Second Edition

Neurobehavioral Genetics

Methods and Applications

Second Edition

Neurobehavioral Genetics

Methods and Applications

Edited by

Byron C. Jones, Ph.D.
Department of Biobehavioral Health
The Pennsylvania State University
University Park, Pennsylvania

Pierre Mormède, D.V.M., Ph.D.
Laboratoire de Neurogénétique et Stress INRA
Université Victor Segalen
Bordeaux, France

Taylor & Francis
Taylor & Francis Group
Boca Raton London New York

CRC is an imprint of the Taylor & Francis Group,
an informa business

CRC Press
Taylor & Francis Group
6000 Broken Sound Parkway NW, Suite 300
Boca Raton, FL 33487-2742

© 2007 by Taylor & Francis Group, LLC
CRC Press is an imprint of Taylor & Francis Group, an Informa business

No claim to original U.S. Government works
Printed in the United States of America on acid-free paper
10 9 8 7 6 5 4 3 2 1

International Standard Book Number-10: 0-8493-1903-X (Hardcover)
International Standard Book Number-13: 978-0-8493-1903-7 (Hardcover)

Visit the Taylor & Francis Web site at
http://www.taylorandfrancis.com

and the CRC Press Web site at
http://www.crcpress.com

Dedication

Between the publishing of our first and second editions, we lost two of our colleagues in the Behavioral Neurogenetics Initiative, Donald J. Nash and Lorraine Flaherty. Both Don and Lori had long, distinguished careers. Don was professor of biology at Colorado State University and Lori was director of the Genomics Institute, chief of the Laboratory of Mammalian Genomics, and director of the Histocompatibility and Paternity Testing Laboratory, Wadsworth Center, New York State Department of Health. Don was president of the Southwestern and Rocky Mountain Division of the American Association for the Advancement of Science for many years. In 2005, Lori received the first Lifetime Achievement Award from the International Behavioral and Neurogenetics Society. Both Don and Lori were active participants in organizing and teaching at our International Summer School on Behavioral Neurogenetics in Europe and North America. In fact, Don was a founding member of the Behavioral Neuro-genetics Initiative (BNI). Both had great and infectious joie de vivre. We even miss Don's outrageous puns. Above all, was their passion for neurobehavioral genetics and bringing this emerging discipline to students and researchers worldwide. The efforts of the BNI have been greatly enriched by the contributions of Lori and Don.

Donald J. Nash
December 20, 1930
January 13, 2002

Lorraine Flaherty
January 8, 1946
February 22, 2006

In Memoriam

Gilbert Gottlieb
1929–2006

Dr. Gilbert Gottlieb was born in Brooklyn, NY in 1929. He earned his Bachelor's (1955) and Master's (1956) degrees from the University of Miami, and his Ph.D. in clinical psychology at Duke University in 1960. This was no ordinary clinical psychology degree because his dissertation was entitled, "The Following Response of Wild and Domestic Ducklings of the Same Species (Anas platyrhynchos)." Thus began an amazing career that spanned several interrelated disciplines (e.g., psychology, zoology, embryology, genetics), the common thread being developmental science and a unique theoretical perspective that he referred to as simply, "the developmental point of view." Instead of reducing development to a simple "interaction" of the organism and its environment, Dr. Gottlieb's view embraced the complexity of organismic development by considering the intricate "transactional" nature of events that occur among all levels of organization (genetic, neural, behavioral, environmental) at every point in time, such that the organism is a "new" organism throughout its transactional developmental trajectory.

This view of development emerged from his early career as a research scientist at the North Carolina Division of Mental Health (Raleigh, NC), where, in 1961, he established the Psychology Laboratory at Dorothea Dix Hospital and pursued both field and laboratory investigations on the acoustic basis of species identification in ducklings. Obtaining baseline data in naturalistic contexts on the kinds of stimuli that organisms normally encounter, as well as the behaviors in which they typically engage, provided a foundation for his elegant and innovative experimental analyses of developmental mechanisms. These studies, which continued for over three decades, revealed the importance of embryonic (as well as early postnatal) experience on the development of species-typical behavior. The empirical and conceptual foundations of this body of work culminated in Dr. Gottlieb's volume, Synthesizing Nature-Nurture (1997, Erlbaum), for which he won the Eleanor Maccoby Book Award of the Developmental Psychology Division of the American Psychological Association.

In 1982, Dr. Gottlieb became the head of the psychology department at the University of North Carolina at Greensboro, where he also held an Excellence Foundation Professorship until his retirement from academia in 1995. That retirement, however, was not a retirement from developmental science, because in that year, he became a research professor and member of the Executive Committee of the Center for Developmental Science at the University of North Carolina at Chapel Hill. In this stimulating multidisciplinary environment, he was able to further develop his theoretical perspectives on the transactional nature of genetic and experiential influences on behavioral development.

Throughout his distinguished career, Dr. Gottlieb received recognition and honors from his peers. He was the guest of the Czechoslovak Academy of Sciences in Prague

(1967) and the USSR Academy of Sciences in Moscow (1989), as well as a Visiting Fellow at The Neurosciences Institute in San Diego (1996). He was also elected Fellow of the Animal Behavior Society and the American Association for the Advancement of Science; and in 1997, the Society for Research in Child Development honored him with the Distinguished Scientific Contributions to Child Development Award. He was supported from 1962 to 2006 by grants from the National Institute of Child Health and Human Development, the National Science Foundation, and the National Institute of Mental Health, all of which enabled him to pursue research questions and theoretical approaches that he recently described to me as "off-the-beaten-track." But he maintained that it is important for scientists to "follow their noses" when so inspired because at the "end of the trail, discoveries will be made."

Dr. Gilbert Gottlieb died on July 13, 2006. Although he did not live to see the publication of this volume, in a taped interview that I conducted with him at his home in Raleigh 1 month before his death, he proclaimed that the chapter contained herein is his most important theoretical contribution to science.

David B. Miller
Department of Psychology
University of Connecticut

Gilbert Gottlieb
1929–2006
Photo Credit: Dr. Marc S. Gottlieb

Preface

This second edition of *Neurobehavioral Genetics: Methods and Applications* is issued about 7 years after the first edition. Since its appearance in 1999, the book has been used as text material for the International Behavioral Neurogenetics Summer School (formerly the French–American Summer School). In September 2005, we held our 12th school in Moscow. This is particularly remarkable because the initial funding, provided by the Scientific Mission of the French Embassy to the U.S. in 1995, was the spark that launched this effort. We will always be thankful to Dr. Jean-Marie Guastavino, attaché for Science and Technology at that time, for arranging for the generous gift from the French government. An alumna of our third school, Dr. Yamima Osher, is author of the chapter on gene–gene interactions.

When we issued the first edition, we knew that the field was developing rapidly, and, in fact, the advances in methods and our knowledge have outstripped even our most optimistic expectations. Methods to examine gene expression, new approaches in bioinformatics, and the rapid growth in knowledge that has followed are remarkable. Major developments in the field of behavioral neurogenetics are the foundation of the Complex Traits Consortium (CTC) and the equally important International Behavioural and Neural Genetics Society (IBANGS). These new societies have provided much needed forums for examination of brain and behavior from a systems genetics perspective. There have been remarkable advances also in single-gene techniques. This second edition has been produced to highlight some of these important advances. We are particularly pleased to include a contribution by Dr. Gilbert Gottlieb, who has been an important and tough critic of behavioral genetics. Fortunately, he has always been kind to behavior geneticists!

Special thanks to our managing editor, Barbara Norwitz, at Taylor & Francis, for her encouragement and guidance.

B.C.J & P.M.
University Park, PA

Editors

Byron C. Jones, Ph.D., is a professor of biobehavioral health and pharmacology at The Pennsylvania State University. Dr. Jones received a B.A. in psychology in 1969 from Eastern Washington University, Cheney, an M.A. in psychology from the University of Arizona, Tucson, in 1972, and a Ph.D. in physiological psychology from the University of Arizona, Tucson, in 1975. He subsequently received post-doctoral training in neuropharmacology at the University of Arizona and in pharmacogenetics at the University of Colorado, Boulder.

Dr. Jones is a member of the Society for Neuroscience, the International Behavioral Neuroscience Society, the Société des Neurosciences, International Brain Research Organization and International Behaviour and Neural Genetics Society. He is a member of the editorial boards for *Pharmacology Biochemistry and Behavior* and *Nutritional Neuroscience*.

Dr. Jones is the author of more than 90 refereed papers, reviews, and chapters. His principal interests are in the pharmacogenetics of drugs that affect the central and peripheral nervous systems.

Pierre Mormède, D.V.M., Ph.D., is director of research at the French National Institute for Agricultural Research (INRA) and the head of the Laboratoire de Neurogénétique et Stress sponsored by INRA and the Université Victor Segalen, Bordeaux, France. Dr. Mormède received a D.V.M. degree from the Veterinary School in Maisons-Alfort, France, and a Ph.D. degree in biology of behaviour from the University Paul Sabatier in Toulouse, France. He subsequently pursued post-doctoral training in molecular neuroendocrinology at the Salk Institute, San Diego, California.

Dr. Mormède is a member of the International Behavioural and Neural Genetics Society, the Société des Neurosciences, and the Société de Neuroendocrinologic Experimentale.

Dr. Mormède is the author of more than 150 refereed papers, reviews, and chapters, and a member of the editorial board of *Psychoneuroendocrinology* and *Genes, Brain and Behavior.* His principal interests are in the psychobiology of stress and drug abuse, and molecular genetics of individual variations in behavioral and neuroendocrine reactivity to environmental constraints.

Contributors

Irmgard Amrein
Institute of Anatomy
University of Zürich-Irchel
Zürich, Switzerland

Rachel Bachner-Melman
Sarah Herzog Memorial Hospital, Givat
  Shaul
Jerusalem, Israel

Amsale T. Belay
Department of Biology
University of Toronto
Mississauga, Canada

John Belknap
Portland Alcohol Research Center
Department of Behavioral
  Neuroscience
Oregon Health & Science University
Portland Veterans Affairs Medical
  Center
Portland, Oregon

Robert H. Belmaker
Division of Psychiatry
Ben Gurion University of the Negev
Beer Sheva, Israel

Jonathan Benjamin
Division of Psychiatry
Ben Gurion University of the Negev
Beer Sheva, Israel

Wade Berrettini
Department of Psychiatry
University of Pennsylvania
Philadelphia, Pennsylvania

Valerie Bolivar
Wadsworth Center
Troy, New York

Guntram Borck
Institut National de la Santé et de la
  Recherche Médicale (INSERM) U781
Hôpital Necker-Enfants Malades
Paris, France

Andrew Canastar
Department of Psychiatry
University of Colorado Health Sciences
  Center
Denver, Colorado

Elissa J. Chesler
Mammalian Genetics and Genomics
  Group
Life Sciences Division
Oak Ridge National Laboratory
Oak Ridge, Tennessee

Laurence Colleaux
INSERM
Hôpital Necker-Enfants Malades
Paris, France

Daniel Comas
Centre National de la Recherche
  Scientifique (CNRS)
Gif-sur-Yvette, France

John C. Crabbe
Portland Alcohol Research Center
Department of Behavioral Neuroscience
Oregon Health & Science University
Portland Veterans Affairs Medical Center
Portland, Oregon

Wim E. Crusio
Laboratoire de Neurosciences
  Cognitives CNRS
Université Bordeaux I
Talence, France

Birgit Dreier
Department of Biochemistry
University of Zürich
Zürich, Switzerland

Richard P. Ebstein
Scheinfeld Center for Human
  Behavioral Genetics
Hebrew University of Jerusalem
Jerusalem, Israel

Lorraine Flaherty
Wadsworth Center
Troy, New York

Philip Gorwood
INSERM
Paris, France

Irving I. Gottesman
Department of Psychiatry
University of Minnesota
Minneapolis, Minnesota

Gilbert Gottlieb
Center for Developmental Science
University of North Carolina
Chapel Hill, North Carolina

Robert Hitzemann
Portland Alcohol Research Center
Department of Behavioral
  Neuroscience
Oregon Health & Science
  University
Portland Veterans Affairs Medical
  Center
Portland, Oregon

Guillaume Isabel
Gènes et Dynamique des Systèmes de
  Mémoire, CNRS
Ecole Supérieure de Physique et de
  Chimie Industrielles
Paris, France

Jean-Marc Jallon
Laboratoire de Neurobiologie de l'
  Apprentissage, Mémoire et
  Communication (NAMC)
Université Paris-Sud XI
Orsay, France

Gert Jansen
Erasmus Medical Center
Rotterdam, the Netherlands

Byron C. Jones
Department of Biobehavioral Health
The Pennsylvania State University
University Park, Pennsylvania

Leslie C. Jones
Interdepartment Graduate Program in
  Neuroscience
The Pennsylvania State University
University Park, Pennsylvania

James E. King
Department of Psychology
University of Arizona
Tuscon, Arizona

Jean-Michel Lassalle
Centre de Recherche sur la
  Cognition Animale
CNRS 5169
Université Paul Sabatier
Toulouse, France

Hans-Peter Lipp
Institute of Anatomy
University of Zürich-Irchel
Zürich, Switzerland

Stephen C. Maxson
Department of Psychology
University of Connecticut
Storrs, Connecticut

Shannon McWeeney
Portland Alcohol Research Center
Department of Behavioral
  Neuroscience
Oregon Health & Science University
Portland, Oregon

Marie-Pierre Moisan
Laboratoire de Neurogénétique et Stress
  INRA
Université Victor Segalen
Bordeaux, France

Florence Molinari
INSERM
Hôpital Necker-Enfants Malades
Paris, France

Pierre Mormède
Laboratoire de Neurogénétique
  et Stress INRA
Université Victor Segalen
Bordeaux, France

Tracy L. Nelson
Department of Health and Exercise
  Science
Colorado State University
Longmont, Colorado

Yamima Osher
Beer Sheva Mental Health Center
Ben Gurion University of the Negev
Beer Sheva, Israel

Jeremy L. Peirce
Department of Anatomy and
  Neurobiology
University of Tennessee
Memphis, Tennessee

Michael F. Pogue-Geile
Department of Psychology
University of Pittsburgh
Pittsburgh, Pennsylvania

Thomas Préat
Gènes et Dynamique des Systèmes de
  Mémoire, CNRS
Ecole Supérieure de Physique et de
  Chimie Industrielles
Paris, France

Richard A. Radcliffe
Department of Pharmaceutical
  Sciences
University of Colorado
Denver, Colorado

André Ramos
Departmento de Biologia Celular,
  Embriologia e Genética
Universidade Federal de Santa
  Catarina
Florianópolis, Brazil

Nicolas Ramoz
INSERM
Paris, France

Laurent Ségalat
Centre de Génétique Moléculaire et
  Cellulaire CNRS
Université Lyon 1 Claude Bernard
Villeurbanne, France

Lutz Slomianka
Institute of Anatomy
University of Zürich-Irchel
Zürich, Switzerland

Marla B. Sokolowski
Department of Biology
University of Toronto
Mississauga, Canada

Peter Sonderegger
Department of Biochemistry
University of Zürich
Zürich, Switzerland

Douglas Wahlsten
Department of Biological Sciences
Great Lakes Institute for Environmental
  Research
University of Windsor
Windsor, Canada

Alexander Weiss
Department of Psychology
University of Edinburgh
Edinburgh, Scotland

Keith E. Whitfield
Department of Biobehavioral Health
The Pennsylvania State University
University Park, Pennsylvania

Robert W. Williams
Department of Anatomy and
  Neurobiology
University of Tennessee
Memphis, Tennessee

David P. Wolfer
Institute of Anatomy
University of Zürich-Irchel
Swiss Federal Institute of Technology
Zürich, Switzerland

Table of Contents

1 A History of Behavior Genetics

Stephen C. Maxson

CONTENTS

SUMMARY

This is a broad brush and highly selective history of behavior genetics. I begin it long ago in prehistoric and ancient times to indicate that even then there was knowledge of and speculation about the inheritance of behavior in animals and humans. I then turn to the contributions of two Victorian cousins, Charles Darwin and Francis Galton. They set the stage for the first century of behavior genetics. I divide the history of first century of behavior genetics into four phases: 1910 to 1952, 1953 to 1970, 1971 to 1985, 1986 to 2005. Within each, I consider the history of behavior genetics by key organisms: nematodes, fruit flies, rodents, other critters, and humans. Throughout, individual contributors to behavior genetics are mentioned. I conclude with some reflections on past conceptual streams of behavior genetics, and on how these streams can and should come together in a comparative genetics of behavior.

1.1 INTRODUCTION

The history of behavior genetics begins long before Darwin's work on evolution or Mendel's work on inheritance. It ends in the very recent past. I hope that this history will be of value in: (1) recognizing who the many contributors to the field were and are, (2) tracing the origins and developments of conceptual themes in behavior genetics, and (3) relating how new methods have contributed again and again to raising questions and enabling answers to them in behavior genetics.

1.2 PRELITERATE AND ANCIENT HISTORY

Humans may have been interested in animals and their own behaviors for as long as there have been *Homo sapiens*, about 150,000 years.[26] For animals, evidence for this is clear by 35,000 years ago. The archeological record shows that humans were hunters and fishers. Both require at least practical knowledge of animal behavior. Furthermore, in France and Spain, there are cave paintings realistically depicting ibex, bison, horses, and other animals; these date from 35,000 years ago. Also, there is ample evidence from anthropology that people in preliterate cultures were and are extremely interested in the behaviors of other humans—watching, recording, and analyzing them. We can only speculate as to whether family resemblances in behavior were noted and theories developed to explain them.

From the beginning of the Neolithic Age (about 12,000 years ago), humans have been domesticating animals. The first domesticated animal may have been the dog. For each domesticated animal, there was subsequently selective breeding for behavioral and other traits. In order to initially domesticate and subsequently selectively breed animals, humans had to recognize that there is variability within a species, that relatives resemble one another, and that breeding like to like produces like. In other words, humans would have to have had some idea about the inheritance of behavior and other traits.

At some point, some humans came to believe that there was inheritance of their own behavioral and other traits. In the sixth century B.C.E., Theognis suggested that well-bred parents have well-bred offspring and that poorly bred parents have poorly bred offspring. Similarly, Plato (427–346 B.C.E.) argued in *The Republic* that matings of the best man to the best woman would produce the best children and that matings of the worst man to the worst woman would produce the worst children. For this reason, both philosophers proposed that humans, like animals, should be selectively bred—with only the well bred or the best having children. This was eugenics long before there was a science of genetics.

There was also an interest in the ancient world in whether the development of behavior was innate or acquired. For example, Aristotle (384–322 B.C.E.) in Greece argued that the brood parasitism of the cuckoo did not depend on post-hatching experience. Similarly, Galen (129–200 C.E.), a Roman physician, proposed, based on an experiment, that the preference of newborn goats for mother's milk did not depend on postnatal experience. Some of the ancient arguments for eugenics may have depended on the belief that if a behavior is inherited, its development is independent of experience. This mistaken idea, which confounds inheritance and innateness, has been a persistent problem in behavior genetics, even until today.

We now leap over almost 2000 years of history to the beginning of a science of behavior genetics.

1.3 TWO VICTORIAN COUSINS

Charles Darwin published *The Origin of Species* in 1859. Darwin argued that the evolution of both adaptations and species was due to past effects of natural selection on heritable variations. He considered complex instincts in chapter 8 of *The Origin of Species*, behavioral sex differences due to sexual selection in chapters 8 to 20 of *The Descent of Man* (1871), animal and human emotions in *The Expression of Emotions in Man and Animals* (1873), and animal mind in chapters 3 and 4 of *The Descent of Man* (1871). In *The Origin of Species*, he also described numerous examples of selective breeding for animal behavior. Selective breeding for animal behaviors would be an important method in the field of behavior genetics. The writings of Darwin on behavioral evolution gave rise to a stream of behavior genetics that is largely concerned with genetics of adaptive behaviors in animals and humans.

The other major stream of behavior genetics is concerned with the causes of variation in human behaviors—especially cognition, personality, psychopathology, and addictions. It is also concerned with the genetics of animal models for these human behaviors. This stream derives largely from the works of Francis Galton, who was Darwin's cousin. Galton initiated three methods for studying behavioral inheritance in humans. These are the family, twin, and adoption methods. The family method and adoption method are discussed in Galton's *Hereditary Genius* (1869). The twin method is discussed in *Inquiry into Human Faculties and Its Development* (1883). Galton also initiated many methods of statistical analysis. Regrettably, he also supported eugenics. Perhaps, he also confounded inheritance and innateness.

1.4 THE FIRST CENTURY OF BEHAVIOR GENETICS

I am going to divide the first century of behavior genetics into four periods: 1910 to 1952, 1953 to 1970, 1971 to 1986, and 1987 to 2005. Within each I will consider the history of behavior genetics by key organisms: the nematode, *C. elegans* from 1973, fruit flies, rodents, other animals, and humans.

1.4.1 1910 TO 1952[12,17]

In 1900, the Mendelian rules of inheritance for the alleles of genes were rediscovered by Correns, DeVries, and von Teschermak, and in 1910, T. H. Morgan proposed the chromosome theory of inheritance. In this theory, each gene is located on one of the chromosomes of its species, and the behavior of the alleles of a gene in gamete formation are due to what happens to chromosomes in gamete formation. Next, it was shown by R. A. Fisher, J. B. S. Haldane, and S. Wright between 1918 and 1921 that the rules of Mendelian genetics would apply to variation and inheritance of quantitative traits like those of concern to Darwin and Galton. The conceptual and methodological stage was now set for this first period in behavior genetics.

1.4.1.1 Fruit Flies[12]

There was a great interest in looking for single-gene effects on fly behavior. This includes the studies of George Wald and Werner Reichard on visual acuity, of Paul Scott on phototropism, of R. S. McEwen on geotaxis and phototaxis, and of A. Sturtevant on mate preference and choice. The objects of study, in many cases, were genes with visible morphological effects on flies, including effects on eye color and shape, body color, and bristle number. There were also studies by H. T. Spieth and Dobzhansky on genetics, mating success, sexual isolation, and evolution. These were both within- and between-species studies.

1.4.1.2 Rodents[17]

In the 1930s and 1940s, R. C. Tryon reported on the selective breeding of two strains of rats that differed in maze learning as indexed by trial errors. The two strains or lines were the "maze brights," with few errors, and the "maze dulls," with many errors. Also, C. S. Hall reported in 1938 on the selective breeding of two lines of rats that differed in emotionality as measured by open-field activity and defecation. Subsequently, selective breeding would be widely used to study the genetics of rat and mouse behaviors.

In 1942, two papers were published showing differences in aggressive behavior among three inbred strains of mice. One was published by John Paul Scott, and the other by Benson Ginsburg and W. C. Allele. In the 1940s, another focus of strain difference in mice was sound-induced, or audiogenic, seizures. Subsequently, there have been a great many studies of strain differences in mouse behaviors. Such strain differences are considered to be evidence for genetic effects on phenotypic variation. Not just mice but rabbits and dogs too were among the organisms studied at Hamilton Station for Behavioral Research, established in 1946 at the Jackson Laboratory. Its full-time or summer staff included John Paul Scott, John Fuller, Benson Ginsburg, Betty Beaman, Calvin Hall, and many others.

For the maze dull and maze bright rats, there were also genotype-by-environment interactions observed for the maze errors. In enriched environments, the error scores of the maze dulls but not the maze brights were decreased, whereas in impoverished environments, the error scores of the maze brights but not the maze dulls were increased.

A genotype-by-environment interaction was also observed for aggression in mice. Scott's most aggressive strain was Ginsburg and Allele's most pacific strain— C57BL/10. When mice of the C57BL/10 strain were transferred from cage to cage by forceps, as Scott did, they were aggressive; and when mice of the C57BL/10 were transferred from cage to cage in a small box, as Ginsburg did, they were pacific. These treatments had no effect on the aggression of the other two strains. Such genotype-by-environment interactions would be found again and again for rodent behaviors and would be an active subject of research up to the present day (see Chapter 20 in this volume).

In 1951, Calvin S. Hall[13] reviewed the literature on the genetics of rat and mouse behaviors and proposed there were three goals for behavior genetics research: (1) to determine whether individual differences in behavior were heritable and to what extent, (2) to determine for each heritable behavior the number of variant genes involved and the chromosomal location of each, and (3) to determine how each of these genes affects the heritable behaviors. These are still among the goals of behavior genetics in animals and humans.

1.4.1.3 Humans[12]

Family and twin studies were conducted for IQ (H. D. Carter, H. H. Neuman, F. N. Freeman, and K. J. Holzinger), personality (M. N. Crook, H. D. Carter, and L. Portenier), schizophrenia (F. J. Kallman), and depression (E. Slater). There were also some adoption studies. The emphasis was on ascertaining whether individual differences were due to genes, environment, or both. Although many studies were consistent with effects of genes on trait variation, definite conclusions were limited by sample size.

Pedigree analysis for single genes with large effects was also developed at this time. Using pedigrees, phenylketonuria was shown by L. Penrose in 1933 to be a single gene recessive trait. It was believed that this was an inborn error of metabolism similar to alkaptonuria as described by A. E. Garrod in 1909.

1.4.2 1953 to 1970[23,25,27]

In 1952, James Watson and Francis Crick described the structure of deoxyribose nucleic acid (DNA) and the implications of that structure for the replication, function, and mutation of genes. Over the next 20 years, the basics of molecular genetics were established, which included an understanding of the two primary functions of DNA. These are its role in determining the amino acid sequence of proteins and in regulating when and where a protein is made as well as how much of it. The stage was now set for Hall's third goal for behavior genetics, namely, identifying and characterizing the effect of DNA and DNA variants on behavior.

In 1958, Benson Ginsburg published *Genes as a Tool in the Study of Behavior*.[11] The first part of this paper was a reflection on the implications of molecular genetics for the study of behavior, and the second part argued that genetics should be used as a tool to analyze animal and human behavior. It could be used as a tool in an evolutionary context to determine natural units of behavior, to study the development of behavior and the respective roles of genes and environment, and to analyze the biological mechanisms of behaviors. This approach would become one of the main tributaries of modern behavior genetics.

The other main tributary of modern behavior genetics focused on the contribution of genetic and environmental factors in individual differences in animal and human behaviors and on identifying and characterizing each of the genes with effects on these individual differences. Also, a prime interest of the research with animals was to develop genetic models relevant to the study of individual differences in human behaviors. In 1960, the literature on this tributary of behavior genetic research on human and animals was reviewed and summarized in the text by J. L. Fuller and R. Thompson. The text was simply titled, *Behavior Genetics*.[12] Developments in quantitative genetics as described in D. S. Falconer's *Introduction to Quantitative Genetics*[11] and in K. Mather's *Biometrical Genetics*[24] were a basis for further advances in this area of behavior genetics.

In 1961 and 1962, there were important meetings on behavior genetics at the Center for Advanced Studies in the Behavioral Sciences at Stanford CA. These were attended by many leaders in the field and in many ways set the agendas for the next decade of behavior genetics research. Some of the presentations at these meetings were published in 1967.[20]

1.4.2.1 Fruit Flies

During this period, there were several selective breeding projects for behaviors in fruit flies. Jerry Hirsch and Niki Erlenmeyer-Kimling initiated bi-directional selection for geotaxis. Selective breeding of these lines has continued to the present day. Now, crosses of the two lines cannot have fertile progeny. Hirsch and Erlenmeyer-Kimling also tested for the contributions of individual chromosome regions to the high and low scores. Similarly, Dobzhansky selectively bred for geotaxis and phototaxis in another species of fruit fly. He reported that very little of the individual differences in these traits was due to genetic variation. About the same time, Aubrey Manning carried out a bi-directional study of mating speed.

During this period, Lee Ehrman initiated her studies of rare male mating advantage in several species of fruit flies. Also, Speiss, Ehrman and Dobzhansky among others continued the research on mating preference and on sexual isolation within and between species.

About 1967 Seymour Benzer first proposed that single gene mutants with large effects could be used to dissect the neural and other mechanisms of behaviors in flies. Later Sydney Brenner made a similar suggestion for the nematode, *C. elegans*. These contributions are described more fully in the next period (1970 to 1986) of behavior genetics. In some ways, this was to be a large substream of the tributary that would use genetics as a tool to analyze behavior as originally proposed by Ginsburg in 1958.

1.4.2.2 Rodents

During this period selective breeding was used to establish new phenotypic lines of rats and mice. These included the Maudsley Emotional and Non-emotional strains (Peter Broadhurst) and the Roman High and Low Avoidance strains (C. Bignami) in rats. Also, mice were selectively bred for high and low activity in an open field (John Defries) and aggression (Geert van Oortmerssen and K. Lagerspetz).

About this time, Jan Bruel suggested in several articles that adaptive traits of mice would have low additive genetic variation and high directional dominance variation. This set the stage for research with diallel F1 crosses and F2 hybrids of inbred strains to determine the additive and dominance variance of many mouse behaviors. This program of research will be further discussed in the next period (1971–1985) of behavior genetics.

Now strain differences were also being used to investigate the neural basis of behaviors. For example, M. R. Rosenzweig, D. Krech, and E. L. Bennett had shown maze dull and maze bright rats differed in brain cholinesterase levels. Subsequently, Thomas Roderick selectively bred for cholinesterase levels in the brains of rats and showed that these strains differed in learning performance.

Also, at this time, research was begun on strain differences in consumption and effects of alcohol (Gerald McClearn). Research on genetics, alcohol, other drugs, and behavior would become a major area of mouse behavior genetics.

1.4.2.3 Dogs

In 1965, John Paul Scott and John L. Fuller published the book, *The Genetics and Social Behavior of the Dog*.[34] This book detailed the findings of nearly 20 years of research at the Jackson Laboratory on the roles of genes and environment in the behavior of five breeds of dogs and their derived crosses. This work remains a critically acclaimed classic on dog behavior and on methods for genetic analysis of social behaviors.

1.4.2.4 Humans

During this period, there were more extensive adoption studies of IQ (M. Skodak and H. M. Skeels, M. Honzig, and J. Sheilds) and schizophrenia (L. Heston, D. Rosenthal, S. S. Kety). About this time, the large Louisville Twin Study was used to assess the contribution of genes and environment to variation in many traits (S. Vandenberg).

Also, during this time, many single genes with large effects on mental retardation were identified, and the behavioral effects of many chromosomal anomalies were also identified. This included the chromosomal basis for Down's and Turner's syndromes. Development of staining techniques enabled the identification of each human chromosome.

1.4.3 1971 to 1985[4,9,14]

In 1967, the Institute for Behavioral Genetics was founded at the University of Colorado. Here many of today's behavioral geneticists received doctoral or postdoctoral training. Subsequently, many more centers of behavior genetics would be established in the U.S. and internationally. In 1971, the Behavior Genetics Association was founded and its journal, *Behavior Genetics*, was first published. Also, new textbooks in behavior genetics were published. These included Gerald McClearn and John Defries's *Introduction to Behavioral Genetics* (1973) and Lee Ehrman and Peter Parsons, *The Genetics of Behavior* (1976).[10] Also a second edition of Fuller and Thompson (now titled, *Foundation of Behavior Genetics*) was published in 1978.[13] In 1974, J. H. F. van Abeelen edited *The Genetics of Behavior*;[36] this had important conceptual, methodological, and substantive reviews of behavior genetics across animals and humans.

1.4.3.1 Nematodes[3]

The research on nematodes began in 1973. The nematode has several advantages for neurobehavioral genetics. These are a small genome (100 Mb), small nervous system of 302 identified and mapped neurons, fast reproductive cycle (three days), and self-fertilization. The major approach has been to expose nematodes to chemical mutagens, to select for single gene mutants with behavioral effects and then to characterize the DNA of the mutant gene. Initially mutants were detected that showed sensory and motor effects.

1.4.3.2 Fruit Flies[18,39]

In 1973, Peter A. Parsons published *Behavioral and Ecological Genetics: A Study in Drosophila*.[30] This summarized much of the findings to date on the genetics of mate choice and mating in relation to fruit fly evolution and speciation. Lee Ehrman, as well as others, continued to work in this area.

A major contributor to finding the initial mutants was Ronald Konopka. This initial work is well summarized in Benzer's biography, *Time, Love and Memory*, by Jonathan Weiner.[39] The long-term goal of this approach was to use genetic mutants to dissect the neural and other biological bases of behavior. About this time, the same approach used by was W. Pak and K. Grossfield to study vision in fruit flies and by many others to investigate the molecular and cell biology of neurons.

1.4.3.3 Rodents[40]

In 1970, Gardner Lindzey and Delbert D. Thiessen edited *Contributions to Behavior-Genetic Analysis: The Mouse as a Prototype*.[22] Each article in it is a review of a major area of research in mouse behavior genetics during in the 1960s, and these articles became the basis for much of mouse behavior genetics research in the 1970s and 1980s.

Selective breeding continued to be used with mice and rats to establish lines differing in behavior and to determine the heritability of the behavior. Some were ones previously described. Others were new. These later included in rats the Syracuse high and low avoidance lines (Bob Brush). Also, David Blizard and others continued the study of the Maudsley Emotional and Non-emotional rats. Blizard also studied behaviors of mice selected for thyroid function. In mice, Carol Lynch selected for lines differing in the size of thermal nests and related her work to the ecology and evolution of mice. Also, Bob Cairns and Kathryn Hood selected mice for male aggression, and Janet Hyde and Patricia Ebert selected mice for female aggression. There was also selection in rats and mice for effects of alcohol and drugs on behavior such as the post-alcohol injection long sleep and short sleep lines. Much of this work in mice was done by John Crabbe, Andrew Smolen, Toni Smolen, Thomas Johnson, Bruce Dudek, as well as many others. Another study selected for high and low brain weights and related these to mouse behaviors (John Fuller and Martin Hahn).

During this period investigators continued to using F1 diallels and F2 dihybrids to assess the adaptive significance of behaviors. The behaviors studied included ultrasounds, open field activity, aggression, mating, and learning. Contributors included Norman Henderson, John Hewitt, Martin Hahn, Tom McGill, and many others.

There was also an interest in finding the individual genes and chromosomes that contributed to variation in mouse behaviors. One approach by Del Theissen and colleagues was to see whether coat color and morphological variants had effects on behavior. They did, but the mechanisms for the effects seemed trivial. More relevant was research described by D. L. Coleman on two spontaneous mutants with effects on feeding, satiety, metabolism, and obesity. These are the *Obese* and *Diabetes* mutant mouse lines. Another approach was to try to use the newly developed recombinant inbred strains to chromosomally map genes with behavioral effects (Basil Eleftheriou and Thomas Seyfried). Because there were not enough chromosomal markers and too few recombinant inbred strains, these initial attempts to map mouse variants with behavioral effects were often not replicated. However, Benson Ginsburg, Stephen Maxson, and their doctoral students successfully used reciprocal F1s and congenic strains to show that the male specific part of the Y chromosome had effects on aggression and mating. The findings on the Y chromosome have been replicated many times (Pierre Roubertoux, Michele Carlier, Frans Sluyter, and Geert van Oortmerssen).

Also, Michele Carlier, Pierre Roubertoux, and their colleagues also began a long-term research program on the interactions of maternal factors and genotype in the development of mouse behaviors.

During this period there were also programs focusing on pharmacology and neuroanatomy to find the connections between genes and behaviors. For example, Kurt Schlesinger and colleagues used drugs to manipulate neurotransmitter systems in relation to audiogenic seizures, and Richard and Cynthia Weimer measured strain differences in cell number and other morphological parameters of the hippocampus in relation to learning and memory. Douglas Whalsten also explored strain differences in corpus colossal morphology in relation to behavior. There were also studies on strain difference in the endocrine systems and behavior such as those by Thomas McGill on male mating, by Jack Vale on male aggression, and by Bruce Svare on female aggression.

1.4.3.4   Canids

In 1970, a study of the evolution and genetics of canids was established at the University of Chicago by Benson Ginsburg. This included research on the social behavior and organization of a pack of wolves and on the inheritance of threat behaviors in dogs (beagles) and coyotes.

1.4.3.5   Humans[19]

Research continued on cognition, personality, and psychopatholgy with family, twin, and adoption methods. There were major conceptual and methodological advances. There were developments in the design and analysis of these studies by David Fulker, John DeFries, John Loehlin, Lindon Eaves, Robert Plomin, and many others. These focused on combining data from at least two of these three, family, twin and adoption studies, to develop mathematical models of genetic and environmental contributions to individual differences. Also a major adoption project was established at the University of Colorado by John DeFries and others, and Thomas Bouchard Jr., Nancy Segal and their colleagues began a long-term intensive study of twins raised apart. In addition, large twin registers were begun in Australia by Nicholas Martin, in Sweden by Nancy Pedersen, and the Netherlands by Dorett Boomsma. Also, the Louisville Twin Study was continued (Ron Wilson).

In this period, pathways from gene to neurological, behavioral or mental effects, such as phenyketonuria, were worked out for many gene variants with large effects.

1.4.4   1986 TO 2005[8,31,32,37]

There were many developments in molecular genetics that had direct impact on behavior genetics. These included: the polymerase chain reaction method, sequencing the genomes of nematodes, flies, mice, and humans, identification of DNA markers on chromosomes of nematodes, flies, mice, and humans for gene mapping, and transgenics and knockouts in mice. These techniques revolutionized all of biology including neural and behavior genetics. Their impact is discussed in detail elsewhere in this volume. Here, I consider some of what was happening in behavior genetics during these years.

Also, in this period, the International Behavioral and Neural Genetics Association or IBANGS (1996) and the International Society for Psychiatric Genetics or ISPG (1992) were established. *Genes, Brain and Behavior* is the journal of IBANGS, and *Neuropsychiatric Genetics* is the journal of ISPG.

1.4.4.1   Nematodes[2]

The genome (100 Mb) of *C. elegans* was sequenced in 1998. During this period, there was continued screening for mutants with effects on olfactory and touch sensory systems, chemotaxis, thermotaxis, biological rhythms, social behavior, learning and memory and behavioral effects of alcohol, cocaine, and nicotine. Many genes with effects on these systems were identified and characterized. In many cases, the effects of the gene can be or are being traced from DNA to protein to neural organization and function to behavior.

1.4.4.2 Fruit Flies[21,35,37]

The genome (160 Mb) of the fruit fly was sequenced in the year 2000. During this period single gene mutants were used to dissect the molecular and neural basis on the circadian clock (Jeff Hall), mating (Jeff Hall, Charalambos P. Kyriacou, and Jean-Marc Jallon), learning and memory (Tim Tully), and effects of alcohol and drugs on behavior (F. E. Wolf and U. Huberlein). There was also an interest in natural genetic variation in fly behavior. One example of this is Marla Sokolowski's studies of the forager gene. Another example is the identification of the genes involved in Hirsch's long-term selection for the high and low geotaxis lines (Ralph Greenspan). This gene identification used a combination of gene expression chips and mutants. Greenspan has also investigated and reviewed the complex interactions of genes with effects on fruit fly behaviors.

1.4.4.3 Bees[33]

The genome (200 Mb) of honeybees is being sequenced. Recent research (Gene Robinson and others) on honeybees provides an excellent example of the association between behavioral development and changes in gene expression and neural structure. Worker bees engage in nursing behavior from birth to about 2 to 3 weeks of age. After that they engage in foraging behavior. This behavioral change is associated with the increase or decrease in expression of many genes in the brain. Two of these genes are period and forager, which were first found in fruit flies.

1.4.4.4 Rodents[7,16,37,38]

The mouse genome (3500 Mb) was sequenced in the year 2002. During this period, the main focus of research has been to identify the genes that cause variation in behavior. There have been four approaches for this. The first is to chromosomally map quantitative trait loci (QTLs). One such study used crosses of the lines selected for open field activity (Jonathan Flint and others.) Another approach used crosses of inbred strains to find QTLs for aggression (Pierre Roubertoux and others). Some studies combined recombinant inbred strains and strain crosses to find QTLs for effects of alcohol and other drugs (John Crabbe, John Belknap, Keri Buck, Thomas Johnson, Byron Jones, and others). In 1991, John C. Crabbe and R. Adron Harris edited *The Genetic Basis of Alcohol and Drug Actions*.[6] There has also been development of large, recombinant, inbred strain sets that have been used to find QTLS for brain structures and expressions of genes in mouse brains (Robert Williams and others). Another approach has used specific gene knockouts and/or transgenics (Crawley). This has been used with many behaviors including sensation, motor skills, emotionality (Rene Hen), feeding (C. H. Vaughn and many others), aggression (Randy Nelson), mating (Sonoko Ogawa and Emilie Rissman), learning and memory (J. Z. Tsien and Jeanne Wehner), and effects of alcohol and drugs (Tamara Philips,

Kari Buck, and many others). Also, there are now tissue or temporal specific knockouts and transgenics.

A related approach is based on spontaneous or induced mutations. As previously described, *Obese* and *Diabetes* are two spontaneous mutations with effects on feeding, satiety and metabolism. Recently, it was shown that the obese mouse has a coding mutation for the polypeptide hormone leptin, and the diabetic mouse has a coding mutation in a leptin receptor (J. M. Friedman and others). The study of these two spontaneous mutants detailing the pathway form DNA to behavior opened wide the door to research on the hormonal and neural mechanisms of feeding and satiety. Also, male mice have been exposed to chemical mutagen and the F1 or F2 progeny of these mice screened to detect mutations with neural or behavioral effects. One of the first induced mouse mutants to be detected was *Clock* which affects circadian rhythms (Joseph Takahashi and others). Other genes with effects on mouse circadian rhythms were found to have fruit fly homologs, such as *Period* and *Timeless*. The study of the *Clock* mutant and variants for *Period* and *Timeless* from DNA to behavior opened up further research on the neural mechanism of the circadian clock in mammals. There are now centers for mouse mutagenesis and behavior at the Jackson Laboratory, Northwestern University, and the University of Tennessee Health Sciences Center. Each has high throughput screens for a range of mouse neural and behavioral traits.

Another approach is to look for differences in gene expression among selected strains of mice. This was done for hippocampal gene expression in SAL (short attack latency) and LAL (long attack latency) strains (D. E. Feldker and others).

Other research in the period has used inbred or selected strain-based genetic correlation to relate brain and behavioral phenotypes. One example is the study of variation in the size of the hippocampal mossy fibers and related cognitive or emotional behaviors (Wim E. Crusio and Hans-Peter Lipp). Another interest has been in the standardization of behavioral tests (Norman Henderson, Martin Hahn, and Stephen Maxson) and phenotypic stability across laboratories (John Crabbe, Douglas Wahlsten, and Bruce Dudek). Furthermore, during this period mouse models for Turner's, Klinefelter's, and Down's syndromes were developed.

One series of fascinating studies has related the genetics of species differences in the behavior of voles focused on oxytocin and vasopressin systems and aggression, parenting and monogamy (Larry J. Young).

1.4.4.5   Humans[1,5,28,29]

The human genome (3500 Mb) was sequenced in the year 2000. During this period path analysis and model fitting were fully developed to analyze the components of individual differences in behavior (David Fulker, John DeFries, Lindon Eaves, Nicholas Martin, Stacy Cherny, and many others). This approach allowed the combined used of family, twin and adoption studies. It was widely applied to cognition (Robert Plomin, Dorett Boomsma, and Sandra Scarr), personality (John Loehlin), personality disorders (Richard Rose, Irwin Waldman, Laura Baker, and Liz DiLalla), sexual orientation (Michael Bailey and Dean Hamer), psychopathology (Peter McGuffin and Irving Gottesman), and addictions (C. R. Cloninger, Kenneth Kendler,

Matt McGue and Andrew Heath). There was also an interest in the relative roles of within family environments, between family environments, genotype–environment correlations and genotype–environment interactions (Robert Plomin and David Rowe) as applied to the same range of human behaviors.

A goal of this period was also to chromosomally map the genes that are involved in the individual differences for a behavioral domain such as cognition. The two approaches to this are linkage analysis and association analysis (Eric Lander and Pak Sham).

The former often uses the affected sib pair method and the latter often uses the transmission disequilibrium test. Both methods make use of nongenic DNA markers and association analysis uses not only nongenic DNA variants, but also DNA variants of known genes. The main problem to date has been replication of findings. For this reason, meta-analysis across studies is often used.

Animal models may also be used in the search for genes with behavioral effects in humans. Animal models can indicate chromosome regions or genes to focus on.

1.5 JANUS: THE PAST AND FUTURE OF BEHAVIOR GENETICS

I believe that there have been at least three conceptual divides in the history of behavior genetics.

The first of these divides traces back to Darwin and Galton. The Darwinian stream has been interested in the genetics of adaptive behavior in animals and humans. Its findings are very relevant to biology in general. The Galtonian stream has been interested in the genetics of human variation, especially cognition, personality, personality disorders, psychopathology, and addictions. Its findings are very relevant to education, employment and medicine.

The second divide has been between those who are interested in using genetics as a tool in the study of behavior and those who are interested in using genetics to analyze the causes of individual differences. The former investigates genetic variants as a way to dissect the developmental and physiological mechanisms of behavior and to find related general principles. The latter is interested in disentangling the genetic and environmental causes of behavior and in finding the genes that cause species behaviors to vary.

The third divide has been between those who study single gene induced mutations or natural variants with large effects and those who study the effects of natural variants of many genes with small effects. In some ways, this divide overlaps that described in the previous paragraph.

These three divides have fragmented behavior genetics. Here I suggest one approach to unifying behavior genetics in its second century. This is a comparative genetics of behavior. In part, this will be based on genetic studies of many more species than the big four of nematodes, fruit flies, mice, and humans. In the first two periods of the first century, a wide range of animals was studied. There should be a return to this practice. A comparative genetics of behavior will also depend on the growing number of projects to sequence the DNA of representatives of major taxa. This now includes not only nematodes, flies, mice and humans but also jellyfish, bees, zebrafish, chickens, pigs, cattle, dogs, cats, monkeys, chimpanzee, and others. Such a comparative genetics should also be gene centered comparing the behavioral effects of homologous genes across species. For example, variants of the gene for MAOA (monoamine oxidase A) have been shown to have similar effects on aggression in mice, monkeys and humans and depend for their effects on specific aspects of the social environment during development.

1.6   A HIGHLY PERSONAL NOTE

I began my career in behavior genetics in 1960. My mentor was Benson Ginsburg. Over the years, I have been privileged and honored to know most of those mentioned in this brief history. All are colleagues and many are friends. I met many of the founders of behavior genetics at the 1961 meeting in Stanford. Others, I have met over the years at annual meetings of the Behavior Genetics Association. I have attended 30 of the 35 meetings. Some I have met at meetings of IBANGs or the Society for Neuroscience. There are many detailed stories that I could tell about each one. But I have tried here to take a wider view of our mutual history. I am sure that there are many behavioral geneticists whom I have not mentioned, especially those working with nematodes, fruit flies, other critters, and humans and perhaps even some of those working with rodents. For this, I apologize to one and all. Regardless, there are so many of you today that I could not mention each and every one of you. Just a list of all the names would fill this history to its page limit. This is one measure of how successful behavior genetics has been in its first 100 years.

REFERENCES

1. Benjamin, J., Ebstein, R. P., and Belmaker, R. H. *Molecular Genetics and Human Personality.* American Psychiatric Publishing Inc., Washington, DC, 2002.
2. Bonn, M. and Villu, Maricq A. Neuronal substrates of complex behaviors in *C. elegans. Ann. Rev. Neurosci.,* 28: 451–501, 2005.
3. Brenner, S. The genetics of behaviour. *Br. Med. Bull.,* 29: 269–271, 1973.
4. Broadhurst, P. L. and Fulker, D. W. Behavioral genetics. *Ann. Rev. Psychol.,* 25: 389–415, 1974.
5. Carey, G. *Human Genetics for the Social Sciences,* Sage Publication, London, 2003.
6. Crabbe, J. C. and Adron Harris, R., Eds., *The Genetic Basis of Alcohol and Drug Actions,* Plenum Press, New York, 1991.
7. Crawley, J. N. *What is Wrong with my Mouse: Behavioral Phenotyping of Transgenic and Knockout Mice.* John Wiley & Sons, New York, 2000.
8. Crusio, W. E. and Gerlai, R. T. *Handbook of Molecular-Genetic Techniques for Brain and Behavior Research.* Elsevier, Amsterdam, 1999.
9. Defries, J. C. and Plomin, R. Behavioral genetics. *Ann. Rev. Psychol.,* 29: 473–515, 1978.
10. Ehrman, L. and Parsons, P. *The Genetics of Behavior.* Sinauer Associates, Sunderland, MA, 1976.

11. Falconer, D. S. *Introduction to Quantitative Genetics.* The Ronald Press, New York, 1960.
12. Fuller, J. and Thompson, W. R. *Behavior Genetics.* John Wiley & Sons, New York, 1960.
13. Fuller, J. and Thompson, W. R. *Foundation of Behaviour Genetics,* The C.V. Mosby Co., St. Louis, 1978.
14. Fuller, J. and Simmel, E. C. *Behavior Genetics: Principles and Applications.* Lawrence Erlbaum Associates Publishers, Hillsdale, NJ, 1983.
15. Ginsburg, B. E. Genetics as a tool in the study of behavior. *Perspectives in Biology and Medicine,* 1: 397–424, 1958.
16. Goldowitz, D., Wahlsten, D. and Wimer, R. E. *Techniques for the Genetic Analysis of Brain and Behavior: Focus on the Mouse.* Elsevier, Amsterdam, 1992.
17. Hall, C. S. The genetics of behavior. In: *Handbook of Experimental Psychology,* Stevens S. S., Ed. John Wiley & Sons, New York, 1951, chap. 9.
18. Hall, J. C., Greenspan, R. J. and Harris, W. A. *Genetic Neurobiology.* MIT Press, Cambridge, MA, 1982.
19. Henderson, N. D. Human behavior genetics. *Ann Rev. Psychol.,* 33: 401–440, 1982.
20. Hirsch, J., Ed. *Behavior-Genetic Analysis.* McGraw-Hill, New York, 1967.
21. Kyriacou, C. P. Single gene mutations in *Drosophila*: What can they tell us about the evolution of sexual behavior? *Genetica,* 116: 197–203, 2002.
22. Lindzey, G. and Thiessen, D. D. Eds., *Contributions to Behavior-Genetic Analaysis: The Mouse as a Prototype,* Appelton Century Crofts, New York, 1970.
23. Lindzey, G., Loehlin, J., Manosevitz, M., and Theissen, D. Behavioral genetics. *Ann. Rev. Psychol.,* 22: 39–94, 1971.
24. Mather, K. *Biometrical Genetics,* Methuen, London, 1949.
25. McClearn, G. E. Behavioral genetics. *Ann. Rev. Genet.,* 4: 417–468, 1970.
26. McClearn, G. E. and DeFries, J. C. *Introduction to Behavioral Genetics.* W.H. Freeman, San Francisco, 1973.
27. McClearn, G. E. and Meredith, W. Behavioral genetics. *Ann. Rev. Psychol.,* 17: 515–550, 1966.
28. McGue, M. and Bouchard Jr., T. J. Genetic and environmental influences on human behavioral differences. *Ann. Rev. Neurosci.,* 21: 1–24, 1998.
29. McGuffin, P., Owen, M. J., and Gottesman, I. J. *Psychiatric Genetics and Genomics.* Oxford University Press, New York, 2002.
30. Parsons, P. A. *Behavioral and Ecological Genetics: A study in Drosophila,* Oxford University Press, New York, 1973.
31. Pfaff, D. W., Berrettini, W., Joh, T. H. and Maxson, S. C. *Genetic Influences on Neural and Behavioral Function,* CRC Press, Boca Raton, FL, 2000.
32. Plomin, R., DeFries, J. C., Craig, I. W., and McGuffin, P. *Behavioral Genetics in the Postgenomic Era.* American Psychological Association, Washington, DC, 2002.
33. Robinson, G. E. and Ben-Shahar, Y. Social behavior and comparative genomics: new genes or new gene regulation. *Genes, Brain, Behav.,* 1: 187–203, 2003.
34. Scott, J. P. and Fuller, J. L. *The Genetics and Social Behavior of the Dog,* University of Chicago Press, 1965.
35. Sokowlowski, M. B. *Drosophila* genetics meets behavior. *Nat. Genet.,* 2: 879–880, 2001.
36. van Abeelen, J. H. F., Ed. *The Genetics of Behavior,* North-Holland, Amsterdam, 1974.
37. Wahlsten, D. Single-gene influences on brain and behavior. *Ann. Rev. Psychol.* 50: 599–624, 1999.

38. Weiner, J. *Time, Lover, and Memory: A Great Biologist and His Quest for the Origins of Behavior.* Alfred A. Knopf, New York, 1999.
39. Whener, J. M., Radcliffe, R. A., and Bowers B. J. Quantitative genetics and mouse behavior. *Ann. Rev. Neurosci.*, 24: 845–867, 2001.
40. Wimer, R. E. and Wimer, C. E. animal behavior genetics: a search for the biological foundations of behavior. *Ann. Rev. Psychol.* 36: 171–218, 1985.

2 Developmental Neurobehavioral Genetics: Development as Explanation

Gilbert Gottlieb

CONTENTS

INTRODUCTION

From the very dawn of human history, there must have been people who wondered how we come to be—not just in the grand religious sense that ancient texts like the Hebrew bible attempt to answer, but also in the more mundane and practical sense of wondering about the mechanisms involved in human (and all animal) development from egg and sperm to full-grown adult. By the time of Aristotle in the fourth century B.C.E., there were two main schools of thought on how we become. In fact, the proper scientific name for the study of individual development was derived from the Greek language: *ontogeny.*

2.1 PREFORMATION AND EPIGENESIS

One school of thought about ontogeny, the *preformationist* school, held that a very tiny version of the complete, adult individual was prepackaged either in the female egg or in the male sperm. (There were actually two schools of preformationist thought: ovists and spermists.) In other words, according to preformationist thinking—whether ovist or spermist—all the parts and organs that an individual will ever have are present from the outset; development is merely the growth of these preexisting parts until they reach their full, adult size. Aristotle's view, on the other hand, gained from his personal observations of the developing chick embryo, was that individual development happens through *epigenesis*; that is, through a series of successive transformations of some sort of early, homogeneous mass that brings the parts and organs of an organism into being. In this view, what happens during development is not just the quantitative growth of preexisting parts, but the gradual qualitative differentiation of parts that then grow as well. Today we still do not completely understand all of the mechanisms that make development happen, but we do know that it is correct to say that individual development—whether in terms of psychology, behavior, physiology, or anatomy—is epigenetic and not preformative.

Nevertheless, some scientists continue to operate under a modern-day version of preformationist thinking when they say that genes govern individual development. The only difference between the ancient idea of the unfolding of preformed parts and the modern idea of the unfolding of the genetic code is that the unfolding, itself, is now understood to be a transformative process—in other words, epigenetic. All biological scientists understand that successive transformations of the forms and functions of tissues occur during development. This is what is meant by epigenetic development. But many scientists seem to think that these transformations are controlled by little, preformed, inherited, autonomous packets of information that we know today as genes. This is a modern-day version of preformationism. As the noted biologist J. S. Haldane[13] observed, the idea of genes as the controllers of individual development only substitutes "an extremely complicated molecular structure" for the ancient idea of the "original miniature adult"(p. 147). I call the idea of epigenetic transformation that is controlled by genes, "predetermined epigenesis." According to this way of thinking, further developmental processes may occur due to environmental influences, but genes and the environment are viewed as separate causal systems rather than as parts of one, integrated system. This idea that there are separate causal systems is the crux of what has become known as the nature–nurture dichotomy.

The advocates of a genetically predetermined epigenesis have often recognized that the nature–nurture dichotomy is a shaky construct because, although based on a recognition that environment does make a difference in development, it lacks an adequate theory of how, when, and where this happens.[22] In the end, only lip service is paid to the idea of gene–environment interactions, and genetic predeterminists say it is genes primarily that make development happen and that the environment plays only a passive, permissive, or supportive role. The formative drive, they say, is governed by genes acting as autonomous agents.

However, as several excellent popular books have already described[16,18] recent research demonstrates that genes are not really unchanging little packets of unfolding information standing on the sidelines of epigenetic processes and stoically directing the developmental game. Instead, as M.-W. Ho has written,[14] they "get variously chopped, rearranged, transposed, and amplified in different cells at different times" (p. 285)—and themselves must be activated in order to make any contribution at all to development. With the supposed role of genes as independent agents of development thus brought into question, the idea of a nature–nurture dichotomy has gotten to be on even shakier ground than before. That is, given that genes need nongenetic inputs to be activated, the nature–nurture dichotomy is undermined at its most basic level.

The aim of this chapter is to replace the nature–nurture dichotomy with a thoroughly integrative, coactional theory of epigenesis that takes into account the role of genes and environments as well as two other domains of influence in the formative process: behavior and the nervous system. The various processes that work together across these four levels of development are what produce a fully formed, behaviorally mature organism. Because according to this theory there is no single factor or influence that causes developmental outcomes to be the way they are, I call the point of view "probabilistic epigenesis." This is the theory that any developmental outcome occurs not as a matter of predetermined necessity but as a result of thoroughly interacting developmental processes and that, if the operating parameters of any of these processes are changed during development, the outcome will also change to one degree or another. The coordination of the necessary influences from the four levels does not occur with absolute fidelity, thus giving a probabilistic character to the process as well as to the outcomes of development. That is why epigenesis is probabilistic, not fixed.

2.2 DEVELOPMENT AS EXPLANATION

The most difficult concept to get across in the understanding of psychology, animal behavior, neuroscience, or neurobehavioral genetics is development-as-explanation. Development is viewed by many as just one more thing we study as psychologists, animal behaviorists, neuroscientists, or geneticists. So, I sometimes get a blank stare or a quizzical look when I propose the idea of development as explanation. For example, some of my colleagues study neuroscience or the brain because in the usual theoretically reductionistic scheme of things, they feel that will allow them to understand the mind; others, for the same reason, study genetics because they feel that will allow them to understand normal and abnormal behavior. My theoretically nonreductionistic scheme (Figure 2.1) for developmental analysis includes the nervous system and genetics, as well as behavior and the physical, social, and cultural aspects of the environment.

These four levels of analysis are mutually influential, so the arrows of influence are top-down as well as bottom-up. This is a fully bidirectional coactional system, with traditionally trained developmental psychologists working at the Behavior↔Environment level, developmental neuropsychologists and neuroscientists at the Neural Activity↔Behavior↔Environment levels, and biologically trained persons beginning now to work on the Genetic Activity↔Neural Activity levels. (For an outstanding example of the latter, see Rampon et al.[21]) Viewed in this way, developmental understanding or explanation is a multilevel affair involving at least culture, society, immediate social and physical environments, anatomy, physiology, hormones, cytoplasm, and genes. Hierarchical, multilevel, or developmental systems analysis is methodologically reductionistic in the sense that biological factors are included to make investigations more complete. However, it is not theoretically

DEVELOPMENTAL ANALYSIS

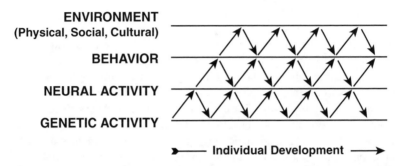

FIGURE 2.1 A developmental–psychobiological systems model to illustrate the comprehensive analysis required for developmental explanation. (Modified from Gottlieb[8] with permission.)

reductionistic in the sense that psychological understanding or explanation does not come from lower levels (or reside in the lower levels), but that both higher and lower levels of analysis are necessary to explain developmental outcomes. The term *coaction* is used to emphasize the multilevel systems nature of developmental analysis.

Any outcome (genetic activity, neural activity, behavior) is an epigenetic outcome of prior developmental processes in the sense that it is an emergent property. The only way to understand the emergence of that outcome, whether genetic, neural, or behavioral, is to study the coactional influences within and between the four levels of analysis that gave rise to the emergence of that outcome. Since all outcomes are emergent, they have a developmental history that explains their emergence. That is why developmental analysis is explanation.

When my model (Figure 2.1) is unpacked to reveal all of its otherwise hidden mysteries, as has been done in Figure 2 by Johnston and Edwards,[15] all the reciprocal intermediate steps between genetic activity, neural activity, behavior and the environment can be visualized. (The issue of time, which is crucial for developmental analysis, is treated elsewhere [pp. 29–30] in Johnston and Edwards's article.) The significance of Figure 2.2 is that it raises the kinds of developmental questions raised by findings that might otherwise be misinterpreted as evidence of a direct link between genes and behavior. Even within the same species there is not just one way to get from genes to behavior.[23]

Figure 2.2 extends and explicates the important features of Figure 2.1 in four ways, as follows from Johnston and Edwards's discussion of time. First, it includes both neural and non-neural components. Certainly, the most thoroughly studied examples of non-neural contributions to structural, physiological, and behavioral development involve hormones; however, development also involves bones, muscles, horns, feathers, and other bodily structures; and all of these need to be taken into account. For example, in Thelen's[25] analysis of infant locomotion, gross morphological changes play a critical role.

Second, the immediate consequences of genetic activity are confined to the cell. Genetic effects on behavioral development must take into account the various

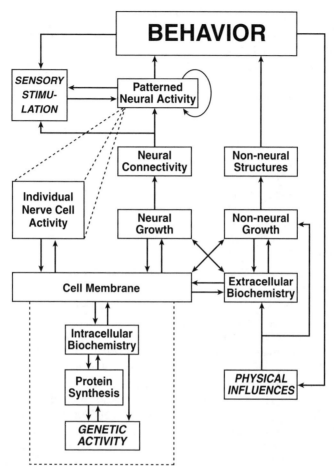

FIGURE 2.2 Completely unpacked model of developmental neurobehavioral genetics, showing all the coacting factors involved in the developmental construction of behavior and the coactions among them. The model includes both neural and non-neural elements, the latter encompassing such influences as hormones (which constitute part of the extracellular biochemistry), bones, muscles, feathers, and so forth. Sensory stimulation is shown to be influenced not only by behavior (as the animal moves about in its environment, both producing and modifying the stimulation it receives), but also by the connectivity of its nervous system (which partly determines its sensitivity to sources of stimulation) and by the current state of neural activity. The elliptical arrow depicts the effects of spontaneous neural activity. All enduring experiential effects on development that have their immediate impact on patterns of neural activity act by modifying events at the cellular level, including patterns of genetic activity. Note that there is no direct connection between genetic activity and behavior; all genetic effects on behavior are mediated through the cell membrane and subsequent coactions among cells and neural networks. Solid lines with arrows represent causal relationships between coacting factors. Dotted lines connecting patterned neural activity to individual nerve cell activity indicate that the latter is nested within the former; the relationship between the two is not causal. (From Johnston and Edwards[15] with the permission of the authors and the American Psychological Association.)

coactions that follow from protein synthesis and its consequences for events at the cell membrane, coactions among cells, and so on. The model treats genes as an integral part of the developing system, rather than placing them outside the system, and sees genes as influencing behavior indirectly, not directly. When a particular gene has been implicated in the development of some behavior, the model would accommodate the identification of various roles that the gene's activity might play in development. For example, in a review of the effects of single-gene mutations on the development of touch receptors in nematodes, Chalfie[5] proposed four developmental roles (generation, specification, function, and maintenance) for the 18 genes that have been implicated so far in touch receptor development. Although Chalfie's taxonomy deals with anatomy rather than behavior, similar taxonomies might be proposed for the development of behavior (e.g., Gottlieb's delineation of the three roles of experience in behavioral and neural development: induction, facilitation, and maintenance[10]). The Johnston–Edwards model (Figure 2.2) has the advantage that it provides an explicit representation of the intervening coactions implied by such taxonomies, even though their coactions may not always be specified. Thus, it indicates the kinds of developmental questions raised by findings that might otherwise be taken as evidence of a direct link between genes and behavior.

Third, when experience has more than a transient effect on behavior, the effect is almost certainly mediated through changes in genetic activity. The model implies that all instances in which experience has been shown to affect behavioral development must involve some change in genetic activity. Developmental theory holds that there can be no genetic effects on behavior independent of the environment and there are probably no environmental effects on behavior independent of genetic activity. The model (Figure 2.2) helps to make this statement more precise by showing the pathway by which experience activates genes through the agency of neural activity, once again supplying a mediating pathway for findings that might otherwise be interpreted as evidence of a direct link between genes and behavior.

Fourth, Johnston and Edwards's model in Figure 2.2 recognizes that nervous system activity needs to be considered at two levels—in terms of neural activity, often involving networks of cells in different anatomical regions, and in terms of the activity of individual nerve cells within which the genes are located. The dotted lines connecting these two boxes in Figure 2.2 indicate that the activity of individual cells is nested within the patterns of activity of cell networks. Individual cell activity neither causes nor is caused by the patterns of activity in cell networks: rather, there are two levels at which neural activity must be analyzed. Such a dual level of analysis does not mean that an account of genetic activity must be given individual cell by individual cell. Rather, ways of describing the developing nervous system both in terms of populations of cells with similar patterns of genetic activity and in terms of networks of cells that participate in behavior must be found. For example, Brennan, Hancock, and Keverne have shown that the immediate-early genes c-*fos* and *zif*-268 (but not c-*jun*) show distinct patterns of both transient induction and persistent induction in the accessory olfactory bulb (AOB) of female mice immediately after mating.[2] The induction of c-*fos* is seen only in the granule cells of the AOB, whereas the *zif*-268 induction occurs in both the granule and mitral cells. The AOB is known to be involved in changes in female olfactory responsiveness to male pheromones following mating, and the differential patterns of gene expression in these two cell types indicate the complexity of the relationships that are likely to exist between neural networks and genetic activity in the AOB.

A further caution to those like myself, who wish to establish correlative links between genetic activity and behavior, is posed by the phenomenon of RNA-editing. The role of genetic activity in producing protein is often depicted in a much too simple unilinear formula: DNA → RNA → Protein, in which DNA establishes the nucleotide structure of mRNA which then translates into the structure of the particular protein resulting from the process. It may come as news to many biologists that Erwin Chargraff[6] could have this to say as early as 1978: "One of the obnoxious dogmas to which it [molecular biology] has given rise—the so-called Central Dogma: DNA makes RNA; RNA makes proteins—is no longer valid." We now know that stretches of DNA are composed of exons (coding portion) and introns (noncoding portion). Before the structure of the protein is composed, nearly 40% of human genes are "alternatively spliced" (i.e., mRNA is edited), not only removing the silent introns but replacing some of the coding exons! This enormously complicates the identification of which genes are actually involved in the making of particular proteins. As stated on p. 1346 of the report *The Sequence of the Human Genome*:[26] "…as was true at the beginning of genome sequencing, ultimately it will be necessary to measure mRNA in specific cell types to demonstrate the presence of a gene." Meanwhile, while we continue to try to correlate genotypes with events at the neural and behavioral levels, we need to remind ourselves of the uncertainty involved. Since much of the genetic analysis in humans and other species involves single nucleotide polymorphisms (SNPs), which are markers for as yet unidentified genes, RNA-editing adds yet a further complication in linking particular genes with particular proteins. It is estimated that one human gene may produce up to five different proteins as a result of alternative splicing.[20]

Before leaving the topic of development as explanation, for the sake of completeness it is essential to mention an often-overlooked factor in developmental neurobehavioral genetics: the role of the *absence* of genetic activity in development. When discussing mental disorders it is common to hear talk of mutations. What is sometimes not distinguished in the absence of genetic activity (deletions) is a frank mutation of a gene in which the usual genetic structure is somehow scrambled, repeated, and/or translocated, thereby resulting in an aberrant protein rather than the mere absence of the normal protein. Talk of "disease genes" is particularly misleading when the culprit is in fact merely the lack of genetic activity and thus the absence of a functional protein essential to normal development as occurs not only in the well-known case of phenylketonuria (PKU) mental retardation but a host of neuropsychiatric disorders such as Prader-Willi syndrome, fragile X syndrome, Williams syndrome, and lissencephaly, among others (see Table 1 in Brodsky and Lombroso,[3] p. 3). If we are to be able to clarify the basic molecular mechanisms in normal and abnormal development, it is necessary to distinguish the role of mutated proteins from the functioning of normal proteins. To discourse about nonexistent "disease genes" when solely deletions are involved only serves to confuse matters in an already very complex situation.

2.3 A CURRENTLY ACCEPTABLE DEVELOPMENTAL NEUROBEHAVIORAL GENETICS: INTERIM SOLUTION

Figure 2.2 portrays a finished or "compleat" developmental neurobehavioral genetics. It is an ideal to be sought and will take many years to achieve. In the meantime we need a model that can be utilized in our day-to-day efforts, so I offer the following, which is based on the notion of the ubiquity of gene-environment coaction documented earlier.[9]

The failure of replications of gene → phenotype associations are legion (extensive review in Wong, Buckle, & Van Tol[27]), the reason being that the environmental (or what I shall call the life experience) component is left out when one merely tries to relate a gene (or genes) to an outcome (e.g., genes for schizophrenia, genes for aggression, genes for novelty-seeking). I am proposing that it is essential to link the gene(s) with a life experience to get a consistent (replicable) result, a proposal that is now receiving some attention in the clinical literature.[17] For example, in a study of the association of a certain genotype (TT genotype) with low concentrations of high density lipoprotein (HDL) cholesterol, it was found that only those TT bearers whose fat intake was at least 30% of their total consumed energy manifested low HDL; the other persons in the sample with the TT genotype did not manifest low HDL.[19] The crucial environmental or life-experience factor here is ingestion of a certain level of particular nutrients. Likewise, in a behavioral example of individuals with a genotype that was associated with low levels of the enzyme monoamine oxidase A (MAOA), it was those who had experienced severe maltreatment in their younger years that were prone to violence in adulthood—those having the same genotype who experienced no maltreatment were unlikely to be violent in adulthood.[4] Eighty-five percent of the males having the low-activity MAOA genotype who were severely maltreated developed some form of antisocial behavior; 15% did not develop antisocial behavior. Obviously, even with the inclusion of the "crucial" life-experience factor, we are still talking about probabilities (epigenesis continues to be probabilistic).

Some of us take it as a given that genes, in and of themselves, cannot produce any neural or behavioral outcome and that gene–environment interaction is a requirement of normal as well as abnormal development. Thus, my probabilistic epigenetic model of developmental outcomes assumes that individuals of the same genotype can have different neural and behavioral outcomes according to the *dissimilarity* of their relevant life experiences, broadly construed. I think this is the basis for the lack of replications among studies that look only at genotypes and attempt to correlate a particular genotype with a certain neural or behavioral outcome without looking for the presence or absence of intervening life experiences that may be crucial to the presence or absence of the outcome. Take the much-studied inhibitory neurotransmitter serotonin. Low levels of serotonin are associated with depression and alcohol abuse in humans. However, correlates of low serotonin are not behaviorally specific (i.e., low serotonin is involved in a number of psychiatric disorders). In rhesus monkeys, low concentrations of serotonin metabolites (collected from cerebral spinal fluid) are associated with higher levels of impulsive aggression and risk taking (e.g., taking long leaps in Suomi's study[24]). Rhesus infants who develop the least secure attachment with their mothers are also the most likely to have deficits in their central serotonin metabolism. Because there is a positive correlation between maternal and infant serotonin level, a genetic deficit could be involved, but it is possible that aberrant maternal care may make a necessary contribution to the serotonin deficit. To shed light on the genetic and interactive aspect, Bennett et al.[1] genotyped the monkeys in Suomi's laboratory for a known polymorphism (long and short allele) in the serotonin transporter gene (5-HTT). The short allele confers low transcriptional efficiency to the 5-HTT gene promoter (relative to the long allele), so low 5-HTT expression may result in lower serotonergic function. However, evidence for this in humans is inconsistent because the necessary life experience correlates have not been examined. In the case of rhesus monkeys, when attempting to correlate the genetic polymorphism to serotonin metabolism, serotonin concentration did not differ as a function of long or short 5-HTT status for mother-reared monkeys, whereas, among peer-reared monkeys, individuals with the short allele had significantly lower serotonin concentrations than those with the long allele.[1] Thus, the lowered serotonin metabolism was not simply a consequence of having the short allele but required the life experience of peer rearing in this instance. This result supports my idea that the inconsistencies in the human literature are likely due to unknown but influential differences in the experiential histories of the populations under study.

Thus, the notion that the short allele of the 5-HTT gene is inevitably associated with a central nervous system (CNS) deficit or defect is not true: the neural outcome depends on the developmental rearing history of the animal, as well as the particular genotype of the animal itself, what has elsewhere been termed "relational causality."[11] The present finding most likely also explains why there are inconsistencies in the human literature in finding anxiety-, depression-, and aggression-related personality traits associated with variations in the serotonin transporter gene.[1] The association, or lack thereof, does not simply reflect genetic causality but developmental-relational causality.

In sum, the chances of linking genotypes to behavioral (and other) outcomes will be vastly improved when crucial intervening life experiences are routinely included in developmental behavioral genetic investigations. This follows from the empirical work reviewed earlier indicating that gene-environment coactions are the rule in developmental investigations.[9]

While gene-environment coactions are a step up from simple single-gene→ outcome associations, single-gene–single-experience coactions are going to continue to be prone to lack of replication because complex behavioral outcomes are no doubt backed by multiple genes (epistasis) and possibly more than one "crucial" life experience. The ultimate solution will be to actually employ more than one relevant gene and more than one relevant life experience if we are eventually to achieve highly replicable findings. For example, the Caspi et al.[4] study of the low MAOA genotype and maltreatment alluded to earlier—a single-gene life experience study— has been replicated at least once[7] and has not been replicated at least once.[12] So, in my opinion, the "interim solution" proposed in the present section of this essay will be best implemented using more than one genetic marker and, whenever possible, more than one life experience if we aspire to a mature developmental neurobehavioral genetics in which replication is the measure of that maturity.

2.4 SUMMARY AND CONCLUSIONS

With the implementation of the psychobiological models presented herein for the analysis of behavioral development at all levels of inquiry, the analytic apparatus is at hand to pursue a developmental neurobehavioral genetics, or development as explanation. Also, adopting a multigene–multilife experience approach should help to overcome the problem of replicability in the field of behavioral genetics.

ACKNOWLEDGMENTS

The author's research and scholarly activities are supported, in part, by National Institute of Mental Health Grant P50-MH-52429 and National Science Foundation Grant BCS-0126475. Some material from my 2003 *Human Development* article[18] is included in this chapter. I thank Brenda Denzler for editorial assistance with the introduction of this chapter.

REFERENCES

1. Bennett, A.J. et al. Early experience and serotonin transporter gene variation interact to influence primate CNS function, *Molecular Psychiatry*, 7: 118–122, 2002.
2. Brennan, P.A., Hancock, D., & Keverne, E.B. The expression of the immediate-early genes *c-fos*, egr-1, and c-jun in the accessory olfactory bulb during the formation of the olfactory memory in mice. *Neuroscience*, 49: 277–284, 1992.
3. Brodsky, M., & Lombroso, P.J. Molecular mechanisms of developmental disorders. *Development and Psychopathology*, 10: 1–20, 1998.
4. Caspi, A. et al. Role of genotype in the cycle of violence in maltreated children. *Science*, 297: 851–854, 2002.
5. Chalfie, M. Touch receptor development and function in *Caenorhabditis elegans, J. Neurobiology*, 24: 1433–1441, 1993.
6. Chargraff, E., *Hericlitian Fire: Sketches from a Life Before Nature*, Rockefeller University Press, New York, 1978, pp. 106–107.
7. Foley, D.L. et al. Childhood adversity, monoamine oxidase a genotype, and risk for conduct disorder. *Archives of General Psychiatry*, 61: 738–744, 2004.
8. Gottlieb, G. *Individual Development and Evolution: The Genesis of Novel Behavior*, Erlbaum, Mahwah, New Jersey, 2002.
9. Gottlieb, G. On making behavioral genetics truly developmental, *Human Development*, 46: 337–355, 2003.
10. Gottlieb, G. The roles of experience in the development of behavior and the nervous system, in *Neural and Behavioral Specificity*, G. Gottlieb, Ed., Academic Press, New York, chap. 2, 1976.
11. Gottlieb, G., & Halpern, C.T. A relational view of causality in normal and abnormal development. *Development and Psychopathology*, 14: 421–435, 2002.
12. Haberstick, B.C. et al. Monoamine oxidase A (MAOA) and antisocial behavior in the presence of childhood and adolescent maltreatment. *Neuropsychiatric Genetics*, 135B:59–64, 2005.
13. Haldane, J.S. *The Philosophical Basis of Biology*, Doubleday, Garden City, New York, 1931.

14. Ho, M.-W. Environment and heredity in development and evolution, in *Beyond Neo-Darwinism: An Introduction to the New Evolutionary Paradigm*, M.-W. Ho & P.T. Saunders, Eds., Academic Press, London, 267–289, 1984.
15. Johnston, T.D., & Edwards, L. Genes, interactions, and the development of behavior. *Psychological Review*, 109: 26–34, 2002.
16. Keller, E.F. *The Century of the Gene*, Harvard University Press, Cambridge, 2000.
17. Merikangas, K.R., & Risch, N. Genomic priorities and public health, *Science*, 302: 599–601, 2003.
18. Moore, D. *The Dependent Gene: The Fallacy of "Nature vs. Nurture,"* Henry Holt, New York, 2002.
19. Ordovas, J.M. et al. Dietary fat intake determines the effect of a common polymorphism in the hepatic lipase gene promoter on high-density lipoprotein metabolism: Evidence of a strong dose effect in the gene–nutrition interaction in the Framingham study, *Circulation*, 106: 2315–2321, 2002.
20. Peters, R.J.G., & Boekholdt, S.M. Gene polymorphisms and the risk of myocardial infarction—an emerging relation, *New England Journal of Medicine*, 347: 1963–1965, 2002.
21. Rampon, C. et al. Enrichment induces structural changes and recovery from nonspatial memory deficits in CA1 NMDAR1-knockout mice, *Nature Neuroscience*, 3: 238–244, 2000.
22. Rutter, M. et al. Testing hypotheses on specific environmental causal effects on behavior, *Psychological Bulletin*, 127: 291–324, 2001.
23. Schaffner, K.F. Genes, behavior, and developmental emergentism: One process indivisible? *International Journal of Behavioral Development*, 24: 5–14, 1998.
24. Suomi, S.J. A biobehavioral perspective on developmental psychopathology, in *Handbook of Developmental Psychopathology*, A.J. Sameroff, M. Lewis, & S.M. Miller, Eds., Kluwer Academic/Plenum, New York, chap. 13, 2000.
25. Thelen, E. Motor development: A new synthesis, *American Psychologist*, 50: 79–95, 1995.
26. Venter, J.C. et al. The sequence of the human genome, *Science*, 291: 304–1351, 2001.
27. Wong, A.H.C., Buckle, C.E., & Van Tol, H.H.M. Polymorphisms in dopamine receptors: What do they tell us? *European J. Pharm.*, 410: 183–203, 2000.

3 Some Basics, Mendelian Traits, Polygenic Traits, Complex Traits

Byron C. Jones

CONTENTS

3.1 INTRODUCTION AND LEVELS OF INVESTIGATION

To some in the functional school of psychology of the early twentieth century, the theory of evolution gave reason to believe that behavior, as much as morphology and biochemistry, served as means of "fitness." Thus, instinct became as valid a criterion for fitness as length of a limb, rate of a biochemical reaction, etc. This led to a circularity of reasoning that gave the ammunition to the behaviorist school of psychology to eschew all things innate as concerns behavior. Accordingly, the environment was the sole source of behaviors, including their differences in expression and development. For many years, the environmentalist view of acquisition and development of behavior dominated the discipline of psychology in America. Even today, the mere mention of genes as influencing behavior elicits skepticism. In Europe, behavioral biologists took a more heredity-friendly view of behavior and its development, and a great deal of debate ensued between the groups. A modern synthesis of the best of the thinking on both sides has lent credence to the notions that (1) genes do in fact influence behavior, (2) genes interact with the environment and with each other, and (3) the actions of genes and their interactions are accessible for study.

Conrad H. Waddington was a developmental biologist who, from the 1930s until the 1960s, helped to elucidate how genes and their environments could influence morphology and biochemical pathways. He asserted that genes exert multiple effects (pleiotropy) and that the immediate environmental conditions, nutrients, hormones, etc., affected gene action and, hence, development. His concept was, *epigenesis + genes = epigenetics*.[10] This useful framework also applies in the examination of neurobehavioral genetics. As we will see in the coming chapters, phenotypes (observable, measurable characteristics), including behaviors, arise as a function of the genotype, the environment, and gene–environment and gene–gene interactions. Another useful conceptualization is that of G.E. McClearn who makes the distinction between the fixed genotype (that which is inherited from the parents) and the momentary-effective genotype, i.e., the summed action of genes that are expressed at a particular time in the life of an organism.[6]

Neurobehavioral genetics is one of the few disciplines that crosses the (artificial) levels of analysis. Thus, we can investigate neurobiology at anatomical and cell-molecular levels, gene expression to behavior. In later chapters, we shall introduce genetic correlational analysis and complex traits analysis, multiple gene techniques, and techniques that place single genes under scrutiny through amplification and nullification.

3.2 FUNDAMENTAL QUESTIONS. WHAT CAUSES WHAT?

Periodically, the popular media may report that some researcher has identified THE GENE for X (fill in the X for yourself). While this may be true for relatively rare, single-gene disorders, such as cystic fibrosis or Huntington's disease, most of the neurobehavioral maladies result from the action of several genes, acting in concert with the environment. Human neurobehavioral diseases such as drug abuse, alcoholism, depression, and most others fall into the latter category. Moreover, multiple genes may operate independently on a particular phenotype or through networks—another of Waddington's insights.[10]

3.3 PHENOTYPES AND PHENOTYPIC SELECTION

Phenotypes are any measurable characteristic of an organism. How phenotypes are operationally defined, however, can be a daunting task. Take, for example, aggression. Aggression is one of the favorite subjects of animal researchers. We are interested in the genetic and environmental underpinnings, yet aggression *per se* is not a phenotype. It is a low-level construct that collects several different behaviors and describes the relationship between pairs (at least) of individuals. The individual behaviors that constitute the larger construct, however, may be suitable phenotypes. A commonly used measure of aggression is the latency to attack. This is the time that it takes a resident mouse to lunge at an intruder mouse after the experimenter has introduced the intruder to the resident mouse's home cage. Another commonly used measure is the number of attacks. The reader can see that there is an inherent danger in naming social behavior phenotypes, i.e., the behavior of the intruder may indeed influence the behavior of the subject and thus the phenotype is partially confounded and less amenable to genetic analysis.

3.4   GENOTYPE–PHENOTYPE RELATIONSHIPS

Sometimes, observing the phenotype reveals all about the genotype concerning that characteristic. Take, for example, the domestic cat with a calico coat color. The calico consists of orange mixed with black, sometimes in distinct spots, sometimes combined with white and sometimes in a complete mélange (tortoise shell). Black and orange alleles of the same gene, located on the X chromosome are codominant and both are expressed. So, by observing calico, one knows that the cat carries one allele of each. One can also discern that barring the XXY karyotype of Kleinfelter's condition, the cat is also female (the Y chromosome does not have a comparable locus, so the tomcat coat color is dependent upon the allele contained on its sole X chromosome). It is the rare case, however, that the behavioral phenotype reveals all about the genotype. Certain kinds of movement may reveal an individual afflicted with Huntington's disease, although there are possible conditions, other than the disease, that may produce similar movement patterns. The lesson here is that most of the neurobehavioral phenotypes of interest in health and disease are influenced by multiple genes, combining algebraically, some increasing the phenotype, and others decreasing, so that merely observing an animal's or human's behavior or neurobiological functions tells us rather little about the genotype. There are other methods, described throughout this volume, that must be employed for that job.

3.5   MENDEL'S LAWS

Gregor Mendel's major contribution to understanding heredity and genetics is that he proposed that inheritance involves acquisition of matter from parents and that the matter could come in different forms. He then proposed two main laws that form the basis of particulate inheritance and basic mathematical models—what we call today *classical genetics*.

3.5.1   LAW OF SEGREGATION

Mendel proposed that trait characteristics are determined by inherited factors (particulate inheritance, if you will) that, for each trait, there is an alternative form. His examples from the garden included green vs. yellow peas, smooth vs. round peas, short vs. long pea vine length, etc. Moreover, these factors come in pairs that may be of the same type or different types (what we call homogeneous or heterogeneous). Finally, during the time of gamete formation, these factors separate to be recombined with like or different types at fertilization. Some of the factor types (what we call alleles) would dominate over the alternative form, so that if an individual plant inherited the factor for wrinkled peas from one parent and the factor for smooth peas from the other parent, the observed type would be smooth. Mendel then termed smooth the dominant trait. From these observations, Mendel produced the familiar mathematical rules to predict outcomes when homogeneous or heterogeneous matings were performed. Thus, mating pure wrinkled with pure smooth would, in the first filial generation, produce all smooth. Mating these first filial (F_1) plants among one another would produce a mix of smooth and wrinkled in the ratio of 3 smooth to 1 wrinkled for the phenotype, whereas the genotypes would be 1(SS):2(SW):1(WW).

3.5.2 LAW OF INDEPENDENT ASSORTMENT

Mendel knew nothing about chromosomes or genes, so it is understandable that he proposed that all characteristics of an organism were inherited independently, i.e., as if there were physically separated particles in the gametes. The fact that many traits are linked together is one of the great contributions to genetic analysis and, in fact, is called *linkage analysis*.

3.6 LINKAGE

During gametogenesis, or more accurately, meiosis, both of the chromosome homologs replicate to give 4 copies of the chromosome—the tetrad. The inner homologous chromosomes can break from physical pressure and reassemble such that a piece of a maternal homolog, containing several genes, splices with the remainder of the paternal homolog and vice versa. So, genes for various characteristics can be "linked." The lineup of genes (alleles) on a single strand homolog is referred to as the haplotype. Now, under certain conditions, a phenotype may inform the researcher about the genetics of a "linked" phenotype.

3.7 MENDELIAN TRAITS

3.7.1 RECESSIVELY TRANSMITTED TRAITS

There are a few human traits that follow apparent single-gene segregation. Sickle cell anemia (named for the characteristic shape of erythrocytes) is well known among individuals of recent sub-Saharan African descent. It is a recessive condition, meaning that both parents must contribute the sickling allele (substitution of valine for glutamic acid at position 6 of the hemoglobin beta-polypeptide). An individual carrying both alleles for sickling has about a 50% chance of living to age 20 without therapy. Remarkably, those individuals carrying sickling and normal alleles (heterozygotes) are resistant to malaria, one of the big killers in Africa and other parts of the world. Thus, the heterozygous configuration confers "fitness"—the ability to pass one's genes on and helps to explain why this allele is preserved in the population.

In neurobehavioral genetics, there are a number of associated recessive traits, although most are rare. One of the best known of these is phenylketonuria or PKU. An estimated 1 in 10,000 individuals among Caucasians (especially from northern Europe) is afflicted and an estimated 1 in 50 individuals is a carrier of the defective allele(s). Affected individuals lack the ability biochemically to convert the amino acid, phenylalanine to tyrosine. Far from being simple, there are numerous mutations in the gene for phenylalanine hydroxylase (PAH, the critical enzyme) that can render the entire enzyme inactive or with substantially reduced activity. As a result, instead of tyrosine (the base amino acid for the catecholamines) the neurotoxin, phenylpyruvic acid is formed and causes severe neurological damage and consequent mental retardation if left untreated. Not surprisingly, the affliction ranges from mild to severe, depending on the type of mutation, but all forms can be treated by eliminating phenylalanine from the diet. The diet is essential during most of development from childhood to adulthood, and there is some indication that it can be abandoned in adulthood, however the timing is in dispute. A recent study reports good results from children staying on the diet for 10 years.[3] Other forms of PKU caused by a deficiency with the PAH cofactors (pterins) are not so easily treated.

3.7.2 DOMINANT TRAITS

These diseases are less common even than recessive trait disorders. Two examples of lethal dominant genetic diseases are Huntington's disease and familial fatal insomnia. Both require only one copy of the allele to produce the disease thus making the probability of contracting the disease when one parent is afflicted 50% compared to 25% when both parents are carriers as is seen with recessively transmitted traits. Most lethal dominant disorders are expressed during embryogenesis; however, in the case of Huntington's disease and familial fatal insomnia, the symptoms of the disease are manifest after the affected individual has attained reproductive age. In 1983, the genetic marker associated with Huntington's disease was located on human chromosome 4 and in 1993, the gene together with its polymorphisms was named IT15.[8] The important polymorphisms consist of cytosine-adenine-guanadine (CAG) repeating sequences with unaffected individuals having 30 or fewer repeats and affected individuals showing 36 to more than 100 repeats. The number of repeats correlates with disease onset and severity such that the longer the repeat sequence, the earlier age of the onset of the disease.

3.8 COMPLEX TRAITS AND ADDITIVE GENETIC EFFECTS

Most of the phenotypes of interest in the study of brain–behavior relations are called complex traits. Complex traits analysis of these phenotypes implies that they are influenced by multiple genes, the environment and their interactions. As evidence for this, when we examine their frequency distributions, we often observe that the shape of the distribution is Gaussian, whereas with single-gene influenced traits,

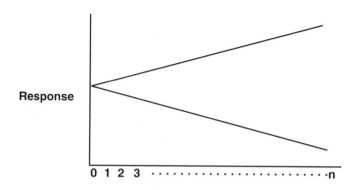

Generation Number

FIGURE 3.1 Theoretical representation of progression of phenotype-based bi-directional selection.

we would expect to see a bimodal distribution. More than 30 years ago, McClearn and Kakihana produced lines of mice selected for high sensitivity and low sensitivity to the hypnotic effects of ethanol.[7] These lines were named long-sleep and short-sleep, respectively. Figure 3.1, is a theoretical representation of how bi-directional selection of a complex trait should progress across generations. Thus alleles that confer increased or decreased sensitivity would accumulate across generations. At the end of more than 40 generations of bi-directional selection, it was concluded that at least four polymorphic genes are involved in sensitivity to the hypnotic effects of ethanol.[5] Thus the genes work together in *additive* fashion in their influence on the phenotype.

3.9 INTERACTIONS

Genes do not operate in isolation. As Waddington[9] proposed, how genes operate in embryogenesis depends in large part on the extra- (or epi-) genetic milieu, e.g., nutrients, toxins, etc. We also know that genes can influence one another to enhance or suppress expression.

3.9.1 GENE–ENVIRONMENT INTERACTIONS

In the example of selection for hypnotic sensitivity to ethanol above, one of the questions addressed the stability of the selected phenotype. In fact—and relevant to the earlier discussion about phenotypes—long-sleep and short-sleep mice differ not only in the time spent sleeping, but in the blood ethanol concentrations (BEC) at which they wake up. Long sleep (LS) wake up at 200–300 mg/dl BEC and short-sleep (SS) wake up at 500–600 mg/dl BEC. Thus, it is clear that the difference between the LS and SS mice is based on differential target tissue (brain) sensitivity. Jones and colleagues[4] investigated the stability of the selected phenotypic difference and performed a simple experiment. Normally, the LS and SS mice are housed in groups of 2–4 with sexes separated except for breeding. How mice are housed—grouped or individually—has a great impact on the central nervous system,[11] so Jones et al. subjected LS and SS male mice to isolation housing following weaning at 21 days of age until 2 months of age. The hypothesis was that isolation housing would decrease the sensitivity of LS mice to alcohol but have no effect on the SS mice. In fact, as predicted, the isolate-housed LS mice slept for a shorter period of time and their BEC was on average 20% higher than their group-housed cohorts. Surprisingly, the isolated SS mice also slept for a shorter period of time, but their BEC was not different from the group-housed animals. The conclusions were (1) the genetic-based mechanisms underlying brain sensitivity to ethanol were altered by isolate housing in LS but not in SS mice and (2) the mechanisms responsible for eliminating ethanol from the blood were changed by isolate housing in the SS but not LS mice.

In 1987, Cloninger published a seminal paper on alcohol-related disorders typologies[1] in humans. According to Cloninger, one type of alcohol-related disorder has a genetic component but is environment limited. That is, manifest only under highly stressful conditions such as a bad marriage, employment or other personal difficulties.

3.9.2 GENE–GENE INTERACTIONS

Gene actions can influence the actions of other genes. This effect is called epistasis. Some of the most salient instances of these interactions can be seen in animals that have been genetically modified, especially mice that have had genes inactivated or "knocked out." The phenotype observed in such animals is dependent upon the genetic background on which the mutation is placed.[2]

3.10 SUMMARY

Basic knowledge about what genes are, what they do, and how they influence neurobiology and behavior has grown exponentially over the past few years. Complex traits analysis is the "new look" in genetics and holds great promise for helping us to see brain–behavior relationships from a systems approach.

REFERENCES

1. Cloninger, C.R. Neurogenetic adaptive mechanisms in alcoholism, *Science,* 1987, 236:410–416.
2. Crawley, J.N. Unusual behavioral phenotypes of inbred mouse strains, *Trends. Neurosci.*, 1996, 19:181–182.
3. Griffiths, P., Paterson, L., and Harvie, A. Neuropsychological effects of subsequent exposure to phenylalanine in adolescents and young adults with early-treated phenylketonuria, *J. Intellect. Disabil. Res.*, 1995, 39:365–372.
4. Jones, B.C., Connell, J.M. and Erwin V.G. Isolate housing alters ethanol sensitivity in long-sleep and short-sleep mice, *Pharmacol. Biochem. Behav.*, 1990, 35:469–472.
5. Markel, P.D., Fulker, D.W., Bennett, B., Corley, R.P., DeFries, J.C., Erwin, V.G., and Johnson, T.E. Quantitative trait loci for ethanol sensitivity in the LS x SS recombinant inbred strains: interval mapping, *Behav. Genet.*, 1996, 26:447–458.

6. McClearn, G.E. Combining molecular and quantitative genetics: decomposing the architecture of lifespan development. In: *Understanding Human Development: Dialogues with Lifespan Psychology.* U. M. Staudinger, and Lindenberger, U., eds., Springer, New York, 2003, 376.

7. McClearn, G.E., and Kakihana, R. Selective breeding for ethanol sensitivity: Short-sleep and long-sleep mice. In: *Development of Animal Models as Pharmacogenetic Tools.* G. E. McClearn, R.A. Deitrich, and V.G. Erwin, eds., 1981, *DHHS Publication No. (ADM)* 81–1133. Washington, DC: U.S. Government Printing Office.

8. The Huntington's Disease Collaborative Research Group, A novel gene containing a trinucleotide repeat that is expanded and unstable on Huntington's disease chromosomes, *Cell*, 1993, 72:971–983.

9. Waddington, C.H. The genetic control of development, *Symp. Soc. Exp. Biol.,* 1949, 145:145–154.

10. Waddington, C.H. *Principles of Development and Differentiation.* The Macmillan Company, New York, 1966, 14, ff.

11. Wilmot, C.A., VanderWende, C. and Spoerlein, M.T. Behavioral and biochemical studies of dopamine receptor sensitivity in differentially housed mice. *Psychopharmacology*, 1986, 89:364–369.

4 An Introduction to Quantitative Genetics

Wim E. Crusio

CONTENTS

SUMMARY

This chapter provides a brief overview of quantitative-genetic theory. Quantitative genetics provides important tools to help elucidate the genetic underpinnings of behavioral and neural phenotypes. This information can then provide substantial insights into the previous evolutionary history of a phenotype, as well as into brain–behavior relationships.

    The most often employed crossbreeding designs are the classical Mendelian cross and the diallel cross. The information rendered by the former is limited to the two parental strains used and cannot be broadly generalized. The principal usefulness of this design is for testing whether a given phenotype is influenced by either one gene or by more genes. The diallel cross renders more generalizable information, the more so if many different strains are used, such as estimates of genetic correlations. To estimate the latter, correlations between inbred strain means may provide a helpful shortcut.

Some commonly encountered mistakes in the interpretation of the results of quantitative-genetic studies are presented and explained.

4.1 INTRODUCTION

Behavior is an animal's way of interacting with its environment and is therefore a prime target for natural selection. Furthermore, as behavior is the output of an animal's nervous system, this indirectly leads to selection pressures on neuronal structures. In consequence, each species' behavior and nervous system have co-evolved in the context of their natural habitat and can be properly comprehended only when their interrelationships are regarded against that background.[4] To arrive at a profound understanding of neurobehavioral traits, one will therefore have to consider problems of causation. Van Abeelen[39] distinguished between the phenogenetic and the phylogenetic aspects of causation. Both aspects deal with the genetic correlates of neurobehavioral traits, the first in a gene-physiological, the second in an evolutionary sense. In other words, neurobehavioral geneticists try to uncover the physiological pathways underlying the expression of a trait and to evaluate its adaptive value for the organism. As I have shown before,[9] quantitative-genetic methods may be employed with profit to address problems related to both aspects of causation.

4.2 PHYLOGENETIC ASPECTS OF CAUSATION

4.2.1 Natural Selection and Genetic Architecture

Selection pressures mold a population's genetic makeup, which subsequently will show traces of this past selection. Therefore, information about the genetic architecture of neurobehavioral traits might permit us to deduce the probable evolutionary history of these traits.[2] With genetic architecture we mean all information pertaining to the effects of genes influencing a particular phenotype in a given population at a given time, including information concerning the presence and size of certain genetic effects, the number of genetic units involved, etc. Generally, however, information about the presence and nature of dominance suffices.[7]

With very few exceptions, natural, nonpathological variation in neurobehavioral phenotypes is polygenically regulated. If dominance is present, we may envisage two different situations: either (uni)directional or ambidirectional dominance. In the first case, dominance acts in the same direction for all genes involved (e.g., for high expression of the trait), whereas in the latter case it acts in one direction for some genes and in the opposite one for others. In its most extreme form, ambidirectional dominance may lead to situations where an F_1 hybrid is exactly intermediate between its parents, despite the presence of strong dominance effects.

Mather[30] distinguished between three kinds of selection: stabilizing, directional, and disruptive. Stabilizing selection favors intermediate expression of the phenotype, directional selection favors either high or low expression, whereas with disruptive selection more than one phenotypic optimum exists. Disruptive selection will lead to di- or polymorphisms, which may be in stable equilibrium or may even lead to breeding isolation and incipient speciation.[37] The most common example of a dimorphism is the existence of two sexes, whereas a possible example of speciation as a consequence of disruptive selection is the explosive adaptive radiation and speciation found among fish species belonging to the family Cichlidae in the great East African lakes.[20]

Selection acts in favor of those genotypes that not only produce the phenotype selected for, but also are capable of producing progeny that differs little from this phenotype. In the end, this results in a population whose mean practically coincides with the optimum. Stabilizing and directional selection have therefore predictably different consequences for the genetic architecture of a trait. Genes for which a dominant allele produces a phenotypic expression opposite to the favored direction will become fixed very rapidly for the recessive allele under directional selection. A similar rapid fixation will then occur for genes for which dominance is absent, i.e., where the heterozygote is intermediate between the two homozygotes. In contrast, selection against recessive alleles is much slower. The result will be that after only a relatively short period of directional selection, the first two types of genes will not contribute to the genetic variation within the population anymore. Those genes where the dominant allele produces the favored phenotypical expression will remain genetically polymorphic for a much longer time, conserving genetic variance. Thus, directional selection leads to situations where dominance is directional, in the same direction as the selection. Stabilizing selection leads to situations where dominance is either absent or ambidirectional. Furthermore, directional selection generally results in lower levels of genetic variation than ambidirectional selection does.[30] The genetic architecture of a trait may be uncovered by using appropriate quantitative genetic methods.

4.2.2 Theoretical Background

At this point, a brief excursion into the field of quantitative genetics is necessary. For the sake of simplicity, considerations of possible interactions between genes (epistatic interactions) will be omitted from the present treatment. Similarly, we will assume that sex-linked genes as well as pre-and postnatal effects are absent. Pertinent references and more technical details for cases where these simplifying assumptions do not hold true may be found in Crusio.[7]

Classical Mendelian analysis studies characters influenced by a limited number of genes, two or three at the most, that are easy to separate into discrete phenotypic classes. Many characters, however, show continuous gradations of expression and are not separable into discrete classes (so-called quantitative traits). The simultaneous actions of a large number of genes (polygenes), combined with phenotypic deviations caused by variations in the environment (environmental variance, V_E), may explain such patterns of phenotypic variation.[19] The genetic contribution to a phenotype can then be divided into two main sources: additive-genetic effects and dominance deviations. In the case of one single gene, with alleles A and a, we may denote the phenotypical values of the three possible genotypes as follows*:

$$AA = m + d_a \qquad Aa = m + h_a \qquad aa = m - d_a \qquad (4.1)$$

The parameter d_a is used to represent half the difference between the homozygotes, h_a designates the deviation of the heterozygote from the midparental point m. Note that in quantitative genetics capital letters are used to indicate increaser alleles, which are not necessarily also the dominant ones. Hence, d_a is positive by definition, whereas h_a may attain all possible values. If we now consider two inbred strains A and B in a situation where many genes affect the phenotype, we may denote the average phenotype of strain A by

$$m + S(d_+) + S(d_-) \qquad (4.2)$$

(shortened to $m + [d]$ for ease of representation), where $S(d_+)$ indicates the summed effects of those genes that are represented by their increaser alleles and $S(d_-)$ indicates the same for decreaser alleles. Parameter m is a constant, reflecting the average environmental effects both strains have in common as well as genetic effects at loci where the strains are fixed for the same alleles. The average phenotype of strain B will then equal $m - [d]$. Similarly, the phenotypic value of an F_1 hybrid between A and B may be written as

$$m + S(h_+) + S(h_-) \qquad (4.3)$$

(shortened to $m + [h]$). It must be noted that $[h]$ is the sum of the dominance deviations of many genes. If these effects are balanced in opposite directions, $[h]$ can be low or zero, even with dominance present. The same applies to $[d]$, of course.

Because variations in a phenotype can be thought of as the summed effects of variations in genotype and environment, plus the interaction and covariation between these two factors, we may express the phenotypic variance P of a population as:

$$P = G + E + G*E + 2cov(g,e) \qquad (4.4)$$

In the controlled situation of animal experiments in the laboratory (but not in the field) the covariance between genotype (g) and environment (e) can be minimized. Further, the absence of genotype–environment interaction means that all individuals, regardless of phenotype, will have similar sensitivities to environmental variations. As a result, genetically homogeneous groups (for example, inbred strains and their F_1 hybrids) should exhibit similar variances. If such is not the case, the effects of ($G*E$) may often be removed by choosing an appropriate measurement scale,[6] leaving

---

* In this chapter, we will follow the notation of Mather and Jinks.[31] In the appendix I present a table comparing this notation with the one followed by Falconer and Mackay.[16]

$$P = G + E \tag{4.5}$$

The genetic component of the variance (G) can, of course, be divided into components due to additive-genetic variation (D) and dominance deviations (H).

We may demonstrate the partitioning of genetic variance into its additive-genetic and dominance components by the example of an F_2 cross between two inbred strains. When only one gene with two alleles influences the phenotype, the expected genetic composition of the F_2 population will be AA, 25%; Aa, 50%; and aa, 25%. From the foregoing, this leads to a phenotypic mean of

$$\tfrac{1}{4}\, d_a + \tfrac{1}{2}\, h_a - \tfrac{1}{4}\, d_a = \tfrac{1}{2}\, h_a \tag{4.6}$$

(m is set at zero by a simple shift of the measurement scale). The sum of squares of deviations from the mean then equals

$$\tfrac{1}{2}\, d_a^2 + \tfrac{1}{4}\, h_a^2 \tag{4.7}$$

In the absence of epistasis and linkage, the contribution of a number of genes (k) to the F_2 variance becomes

$$\tfrac{1}{2} \sum_{i=1}^{k} d_i^2 + \tfrac{1}{4} \sum_{i=1}^{k} h_i^2 \tag{4.8}$$

shortened to

$$\tfrac{1}{2}\, D + \tfrac{1}{4}\, H \tag{4.9}$$

for ease of representation. The total phenotypical variance of an F_2 is thus

$$V_P = \tfrac{1}{2}\, D + \tfrac{1}{4}\, H + E \tag{4.10}$$

By using groups with different genetic compositions it will be possible to obtain estimates for the three parameters D, H, and E. From these parameters we may then estimate the proportion of the phenotypical variance due to additive-genetic effects, the heritability in the narrow sense

$$h_n^2 = (\tfrac{1}{2}\, D)/V_P \tag{4.11}$$

and the proportion of the phenotypical variance due to all genetic effects, the heritability in the broad sense

$$h_b^2 = (\tfrac{1}{2}\, D + \tfrac{1}{4}\, H)/V_P \tag{4.12}$$

When investigating populations other than an F_2 between two inbred strains (such as a diallel cross), allele frequencies need not be identical. Using u to indicate the frequency of the increaser allele and v as the frequency of the decreaser allele (with $u = 1-v$) we may amend the definitions of D and H as follows:

$$D = \sum_{i=1}^{k} 4u_i v_i d_i^2 \qquad (4.13)$$

and

$$H = \sum_{i=1}^{k} 4u_i v_i h_i^2 \qquad (4.14)$$

(Formally, this definition of H should be called H_1, to distinguish it from H_2, the other one of the two diallel forms of H.[31])

It should be noted that because D and H represent summations of the squared effects of single genes, they can only be zero if additive-genetic effects or dominance, respectively, are absent. This is in obvious contrast to $[d]$ and $[h]$.

The crossbreeding designs employed most often in neurobehavioral genetic studies are the classical Mendelian cross and the diallel cross (for other possible designs, see Chapter 10 and Crusio[7]). The former consists of two inbred strains and, at least, their F_1 and F_2, often supplemented with backcrosses of the F_1 with both parentals. The latter consists of a number of inbred strains (at least 3) that are crossed in all possible combinations. The two designs render different types of information on the genetic architecture of a trait. The classical cross permits very detailed genetic analyses and the detection of very small genetic effects, but on a very restricted sample of two inbred strains, only. Furthermore, it is very hard and almost always outright impossible to distinguish between directional and ambidirectional dominance when using this design, because a significant parameter $[h]$ only indicates that dominance is present and, at least, not completely balanced. In fact, a classical Mendelian cross is, generally speaking, only useful if one wants to establish whether the difference between two inbred strains is determined by either one gene or by more genes. In contrast, the analysis of a diallel cross renders information on a larger genetic sample and is therefore much more generalizable, but this carries a price in that the information obtained is less detailed. A great advantage, however, is the possibility to distinguish between ambi- and unidirectional dominance. To uncover the genetic architecture of a trait, the diallel cross will be nearly always the design of choice.

When results from different crosses are available, it should be realized that *comparing* them is not very informative. Obviously, different results may be obtained depending on the genetic makeup of the parental strains used, especially if the crossbreeding design employed has a low generalizability (e.g., the classical cross). However, such results may be *combined* in order to provide a more complete and generalized picture of the genetic architecture of a trait. For instance, if one cross-breeding experiment indicates dominance in the direction of, say, high expression of the trait, but another cross indicates dominance in the opposite direction, then this constitutes *prima facie* evidence for ambidirectional dominance. In fact, the presence of directional dominance may only be inferred if all available evidence indicates that dominance is acting in the same direction.

4.2.3 EXAMPLES

Crusio and van Abeelen[14] addressed the question of what exactly is the adaptive value of various mouse exploratory behaviors carried out in novel surroundings. As one result of exploration is the collection of new, or the updating of previously acquired, information, we argued that, if an animal enters a completely novel environment, it is obviously of prime interest to collect as much information as possible in a short time. On the other hand, high exploration levels will render the animal more vulnerable to predation. Taking all together, we hypothesized an evolutionary history of stabilizing selection for exploration. This hypothesis was subsequently confirmed by the results of several crossbreeding experiments[11,14] that revealed genetic architectures comprising additive genetic variation and/or ambidirectional dominance for most behaviors displayed in an open field. The above reasoning is, of course, not specific for the mouse species. Indeed, Gerlai et al.[21] found similar genetic architectures for exploration in an open field in a diallel cross between inbred strains of paradise fish, *Macropodus operculatus*.

4.3 PHENOGENETIC ASPECTS OF CAUSATION

4.3.1 THE CORRELATIONAL APPROACH: BRAIN LESIONS AND THE LOCALITY ASSUMPTION

When a behavioral neuroscientist wants to investigate the function of some brain system, he or she often will do so by manipulating the system in question. Brain lesions or pharmacological interventions to impair the functioning of the structure of interest are among the most-often-used techniques. Some time ago, Farah[18] reviewed the problems connected with the use of the so-called locality assumption, that more or less equalizes the function of an impaired structure with the defects exhibited by the damaged brain and which is almost always invoked to interpret the results of interventionist studies. In an elegant way, Farah[18] provided evidence that this reasoning may lead to false conclusions. An additional disadvantage of interventionist studies is related to the fact that large interindividual differences in brain structure exist. This heritable variation of the brain is an aspect that many neuroscientists tend to ignore, most likely at their own peril. For example, widely divergent behavioral effects of septal,[15,17] or limbic-system lesions,[1] and pharmacological interventions in the hippocampus[40] have been reported in mice, depending on which particular inbred strain was being used.

It appears that, in the field of neurobehavioral genetics, an alternative approach that does not suffer from these drawbacks exists: using genetic methods exploiting naturally occurring individual differences as a tool for understanding brain function. No brain is like another and every individual behaves differently. The assumption that there is a link between the variability of the brain and individual talents and propensities seems quite plausible. This approach differs from the usual one in neuropsychology in two important aspects. First, no subjects are studied that, by accident or by design, have impaired or damaged brains. Rather, all subjects fall within the range of normal, nonpathological variation (provided animals carrying deleterious neurological mutations are excluded). Second, instead of comparing a damaged group with normal controls, we study a whole range of subjects and try to correlate variation at the behavioral level with that at the neuronal level.

This noninvasive strategy is reminiscent of the phrenological approach propagated by Franz Josef Gall (1758–1828); Lipp has coined the name "microphrenology" for it.[29] It appears that, as long as variation in one neuronal structure is independent of that in another, there will be no need for a locality assumption to interpret results of experiments carried out along these lines. Especially when used in combination with methods permitting the estimation of genetic correlations,[7,8] this strategy yields a very powerful approach.

4.3.2 GENETIC CORRELATIONS

A weakness inherent in correlational studies is that a phenotypical correlation between characters does not necessarily reflect a functional relationship. On the other hand, if two independent processes, one causing a positive relationship, the other causing a negative relationship, act simultaneously upon two characters, the effects may cancel each other so that no detectable correlation can emerge.

These problems can largely be avoided by looking at the *genetic* correlations, that is, at correlations between the genetic effects that influence certain characters. Such correlations are caused either by genes with pleiotropic effects or by a linkage disequilibrium. With linkage disequilibrium we mean situations where certain allele combinations at closely linked genes are more frequent than might be expected based on chance. This will occur, for instance, in F_2 crosses between two inbred strains (or populations derived from them, such as Recombinant Inbred or Recombinant Congenic Strains).

By using inbred strains that are only distantly related, the probability that a linkage disequilibrium occurs may be minimized so that a possible genetic correlation will most probably be caused by pleiotropy, that is, there exist one or more genes that influence both characters simultaneously. Thus, for these characters, at least part of the physiological pathways leading from genotype to phenotype must be shared and a causal, perhaps also functional, relationship must exist. It is this special property that makes the genetic-correlational approach such a uniquely valuable addition to the behavioral neuroscientists' toolbox. It should perhaps be noted at this point (as stated by Carey[5] and to be seen easily from the equations below) that the inverse need not be true; that is, the genetical correlation can still be low or even zero although pleiotropic genes are present.

4.3.3 THEORETICAL BACKGROUND

In the univariate analysis, we may partition the phenotypic variation into its components E, D, and H, the environmental, additive-genetic, and dominance contributions. In the bivariate analysis, the covariation between two traits x and y is partitioned into its equivalent components E_{xy}, D_{xy}, and H_{xy}. We may define the latter two parameters as:

$$D_{xy} = \sum_{i=1}^{k} 4u_i v_i d_{xi} d_{yi} \tag{4.15}$$

and

$$H_{xy} = \sum_{i=1}^{k} 4u_i v_i h_{xi} h_{yi} \tag{4.16}$$

where d_{xi} and d_{yi} are the additive-genetic effects of the ith gene on characters x and y, respectively, and h_{xi} and h_{yi} are the respective dominance deviations due to the ith gene. Evidently, only genes that have effects on *both* of the characters x and y contribute to the genetic covariance terms, whereas all genes that affect either x or y contribute only to the respective genetic variance terms. Combining these components of the covariance with the components of the variance obtained in the univariate analyses we may estimate genetic correlations as follows:

$$r_D = D_{xy} / \sqrt{D_x D_y} \tag{4.17}$$

and

$$r_H = H_{xy} / \sqrt{H_x H_y} \tag{4.18}$$

As is the case with normal correlations, genetic correlations are bound by -1 and 1. If a genetic correlation equals unity, then all genes affecting character x also affect character y with gene effects on both characters being completely proportional. A genetic correlation will become zero only in case no gene at all affects both characters simultaneously or in some balanced cases. For instance, if the effects of genes on character x are uncorrelated to the effects on character y (so that some genes influence both characters in the same direction, whereas others do so in opposite directions[5]). Obviously, similar observations can be made about r_E, the correlation between environmental effects on two phenotypes x and y.

In principle, every breeding design allowing the partitioning of variation also enables one to partition covariation. However, in practice some designs turn out to be not very well-suited to estimating genetical correlations. Especially the classical cross is very problematic in this respect and although many examples exist in the literature in which authors claim to have analyzed genetic correlations with this design, none really have done so. The problem appears to be due mainly to the frequent occurrence of the phenomenon, first observed by Tryon,[38] that the variance of a segregating F_2 population is not significantly larger than those of nonsegregating populations (in extreme cases, it will even be smaller). Several possible explanations have been brought forward. First, Hall[23] attributed the Tryon effect to an insufficient degree of inbreeding of Tryon's selected (but not inbred) strains. Of course, this would enlarge the genetic variation within the parental and F_1 generations, but one would still expect the F_2 to have a somewhat larger variance. In addition, the Tryon effect has since also been observed in crosses between highly inbred strains. A second explanation was presented by Hirsch.[26] He argued that most phenotypes are influenced by more than one gene. If we take the rat, with a karyotype of 21 chromosome pairs, as an example and, for simplicity, treat these chromosomes as major indivisible genes, one can see that this organism can produce 2^{21} different kinds of gametes, leading to 3^{21} $(= 1.05 \times 10^{10})$ different possible genotypes. In reality, this number will be even larger because chromosomes are not indivisible. Obviously, no experiment can take from an F_2 generation a sample large enough to have all these genotypes represented and Hirsch assumed this sampling effect to lower the observed F_2 variances below expected levels. Tellegen[36] quite correctly countered that as long as the sampling from the F_2 is random, an unbiased estimate of the population variance should be obtained. In addition, it can easily be seen that even if Hirsch's reasoning were correct, the F_2 variance would still be expected to significantly exceed that of nonsegregating generations, being the sum of environmentally induced variation and, in his reasoning, at least some genetic variation.

Bruell[3] had observed that the amount of variance caused by segregation in the F_2 increases if gene effects are larger and decreases if more genes influence the phenotype studied (cf. Equations 4.13 and 4.14). If environmental influences on the phenotype are large, an extremely large sample would be needed to detect the difference in variance between the F_2 and F_1 populations at a sufficient level of significance in situations with many genes and relatively small gene effects. As sample sizes are limited by considerations of time, money, and space, while environmental influences on behavioral characters are usually very pronounced, one should normally expect F_2 variances not to differ significantly from F_1 variances. Due to sampling error, they may then even be smaller, although usually not significantly so. Homeostatic processes may be responsible if the latter situation occurs.[27]

The problem therefore boils down to one of statistical power. It should be recalled here that larger sample sizes are needed to obtain accurate estimates of the variance of a population than for estimating its mean. By analogy, this also goes for covariances. In sum, unless rather huge sample sizes are used, classical crosses will generally lack the statistical power needed to accurately estimate genetical correlations.

Another reason that the classical cross is less suited for bi- and multivariate studies is the fact that results are not generalizable, but based on a restricted sample of two inbred strains only: even if genetic correlations are estimated correctly with this design (which happens only rarely; see Hayman[24] for appropriate statistical methods), there exists a non-negligible probability that they would be due to a linkage disequilibrium instead of pleiotropy. Of course, it is exactly the latter property that researchers employ when localizing genes. Note, however, that this is a special case in which one character, the molecular marker, is completely determined by the genotype (see also the following section).

A more suitable method is the diallel cross, for which a bivariate extension is available,[8] whereas an interesting shortcut is offered by using a panel of inbred strains. Correlations between strain means either permit the estimation of additive-genetic correlations,[25] or provide a direct lower-bound estimate of additive-genetic correlations (if the traits to be correlated have been measured in different individuals from these strains).

From the foregoing, it may easily be seen that the variance of the means of a set of inbred strains equals

$$D + E/n \tag{4.19}$$

where n is the harmonic mean of the number of subjects per strain. The covariance between the means obtained for two characters x and y can then be expressed as

$$D_{xy} + E_{xy}/n \tag{4.20a}$$

in case both characters are being measured on the same individuals or as

$$D_{xy} \tag{4.20b}$$

in case characters x and y are being measured on different individuals from the same strains. The correlation between the strain means in these two situations will now equal

$$(D_{xy} + E_{xy}/n)/ \sqrt{(D_x + E_x/n)(D_y + E_y/n)} \tag{4.21a}$$

or

$$D_{xy}/ \sqrt{(D_x + E_x/n)(D_y + E_y/n)} \tag{4.21b}$$

respectively. Especially if environmental effects are small and large numbers of subjects are being used, the correlation between inbred strain means will approach the genetical correlation. In addition, it can easily be seen that Equation 4.21b will always render a lower-bound estimate of the genetical correlation, even if n is small or if environmental effects are large. It should be realized that Equation 4.21a may render a significant correlation even in the complete absence of any genetical effects. In the latter case, Equation 4.21a reduces to the environmental correlation. By using the within-strain variation and covariation as estimates of the environmental variances and covariances, respectively, Equations 4.19 and 4.20 render unbiased estimates of environmental and genetical correlations for both cases.[25]

Recombinant inbred strains (RISs) have sometimes also been used to estimate genetic correlations between phenotypes, using the above-described methods for correlations using ordinary inbred strains. It should be realized, however, that there is a considerable risk that any genetic correlations thus found will be due to a linkage disequilibrium because RISs have been derived from an F_2 between two inbred strains. As was the case with the classical cross itself, this property is of course of interest for researchers hoping to localize quantitative trait loci (QTL; but see Section 4.3.4.3). In fact, a genetic correlation obtained with RISs implies that both characters under study are influenced by linked genes, i.e., that map to the same chromosomal segments, at least in part. Only if previous evidence of the existence of a genetic correlation has been obtained with other methods (such as a screen of normal, unrelated inbred strains), can a genetic correlation obtained from an RI study be considered evidence of the localization of a QTL caused by a common gene.

Except in the case of correlations between inbred strain means, testing the significance of genetic correlations is often problematic and the power of available tests is not well known. Fortunately, when the environmental and genetic correlations are used as input for further, multivariate, analyses, the possible significance or lack thereof of an individual correlation is no longer very important.

4.3.4 EXAMPLES

4.3.4.1 Rearing Behavior in an Open Field and Hippocampal Mossy Fibers in Mice

Crusio et al.[12] carried out a diallel cross study in which 5 different inbred strains were crossed in all possible ways. In 150 male mice from the 25 resulting crosses they measured the rearing-up frequency during a 20 min session in an open field and the extent of the hippocampal intra- and infrapyramidal mossy fiber (IIPMF) projection. They obtained a marginally significant phenotypical correlation of 0.138 (df = 148, $0.05 < P < 0.10$). Ordinarily, one would take such a result as evidence that variation in the size of the IIPMF is not related to behavioral variation. However, a quantitative-genetic partitioning of the covariation showed that the genetic correlation was quite sizable: 0.479. The low phenotypical correlation was explained by the modest heritability for rearing (0.25 vs. 0.53 for the IIPMF[10]) and by the fact that the (low) environmental correlation had a sign opposite to that of the genetic correlation. The genetic relationship between rearing, on the one hand, and the IIPMF, on the other hand, was confirmed by the finding that a line selected for high rearing frequency had larger IIPMF projections than a line selected for low rearing frequencies,[13] exactly as would be predicted from a positive genetic correlation between these characters. From these data, it was concluded that the IIPMF plays an important role in the regulation of open-field rearing.[9]

4.3.4.2 Nerve Conduction Velocity and IQ in Man

Lately, human behavior geneticists have also started to use genetic correlations to uncover brain–behavior relationships, especially the very active group around Dorret Boomsma at the Free University of Amsterdam (the Netherlands). In a recent experiment they used the twin method (see Chapter 12) to examine the possible existence of a genetic correlation between speed-of-information processing (SIP) and IQ.[35] It was postulated that SIP, as derived from reaction times on experimental tasks, measures the efficiency with which subjects can perform basic cognitive operations underlying a wide range of intellectual abilities. Phenotypic correlations generally range from –0.2 to –0.4. Rijsdijk et al.[35] showed that genetic correlations also fell in this range at ages 16 and 18 years, whereas environmental correlations were essentially zero. A common, heritable biological basis underlying the SIP–IQ relationship is thus very probable.

4.3.4.3   Localization of QTL

Currently, RISs are widely used as a tool to localize QTL. Unfortunately, problems of statistical power (often also due to multiple testing) lead to many false positives and negatives, as illustrated by the studies of Mathis et al.[32] and Gershenfeld et al.[22] In the first study, a number of QTLs associated with open-field behavior were "identified" using a large set of RIS between the inbred mouse strains C57BL/6J and A/J. In the second study, an F_2 generation between these same inbred strains was studied, again leading to the "identification" of a number of QTLs for the same behavioral phenotypes. None of the QTL found was common to both studies, illustrating the problem of both types of statistical errors inherent with the use of recombinant inbred strains. Nevertheless, there have been a few instances where QTL have been replicated within and across laboratories.[28,33] Recently, recombinant strain sets have become much larger through the development of new, large sets as well as of additional strains for existing sets. This development may expect to lead to a vast increase of power and a sizeable reduction of type I and type II errors.

4.4   SOME COMMON MISAPPLICATIONS

A final point of caution is warranted here. Quantitative-genetic methods are not only used, but unfortunately, also regularly abused. One problematic issue already addressed above concerns whether information obtained at one level (on the components of means, say) can render information about another level (on the components of variance or covariance, for instance). This is sometimes the case, sometimes not. A few other frequently made mistakes are mentioned here.

1. *If some genetic effects are found for one character but not for another, this implies that these characters are influenced by different genes.* Wrong! The equations given above should already make it abundantly clear that this is not true. An example may illustrate this.

Albinism in mice, for instance, is a character that is completely recessive as far as coat color is concerned. A quantitative-genetic analysis would indicate the significant presence of both *d* and *h* of equal size, in such a way that the heterozygote would completely resemble the non-albino homozygote. However, if we now would perform a quantitative-genetic analysis of the phenotype "activity of the enzyme tyrosinase," we would find that dominance is completely absent for this character, the heterozygote being completely intermediate between the two homozygotes. Still, as we know very well, only one and the same gene is involved here, which acts as a recessive on the level of coat color, but shows intermediate inheritance on the level of the activity of the responsible enzyme.

In fact, only the presence of a genetic correlation between two phenotypes provides evidence that (a) gene(s) is (are) simultaneously influencing both. As was pointed out above, the reverse need not be true.

2. *IF TWO CHARACTERS ARE CORRELATED BETWEEN TWO PARENTAL STRAINS BUT NOT IN THEIR* F_2, *THEY SEGREGATE INDEPENDENTLY AND, HENCE, ARE INFLUENCED BY DIFFERENT GENES* (IN OTHER WORDS, THERE IS NO *GENETIC* CORRELATION BETWEEN THEM). WRONG! IN FACT, THIS OBSERVATION MAY BE TRUE, BUT IN ONLY ONE EXCEPTIONAL SITUATION: IF AT LEAST ONE OF THE CHARACTERS WE ARE DEALING WITH (FOR INSTANCE, A MOLECULAR-GENETIC MARKER) IS COMPLETELY DETERMINED BY THE GENOTYPE. THIS LATTER CONDITION IS CALLED COMPLETE PENETRANCE OR, IN QUANTITATIVE-GENETIC TERMS, THE HERITABILITY IN THE BROAD SENSE IS SAID TO EQUAL 1. FOR BEHAVIORAL AND NEURAL PHENOTYPES, THIS CONDITION ALMOST NEVER OCCURS. A FEW FURTHER OBSERVATIONS SHOULD BE MADE.

First, in all situations (including the above one), the *phenotypical* correlation within an F_2 generation will be a function of the heritabilities of both characters and the sizes and signs of the genetic and environmental correlations between them. If, to simplify the equation, we suppose dominance effects to be absent for both characters x and y, then

$$r_P = h_x h_y r_D + e_x e_y r_E \qquad (4.22)$$

where r_P is the phenotypical correlation, h_x and h_y are the square roots of the narrow-sense heritabilities of characters x and y, e_x and e_y are the square roots of the "environmentalities" (the proportion of the phenotypical variance due to the environment; $e_x^2 = 1 - h_x^2$), and r_E is the environmental correlation.

Equation 4.22 has a number of important implications. For instance, the size and sign of r_P evidently do not render any information at all about the size and sign of r_D. It can easily be seen that a significant phenotypical correlation may even be completely absent (r_P not significantly different from zero) in case r_D and r_E have opposite signs and the absolute value of $h_x h_y r_D$ comes close to that of $e_x e_y r_E$. In recent years, experiments attempting to determine the locations of polygenes, or QTL, have become ever more popular. In such experiments one character, the molecular-genetic marker whose possible linkage to a putative QTL is being tested, will have a heritability of 1 and zero environmentality. In such a case, Equation 4.22 reduces to

$$r_P = h_x r_D \qquad (4.23)$$

In this case, the additive-genetic correlation r_D is equivalent to the square root of the proportion of the genetic variance explained by the QTL under study. This in turn, is a function of the distance between the genetic marker and the QTL and the relative effect of the QTL compared to other genes. An important implication of this equation is that there exists an upper bound to the correlation between a behavioral or neural phenotype, on the one hand, and a marker locus, on the other hand, *even if this marker locus would not just be linked to the hypothetical QTL but actually be identical with it and even if there is only one single gene influencing the phenotype.* In the latter case, r_D would reduce to 1 and the phenotypical correlation would be expected to equal the heritability. Equation 4.23 explains why, especially when heritabilities are low or numbers of QTL are large (and also because of the lack of statistical power of estimates of variance and covariance in F_2 populations referred to in Section 4.3.3), the power to reliably detect QTL is often very low, leading to many false positives and negatives.[34]

Second, note that in the above erroneous statement the words *between two parental strains* were used. The correlation *within* such strains has obviously no bearing at all on the eventual presence or absence of genetic correlations. As all individuals within inbred strains have the same genotype, any correlations occurring within such a strain are of environmental origin, of course. This is not to say, of course, that at any given point in time, two individuals belonging to the same inbred strain may not have different profiles of gene expression. If such is the case, however, then such differences in gene expression itself must be due to environmental influences (assuming, of course, that both individuals are at similar stages of development).

3. *To determine the heritability of a character one should carry out a selection study.*

This proposition is perhaps formally not incorrect (heritabilities *can* be derived from selection studies), but it contains in fact two conceptual errors. The first one is that there is some information to be gained from a heritability coefficient. Actually, except as an intermediate step in estimating genetic correlations, heritabilities do not have any intrinsic value. The only interesting facts about heritabilities are whether they differ significantly from zero (meaning that there is significant genetic variation) or from unity (meaning that there is significant environmental variation). There is one further use of heritability estimates, which is that their size predicts the eventual effects of selection pressures (whether artificial or natural[16]). The second conceptual error derives from this fact: once a selection study has been carried out and selection has led to the successful establishment of divergent lines, knowledge about heritability become more or less useless: it is not interesting any more to determine whether selection might be successful! Thus, there is only a single situation in which one would perform a selection experiment to estimate heritabilities: when one wishes to estimate genetic correlations and for some reason inbred strains are not available.

4.5 CONCLUSION

In conclusion, if the above pitfalls are avoided, quantitative-genetic experiments can render valuable information on the genetic architecture of a trait. In addition, they can provide information about the multivariate genetic structure of complexes of traits. Because of this last property, quantitative genetics may serve as a valuable additional tool in the neuroscientist's arsenal and may greatly enhance our understanding of the genetic and neural mechanisms underlying individual differences in behavior.

ACKNOWLEDGMENTS

This chapter is dedicated to the memory of my teacher, mentor, and dear friend, Hans van Abeelen (1936–1998).

REFERENCES

1. AMMASSARI-TEULE M, FAGIOLI S, ROSSI-ARNAUD C: Genotype-dependent involvement of limbic areas in spatial learning and postlesion recovery. *Physiol Behav*, 1992, 52:505–510.
2. BROADHURST PL, JINKS JL: What genetical architecture can tell us about the natural selection of behavioural traits. In JHF van Abeelen (Ed): *The Genetics of Behaviour*. North-Holland, Amsterdam, 1974, pp 43–63.
3. BRUELL JH: Dominance and segregation in the inheritance of quantitative behavior in mice. In EL Bliss (Ed): *Roots of behavior*. Hoeber and Harper, New York, 1962, pp 48–67.
4. CAMHI JM: Neural mechanisms of behavior. *Curr Opinion Neurobiol*, 1993, 3:1011–1019.
5. CAREY G: Inference about genetic correlations. *Behav Genet*, 1988, 18:329–338.
6. CRUSIO WE: HOMAL: A computer program for selecting adequate data transformations, *J Hered*, 1990, 81:173.
7. CRUSIO WE: Quantitative Genetics. In D Goldowitz, D Wahlsten, R Wimer (Eds): *Techniques for the Genetic Analysis of Brain and Behavior: Focus on the Mouse. Techniques in the Behavioral and Neural Sciences, Volume 8*, Elsevier, Amsterdam, 1992, pp 231–250.
8. CRUSIO WE: Bi- and multivariate analyses of diallel crosses: A tool for the genetic dissection of neurobehavioral phenotypes. *Behav Genet*, 1993, 23:59–67.
9. CRUSIO WE: Natural selection on hippocampal circuitry underlying exploratory behaviour in mice: Quantitative-genetic analysis. In E Alleva, A Fasolo, H-P Lipp, L Nadel, L Ricceri (Eds): *Behavioural Brain Research in Naturalistic and Seminaturalistic Settings*. NATO Advanced Study Institutes Series D, Behavioural and Social Sciences, Kluwer Academic Press, Dordrecht, Pays-Bas, 1995, pp 323–342.
10. CRUSIO WE, GENTHNER-GRIMM G, SCHWEGLER H: A quantitative-genetic analysis of hippocampal variation in the mouse. *J Neurogenet*, 1986, 3:203–214.
11. CRUSIO WE, SCHWEGLER H, VAN ABEELEN JHF: Behavioral responses to novelty and structural variation of the hippocampus in mice. I. Quantitative-genetic analysis of behavior in the open-field. *Behav Brain Res*, 1989, 32:75–80.
12. CRUSIO WE, SCHWEGLER H, VAN ABEELEN JHF: Behavioral responses to novelty and structural variation of the hippocampus in mice. II. Multivariate genetic analysis. *Behav Brain Res*, 1989, 32:81–88.
13. CRUSIO WE, SCHWEGLER H, BRUST I, VAN ABEELEN JHF: Genetic selection for novelty-induced rearing behavior in mice produces changes in hippocampal mossy fiber distributions. *J Neurogenet*, 1989, 5:87–93.
14. CRUSIO WE, VAN ABEELEN JHF: The genetic architecture of behavioural responses to novelty in mice. *Heredity*, 1986, 56:55–63.
15. DONOVICK PJ, BURRIGHT RG, FANELLI RJ, ENGELLENNER WJ: Septal lesions and avoidance behavior: Genetic, neurochemical and behavioral considerations. *Physiol Behav*, 1981, 26:495–507.

16. FALCONER DS, MACKAY TFC: *Introduction to Quantitative Genetics*. 4th Edition. Longman Group Ltd, Harlow, Essex, 1996.
17. FANELLI RJ, BURRIGHT RJ, DONOVICK PJ: A multivariate approach to the analysis of genetic and septal lesion effects on maze performance in mice. *Behav Neurosci*, 1983, 97:354–369.
18. FARAH MJ: Neuropsychological inference with an interactive brain: A critique of the "locality" assumption. *Behav Brain Sci*, 1994, 17:43–104.
19. FISHER RA: The correlation between relatives on the supposition of Mendelian inheritance. *Trans R Soc Edinb*, 1918, 52:399–433.
20. FRYER G, ILES TD: *The Cichlid Fishes of the Great Lakes of Africa: Their Biology and Evolution*. Oliver and Boyd, Edinburgh, 1972.
21. GERLAI R, CRUSIO WE, CSÁNYI, V: Inheritance of species-specific behaviors in the paradise fish (*Macropodus opercularis*): A diallel study. *Behav Genet*, 1990, 20:487–498.
22. GERSHENFELD HK, NEUMANN PE, MATHIS C, CRAWLEY JN, LI X, PAUL SM: Mapping quantitative trait loci for open-field behavior in mice. *Behav Genet*, 1997, 27:201–210.
23. HALL CS: The genetics of behavior. In SS Stevens (Ed): *Handbook of Experimental Psychology*, John Wiley & Sons, New York, 1951, pp 304–329.
24. HAYMAN BI: Maximum likelihood estimation of genetic components of variation. *Biometrics*, 1960, 16:369–381.
25. HEGMANN JP, POSSIDENTE B: Estimating genetic correlations from inbred strains. *Behav Genet*, 1981, 11:103–114.
26. HIRSCH J: Behavior-genetic, or "experimental," analysis: The challenge of science versus the lure of technology. *Am Psychol*, 1967, 22:118–130.
27. HYDE JS: Genetic homeostasis and behavior: Analysis, data, and theory. *Behav Genet*, 1973, 3:233–245.
28. JONES BC, TARANTINO LM, RODRIGUEZ, LA, REED CL, MCCLEARN GE, PLOMIN R, ERWIN VG: Quantitative-trait loci analysis of cocaine-related behaviours and neurochemistry. *Pharmacogenet*, 1999, 9:607–617.
29. LIPP HP, SCHWEGLER H, CRUSIO WE, WOLFER D, LEISINGER-TRIGONA M C, HEIMRICH B, DRISCOLL P: Using genetically-defined rodent strains for the identification of hippocampal traits relevant for two way avoidance learning: A non-invasive approach. *Experientia*, 1989, 45:845–859.
30. MATHER K: *The Genetical Structure of Populations*. Chapman & Hall, London, 1973.
31. MATHER K, JINKS JL: *Biometrical Genetics, 3rd ed.*, Chapman & Hall, London, 1982.
32. MATHIS C, NEUMANN PE, GERSHENFELD H, PAUL SM, CRAWLEY JN: Genetic analysis of anxiety-related behaviors and responses to benzodiazepine-related drugs in AXB and BXA recombinant inbred mouse strains. *Behav Genet*, 1995, 25:557–568.
33. MCCLEARN GE, TARANTINO LM, RODRIGUEZ LA, JONES BC, BLIZARD DA, PLOMIN R: Genotypic selection provides experimental confirmation for an alcohol consumption quantitative trait locus in mouse. *Molec Psychiat*, 1997, 2:486–489.
34. MELCHINGER AE, UTZ FH, SCHÖN CC: Quantitative trait locus (QTL) mapping using different testers and independent population samples in maize reveals low power of QTL detection and large bias in estimates of QTL effects. *Genetics*, 1998, 149:383–403.

35. RIJSDIJK FV, VERNON PA, BOOMSMA DI: The genetic basis of the relation between speed-or-information-processing and IQ. *Behav Brain Res*, 1998, 95:77–84.
36. TELLEGEN A: Note on the Tryon effect. *Am Psychol*, 23:585–587.
37. THODAY JM: Disruptive selection. *Proc R Soc Lond B*, 1972, 182:109–143.
38. TRYON RC: Genetic differences in maze learning ability in rats. *Yrbk Natl Soc Stud Educ*, 1940, 39:111–119.
39. VAN ABEELEN JHF: Ethology and the genetic foundations of animal behavior. In JR Royce, LP Mos (Eds): *Theoretical Advances in Behavior Genetics*. Sijthoff & Noordhoff, Alphen aan den Rijn, 1979, pp 101–112.
40. VAN ABEELEN JHF: Genetic control of hippocampal cholinergic and dynorphinergic mechanisms regulating novelty-induced exploratory behavior in house mice. *Experientia*, 1989, 45:839–845.

APPENDIX

Comparison of the Notation Systems Used by Mather and Jinks[31] and Falconer and Mackay[16]

Description of Statistic	Mather and Jinks	Falconer
Additive-genetic effect	d	a
Dominance deviation	h	d
Additive-genetic variance	D	$2V_A$
Dominance variance	H	$4V_D$
Environmental variance	E	V_E
Genotypic variance	G	V_G
Phenotypic variance	P	V_P
Additive-genetic correlation	r_D	r_A
Dominance correlation	r_H	r_D
Environmental correlation	r_E	r_E
Phenotypic correlation	r_P	r_P

5 From QTL Detection to Gene Identification

Marie-Pierre Moisan

CONTENTS

SUMMARY

Although quantitative trait loci (QTL) mapping analysis has become very popular, the final goal to positionally clone and identify the relevant gene(s) remains difficult for the genetic dissection of complex traits. This chapter reviews the different approaches that have been used thus far. First, various breeding strategies and haplotype mapping have been developed in order to narrow down the QTL interval. Once high resolution mapping of the QTL is achieved, positional cloning of the relevant gene can be performed. Until a few years ago, a contig of DNA clones covering the QTL interval was constructed and several techniques were used in order to extract the coding sequences of the contig. This is now obsolete in human, mouse and rat species for which genomes are sequenced. Progress in genome annotation and techniques based on microarrays gene expression analysis are used to pull out candidate genes from the fine-mapped QTL interval. Finally, sequence analysis provides candidate mutations that then need to be validated by functional studies.

5.1 INTRODUCTION

Recent progress in genetic strategies and technologies has allowed for the dissection of complex traits in mice, rats and humans. Behavior geneticists took part in this new research field and a high number of QTL for behavioral traits has now been reported in mice[10,12,13,24,26,40,54,61,62] and in rats.[23,42,43,52,57]

Many of these studies are conducted in inbred animals. The advantages are multiple: (1) the environment is controlled, thus its effect on the phenotype is minimized, (2) problems of genetic heterogeneity (different genes causing the same phenotype) are avoided, (3) an adequate number of progeny and type of crosses are available permitting robust statistical testing. The first behavioral QTL mapping studies used recombinant inbred strains (RISs). The main advantage of using these strains is that behavioral measures are conducted on groups of animals of each strain thus strain means and variances become the unit of analysis. This is important for phenotypes such as behavioral traits that are easily influenced by environmental noise. However, because only a limited number of RISs are available from the same panel, mapping resolution is often very low (unless the QTL effect is strong); therefore, QTL detection using RISs is usually replicated on an F_2 population for validation. Although many successful F_2 and backcross QTL mapping analyses have been reported, very few causative gene variants have been identified (see Korstanje[34] for a 2002 report), as the route from QTL detection to gene identification is long and uncertain. This review will attempt to report various strategies that have been used thus far for complex traits with their successes and limitations and the future prospects of this continuously evolving discipline.

5.2 HIGH-RESOLUTION MAPPING

Once a QTL is detected with a significant log of odds (LOD) score (consensus guidelines require LOD > 4.3),[35] the next step is to saturate the chromosomal area with polymorphic DNA markers in order to narrow the confidence interval as much as possible. However at this stage in an F_2 or backcross population, the chromosomal region containing a QTL is still very large, typically 5 to 20 cM. Therefore additional experimental strategies must be undertaken to improve the mapping resolution before moving to a candidate gene approach or positional cloning of the relevant gene.

5.2.1 CONGENIC STRAINS AND SUBSTRAINS

An approach that has been used for many complex traits such as diabetes, hypertension epilepsy and behavior is the construction of congenic strains in which a mapped QTL is isolated from the others present in the parental strain, and then substrains are produced to obtain a higher resolution of its localization.

To obtain a congenic strain, DNA of a QTL region is moved from one parental strain (donor strain) to another (recipient strain). This is achieved first by an F_1 mating of the two parental strains followed by successive backcrosses to the recipient strain (Figure 5.1). Each backcross results in a 50% loss of donor strain alleles, on average. Thus, progressively the genome of the hybrids become of the recipient type

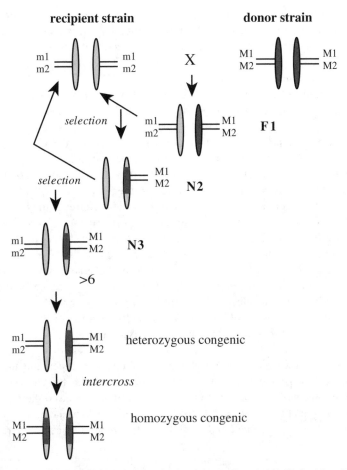

FIGURE 5.1 Congenic strains. A QTL contained in the interval delimited by the markers M1 and M2 is introgressed from a donor strain into a recipient strain by marker-assisted selection. N2 is the first backcross generation.

except at the QTL because at each backcross generation animals are genotyped with genetic markers flanking the QTL. Heterozygous individuals that have retained the donor strain alleles (for example M1 and M2) at the QTL are selected and mated to the recipient strain thus producing the next backcross generation. To speed up the congenic process, individuals of each generation are also genotyped with a set of genetic markers covering the rest of the genome. This time one looks for animals that are homozygous for the recipient alleles at every loci except the QTL. This strategy is referred to as marker-assisted selection (MAS). Using this strategy, full congenics could be obtained in 4 to 6 generations of backcrosses instead of eight backcrosses if no counter-selection on the other chromosome is done.[29] Once het-erozygous full congenics (M1m1, M2m2) are obtained, they are intercrossed in order to obtain homozygous animals (M1M1, M2M2), theoretically one fourth of the progeny. Finally, two homozygous congenic animals are selected, based on their phenotypic differences with the recipient strains, and bred brother–sister to fix the donor strain alleles in the recipient strain background. Alternatively, a heterozygous congenic male is mated to several females of the recipient strain in order to obtain identical congenic animals to be then intercrossed. This procedure takes about two years to complete in a mouse or rat.

Theoretically, a pure congenic strain differs from the parental strain only in the QTL area; thus, rigorous phenotypic comparisons between congenic and parental strains can be made such that: (1) the existence of the QTL is verified (congruent variation in phenotype), (2) the impact of a given QTL on the phenotype can be measured and, (3) interaction between various QTLs can be tested.

Once the existence of a QTL has been verified by the development of a congenic strain, the QTL interval may be further refined by constructing substrains (Figure 5.2). A congenic strain that shows the expected difference in phenotype from the recipient strain is crossed to the latter and the offsprings are backcrossed to the recipient strain, producing a large segregating population. Using a dense genetic map, individuals recombinant at various places throughout the QTL region are then selected. Again, homozygous animals for the desired crossover chromosome must be obtained (by mating 2 heterozygotes) and these homozygotes are bred to fix the genotype in the congenic substrain. Theoretically phenotypic analysis of these various congenic substrains should lead to a refined localization of the QTL.

There are, however, several limitations to this strategy:

1. There is always an unknown amount of unselected genome that may remain associated within the congenic strain and bias the analysis.
2. The effects of such isolated QTL are reduced compared to their effect in the original F_2 or backcross. Indeed, the genetic background of the recipient strain is very important and some QTL may require epistatic interactions

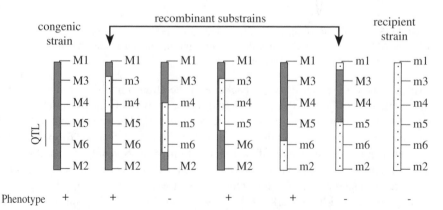

FIGURE 5.2 Fine mapping using congenic substrains. A QTL was localized between markers M1 and M2 in a congenic strain. Phenotypic analysis of congenic substrains recombinant within the M1-M2 interval has allowed to fine-map the QTL to the M5-M6 chromosomal segment.

with other QTL to be expressed (e.g., *Idd18*, a QTL contributing to the development of insulin-dependent diabetes mellitus in the NOD mouse,[50]). In some cases, the phenotypic effects of a single QTL may be too small to be detected (e.g., epilepsy QTL *El3*),[25] or it is reduced to a stage that is hardly reliable (e.g., the congenic substrain for *El2* locus shows a within-strain variance that is close to that of interstrain crosses).[25,36]

3. Many congenic mapping studies have led to the dissection of a QTL (localized by segregation analysis) into several linked QTLs.[50,51] Sometimes these linked QTLs have an opposite effect on the phenotype,[21,63] thus increasing the difficulty of identifying the gene(s) involved.

4. A dense genetic map of polymorphic DNA markers must be available in the QTL area in order to follow properly the introgression of the donor strain chromosomal segment and also to construct substrains.

5. When constructing congenic substrains, the number of animals that must be screened to isolate recombinants over short distances may be excessively large.

Despite these limitations, many QTLs have been fine-mapped by means of congenic strains. As examples, congenic strains with the expected phenotypic effect were obtained for epilepsy,[36] for alcohol and drug–related phenotypes,[8,22] or for saccharin preference.[5,6,38]

5.2.2 ALTERNATIVE STRATEGIES

Recombinant congenic strains (RCS) is an approach related to congenics except that nonlinked genes controlling a complex trait are separated before their localization by linkage analysis. Such strains are obtained by 2 generations of backcrossing followed by 20 generations of inbreeding. In this manner, each RCS of a panel contains, on average, 87.5% genes of a common background and 12.5% genes of a common donor strain. For example, the Ccs series, developed by Demant and Hart[18] contains 20 homozygous RCS derived from the parental strains BALB/c and STS.

Using this panel of RCS, P. Demant's group was able to fine map a colon tumor susceptibility gene (*Scc1*) to a 2.4 cM region thus opening the way to its positional cloning. *Scc1* was first mapped by linkage analysis to a 31 cM interval in an F_2 cross between BALB/c and a colon tumor susceptible RCS, Ccs-19. To achieve a higher resolution, Css-19 mice were crossed again to BALB/c in order to obtain a new segregating population. Sixty-eight F_2 hybrids were genotyped with closely linked markers covering the *Scc1* interval. F_2 animals who showed crossings-over between Ccs-19 and BALB/c alleles within the critical interval were selected. These recombinants were then backcrossed to BALB/c producing mice that had crossings-over only on one chromosome. These heterozygous recombinants were then tested for colon tumor susceptibility, which is the dominant phenotype. As in congenic substrains, phenotypic comparison of the various recombinants allowed for fine mapping of *Scc1*. Two additional backcross generations of some recombinants were performed to produce new types of recombinants and to finally reach the 2.4 cM resolution.[41]

Chromosomal substitution strains or consomic strains are strains in which an entire chromosome is introgressed into the isogenic background of another inbred strain using marker-assisted selection. Although such a panel takes several years to develop, many advantages are linked to it: (1) QTL are assigned to chromosomes by simply phenotyping the panel of strains with substituted chromosomes, (2) fine-mapping of the QTL is obtained by crossing the consomic strain containing the QTL with the parental recipient strain, with a much higher power than intercrossing the 2 parental strains of interest, (3) congenic strains over a narrow region are rapidly developed from a consomic strain containing the QTL of interest and (4) replicative and longitudinal studies are possible. A panel of 44 consomic rat strains from the BN and Dahl salt-sensitive parental strains is under development at the Medical College of Wisconsin.[15] The Physgen project has the goal to characterize the panel of consomic rats using 214 phenotypes specific to heart, lung, kidney, vasculature, and blood function in response to environmental stressors (hypoxia, exercise, salt intake) and to link these traits to the genomes of the mouse and human by comparative mapping (http://pga.mcw.edu). Panels of consomic strains have also been constructed in mice.[45]

Interval-specific congenic strains (ISCS) are described as a means to map QTL to a 1 cM interval. The strategy is a shortcut to congenic mapping, in that various recombinant animals throughout the QTL interval are selected from an F_2 population and they are then backcrossed to the recipient strain with marker-assisted selection to retain the specific donor strain interval. Counter-selection for the rest of the genome is done only for chromosomal intervals known to contain other interacting QTL. This strategy was used by Bennet and colleagues[9,22] thus yielding resolution for an alcohol-related QTL interval to 3.7 cM.

Advanced intercross lines (AIL) are proposed as a resource for fine genetic mapping. These lines are produced from a standard F_2 population by randomly and sequentially intercrossing each generation inter se (Figure 5.3). Consequently, animals are recombinants over shorter and shorter chromosomal segments thus increasing the mapping resolution. This strategy has been used to finely map trypanosomiasis resistance QTL in mice.[32] Recently, a new set of recombinant inbred (RI) strains was generated from two independent advanced intercrosses between C57BL/6J and DBA/2J mouse strains.[49] Compared to standard RI strains, this AIL-based panel has approximately twice as many recombinations as classical panels generated from an F_2 population.

Heterogeneous stocks derived from inbred mouse strains have been used to fine-map QTL. This approach takes advantage of the fact that the genealogy of the heterogeneous stock is known, coming from an eight-way cross between eight known inbred strains, and the high number of filial generations of one current stock (>58 generations) allows for high resolution of QTL. To fine-map an anxiety-related QTL, Talbot et al.[56] had estimated genotype–phenotype associations by ANOVA in ~750 mice from the heterogeneous stock to detect QTL. Then, multiple regression was used to establish the direction of effect of each allele and to infer the strain origins of the QTL. Finally, haplotyping of each strain (see below) in the QTL interval allowed to refine the QTL to 0.8 cM. The same strategy was used to fine-map ethanol-induced locomotor activity.[19] To learn about other genetic strategies, the reader is directed to a review by A. Darvasi.[17]

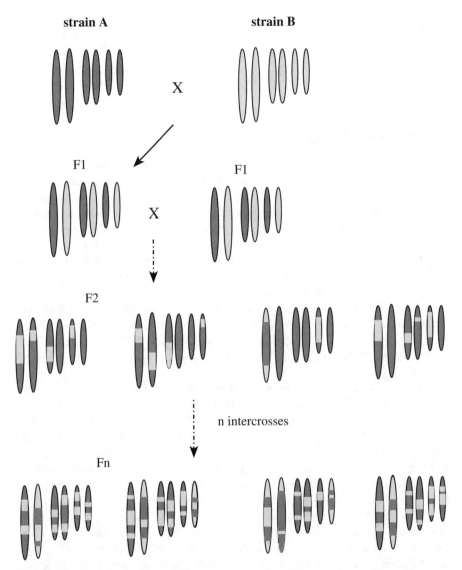

strain A   **strain B**

X

F1   F1

X

F2

n intercrosses

Fn

FIGURE 5.3 Advanced intercross lines (AIL). A standard F_2 is produced, then each generation is intercrossed with itself. Inbreeding is avoided as much as possible.

A few years ago, a community of geneticists working on complex traits formed
the Complex Traits Consortium (CTC). Part of their mission was to discuss new
resources needed for QTL fine-mapping and identification. As a result of their
conversations, 1000 new lines of advanced, multi-parental recombinant inbred mice
are in the process of development under the operational name, Collaborative Cross.
These lines are being developed from an initial panel of eight inbred strains that
have been outcrossed for four generations and then followed by inbreeding, now in
progress. The main advantage of the resource will be the high mapping resolution

and cumulative, integrated data from the various participating laboratories.[58] Funding for this endeavor has been obtained from the Ellison Foundation, and progress of the Collaborative Cross and other CTC projects can be monitored at www.complextrait.org.

Should one of the aforementioned strategies work, the next step is positional cloning of the gene(s) involved.

5.3 POSITIONAL CLONING

Once a QTL interval is isolated, if possible fixed in a homozygous fashion in a strain, and its size sufficiently reduced, *physical mapping* is undertaken prior to cloning of the relevant gene. For sequenced genomes, this step is no longer necessary, as all the information can be found in databases. However, I shall briefly describe the methodologies that were used before whole genome sequences were available as they are still used in nonsequenced species of interest for complex traits.

Physical mapping implies coverage of the chromosomal area defined by linkage studies with a set of overlapping DNA clones. In order to reduce the number of clones to analyze, large-insert vectors are used such as YACs (yeast artificial chromosomes),[11] PACs (P1-derived artificial chromosomes)[31] or BACs (bacterial artificial chromosomes).[55] These vectors can integrate DNA inserts of several hundred kilobases (kb) (PACs and BACs) up to 2 megabases (Mb) for YACs; however, the latter often display chimerism, that is integration of two or more noncontiguous DNA inserts.

The first step involves screening of genomic libraries of these large-insert vectors using the DNA markers contained in the QTL interval. At this stage, even nonpolymorphic markers such as sequence tag sites (STSs) can be used. STSs are short sequences of DNA obtained by random sequencing of genomic DNA. From their sequence, oligonucleotides are derived so that STSs are easily amplifiable by PCR and their position can be specifically determined.[46] To "walk" across the interval, both ends of the first isolated clones are sequenced and new STSs are derived to extract overlapping clones from the same libraries.

Clones are then ordered by analyzing the STS they have in common such that at the end, a set of contiguous clones called a "contig" is constructed (Figure 5.4). High throughput technology is available to rapidly construct contigs. Thus, large-insert libraries are arranged in super pools ready for PCR analysis. The STS databases and their localization are available via Internet; robots are used to replicate these libraries rapidly and efficiently, and to extract DNA.

Contigs are constructed that cover a region as large as 10 Mb which roughly corresponds to 5–10 cM depending on the species. Such big contigs are used to find new polymorphic markers that can then be tested in congenic substrains to reduce the QTL interval (see for example Denny et al.[20]). However, some chromosomal regions may contain as much as one gene every 30 kb, thus positional cloning of gene(s) responsible for the mutant phenotype from the contig is started once the interval is restricted to around 1 cM.

In species for which the genome is sequenced, there is no need to construct a contig of DNA clones in the QTL interval. New markers are found by simply

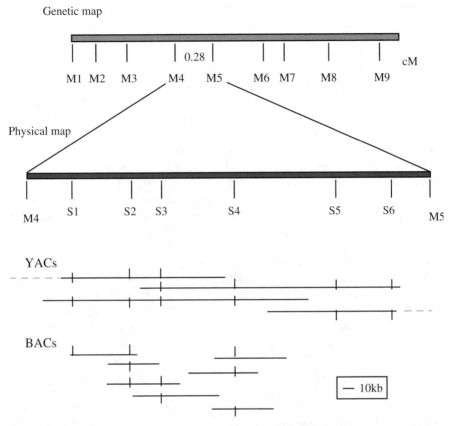

FIGURE 5.4 Construction of a contig. M1-M9 are genetic markers. S1-S6 are sequence tag sites.

consulting the genome sequence corresponding to the QTL interval in databases [National Center for Biotechnology Information (NCBI) genome browser, Ensembl].

5.3.1 Pure Positional Cloning

Identification of genes responsible for a phenotype requires the isolation of expressed sequences from cloned DNA. Again, when the genome has not been sequenced, various techniques have been developed and good reviews on the subject are available.[44,47,48] I shall describe them only briefly.

Evolutionary conservation: Because coding sequences are more conserved than are noncoding sequences, candidate DNA fragments from the contig are hybridized to Southern blots of genomic DNA from multiple species (also called zoo blot). DNA fragments that hybridize to several species are selected for further investigations as these hybridizations indicate that these clones contain an evolutionary conserved sequence.

CpG islands: This strategy is based on the observation that around 60% of genes are associated in their 5'untranslated region with short stretches of DNA with a high frequency of the dinucleotide CpG (i.e., cytosine connected by a 3'-5' phosphodiester bond to a guanine). These unmethylated CpG-rich sequences are called CpG islands. They are detected easily by cleavage with rare-cutting restriction enzymes of the contig DNA clones.

cDNA selection: In this approach candidate DNA fragments are hybridized to amplified cDNA libraries. After removal of nonspecific binding, the cDNA inserts that hybridize specifically to the cloned genomic DNA are eluted and amplified again by PCR. Two or three rounds of this process are performed before cloning the selected cDNAs. Several variations in the cDNA selection protocol have been used: the cloned genomic DNAs are either immobilized on nylon membrane or the hybridization is performed in solution. A biotin-streptavidin system and magnetic beads that capture the cDNA inserts have also been designed.

Exon trapping: This technique relies on splice sites flanking coding sequences and necessary for RNA splicing. Candidate DNA fragments are cloned into a vector that contains all the necessary sequences to produce mRNA. This recombinant vector is then transfected for expression in cells and the mRNA produced from the clones is identified by RT-PCR. If the DNA insert contains an exon with splice sites this exon will "be trapped" into this chimeric mRNA which can then be cloned and sequenced.

Direct sequencing: The availability of high-throughput automated sequencing machines, combined with advances in computer algorithms, has favored the development of sequence-based methods for searching exons in cloned genomic DNA. "Shotgun" sequencing approaches (that is cloning of an entire DNA population from randomly generated fragments) from the contig clones is often used. Programs devised to find potential coding sequences include GRAIL[59] and BLAST.[3,27]

So far, most positionally cloned genes have been identified by a combination of all of these techniques.[48] For the positional cloning of the mouse circadian *Clock* gene, a >4 Mb contig was constructed.[33] The techniques that have been used to identify the gene include: (1) the search for CpG islands, (2) cDNA selection of suprachiasmatic nucleus cDNA using one BAC, and (3) shotgun sequencing of subclones from two BACs of the contig. Selected candidate sequences issued from these approaches were used as probes on Northern blots of RNA from hypothalamus and eye from the mutant and wild type mice. Using one of the clones, RNA differences of abundance and size were detected between the two types of mice. The complete cDNA sequence of this gene was obtained with 11 overlapping cDNA clones isolated with the approaches mentioned above. This gene named *Clock* spans almost 100,000 bp of genomic DNA and contains 24 exons. Further evidence for the action of *Clock* gene in the circadian rhythm in mutant mice was provided by a BAC rescue experiment. A BAC clone from the critical region of the contig was microinjected into mutant *Clock* mouse oocytes. Its expression in the adult transgenic mouse permitted *in vivo* complementation of the loss-of-rhythm phenotype;[4] i.e., the mice showed recovery of the circadian activity cycle. Similarly, the mouse saccharin preference locus was positionally cloned after reducing the QTL interval by congenics to 0.7 cM and constructing a BAC contig of the region. The 0.7 cM QTL interval had a physical size of 194 kb and contained 12 predicted genes. Among them, the Tas1r3 gene stood out as a good candidate as its protein is a G-protein coupled receptor with homology to taste receptor protein, T1R1.[6]

5.3.2 POSITIONAL CANDIDATE APPROACH

5.3.2.1 Mapping of Candidate Genes in the QTL

In a 1995 review, Francis Collins[14] made the now verified prediction that the "positional candidate" strategy would overtake pure positional cloning. This involves combining fine scale mapping and survey of candidate genes within the interval. In species not yet sequenced, efforts are underway to improve the transcript maps in which coding sequences are localized precisely. Besides cDNA sequences already identified and mapped, short, PCR-amplifiable partial cDNA sequences called EST (expressed sequence tag, a subtype of STS) have been generated for this purpose and stored in a public database (dbEST, NCBI). Fine mapping of these coding sequences has been possible with the advent of radiation hybrid panels. *Radiation hybrid mapping* is a somatic cell genetic approach in which diploid cells of the species studied are exposed to a lethal dose of x-rays, breaking each chromosome into small fragments. These irradiated cells are then rescued by fusion to non-irradiated hamster cells. Each independent radiation hybrid clone retains between 15 and 20% of the irradiated cell and the retention of these random fragments of the species chromosomes is stable without selection. The mapping strategy employs the frequency of x-ray breakage between two markers as a statistical measure of distance between these markers. Therefore any sequence that can be distinguished from hamster DNA can be mapped using this approach without need for polymorphisms.[16] These tools have been widely used in humans, mice and rats before their genome was sequenced and remains very useful for other species.

Thus, the first step after establishing a contig, is a computer exercise in which a query is made to select ESTs mapped in a given chromosomal subregion. These ESTs are then placed on the contig simply by PCR amplification of the contig DNA clones using primers specific for a given EST. A considerable amount of information can be extracted from an EST; for example, homologies with known cDNA sequences can be obtained (for example using the BLAST program). In turn, these data may point to a gene already cloned or to a gene family, providing important clues about the function of the gene and thus its relevance to the disease or mutant phenotype. If an EST does not display homology to any known sequence, it could be used as a hybridization probe to examine its tissue expression and to clone a longer fragment of the cDNA in question. As outlined by Ballabio,[7] the features of candidate genes—including sequence domains, expression pattern, imprinted expression, sequence instability (such as nucleotide triplet expansion) and developmental expression—are compared to the disease or phenotype features in order to select the strongest positional candidates. Finally, variants of that gene are screened for by comparing affected individuals to controls.

An example of this approach is provided by the cloning of genes for early-onset familial Alzheimer's disease. The AD3 locus was mapped to chromosome 14 by linkage analysis and a contig of >5 Mb was constructed around this region. Expressed sequences encoded within the critical interval were then isolated by hybridization of brain cDNA to the contig YAC clones. After selection based upon map location and evolutionary conservation (see above), 19 cDNA clones were kept. The sequences of these cDNA clones were compared in affected and normal post-mortem brain tissue by reverse-transcriptase PCR, leading to the identification of a novel gene, presenilin-1 (PS-1) within the AD3 locus.[53] Shortly after this study, the presenilin-2 gene was identified on chromosome 1. This finding was based on the location of an EST within the candidate region, defined by linkage studies, that has a high sequence homology with PS-1.[37]

5.3.2.2 Combining QTL Mapping and Expression Profiling

The availability of genome sequences, together with the advent of DNA microarrays or "chips," has favored the search for positional candidates through differential expression analysis. The first time this strategy combining QTL mapping and large-scale expression analysis was used, it was in a rat model of insulin-resistance.[2] cDNA molecules obtained from reverse-transcribed RNA extracted from adipose tissue (the target tissue) of the two parental strains and a congenic strain for the detected QTL were hybridized pairwise to a DNA chip of 10,000 arrayed rat cDNA. Around 70 genes were found to be differentially regulated between the parental and congenic strains. Among them, there was a biological candidate that was then mapped to the expected chromosome region. Further analysis of this candidate showed that it maps at the peak of the QTL and displays functional sequence variation between the parental strains. Similarly, α-synuclein has been proposed as the causative gene of a QTL associated to alcohol preference in a rat model following QTL mapping combined to microarray expression analysis in various brain regions.[39] In addition to its strengths, this approach has several limitations: (1) the causative gene may not necessarily vary in expression between the parental strains (or subject and control in human studies); (2) the difference in expression may be too small to be detected by microarray analysis. Indeed, this is of particular concern for behavioral traits in which the brain is the target tissue and a small difference in gene expression may have an important functional impact; (3) it is not always obvious which target tissue to choose in which the causative gene will be differentially regulated; and (4) the gene under investigation is not represented on the chip; this problem should be resolved with progress in genome annotation and availability of pangenomic chips.

5.3.2.3 Haplotype Mapping

An haplotype is a set of linked alleles on one chromosome or a segment of a chromosome. Single nucleotide polymorphisms (SNPs) are the polymorphisms most often used to define a haplotype because they are stable over many generations and can be traced back to a common ancestral population. Recent studies have suggested that genomes are composed of haplotype blocks with regions of high linkage disequilibrium (the nonrandom association of alleles) separated by a region of a high number of recombination events. Identification of such haplotype blocks is useful for association studies in humans by comparing the frequency of haplotypes in affected vs. control individuals.[60] In mouse and rat inbred strains, the situation is even more favorable because all inbred strains are derived from a limited number of ancestral populations. Thus, genomes of inbred strains appear to be haplotype blocks mosaics derived from few original strains. This genome structure is exploited to reduce the number of high-priority candidates genes in a QTL interval by concentrating on genes that are in divergent haplotype blocks between the two strains that have served to detect the QTL. Even more rewarding is to compare haplotypes of a QTL region detected in various crosses using different strains. Such a multiple crossmapping strategy has been applied and combined to expression profiling and sequence analysis, leading to the identification of a strong candidate gene for a QTL associated with basal locomotor activity.[30]

5.4 CONCLUSION

Although still challenging, identification of genes underlying QTL is becoming easier. Completion of genome sequences in human and model organisms, advances in genomic and statistical techniques, and production of new genetic resources in mice and rats are the main components of this progress. Nowadays, the tendency is to set up a large-scale project integrating advanced genetic resources, in-depth phenotyping and high-throughput genomics, such as in the *Physgen* project in rat or the *Collaborative Cross** in mice. Two tasks demanding a great deal of effort remain to complete a QTL gene identification study. The first is to prove the involvement of the candidate gene and then to find the causative mutation(s). The best approach in model organisms is to introduce the variant allele in the control strain by means of knock-in technology. So far, however, this is only possible in one strain of mice and not yet in rats. Thus, as stated and detailed by others,[1,28] accumulation of evidence converging to a candidate gene, including functional *in vitro* and *in vivo* studies, is generally accepted.

REFERENCES

1. Abiola O, Angel JM, Avner P, Bachmanov AA, Belknap JK, Bennett B, Blankenhorn EP, Blizard DA, Bolivar V, Brockmann GA, Buck KJ, Bureau JF, Casley WL, Chesler EJ, Cheverud JM, Churchill GA, Cook M, Crabbe JC, Crusio WE, Darvasi A, de Haan G, Dermant P, Doerge RW, Elliot RW, Farber CR, Flaherty L, Flint J, Gershenfeld H, Gibson JP, Gu J, Gu W, Himmelbauer H, Hitzemann R, Hsu HC, Hunter K, Iraqi FF, Jansen RC, Johnson TE, Jones BC, Kempermann G, Lammert F, Lu L, Manly KF, Matthews DB, Medrano JF, Mehrabian M, Mittlemann G, Mock BA, Mogil JS, Montagutelli X, Morahan G, Mountz JD, Nagase H, Nowakowski RS, O'Hara BF, Osadchuk AV, Paigen B, Palmer AA, Peirce JL, Pomp D, Rosemann M, Rosen GD, Schalkwyk LC, Seltzer Z, Settle S, Shimomura K, Shou S, Sikela JM, Siracusa LD, Spearow JL, Teuscher C, Threadgill DW, Toth LA, Toye AA, Vadasz C, Van Zant G, Wakeland E, Williams RW, Zhang HG, Zou F. The nature and identification of quantitative trait loci: a community's view. *Nat Rev Genet* 4:911–916, 2003.

---

* At printing, the Collaborative Cross is operative and had received generous support from the Ellison Foundation. For further information consult http://atlas.utmem.edu/pipermail/ctc/2005-October/000284.html.

2. Aitman TJ, Glazier AM, Wallace CA, Cooper LD, Norsworthy PJ, Wahid FN, al
   Majali KM, Trembling PM, Mann CJ, Shoulders CC, Graf D, St. Lezin E, Kurtz TW,
   Kren V, Pravenec M, Ibrahimi A, Abumrad NA, Stanton LW, Scott J. Identification
   of Cd36 (Fat) as an insulin-resistance gene causing defective fatty acid and glucose
   metabolism in hypertensive rats [see comments]. *Nat Gen*et 21:76–83, 1999.
3. Altschul SF, Lipman DJ. Protein database searches for multiple alignments. *Proced
   Nat Acad Sci USA* 87[14], 5509–5513, 1990.
4. Antoch MP, Song EJ, Chang AM, Vitaterna MH, Zhao Y, Wilsbacher LD, Sangoram
   AM, King DP, Pinto LH, Takahashi JS. Functional identification of the mouse circadian clock gene by transgenic BAC rescue. *Cell* 89, 655–667, 1997.
5. Bachmanov AA, Li X, Li S, Neira M, Beauchamp GK, Azen EA High-resolution
   genetic mapping of the sucrose octaacetate taste aversion (Soa) locus on mouse
   Chromosome 6. *Mamm Genome* 12:695–699, 2001.
6. Bachmanov AA, Li X, Reed DR, Ohmen JD, Li S, Chen Z, Tordoff MG, de Jong
   PJ, Wu C, West DB, Chatterjee A, Ross DA, Beauchamp GK. Positional cloning of
   the mouse saccharin preference (Sac) locus. *Chem Senses* 26:925–933, 2001.
7. Ballabio A. The rise and fall position cloning? *Nat Genet* 3:277–281, 1993.
8. Bennett B. Congenic strains developed for alcohol- and drug-related phenotypes.
   Pharmacol Biochem Behav 67:671–681, 2000.
9. Bennett B, Beeson M, Gordon L, Carosone-Link P, Johnson TE. Genetic dissection
   of quantitative trait loci specifying sedative/hypnotic sensitivity to ethanol: mapping
   with interval-specific congenic recombinant lines. *Alcohol Clin Exp Res*
   26:1615–1624, 2002.
10. Berrettini WH, Ferraro TN, Alexander RC, Buchberg AM, Vogel WH. Quantitative
    trait loci mapping of three loci controlling morphine preference using inbred mouse
    strains. *Nat Genet* 7:54–58, 1994.
11. Burke DT, Carle GF, Olson MV. Cloning of large segments of exogenous DNA into
    yeast by means of articicial chromosome vectors. *Science* 236:806–812, 1987.
12. Caldarone B, Saavedra C, Tartaglia K, Wehner JM, Dudek BC, Flaherty L. Quanti-
    tative trait loci analysis affecting contextual conditioning in mice. *Nat Genet*
    17:335–337, 1997.
13. Carr LG, Foroud T, Bice P, Gobbett T, Ivashina J, Edenberg H, Lumeng L, Li TK.
    A quantitative trait locus for alcohol consumption in selectively bred rat lines. *Alcohol
    Clin Exp Res* 22:884–887, 1998.
14. Collins FS. Positional cloning moves from perditional to traditional. *Nat Genet*
    9:347–350, 1995.
15. Cowley AW, Jr., Roman RJ, Jacob HJ Application of chromosomal substitution
    techniques in gene-function discovery. *J Physiol* 554:46–55, 2004.
16. Cox DR, Burmeister M, Price ER, Kim S, Myers RM. Radiation hybrid mapping:
    a somatic cell genetic method for constructing high-resolution maps of mammalian
    chromosomes. *Science* 250:245–250, 1990.
17. Darvasi A. Experimental strategies for the genetic dissection of complex traits in
    animal models. *Nat Genet* 18:19–24, 1998.
18. Demant P, Hart AA. Recombinant congenic strains—a new tool for analyzing
    genetic traits determined by more than one gene. *Immunogenetics* 24:416–422,
    1986.
19. Demarest K, Koyner J, McCaughran J, Jr., Cipp L, Hitzemann R. Further character-
    ization and high-resolution mapping of quantitative trait loci for ethanol-induced
    locomotor activity. *Behav Genet* 31:79–91, 2001.

20. Denny P, Lord CJ, Hill NJ, Goy JV, Levy ER, Podolin PL, Peterson LB, Wicker LS, Todd JA, Lyons PA. Mapping of the IDDM locus Idd3 to a 0.35-cM interval containing the interleukin-2 gene. *Diabetes* 46:695–700, 1997.

21. Dukhanina OI, Dene HY, Deng AY, Choi CR, Hoebee B, Rapp JP. Linkage map and congenic strains to localize blood pressure QTL on rat chromosome 10. *Mamm Genome* 8:229–235, 1997.

22. Ehringer MA, Thompson J, Conroy O, Yang F, Hink R, Bennett B, Johnson TE, Sikela JM. Fine mapping of polymorphic alcohol-related quantitative trait loci candidate genes using interval-specific congenic recombinant mice. *Alcohol Clin Exp Res* 26:1603–1608, 2002.

23. Fernandez-Teruel A, Escorihuela RM, Gray JA, Aguilar R, Gil L, Gimenez-Llort L, Tobena A, Bhomra A, Nicod A, Mott R, Driscoll P, Dawson GR, Flint J. A quantitative trait locus influencing anxiety in the laboratory rat. *Genome Res* 12:618–626, 2002.

24. Flint J, Corley R, DeFries JC, Fulker DW, Gray JA, Miller S, Collins AC. A simple genetic basis for a complex psychological trait in laboratory mice. *Science* 269:1432–1435, 1995.

25. Frankel WN, Johnson EW, Lutz CM. Congenic strains reveal effects of the epilepsy quantitative trait locus, El2, separate from other El loci. *Mamm Genome* 6:839–843, 1995.

26. Gershenfeld HK, Paul SM. Mapping quantitative trait loci for fear-like behaviors in mice. *Genomics* 46:1–8, 1997.

27. Gish W, States DJ. Identification of protein coding regions by database similarity search. *Nat Genets* 3:266–272, 1993.

28. Glazier AM, Nadeau JH, Aitman TJ. Finding genes that underlie complex traits. *Science* 298:2345–2349, 2002

29. Gould KA, Dietrich WF, Borenstein N, Lander ES, Dove WF. Mom1 is a semidominant modifier of intestinal adenoma size and multiplicity in Min/+ mice. *Genetics* 144:1769–1776, 1996.

30. Hitzemann R, Malmanger B, Reed C, Lawler M, Hitzemann B, Coulombe S, Buck K, Rademacher B, Walter N, Polyakov Y, Sikela J, Gensler B, Burgers S, Williams RW, Manly K, Flint J, Talbot C. A strategy for the integration of QTL, gene expression, and sequence analyses. *Mamm Genome* 14:733–747, 2003.

31. Ioannou PA, Amemiya CT, Garnes J, Kroisel PM, Shizuya H, Chen C, Batzer MA, DeJong PJ. A new bacteriophage P1-derived vector for the propagation of large human DNA fragments. *Nat Genet* 6:84–89, 1994.

32. Iraqi F, Clapcott SJ, Kumari P, Haley CS, Kemp SJ, Teale AJ. Fine mapping of trypanosomiasis resistance loci in murine advanced intercross lines. *Mamm Genome* 11:645–648, 2000.

33. King DP, Zhao Y, Sangoram AM, Wilsbacher LD, Tanaka M, Antoch MP, Steeves TDL, Vitaterna MH, Kornhauser JM, Lowrey PL, Turek FW, Takahashi JS. Positional cloning of the mouse circadian clock gene. *Cell* 89:641–653, 1997.

34. Korstanje R, Paigen B. From QTL to gene: the harvest begins. *Nat Genet* 31:235–236, 2002.

35. Lander E, Kruglyak L. Genetic dissection of complex traits: guidelines for interpreting and reporting linkage results. *Nat Genet* 11:241–247, 1995.

36. Legare ME, Bartlett FS, Frankel WN. A major effect QTL determined by multiple genes in epileptic EL mice. *Genome Res* 10:42–48, 2000.

37. Levy-Lahad E, Wasco W, Poorkaj P, Romano DM, Oshima J, Pettingell CY, Jondro PD, Schmidt SD, Wang K, Crowley AC, Fu YH, Guenette SY, Galas D, Nemens E,

Wijsman EM, Bird TD, Schellenberg GD, Tanzi RE. Candidate gene for the chromosome 1 familial Alzheimer's disease locus. *Science* 269:973–977, 1995.

38. Li X, Inoue M, Reed DR, Huque T, Puchalski RB, Tordoff MG, Ninomiya Y, Beauchamp GK, Bachmanov AA. High-resolution genetic mapping of the saccharin preference locus (Sac) and the putative sweet taste receptor (T1R1) gene (Gpr70) to mouse distal Chromosome 4. *Mamm Genome* 12:13–16, 2001.

39. Liang T, Spence J, Liu L, Strother WN, Chang HW, Ellison JA, Lumeng L, Li TK, Foroud T, Carr LG. alpha-Synuclein maps to a quantitative trait locus for alcohol preference and is differentially expressed in alcohol-preferring and -nonpreferring rats. *Proc Natl Acad Sci U S A* 100:4690–4695, 2003.

40. Melo JA, Shendure J, Pociask K, Silver LM. Identification of sex-specific quantitative trait loci controlling alcohol preference in C57BL/6 mice. *Nat Genet* 13:147–153, 1996.

41. Moen CJ, Groot PC, Hart AA, Snoek M, Demant P. Fine mapping of colon tumor susceptibility (Scc) genes in the mouse, different from the genes known to be somatically mutated in colon. *Proc Natl Acad Sci U S A* 93:1082–1086, 1996.

42. Moisan MP, Courvoisier H, Bihoreau MT, Gauguier D, Hendley ED, Lathrop M, James MR, Mormède P. A major quantitative trait locus influences hyperactivity in the WKHA rat. *Nat Genet* 14:471–473, 1996.

43. Moisan MP, Llamas B, Cook MN, Mormede P. Further dissection of a genomic locus associated with behavioral activity in the Wistar-Kyoto hyperactive rat, an animal model of hyperkinesis. *Mol Psychiatry* 8:348–352, 2003.

44. Monaco AP. Isolation of genes from cloned DNA. *Current opinion in Genetics and Development* 4:360–365, 1994.

45. Nadeau JH, Singer JB, Matin A, Lander ES. Analysing complex genetic traits with chromosome substitution strains. *Nat Genet* 24:221–225, 2000.

46. Olson M, Hood L, Cantor C, Botstein D. A common language for physical mapping of the human genome. *Science* 245:1434–1435, 1989.

47. Parimoo S, Patanjali SR, Kolluri R, Xu H, Wei H, Weissman SM. cDNA selection and other approaches in positional cloning. *Analyt Biochem* 228:1–17, 1995.

48. Parrish JE, Nelson DL. Methods for finding genes a major rate-limiting step in positional cloning. *GATA* 10:29–41, 1993.

49. Peirce JL, Lu L, Gu J, Silver LM, Williams RW. A new set of BXD recombinant inbred lines from advanced intercross populations in mice. *BMC Genet* 5:7, 2004.

50. Podolin PL, Denny P, Armitage N, Lord CJ, Hill NJ, Levy ER, Peterson LB, Todd JA, Wicker LS, Lyons PA. Localization of two insulin-dependent diabetes (*Idd*) genes to the *Idd10* region on mouse chromosome 3. *Mamma Genome* 9:283–286, 1998.

51. Podolin PL, Denny P, Lord CJ, Hill NJ, Todd JA, Peterson LB, Wicker LS, Lyons PA. Congenic mapping of the insulin-dependent diabetes (Idd) gene, Idd10, localizes two genes mediating the Idd10 effect and eliminates the candidate Fcgr1. *J Immunol* 159:1835–1843, 1997.

52. Ramos A, Moisan MP, Chaouloff F, Mormede P. Identification of female-specific QTLs affecting an emotionality-related behavior in rats. *Mol Psychiatry* 4:453–462, 1999.

53. Sherrington R, Rogaev EI, Liang Y, Rogaeva EA, Levesque G, Ikeda M, Chi H, Lin C, Li G, Holman K, Tsuda T, Mar L, Foncin JF, Bruni AC, Montesi MP, Sorbi S, Rainero I, Pinessi L, Nee L, Chumakov I, Pollen D, Brookes A, Sanseau P, Polinsky RJ, Wasco W, DaSilva HAR, Haines JL, Pericak-Vance MA, Tanzi RE, Roses AD, Fraser PE, Rommens JM, St George-Hyslop PH. Cloning of a gene bearing missense mutations in early-onset familial Alzheimer's disease. *Nature* 375:754–760, 1995.

54. Shimomura K, Low-Zeddies SS, King DP, Steeves TD, Whiteley A, Kushla J, Zemenides PD, Lin A, Vitaterna MH, Churchill GA, Takahashi JS. Genome-wide epistatic interaction analysis reveals complex genetic determinants of circadian behavior in mice. *Genome Res* 11:959–980, 2001.
55. Shizuya H, Birren B, Kim UJ, Mancino V, Slepak T, Tachiiri Y, Simon M. Cloning and stable maintenance of 300-kilobase-pair fragments of human DNA in Escherichia coli using an F-factor-based vector. *Proc Nat Acad Sci USA* 89:8794–8797, 1992.
56. Talbot CJ, Nicod A, Cherny SS, Fulker DW, Collins AC, Flint J. High-resolution mapping of quantitative trait loci in outbred mice. *Nat Genet* 21:305–308, 1999.
57. Terenina-Rigaldie E, Moisan MP, Colas A, Beauge F, Shah KV, Jones BC, Mormede P. Genetics of behaviour: phenotypic and molecular study of rats derived from high- and low-alcohol consuming lines. *Pharmacogenetics* 13:543–554, 2003.
58. Threadgill DW, Hunter KW, Williams RW. Genetic dissection of complex and quantitative traits: from fantasy to reality via a community effort. *Mamm Genome* 13:175–178, 2002.
59. Uberbacher EC, Mural RJ. Locating protein-coding regions in human DNA sequences by a multiple sensor-neural network approach. *Proc Natl Acad Sci USA* 88:11261–11265, 1991.
60. Wall JD, Pritchard JK. Haplotype blocks and linkage disequilibrium in the human genome. *Nat Rev Genet* 4:587–597, 2003.
61. Wehner JM, Radcliffe RA, Rossmann ST, Christensen SC, Rasmussen DL, Fulker DW, Wiles M. Quantitative trait locus analysis of contextual fear conditioning in mice. *Nat Genet* 17:331–334, 1997.
62. Yoshikawa T, Watanabe A, Ishitsuka Y, Nakaya A, Nakatani N. Identification of multiple genetic loci linked to the propensity for "behavioral despair" in mice. *Genome Res* 12:357–366, 2002.
63. Zhang QY, Dene HY, Deng AY, Garrett MR, Jacob HJ, Rapp JP. Interval mapping and congenic strains for a blood pressure QTL on rat chromosome 13. *Mamm Genome* 8:636–641, 1997.

6 Gene Expression

Richard A. Radcliffe

CONTENTS

INTRODUCTION

Gene expression, or *transcription*, is an important element of "The Central Dogma," the phrase used by Francis Crick[1] to describe what was and still is thought to be the basic and only flow of genetic information in a cell: DNA is *transcribed* to messenger RNA (mRNA) which is then *translated* to protein. In this scheme, only nucleic acids actually contain information that is encoded in the specific sequence of nucleotides that make up the DNA or RNA. Cells can construct proteins from the information contained in nucleic acids, but the transfer of information is one way; cells do not have the capacity to reproduce nucleic acid sequence information from proteins alone. While it is known that RNA performs more than just information functions,[2] proteins are still thought to be the primary working molecules in a cell carrying out diverse functions related to energy management, immune response, structure, transport and storage, signal transduction, cell division, etc. Thus, implicit in the Central Dogma are mechanisms for the tightly controlled regulation of mRNA and thus protein amount, one of several mechanisms through which the cell maintains strict control over the final activity of a protein. On a grander scale, the integrated transcriptional activity of the entire *genome* (the entire DNA sequence in an organism) to a great extent determines the nature of a cell and its interactions with other cells in an organism or with the environment.

Transcription is an essential mechanism throughout the development of a multicellular organism. The coordinate induction or repression of the thousands of genes found in the organism's genome results in a wide array of cell phenotypes through the process of cell differentiation. Mature cells, including highly specialized post-mitotic cells such as neurons, continue to make use of transcription throughout their life. One reason for this is that active transcription is required to replace proteins all of which have a finite duration, at least to the extent that is known, and therefore must be replaced throughout the life of the cell. Cells also need to adapt to a dynamic environment necessitating variable levels of some proteins. Environmental adaptation through transcription is perhaps most highly developed in neurons of the central nervous system (CNS) in which one of the most important and interesting functions of the CNS takes place: learning and memory.[3] The CNS is constantly being bombarded with sensory information, some of which is critical and must be dealt with immediately, but most of which is extraneous and can be ignored. In either case, processes of learning and memory aid the organism in distinguishing into which category the incoming information belongs. These processes cannot take place in the absence of a functioning transcription apparatus.

The fundamentals of transcription are essentially the same in all eukaryotic cells, but the timing and nature of the regulatory events that determine which genes are actively being transcribed vary from gene to gene. This chapter describes the fundamentals of transcription as well as the basic principles of transcriptional regulation. Also included is a discussion of sources of variation in mRNA levels and a brief description of some standard methods for measuring mRNA amounts from cell or tissue samples.

6.1 THE GENE

The molecular mass of a eukaryotic chromosome is composed of approximately 50% DNA, with the other half made up of proteins. DNA with its associated proteins is called *chromatin*. *Histones*, the majority of proteins found in chromatin, compact, organize, and protect the DNA. There are four histone subunits that assemble into a flat, round octamer around which DNA wraps approximately 1.65 times. This complex of histone proteins and the associated approximately 147 base-pair (bp; a nucleotide and its compliment in native double-stranded DNA) length of DNA is called a *nucleosome*. Nucleosomes are connected to one another by a length of linker DNA (20–60 bp) and are found along the entire length of DNA except in areas where the DNA is actively engaged by various nonhistone proteins. Nonhistone proteins are much less abundant than histones and are involved in the regulation and execution of transcription, DNA repair, recombination, and replication.[4,5]

The protein fraction of chromatin is very important for its proper functioning, but it is the nucleotide sequence in DNA that is the critical information-storage component, with the *gene* as the elemental unit of information (see Figure 6.1). A gene is generally considered to be the *coding region* portion of a particular DNA sequence. This is the sequence that is transcribed into mRNA and that contains the specific information from which the primary structure of the gene product is determined; i.e., the protein. Just as important, however, is the *regulatory region* of a gene which is composed of noncoding DNA sequence. The regulatory region interacts with a large number of proteins that dictate when and where the gene will be transcribed and can include sequence found many thousands of bp away from the

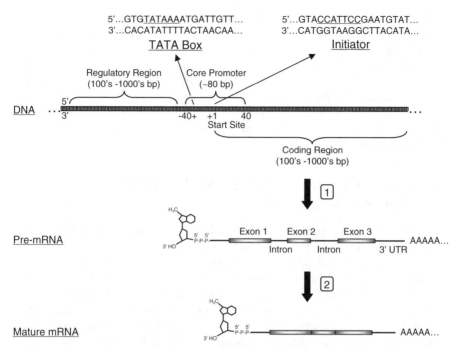

5'...GTGTATAAAATGATTGTT...   5'...GTACCATTCCGAATGTAT...
3'...CACATATTTTACTAACAA...   3'...CATGGTAAGGCTTACATA...
        TATA Box   Initiator

FIGURE 6.1 Basic structure and transcription of a hypothetical gene. Two binding elements are shown, the TATA box and the initiator, along with their complementary DNA strand (sequence on either side of the element is arbitrary). During transcription, the 5' end of the nascent mRNA is capped with methyl-guanine and the 3' end is polyadenylated (step 1). Introns are removed from the pre-mRNA in the final processing step toward the mature mRNA molecule (step 2). In a very small number of cases, the mRNA is further processed with RNA editing (not shown). Transcription takes place in the nucleus followed by transport of the mature mRNA out of the nucleus to other cellular compartments where translational occurs.

coding region. The regulatory region can be further subdivided into the *promoter* (the location on the DNA where the transcriptional machinery binds) and the *regulator binding sites* (individual sites on the DNA to which proteins bind; also called *binding elements*). Regulator binding sites that promote transcription are often clustered in regions known as *enhancers*. Similarly, clusters of binding sites that suppress transcription when bound by regulatory proteins are called *silencers*.[6]

The basis for the specificity of interactions between nucleic acids or between nucleic acids and proteins is in the linear sequence of the four nucleotide bases that are found in double-stranded DNA or single-stranded RNA: adenine, guanine, cytosine, and thymine (uracil in RNA). This was the fundamental discovery made by Watson and Crick some 50 years ago.[7] It is the unidirectional order of these bases (specified with relation to the 5' and 3' phosphodiester linkages between ribose or deoxyribose sugar molecules in the RNA or DNA backbone) and their two- and three-dimensional relationships to one another that contain specific information about the structure of proteins and their temporal and spatial expression. Many different types of regulatory and enzymatic proteins are able to "read" the nucleic acid–based information in an integrated fashion and efficiently accomplish accurate expression of the gene product. A crucial aspect of the information-carrying capacity of nucleic acids is that it is easily transmittable in a manner that accurately preserves the sequence information. Thus, when a cell divides, the daughter cells each contain an exact copy of DNA. Similarly, during transcription, the mRNA is synthesized in a precise rendition of the coding portion of the gene. This is a result of the *complementary* nature of DNA, often referred to as *Watson–Crick base-pairing*, wherein an adenine (A) is always paired to a thymine (T; uracil in mRNA – U) and a guanine (G) is always paired to a cytosine (C).

6.2 TRANSCRIPTION: BASIC MECHANISM OF MRNA BIOGENESIS

Synthesis of mRNA proceeds through three basic phases: initiation, elongation, and termination. Initiation occurs when the histone-bound DNA is made accessible, the transcriptional machinery binds to the appropriate location on the DNA, and synthesis of the complementary strand of RNA commences. During the elongation phase, the nascent RNA is lengthened and proofreading functions take place to ensure that the message is error-free. The last phase of transcription, termination, occurs when the new RNA molecule is released thus causing transcription to end. This is followed by detachment of the transcriptional machinery from the DNA. Often transcription is taking place simultaneously at multiple sites on the same gene such that many identical mRNA molecules are being synthesized simultaneously (see Figure 6.2).

Single-stranded mRNA is synthesized from and complementary to the 3'→5' strand of the DNA; thus, the mRNA sequence is identical to the 5'→3' DNA strand. Transcription starts at the location on the DNA that corresponds to what will be the 5' terminus of the single-stranded mRNA molecule, often called the *transcript*. Approximately 40 nucleotides on either side of the start site constitute the *core promoter*. The core promoter contains the minimal amount of sequence information required for the initiation of transcription. There are four specific sequence motifs located in the core promoter; any given gene typically includes only two or three of these elements. Additional elements located within 200 bp of the start site are referred to as *promoter-proximal elements*. Classes of proteins called *transcription factors* are able to recognize the promoter sequences and by doing so facilitate the binding and activation of the RNA-synthesizing enzyme *RNA polymerase*. Some important transcription factors and their binding element consensus sequences are shown in Table 6.1. Additional elements, some as far as 100,000 bp upstream of the start site, may also contribute to the induction or suppression of transcription. Thus, certain combinations of transcription factors are required to initiate transcription and it is the cooperative activity of the specific combination of factors that are present in a cell that determines the temporal and spatial expression of a given gene.[8–10]

Eukaryotic cells have three distinct polymerases for the synthesis of RNA: RNA Pol I, II, and III. Pol II is responsible for the transcription of all protein-coding genes; Pol I and III transcribe specialized RNA molecules including transfer RNA (tRNA) and ribosomal RNA (rRNA) and will not be considered here. Pol II forms

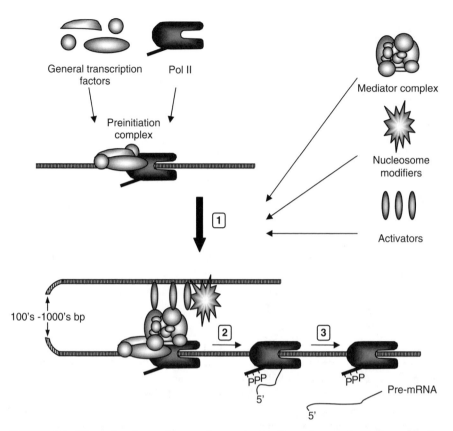

FIGURE 6.2 The molecular machinery of transcription. General transcription factors facilitate the binding of Pol II onto the promoter; this is the preinitiation complex. *In vivo*, many other components are needed to initiate transcription including the mediator complex, nucleosome modifiers, and activators (step 1). Phosphorylation of the Pol II tail by TFIIH causes it to be released from the preinitiation complex and begin transcription (step 2). Sequence signals in the transcript cause it to be released from Pol II and polyadenylated, thus terminating transcription (step 3).

a *pre-initiation complex* with numerous *general transcription factors*—proteins that aid Pol II in the recognition of and binding to the core promoter (see Table 6.1). It is thought that the general transcription factors assemble on the core promoter in a step-wise fashion in such a way as to denature the DNA and ultimately allow the binding of Pol II in the appropriate location. Pol II wraps around a single strand of the DNA duplex (the 3'→5' strand) and uses the information contained in the DNA sequence to consecutively add nucleotides to a growing RNA chain that shows Watson–Crick complementarity to the template DNA. The preinitiation complex is the minimal requirement for transcription to proceed on naked DNA *in vitro*. *In vivo*, however, nucleosomal DNA is generally inaccessible to nonhistone proteins. Thus, initiation of transcription can only take place when an inaccessible target region of the genome is made accessible by the activity of other transcription factors

TABLE 6.1
Common Promoter Elements

Promoter	Location	Position[a]	Binding Element Consensus Sequence[b]	Transcription Factor[c]
Initiator	Core promoter element	+1	PyPyAN(T/A)PyPy	TBP
TATA box	Core promoter element	−35 to −20	TATAAA	TBP
CAAT box	Promoter-proximal element	−200 to −70	CCAAT	CBF, NF1, C/EBP
GC box	Promoter-proximal element	−200 to −70	GGGCGG	SP1

[a] Number of nucleotides from the start site with negative numbers indicating in the upstream direction; i.e., toward the 5' end.

[b] The *consensus sequence* is the ideal sequence for the most effective interaction with its factor; binding elements can often tolerate slight differences from the consensus. Py = pyrimidine (C or T); N = any (A, C, T, G).

[c] TBP: TATA binding protein; CBF: CAAT binding protein; NF1: nuclear factor 1; C/EBP: CAAT/enhancer binding protein.

that facilitate the binding of the preinitiation complex to the core promoter. These factors, which include the mediator, a complex of more than 20 subunits, regulatory proteins, and nucleosome-modifying proteins, are engaged in various activities including covalent modification of histones and Pol II recruitment (see below).[10,11]

Once the preinitiation complex has formed at the core promoter, Pol II begins transcription and attempts to separate itself from the complex. Several short transcripts are synthesized, but the polymerase cannot enter into the elongation phase until its C-terminal tail, which is composed of a long string of Tyr, Ser, Pro, and Thr residues, is phosphorylated by the kinase activity of the factor TFIIH and other kinases. Once the tail has been phosphorylated to a certain level, affinity of Pol II for the general transcription factors decreases considerably allowing the polymerase to move away from the core promoter and begin synthesizing a full-length transcript during the elongation phase. Phosphorylation and dephosphorylation of the C-terminal tail are important regulatory mechanisms of Pol II activity.[11–13]

As Pol II enters into the elongation phase, the general transcription factors are replaced with a new set of factors that bind favorably to the phosphorylated form of the C-terminal tail. The new factors fulfill three basic functions: (1) promote continued RNA synthesis by stimulating Pol II activity; (2) proofread and correct the nascent RNA to ensure exact complementarity to the DNA template; and (3) process the emerging RNA molecule with several types of covalent modifications primarily for stabilization. Many of these factors bind to the C-terminal tail of Pol II. As the RNA is synthesized, its 5' end emerges from the polymerase near the C-terminal tail. Thus, the factors are in the ideal position for gaining access to the newly synthesized RNA.[11–13]

P-TEFb is a central factor recruited to the C-terminal tail early in the elongation phase. Along with other factors, it stimulates the elongation of the RNA molecule by Pol II. P-TEFb has inherent kinase activity and it phosphorylates additional residues on the Pol II tail. This activity aids in the recruitment of the elongation factors hSPT5, TAT-SF1, and TFIIS, all of which promote the continued elongation of the transcript. Pol II does not synthesize RNA at a constant rate; rather, the rate is dependent on the specific sequence being transcribed and some sequences will cause the polymerase to come to a near complete stop. One of the functions of TFIIS is to smooth the progress of Pol II through these more involved sequences so that it progresses at a faster overall rate. Combined with the nuclease activity of Pol II, TFIIS also aids in the detection and replacement of mismatched nucleotides. This proofreading capability is not as efficient as that which occurs during DNA synthesis. Approximately 1 in 10^7 mismatched nucleotides escape detection during DNA replication, but transcription is somewhat less accurate to approximately 1 in 10^4 to 10^5 errors are made during transcription. Of course, the consequences of a DNA mutation are potentially much more severe than the occasional single mutant protein translated from a mismatched mRNA sequence.[11–13]

When the new transcript reaches a length of 20 to 40 nucleotides, the 5' end is modified with the addition of a methylated guanine, known as *capping*. Guanine is added in an unusual 5'-5' triphosphate linkage and then a methyl group is added to the nitrogen in the 7' position of the guanine moiety (see Figure 6.2). The cap serves as a signal for other processes including recruitment of the translation machinery. As the transcript reaches full-length, a second processing step takes place, *polyadenylation*, that coincides with termination of transcription. CPSF and CstF are proteins that have been recruited to the phosphorylated tail of Pol II to separate the nascent mRNA from the RNA that is still being made and to facilitate polyadenylation. Transcription of the *poly-A signal sequence* promotes release of CPSF and CstF from the Pol II onto the new transcript. These factors cause the mRNA to be cleaved from the still growing RNA strand and recruit *poly-A polymerase* which enzymatically adds approximately 200 adenine molecules (from ATP) to the 3' end of the transcript. Pol II continues to synthesize complementary RNA sometimes as long as several hundred nucleotides, but eventually Pol II dissociates from the DNA and releases the aberrant RNA molecule which is rapidly degraded. The signals that cause dissociation of the polymerase may have to do with either the absence of a 5' guanine cap and/or with the transfer of CPSF and CstF from the tail of Pol II, but the details of this have not been entirely worked out.[12,13]

For most eukaryotic genes, the *primary transcript*, or *pre-mRNA*, requires one additional processing step, *splicing*, before it is mature and ready for the translation of protein. The pre-mRNA usually consists of alternating segments known as *exons* and *introns* that specify coding and noncoding portions of the gene, respectively. All of the exons in a gene (more than 300 in some cases) must be joined in a contiguous fashion in order for the transcript to be appropriately translated into the protein product. Thus, introns are removed and exons are spliced together in a reaction catalyzed by a large protein-RNA complex known as the *spliceosome*. The spliceosome is directed by specific sequences at the exon/intron border which guide the splicing reaction. Some primary transcripts have the capacity to be spliced in different configurations. This *alternative splicing* provides a method by which a single gene can produce many, often related, protein products. Alternative splicing can be either *constitutive* or *regulated*; i.e., alternate transcripts are made on a regular basis or they are made as dictated by the needs of the cell.[12,13]

RNA editing is another process that can change one or more nucleotides in a transcript such that the protein product is changed either in sequence or in size.[14] RNA editing does not occur for all transcripts and it typically takes place in a cell or tissue specific manner. An important neuronal gene that undergoes RNA editing is the Glur2 subunit of the AMPA-type glutamate receptors, ligand-gated cation channels. A specific adenosine is deaminated to inosine in Glur2 pre-mRNA.[15] The adenosine is part of a codon (a nucleotide triplet that codes for an amino acid) that normally codes for a glutamine residue, but the inosine-containing codon codes for arginine. This amino acid switch has profound effects on the function of the channel, especially with regard to calcium conductance. The editing appears to be critical because mutant mice that lacked the ability to make the amino acid conversion were found to have severe neurological problems and die within just a few weeks after birth.[16]

Messenger RNA is transported out of the nucleus into the cytoplasm where it is directed to the appropriate location for translation, but only capped, polyadenylated, fully spliced mature mRNA is transported. The transport signal consists of the various proteins that have accumulated on the mRNA since the beginning of its synthesis. Once out of the nucleus, the proteins are shed and transported back into the nucleus where they are recycled for continued transcription. Some proteins are translated and then transported or diffused to their final destination, but in many cases, the mRNA is transported to specific sites and translated locally. This is of special interest in the nervous system in which mRNAs related to cytoskeletal structure, signaling, and even transcriptional regulation have been detected in dendrites and axons where the machinery for translation is also found and in which local translation does appear to take place.[17]

6.3 REGULATION OF TRANSCRIPTION

Regulation of transcription is one method through which the cell controls protein amount, although it is important to bear in mind that there is not always a direct correspondence between mRNA levels and protein levels and also that there are numerous other mechanisms for controlling the amount and activity of a protein.[18] Transcriptional regulation acts primarily at the level of initiation. Regulation also occurs during elongation, termination, and RNA splicing, but these will not be discussed here. The importance of gene regulation lies in the notion that for a cell to function properly, only a specific subset of proteins is required at any particular time. Thus, mRNA for a given gene is typically expressed in a highly regulated tissue- and time-dependent manner. In any particular cell, some genes are exclusively transcribed during certain periods of early development and others only later. Still other mRNAs are expressed throughout the life of a cell or only in response to intrinsic or extrinsic signals. Whether a particular mRNA is transcribed is dictated by dynamic and concerted interactions between DNA and a wide variety of transcription factors and other regulatory proteins.

Transcription factors can either facilitate or suppress transcription and are either *constitutive* or *inducible*. Constitutive transcription factors are present in the cell at all times and are typically activated through signal transduction pathways. Inducible transcription factors are not usually present in the cell, but are transcribed and then translated following some kind of intra- or extracellular signal. Positive transcription factors are known as *activators*; negative factors are called *repressors*. Transcription factors bind with high specificity to specific DNA sequences in the promoter or elsewhere in the regulatory region of genes and interact with other proteins to influence transcription. The protein–protein interactions are of two general types: *recruitment*, in which a protein facilitates the *cooperative binding* of one or more other proteins to a specific genomic location; and *allostery*, in which one protein triggers a conformational change in another eliciting some kind of functional modification in the second protein. Often both types of interactions take place. In many cases, DNA-binding proteins bind to adjacent sites on the DNA, but this is not always true. Interacting proteins may bind at very long distances from one another and are able to make physical contact because of DNA's ability to loop back on itself. Indeed, there are specific DNA-binding proteins that facilitate the bending of DNA so that distant sites can be brought in close proximity to one another.[19,20]

Transcription factors typically have distinct DNA-binding domains and protein-binding domains. The DNA-binding domains have high affinity for specific DNA nucleotide sequences and often bind to DNA as dimers with an α-helix functioning as the DNA recognition site. Other parts of the DNA-binding domain interact with the DNA backbone to align the DNA recognition helix in the proper position in the major groove of DNA. Some of the more common DNA-binding domains found in transcription factors include homeodomain proteins, zinc-containing DNA-binding domains (includes zinc finger proteins), leucine zippers, and helix-loop-helix proteins.[21]

One of the important functions of activator proteins is to recruit Pol II to the promoter. Typically this is not a direct interaction with Pol II, but rather recruitment takes place indirectly through interactions with preformed polymerase-associated protein complexes, such as mediator or TFIID, which are often already associated with Pol II. Often the concerted action of many activators recruits the transcriptional machinery. For a simple example, consider two distinct activators that contact mediator at different sites. Either one alone provides insufficient binding energy, but the combined binding energy of the two is able to effectively recruit the complex. Transcriptional initiation often requires the presence of many activators that work together in such an integrated manner. Each gene does not have its own specific activator; rather, the combined activity of a group of activators present in the right combination and at the right time is required for gene expression. This kind of *combinatorial control* is a key element for the appropriate cell-, time-, and response-dependent expression of many genes.[8]

The DNA in chromatin is generally inaccessible for transcription because promoters are tightly bound by histone proteins. Some activators recruit specific proteins that alter nucleosomes so that the transcriptional machinery can access the gene. This occurs in one of two ways: *nucleosome remodeling* or through *covalent modifications*. Activators recruit nucleosome remodeling complexes that change the nature of DNA-histone interactions. These complexes can free the promoter either by moving the position of the histone relative to the promoter or by causing the promoter to become less tightly bound to the histone. Either situation promotes the binding of Pol II and its associated proteins. The second type of alteration involves reversible covalent modifications to histone proteins. Activators recruit *histone acety-lases* that add acetyl groups to the N-terminal tails of histones. The tails are rich in lysine residues whose positive charge normally facilitates tight binding of the histone to the negatively charged phosphate portion of the DNA backbone. Acetylation effectively neutralizes the tail's positive charge causing the nucleosome structure to relax facilitating the binding of the transcriptional machinery to the promoter. In addition, certain proteins and protein complexes such as TFIID have specific domains known as *bromodomains* that bind with high affinity to acetylated histones further promoting the initiation of transcription. Remodeling and histone modification typically act in combination to promote transcription, sometimes the activity of one enhancing that of the other.[5,22]

Histone tails can also be modified with the addition of methyl groups by *histone methylases*. This can occur in an analogous manner as acetylation, but histone methylation is usually associated with inhibition of transcription. Thus, *chromo-domain*-containing proteins interact with methylated histones to prevent transcription by inhibiting the binding of transcriptional activators. Methylation also recruits proteins with methylase activity that propagate methylation locally to unmodified histones (see below). The covalent modification of histone tails is a dynamic process and it is believed that it is the overall pattern of histone modifications that is an important factor in transcriptional regulation.[5,22]

As a result of the integrated activity of many activators and the proteins with which they interact, only certain genes are transcribed at any given time. Transcriptional *repressors* also contribute to the transcriptional status of a gene. In general, repressors behave very similarly to activators, i.e., they bind with high affinity to specific DNA sequences and also bind to other proteins, but with the end result being the inhibition of initiation. A repressor can bind to a specific DNA site allowing it to interact with the transcriptional machinery thereby preventing initiation through, for example, allosteric modulation. Repressors can also inhibit transcription more indirectly by preventing the binding of activators either to DNA or to the proteins with which the activators interact. Repressors can also prevent transcription through the recruitment of *histone deacetylases*. These enzymes remove acetyl groups from the N-terminal tails of histones thereby restoring nucleosome structure.[23]

Messenger RNA is ultimately degraded through the sequential activity of specific decay enzymes found localized to discrete cytoplasmic bodies. The initiating step of decay is the deadenylation of the mRNA. From there, it proceeds through one of two known pathways that end in its digestion by the activity of either 5'→3' or 3'→5' exonucleases. As might be expected, the decay of mRNA is highly regulated, a consequence of which is that the mRNA for constitutively expressed genes tends to have a longer half-life than for genes that are expressed in response to external signals. The signal for degradation is generally, but not exclusively, found in the *3' untranslated region* (3'UTR) of the mature mRNA. There are specific elements in the 3'UTR whose bound regulatory proteins recruit the decay enzymes. Conversely, certain regulatory proteins are able to stabilize an mRNA, prolonging its life in the cell. The fate of a transcript is determined by the localization, levels, or regulated activity of the decay enzymes and/or the regulatory proteins, some of which are influenced by external signals. Thus, the amount of any given transcript is the combined result of its transcription rate and its decay rate both of which are highly regulated processes.[24]

6.4 TRANSCRIPTIONAL REGULATION THROUGH SIGNAL TRANSDUCTION

Whether a gene is transcribed is dependent on the coordinate activity of many proteins: activators, repressors, mediator, TFIID, Pol II, etc. How are all of these proteins brought together at the right time and in the right place? In prokaryotes, environmental signals can directly influence the activity of activators or repressors. For example, the sugar lactose binds to the Lac repressor preventing its binding to the *lacZ* gene, the product of which is β-galactosidase. β-galactosidase cleaves lactose into glucose and galactose, the preferred energy source. Thus, in the presence of lactose, *lacZ* is "derepressed" and β-galactosidase is synthesized for the metabolism of lactose, but in the absence of lactose, the Lac repressor binds to *lacZ* preventing transcription. This efficient scheme ensures that the cell only makes the enzyme when it is needed. At some level, all cells respond to environmental signals, but this sort of simple, direct mechanism is generally not the case in multicellular eukaryotic organisms. Rather, transcription is often mediated indirectly through *signal transduction pathways*.

Signal transduction pathways are activated by the binding of a signal molecule *ligand* to a cell surface *receptor*, a protein or protein complex that is embedded in the cell membrane with one surface exposed on the extracellular side and a second surface exposed on the intracellular side. Signal molecules include neurotransmitters, steroid hormones, neuropeptides, modified amino acids, and other kinds of molecules that are released from other cells in response to external or internal signals. The binding of the signal molecule on the extracellular surface causes a conformational change in the receptor which then triggers a cascade of intracellular events ultimately leading to some sort of transcriptional regulatory activity. The signal transduction pathway can be simple or it can be very elaborate and complex. The cAMP response element binding (CREB) protein system is presented as a simple example of signal-mediated activation of transcription and as an illustration of many of the principles described above (see Figure 6.3).[25]

CREB is a constitutive transcription factor and the prototypical member of the CREB family that includes CREM (cAMP response element modulator) and ATF-1 (activating transcription factor 1). CREB itself has at least five splice variants that can act as either activators or repressors, dependent on the exon configuration. CREB variants that are activators all contain three important domains: bZIP which mediates DNA binding to the cAMP response element (CRE) and also mediates CREB dimerization; Q2/CAD (constitutive active domain) which interacts with the transcriptional machinery; and the kinase inducible domain (KID) which contains a serine residue (Ser-133) whose phosphorylation state is a critical determinant of CREB's activity level.[26,27]

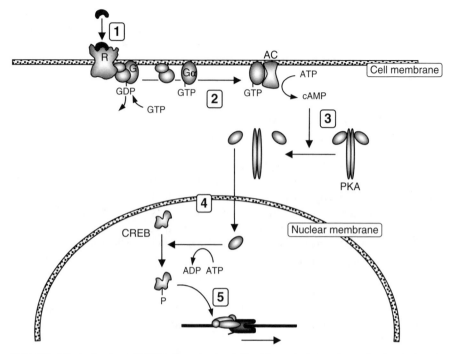

FIGURE 6.3 Activation of CREB through stimulation of a typical signal transduction pathway. The process is initiated by the binding of a small molecule ligand to a cell surface receptor (R) embedded in the cell membrane (step 1). A subsequent conformational change in the receptor causes guanosine-diphosphate (GDP) to be replaced with guanosine-triphosphate (GTP) on the trimeric G-protein (G) and its GTP-containing α subunit to become dissociated (step 2). The α/GTP complex activates the enzyme adenylyl cyclase (AC) which catalyzes the formation of cyclic-AMP (cAMP) from ATP. cAMP triggers the dissociation of the catalytic subunit of protein kinase A (PKA) which then diffuses into the nucleus and phosphorylates CREB (steps 3 and 4). Phosphorylated CREB now binds to CRE on the DNA and promotes transcription through its interactions with the preinitiation complex and CREB binding protein (CBP) (step 5).

Unphosphorylated CREB resides in the nucleus of many different cell types. In this state, it is thought to be able to stimulate a very low level of basal transcription of CRE-containing genes through its interaction with the transcriptional machinery. Stimulus-induced phosphorylation of Ser-133 increases CREB-mediated transcription manyfold. It has become clear that numerous pathways can lead to the phosphorylation of Ser-133, but the first described and perhaps most prominent is through activation of adenylyl cyclase (AC). Ligand binding to any one of a number of membrane-bound G-protein-coupled receptors activates AC through the G-protein mediated hydrolysis of GTP. Activated AC catalyzes the formation of cAMP from ATP. cAMP interacts with protein kinase A (PKA) stimulating the release of its catalytic subunit which then migrates into the nucleus and phosphorylates Ser-133. Phosphorylated CREB has very high affinity for CREB binding protein (CBP) and its paralogue p300. Binding of either of these two proteins to CREB promotes transcription through two mechanisms: stabilization of the preinitiation complex and the intrinsic histone acetylase activity of CBP. CREB activity is terminated by the removal of the Ser-133 phosphate through the action of a phosphatase, itself activated through signal transduction pathways.[26,27]

CREB-mediated transcription peaks at about 1 hour after stimulation and then slowly declines to basal levels after about 4 hours. The rate-limiting step is the movement of the PKA catalytic subunit into the nucleus. Some of the first genes to be transcribed following stimulation are the so-called *immediate early genes* (IEGs). Many of these are themselves transcription factors including c-fos, ATF-3, JunD, Nurr1, and CREB itself.[26,27]

The basic mechanism of CREB activation described above is well supported, but there are unresolved questions related to CREB signal discrimination. In addition to AC signaling, CREB can be activated by many signals and in response to a wide variety of physiological stimuli. Moreover, a potentially large number of genes are CREB-responsive, but under a given stimulus or in a particular cell type, only a small set of specific genes are transcribed in response to CREB activation. What accounts for this specificity? This is likely determined by the presence of other activators or repressors and possibly also by the phosphorylation state of other sites on CREB. Herein is an example of combinatorial control whose complexity belies what was once thought to be a simple model of transcriptional regulation.[26,27]

6.5 RNA INTERFERENCE

RNA interference (RNAi) through *microRNA*-mediated mechanisms is a novel form of post-transcriptional gene regulation. RNAi has profound implications not only from a basic biological perspective, but also for its exploitation as an experimental tool and its potential therapeutic application in disease. *MicroRNAs* (miRNAs) are small RNA molecules (approximately 22 nucleotides) that are complementary to the sequence of mRNA targets and are thought to be an evolutionary by-product of a cellular defense mechanism directed against viral pathogens.[28] MicroRNAs are thought to be synthesized in a manner that is similar if not identical to mechanisms of mRNA synthesis.[29] Approximately 300 miRNAs have been definitively identified in plants, nematodes, fruit flies, fish, rodents, and humans, and it has been estimated that there are as many as 255 human *miRNA* genes.[30] The few miRNA targets that have been characterized are involved in developmental patterning and timing,[31] but in general little is known of miRNA targets, especially in animals. Dysfunction of miRNA processes have been implicated in a limited number of human pathologies,[32] but miRNA research is an emerging field and the ubiquity of this observation has yet to be determined.

Many miRNA genes are *polycistronic* (tightly clustered in a single genomic region and transcribed together), but quite a few are embedded within the introns of their target genes.[29] Clustered miRNAs are often, but not always functionally related. Similarly, functionally related miRNA genes are sometimes physically distant to one another within the genome. Some miRNAs are known to be synthesized by Pol II and the normal mRNA transcriptional machinery suggesting that regulation of miRNA expression occurs in conjunction with or similarly to that of mRNA.[29] However, little is currently known about the regulation of miRNA expression. miRNAs are initially synthesized as a single-stranded RNA precursor of about 600 nucleotides in length known as *pri*-miRNA. A portion of each of the 5' and 3' ends are cleaved off by the *Drosha* endonuclease leaving behind an approximately 75-nucleotide intermediate (now called *pre*-miRNA) that hybridizes to itself by Watson–Crick base-pairing into a *stem-loop* structure in which the miRNA segment forms the double-stranded stem. This structure is transported out of the nucleus and into the cytoplasm where it interacts with another endonuclease known as *Dicer* and possibly with members of the *Argonaute* protein family. Dicer cleaves off the loop of the pre-miRNA to complete the processing of the now mature approximately 22-nuleotide double-stranded miRNA molecule.[33]

Following Dicer processing, miRNA becomes incorporated into the ribonucleoprotein *RNA-induced silencing complex* (RISC; note that many authors refer to this complex as miRNP). RISC contains numerous proteins including one of the Argonaute family members around which the miRNA binds very tightly. Only one strand of the miRNA duplex is incorporated, the strand that shows complementarity to the target mRNA; the other strand is identified and degraded by RISC by unknown mechanisms. RISC binds the target mRNA and inhibits its further activity by one of two methods. If there is perfect complementarity between the miRNA and the target transcript, the mRNA is degraded. If there are mismatches between the two RNA molecules, translation is prevented. The degree of complementarity is probably not the sole determinant of the specific mechanism of interference, but either situation concludes with the same result: protein expression is inhibited.[33]

6.6 GENE SILENCING

Large sections of DNA can be "turned off" through the concerted activity of methylases, deacetylases, and other proteins. *Gene silencing* is associated with protein-dense genomic regions known as *heterochromatin*. One such area is found at the end of chromosomes – the *telomere*. A complex of proteins that includes Sir 1, 2, 3, and 4 (silent information regulators) is recruited by proteins that bind to specific sequences in the telomere. The Sir proteins bind to unacetylated histones and then deacetylate nearby histones thus propagating a growing heterochromatin complex. Transcription is inhibited as a result of the inability of the transcription machinery or activators to bind to the complex. Eventually histones are encountered that have been modified in certain ways that prevent the binding or deacetylation activity of the Sir proteins. Heterochromatin is thought to make up approximately 50% of the genome and most of it is located at the telomeres or centromeres, but is also found elsewhere throughout the genome.[34,35]

Gene silencing can also occur through covalent modifications on the DNA itself. During development or in other situations, an initiating signal induces certain genes to be completely shut off. This is accomplished through methylation of the DNA. *DNA methyltransferase* adds a methyl group onto the ring structure of the nucleotide cytosine. Proteins that recognize methylcytosine bind with high affinity and recruit histone remodeling and modifying proteins whose activity results in the complete inhibition of transcription on that gene.[36]

During cell division, it is usually necessary to preserve the status of silenced genes in daughter cells. This implies that the 50% of newly synthesized histones and DNA must be properly modified, but in the absence of the original initiating signal. Methylated histones will remain in the correct position within the chromatin of daughter cell DNA distributed approximately equally between the two cells. These modified histones subsequently recruit chromodomain-containing proteins which include histone methylase. Methylation then proceeds on the nearby unmethylated histones. A similar mechanism occurs for methylated DNA. A specific enzyme, *maintenance methylase*, recognizes the partially methylated DNA and completes local methylation of the new strand. This triggers protein recruitment and gene silencing as described above. This is the basis of a phenomenon known as *imprinting* in which a gene *allele* is silenced selectively on either the maternal or paternal chromosome and is transmitted to the offspring. Preservation of silenced genes or alleles during mitosis, meiosis, or fertilization is a form of DNA modification that is heritable in the absence of DNA mutation and that does not require the original initiating signal. This general phenomenon is referred to as *epigenetic regulation*.[37,38]

6.7 SOURCES OF VARIATION IN GENE EXPRESSION

Transcription can be influenced by a wide array of stimuli including exposure to exogenous substances, external sensory stimulation, and in response to pathological conditions. Transcriptional regulation by these sources facilitates an organism's response and adaptation to an ever-changing internal and external environment. For example, acute exposure to many psychoactive substances such as alcohol, cocaine, morphine, and other drugs of abuse, stimulates the phosphorylation of CREB leading to a cascade of gene expression events.[39] Most notably, there is an increase in the transcription of inducible transcription factors, also known as *immediate early genes* (IEGs) including *c-Fos*, *FosB*, *c-Jun*, *JunB*, *JunD*, *Egr1*, *Egr2*, *Zif268*, and others. The increased mRNA is often translated into increased protein which then induces late-response target genes. This phenomenon has been exploited in model organisms to map regions of the brain in which the drugs are active, as well as to learn something about the neurobiology of psychoactive drug responses. Different drugs do not all show similar patterns of response across IEGs, time, or brain region. For example, cocaine induces *c-Fos* in a number of brain regions including a very robust induction in the striatum.[40] This is not too surprising because the striatum is a major dopamine output area and cocaine stimulates increased dopamine activity through its action as an indirect dopamine agonist. However, alcohol, which stimulates dopamine release in the striatum does not or only weakly induces *c-Fos* transcription in this brain structure.[41,42] On the other hand, *Egr1* is significantly induced in the striatum after either drug.[39,43] It is believed that these kinds of IEG cascades in the reward and motivation circuitry of the brain contribute to the pathological state of addiction;[44] however, a clear understanding of how these events do so has not yet been achieved.

Within a heterogeneous population, variation in transcription is also attributable to genetic factors; i.e., sequence *polymorphisms* in a gene. It is easy to imagine that a DNA sequence difference in a promoter or binding element might cause differential transcription factor binding affinity, and therefore result in a differential rate or magnitude of transcription. Sequence variation can also affect other pretranslation processes such as splicing and mRNA stability. These kinds of genetic influences on gene expression are undoubtedly an important source of genetic variation for behavior and other complex traits. Indeed, gene expression itself can be considered as a complex or quantitative trait just as can virtually all measures of behavior or physiology. This simply means that the expression level of a given gene in a heterogeneous population is continuously distributed due to influences from both environmental and genetic sources. Like other complex traits, the genetic component of the expression variance is often under the control of multiple genes. As such, it is possible to map *quantitative trait loci* (QTL) for the expression of a gene (see Chapter 4). QTLs are regions of the genome that contain two or more DNA sequence variants or alleles (within a population) that mediate the genetic component of variation for the trait of interest; in this case, gene expression. An expression QTL can be either *cis-* or *trans-* acting. In the former case, the QTL is tightly linked to the gene itself, suggesting that a polymorphism within a *cis* regulatory element of the gene impacts its own expression, while in the latter case, the QTL is found physically distant from the gene and modulates expression through any one of a number of possible interactions. Single gene mutations can affect the expression of many genes through *trans* effects,[45-47] but the discovery of a *cis*-QTL in a more genetically complex population is especially informative because this immediately points to a candidate gene for any other trait that has had a QTL mapped to the same genomic region. Standard mRNA quantification methods can be used to map gene expression for a single gene at a time or gene expression microarrays can be used to map QTLs for thousands of genes simultaneously (see below). This general strategy was used in combination with other methods to identify a specific polymorphism in the regulatory region of mouse *Kcnj9* (a G-protein inwardly rectifying potassium channel) that appeared to be responsible for differential regulation of the gene. Perhaps more importantly, a compelling argument was made that this *Kcnj9* polymorphism was, in fact, a QTL for the behavioral trait of basal locomotor activity.[48]

6.8 MEASURING GENE EXPRESSION

Measurement of the expression level of a gene can begin to reveal clues about its general regulation and about its involvement in a cell's response to external perturbations. Thus, techniques have been developed to accurately determine the amount of a given transcript from a cell or tissue sample. The most common methods rely on the properties of Watson–Crick base-pairing making it very easy to design probes that are highly selective for a specific transcript. However, members of gene families typically have at least partial sequence similarities and therefore care must be taken in some situations to ensure that only the gene of interest is being measured rather than the gene in addition to one or more related genes with similar sequences. The methods are generally not difficult to perform, but certain precautions must be taken in handling any kind of RNA because of the ubiquity of *ribonucleases* (RNA-specific degrading enzymes) in both prokaryotic and eukaryotic organisms. Several commonly used methods of mRNA quantification are briefly described.

The standard workhorse method of quantification for a single mRNA species is the *Northern blot*. Total RNA is isolated from cultured cells or a tissue sample using one of several methods.[49] The RNA sample can be further fractionated into purified polyadenylated RNA (i.e., pure mRNA), but this is not always necessary. Following isolation, mRNA of different molecular weights (i.e., lengths) is separated by polyacrylamide gel electrophoresis (PAGE) using standard methods. The separated bands are transferred to a nylon membrane and incubated in the presence of a nucleic acid probe—a strand of radiolabeled DNA that is complementary to the transcript of interest. The membrane is then exposed to x-ray film. If all goes well, a single band is revealed on the film from which the amount of target mRNA in the sample is quantified using image densitometry. Typically the membrane is stripped and the process repeated with a probe for a so-called "housekeeping gene" to control for loading on the gel and other factors that could potentially introduce technical error to quantification of the band. Housekeeping genes are genes that are thought to remain at a constant level across experimental conditions and include the actins, glyceraldehyde-3-phosphate dehydrogenase (G3PDH), the tubulins, cyclophilin, hypoxanthine phosphoribosyltransferase (HRPT), and others, as well as the 28S and 18S ribosomal RNAs.[50] However, the consistency of these genes' expression level must be verified when possible because they often do change across experimental conditions. Northern blotting can be reasonably quantitative, but the many steps involved and the compressed dynamic range of densitometric readings reduces quantitative precision somewhat.

The Taqman-based *quantitative reverse-transcriptase polymerase chain reaction* (RT-PCR) is becoming increasingly popular for the quantification of single-gene mRNA concentration. It is highly selective for the gene of interest, extremely quantitative, and can measure very small amounts of transcript; in theory, just a single mRNA molecule.[51–53] PCR primers are designed to bracket a fairly short (30–50 nucleotides) sequence within a region unique to the target mRNA. A probe of 18 to 22 nucleotides in length is designed to be complementary to the sequence between the primers. The 5' end of the probe is covalently tagged with a fluorescing reporter dye (e.g., FAM) and a quenching dye is added to the 3' end (e.g., TAMRA). Standard PCR is performed in the presence of the RNA sample, the primers, the probe, and a polymerase that has three important enzymatic capabilities: *reverse transcriptase* activity (synthesis of a single-stranded DNA molecule that is complementary to the single-stranded mRNA template), DNA polymerase activity, and 5'→3' exonuclease activity. The reaction is conducted in a specialized thermocycler that is capable of measuring fluorescence at the end of each PCR cycle. Normal PCR amplification takes place with the mRNA as the initial template. During the hybridization step of each cycle, the probe binds to newly synthesized DNA product and is then digested by the exonuclease activity of the polymerase during the extension step. This effectively separates the reporter dye from the quencher dye allowing the reporter to be detected by the thermocycler. Thus, fluorescence increases as the product increases with a rate of increase that is dependent on the amount of mRNA template initially present. The number of PCR cycles it takes to reach a certain level of detected fluorescence (the *threshold cycle*) is a quantitative measure of how much target mRNA was present.[54] If a clone of the target transcript is available, extremely accurate quantification can be achieved using a standard curve. Otherwise, the assay will include a primer/probe set for a housekeeping gene that is amplified in the same reaction vessel and the threshold cycle of the target is expressed as a ratio to that of the housekeeping gene mRNA. There are other RT-PCR-based expression assays,[55] but the Taqman version possesses the combined benefits of high specificity, sensitivity, and quantitative precision.

In some cases, it is desirable to not only quantify a transcript of interest, but also to determine its ultrastructural localization. With either of the above two methods, localization is only as good as the dissection procedure. This is often unacceptable in a brain where fine structure is quite complex. Thus, the method of *in situ hybridization* has been devised to measure mRNA from a section of an intact tissue sample that can be examined at high resolution, down to the level of individual cells. Typically, the tissue is frozen at the time of sacrifice, sliced on a cryostat, and the slices mounted onto glass microscope slides. The tissue is fixed in such a way as to preserve the structural and chemical nature of the tissue and to make it permeable to nucleic acid probes. RNA or DNA probes complementary to the target transcript are labeled with a radioisotope during synthesis. Other labels can be used (e.g., fluorescent), but these typically are not as sensitive as radioisotopes. The mounted tissue is incubated in the presence of the probe during which time it hybridizes selectively to the target transcript. The slide is exposed to an x-ray film and selected areas from the subsequent image can be quantified using densitometry techniques. This method can be considered only semiquantitative because of the many sources of error that can be introduced during the procedure.[56]

Over the past decade, methods have been developed to quantify thousands of different transcripts from the same RNA sample making it possible to measure the level of every single transcript from a single cell culture or tissue sample. Thus, control cells can be compared to various kinds of experimental groups (e.g., drug-treated, diseased, cells in a particular stage of growth, etc.) and the genes that respond to the experimental condition can then be identified. Often referred to as *gene expression profiling*, these methods have been employed as an unbiased approach applied with many goals in mind including the elucidation of biochemical or neuronal pathways, the identification of coregulated networks of gene expression, the determination of patterns of recognition as predictors of drug responses or pathologic states, and in gene discovery applications as described above. A variety of methodologies have been developed including *differential display,*[57] *serial analysis of gene expression* (SAGE[58]), and the solid platform *microarray* of which several different types have been devised and is by far the most commonly used technique for expression profiling. There are a number of commercially available microarrays and many individual laboratories possess the instrumentation for producing their own in-house microarrays. Perhaps the most well-developed commercial microarray system is that offered by Affymetrix. The Affymetrix system will be used to illustrate the general principles of expression profiling using microarray methodologies.

Total RNA is isolated using standard methods and then amplified and biotin-labeled through a series of enzymatic reactions and purification steps.[49,59] This biotin-labeled complementary RNA (cRNA) is fragmented and hybridized to the Affymetrix GeneChip® microarray (Affymetrix, Santa Clara, CA). Affymetrix uses a lithographic manufacturing process to "grow" short oligonucleotides (nucleic acid sequences of 25 nucleotides in length) in a specific location, or array "cell," on a solid support. Each cell contains hundreds of thousands of identical oligonucleotides. Each organism's mRNAs are represented by 22 such spots that consist of a series of 11 "perfect match" (PM) and "mismatch" (MM) pairs, known collectively as a probe set. The PM sequence is perfectly complementary to a unique sequence within the target mRNA, while its corresponding MM is identical but with a single altered nucleotide. The theory is that the biotin-labeled material will bind with high affinity and specificity to only the PM sequences, but not the MM sequences because of the mismatched nucleotide. Thus the MM probes control for background and nonspecific binding. In reality, some of the PM/MM pairs do not always perform as desired, but the system generally works well when the entire probe set is considered. Following hybridization, the microarray is bathed in a solution containing phycoerythrin-streptavidin. Streptavidin binds with very high affinity to the hybridized biotinylated cRNA molecules and phycoerythrin is a broad spectrum fluorophore. The fluorescence intensity of each individual cell is determined with the use of a scanning laser and the intensity corresponds directly to the amount of labeled cRNA bound to the oligonucleotides in the cell. One of various analytical methods is used to combine the fluorescence values from all cells in a probe set to generate a single value of transcript abundance.[60–62] Many thousands of probe sets can be manufactured onto a single array meaning that a substantial proportion if not all of an organism's transcripts can be represented on a single microarray. Most other types of microarrays use only a single cell per mRNA target, do not use the PM/MM concept, and are manufactured differently, but otherwise are very similar in basic design and overall theory of operation.[63]

The wet laboratory procedures for expression profiling are relatively straightforward, but the analysis can be complex in design and interpretation. Typically, an experiment will go through three levels of analysis. First, procedures are used to determine if a gene was actually present in the sample taking into account such factors as the intensity of a given cell and nonspecific background fluorescence. Second, statistical analyses are conducted to identify those genes that have changed as a result of the experimental condition(s). This step can be problematic because of multiple testing issues. Strictly speaking, the number of genes being examined on a microarray is equivalent to the number of hypotheses being tested. Therefore, the nominal P value in a standard statistical test would have to be extremely low for an expression difference to be considered significantly different and many true differences would be rejected because of this. On the other hand, many false positives would be accepted as true if the statistical criteria are not stringent enough. Currently, methods such as the Benjamini–Hochberg False Discovery Rate (FDR[64]) are widely accepted to control for error rates in microarray experiments. Rather than a strict cutoff, the FDR uses a sort of sliding scale that allows the experimenter to dictate the extent of error in an experiment. Finally, with a set of genes in hand, higher order analyses such as clustering and biological pathway analyses are used to identify coregulated genes and the biological networks that they fit into.[65] The computational tools and bioinformatics resources for these kinds of analyses can still be considered fairly immature, but are constantly being improved. It will be the next generation of tools and the scientists trained to use them that will start to really tap into the power of this experimental strategy.

REFERENCES

1. Crick, F.H.C., On protein synthesis, in *Symposium of the Society for Experimental Biology XII*, Academic Press, New York, 1958, 153.
2. Cech, T.R., The efficiency and versatility of catalytic RNA: implications for an RNA world, *Gene*, 135, 33, 1993.
3. Kaczmarek, L., Molecular biology of vertebrate learning: is c-fos a new beginning?, *J. Neurosci. Res.*, 34, 377, 1993.
4. Belmont, A.S., Dietzel, S., Nye, A.C., Strukov, Y.G., and Tumbar, T., Large-scale chromatin structure and function, *Curr. Opin. Cell Biol.*, 11, 307, 1999.
5. Wolffe, A.P., Chromatin structure and the regulation of transcription, in *Transcription Factors*, Locker, J., Ed., BIOS Scientific Publishers Ltd., Oxford, 2001, chap. 3.
6. Watson, J.D. et al., *Molecular Biology of the Gene*, 5th ed., Cold Spring Harbor Laboratory Press, San Francisco, 2004.
7. Watson, J.D. and Crick, F.H.C., Genetical implications of the structure of deoxyribonucleic acid, *Nature*, 171, 964, 1953.
8. Courey, A.J., Regulatory transcription factors and cis-regulatory regions, in *Transcription Factors*, Locker, J., Ed., BIOS Scientific Publishers Ltd., Oxford, 2001, chap. 2.
9. Garvie, C.W. and Wolberger, C., Recognition of specific DNA sequences, *Mol. Cell*, 8, 937, 2001.
10. Young, B.A., Gruber, T.M., and Gross, C.A., Views of transcription initiation, *Cell*, 109, 417, 2002.
11. Pugh, B.F., RNA polymerase II transcription machinery, in *Transcription Factors*, Locker, J., Ed., BIOS Scientific Publishers Ltd., Oxford, 2001, chap. 1.
12. Maniatis, T. and Reed R., An extensive network of coupling among gene expression machines, *Nature*, 416, 499, 2002.
13. Howe, K.J., RNA polymerase II conducts a symphony of pre-mRNA processing activities, *Biochim. Biophys. Acta*, 1577, 308, 2002.
14. Schmauss, C. and Howe, J.R., RNA editing of neurotransmitter receptors in the mammalian brain, *Sci. STKE*, 2002, pe26, 2002.
15. Sommer, B. et al., RNA editing in brain controls a determinant of ion flow in glutamate-gated channels, *Cell*, 67, 11, 1991.
16. Brusa, R. et al., Early-onset epilepsy and postnatal lethality associated with an editing-deficient GluR-B allele in mice, *Science*, 270, 1677, 1995.
17. Job, C. and Eberwine, J., Localization and translation of mRNA in dendrites and axons, *Nat. Rev. Neurosci.*, 2, 889, 2001.
18. Anderson, L. and Seilhammer. J., A comparsion of selected mRNA and protein abundances in human liver, *Electrophoresis*, 18, 533, 1997.
19. Lefstin, J.A. and Yamamoto, K.R., Allosteric effects of DNA on transcriptional regulators, *Nature*, 392, 885, 1998.
20. Ptashne, M. and Gann, A., Transcriptional activation by recruitment, *Nature*, 386, 569, 1997.
21. Fairall, L. and Schwabe, J.W.R., DNA binding by transcription factors, in *Transcription Factors*, Locker, J., Ed., BIOS Scientific Publishers Ltd., Oxford, 2001, chap. 4.

22. Fry, C.J. and Peterson, C.L., Chromatin remodeling enzymes: who's on first?, *Curr. Biol.*, 11, R185, 2001.

23. Maldonado, E., Hampsey, M., and Reinberg, D., Repression: targeting the heart of the matter, *Cell*, 99, 455, 1999.

24. Wilusz, C.J. and Wilusz, J., Bringing the role of mRNA decay in the control of gene expression into focus, *Trends Genet.*, 20, 491, 2004.

25. Shore, P. and Sharrocks, A.D., Regulation of transcription by extracellular signals, in *Transcription Factors*, Locker, J., Ed., BIOS Scientific Publishers Ltd., Oxford, 2001, chap. 6.

26. Mayr, B. and Montminy, M., Transcriptional regulation by the phosphorylation-dependent factor CREB, *Nat. Rev. Mol. Cell Biol.*, 2, 599, 2001.

27. Lonze, B.E. and Ginty, D.D., Function and regulation of CREB family transcription factors in the nervous system, *Neuron*, 35, 605, 2002.

28. Waterhouse, P.M., Wang, M.B., and Lough, T., Gene silencing as an adaptive defence against viruses, *Nature*, 411, 834, 2001.

29. Bartel, D.P., MicroRNAs: genomics, biogenesis, mechanism, and function, *Cell*, 116, 281, 2004.

30. Lim, L.P. et al., Vertebrate microRNA genes, *Science*, 299, 1540, 2003.

31. Ambros, V., The functions of animal microRNAs, *Nature*, 431, 350, 2004.

32. Nelson, P. et al., The microRNA world: small is mighty, *Trends Biochem. Sci.*, 28, 534, 2003.

33. Meister, G. and Tuschl, T., Mechanisms of gene silencing by double-stranded RNA, *Nature*, 431, 343, 2004.

34. Gartenberg, M.R., The Sir proteins of *Saccharomyces cerevisiae*: mediators of transcriptional silencing and much more, *Curr. Opin. Microbiol.*, 3, 132, 2000.

35. Gottschling, D.E., Gene silencing: two faces of SIR2, *Curr. Biol.*, 10, R708, 2000.

36. Martienssen, R.A. and Colot, V., DNA methylation and epigenetic inheritance in plants and filamentous fungi, *Science*, 293, 1070, 2001.

37. Richards, E.J. and Elgin, S.C., Epigenetic codes for heterochromatin formation and silencing: rounding up the usual suspects, *Cell*, 108, 489, 2002.

38. Tilghman, S.M., The sins of the fathers and mothers: genomic imprinting in mammalian development, *Cell*, 96, 185, 1999.

39. Ryabinin, A.E., ITF mapping after drugs of abuse: pharmacological versus perceptional effects, *Acta Neurobiol. Exp.*, 60, 547, 2000.

40. Graybiel, A.M., Moratalla, R., and Robertson H.A., Amphetamine and cocaine induce drug-specific activation of the c-fos gene in striosome-matrix compartments and limbic subdivisions of the striatum, *P. Nat. Acad. Sci. U.S.A.*, 87, 6912, 1990.

41. Hitzemann, B. and Hitzemann, R., Genetics ethanol and the Fos response: a comparison of the C57BL/6J and DBA/2J inbred mouse strains, *Alcohol. Clin. Exp. Res.*, 21, 1497, 1997.

42. Ryabinin, A.E. et al., Differential sensitivity of c-Fos expression in hippocampus and other brain regions to moderate and low doses of alcohol, *Mol. Psychiatr.*, 2, 32, 1997.

43. Jouvert, P. et al., Differential rat brain expression of EGR proteins and of the transcriptional corepressor NAB in response to acute or chronic cocaine administration. *Neuromol. Med.*, 1, 137, 2002.

44. Nestler, E.J., Molecular basis of long-term plasticity underlying addiction, *Nat. Rev. Neurosci.*, 2, 119, 2001.

45. Monti, J. et al., Expression analysis using oligonucleotide microarrays in mice lacking bradykinin type 2 receptors, *Hypertension*, 38, E1, 2001.

46. Yoshihara, T. et al., Differential expression of inflammation- and apoptosis-related genes in spinal cords of a mutant SOD1 transgenic mouse model of familial amyotrophic lateral sclerosis, *J. Neurochem.*, 80, 158, 2002.

47. Cadet, J.L. et al., Temporal profiling of methamphetamine-induced changes in gene expression in the mouse brain: evidence from cDNA array, *Synapse*, 41, 40, 2001.

48. Hitzemann, R. et al., A strategy for the integration of QTL, gene expression, and sequence analyses, *Mamm. Genome*, 14, 733, 2003.

49. Mannhalter, C., Koizar, D., and Mitterbauer, G. Evaluation of RNA isolation methods and reference genes for RT-PCR analyses of rare target RNA, *Clin. Chem. Lab. Med.*, 38, 171, 2000.

50. Thellin, O. et al., Housekeeping genes as internal standards: use and limits, *J. Biotechnol.*, 75, 291, 1999.

51. Giulietti, A. et al., An overview of real-time quantitative PCR: applications to quantify cytokine gene expression, *Methods*, 25386, 2001.

52. Applied Biosystems, User Bulletin #2, Relative Quantitation of Gene Expression: ABI PRISM 7700 Sequence Detection System, 2001.

53. Bustin, S.A., Absolute quantification of mRNA using real-time reverse transcription polymerase chain reaction assays, *J. Mol. Endocrinol.*, 25, 169, 2000.

54. Livak, K.J. and Schmittgen, T.D., Analysis of relative gene expression data using real-time quantitative PCR and the 2(-Delta Delta C(T)) Method, *Methods*, 25, 402, 2001.

55. Ponchel, F. et al., Real-time PCR based on SYBR-Green I fluorescence: an alternative to the TaqMan assay for a relative quantification of gene rearrangements, gene amplifications and micro gene deletions, *BMC Biotechnol.*, 3, 18, 2003.

56. Ziolkowska, B. and Przewlocki, R., Methods used in inducible transcription factor studies: focus on mRNA, in *Immediate Early Genes and Inducible Transcription Factors in Mapping of the Central Nervous System Function and Dysfunction*, Kaczmarek, L., and Roobertson, H.A., Eds., Elsevier Science, Amsterdam, 2002, chap. 1.

57. Bartlett, J.M., Differential display: a technical overview, *Methods Mol. Biol.*, 226, 217, 2003.

58. Powell, J., SAGE. The serial analysis of gene expression, *Methods Mol. Biol.*, 99, 297, 2000.

59. Affymetrix, *GeneChip® Expression Analysis Technical Manual*, 2004.

60. Zhang, L., Miles, M.F., and Aldape, K.D., A model of molecular interactions on short oligonucleotide microarrays, *Nat. Biotechnol.*, 21, 818, 2003.

61. Irizarry, R.A. et al., Exploration, normalization, and summaries of high density oligonucleotide array probe level data, *Biostatistics*, 4, 249, 2003.

62. Han, E.S. et al., Reproducibility, sources of variability, pooling, and sample size: important considerations for the design of high-density oligonucleotide array experiments, *J. Gerontol. A Biol.*, 59, 306, 2004.

63. Holloway, A.J. et al., Options available—from start to finish—for obtaining data from DNA microarrays II, *Nat. Genet.*, 32 Suppl., 481, 2002.

64. Benjamini, Y. and Hochberg, Y. Controlling the false discovery rate: a practical and powerful approach to multiple testing, *J. Roy. Stat. Soc. B*, 57, 289, 1995.

65. Slonim, D.K., From patterns to pathways: gene expression data analysis comes of age, *Nat. Genet.*, 32 Suppl., 502, 2002.

7 Bioinformatics of Behavior

Elissa J. Chesler

CONTENTS

SUMMARY

This chapter is intended to provide a brief introduction to biological databases for two major purposes—first, to familiarize readers with the structure and design of databases for use in their own laboratories, and second, to illustrate examples of public biological databases and approaches that have grown from early bioinformatic methods.

7.1 WHAT IS BIOINFORMATICS?

Bioinformatics is a rapidly developing field that originated with the need for integration, analysis, and dissemination of rapidly accumulating nucleic acid and protein sequence data. With the massive influx of data from genome sequencing efforts, tools for the compilation, storage and assembly were devised. Phylogenic methods were developed to examine the relationships among species through the comparison of sequence data. Though biologists now had the string of letters, we only knew a few of the words and sentences. Annotation was required to parse the genome sequence and determine the identity and location of genes, their start and stop codons, introns, exons, enhancers, promoters, and other features. The mouse genome sequence was largely completed in 2002,[1] but the accumulation and annotation of nucleic acid sequence data continues today. The location of polymorphisms and, increasingly, their distribution patterns among strains are being evaluated. The genome, now largely anchored to the C57BL/6J strain, is being generalized across members of the species, and the study of haplotype blocks[2-4] and linkage disequilibrium is bringing new understanding to the interaction and selection pressures on genomes within a species. Newly discovered genome features such as micro-RNAs[5,6] and other functional noncoding structures are being annotated.

As behavioral geneticists we attempt to perform a second, much higher-order annotation of the genome. A major goal of neurobehavioral genetic analysis is to find genome locations, genes and polymorphisms that influence the variability in behavioral traits. This brings up an important distinction between the search for genes underlying behavioral variation and the genomic analysis of behavior. The former endeavor is the attempt to identify the actual polymorphisms that are segregating in a natural population that are responsible for the heritable variation in a behavioral phenotype—identifying the gene in a Mendelian sense, as a mode of inheritance. The latter effort uses genomic technology such as genome wide mutagenesis and microarray methods to determine which genes and gene products are involved in the effector pathway, regardless of whether the pathway members are polymorphic. Both of these efforts have been fruitful in identifying novel gene-to-behavior relations.

A third emerging application in bioinformatics will be discussed in the present chapter—the integration of data resources to address neurobehavioral questions. In much the same way that functional MRI has aided in the identification of the shared biological substrates for cognitive processes, we can now use genomic technology to address hypotheses concerning our understanding of the relationships among behavioral traits, at the same time identifying sources of individual differences in the underlying processes. The approach involves the integration of systems-level trait exploration with knowledge of the genome. Both of these endeavors make use of the tremendous array of biological databases that have been developed over the past few decades.

It is not possible to present all of these databases or even a large fraction of them in this chapter. Those most relevant to readers of this volume have been compiled at the Neuroscience Database Gateway (NDG) by the Brain Information Group (BIG) of the Society For Neuroscience, http://big.sfn.org/NDG/site/.[7] Some basic underlying concepts and examples are presented, with the intention of providing the reader with a framework for understanding the utility of these dynamic resources. Readers are urged to engage in deep exploration of these databases, starting with genes or traits of interest. Most are equipped with online tutorials and extensive documentation. The process of operant conditioning is also quite an effective approach to training. Click and you shall receive.

7.2 DATABASE DESIGN

7.2.1 DATA STORAGE BEYOND THE SPREADSHEET: LABORATORY INFORMATION MANAGEMENT SYSTEMS (LIMS)

The integration of massive amounts of biological data requires a shift in data management practices for most laboratories. It is more effective to archive raw data from a laboratory in a relational database system than a set of spreadsheets. Though spreadsheets can be very sophisticated and often have sufficient storage capacity for the average laboratory, their applications are limited. Most users continue to employ them for the relative ease with which formulae can be entered and summary statistics computed. However, this is a dubious benefit. Beyond the most simple descriptive statistics, analysis using the most common spreadsheet software is not recommended,[8] and it is best to store data in a format amenable to more sophisticated analyses. Spreadsheets do not provide a log of changes, can be erroneously modified with ease, readily become desynchronized with one another and with external resources, and are difficult to share without spawning a proliferation of multiple documents. In contrast, many simultaneous users in multiple sites can share a single simple database management system (DBMS).

DBMSs are often referred to as laboratory information management systems (LIMS) when customized for scientific laboratory applications. Though costly commercial systems are available, most typical laboratories will find that a simple database system such as Filemaker Pro® (Filemaker, Inc.) or Access® (Microsoft) will be sufficient. These can be used to track individual subject data (all the way from breeding and animal colony climate conditions) through bio-behavioral assay and higher order statistical results derived from them. This type of archiving allows the ability to store, retrieve and analyze large sets of data from multiple laboratories for retrospective meta-analysis[9] in addition to simple laboratory sample tracking and error checking. Using open data base connectivity (ODBC), most statistical packages can query these databases for customization of data tables and views needed for routine analyses.

7.2.2 Data Modeling

A brief introduction to some essential concepts in data modeling is presented as a framework for understanding data storage and database design. The transition from simple spreadsheets to database systems can be achieved as a progression through an orderly series of growth from paper archives and spreadsheets to high-capacity open-source relational databases as the needs of the laboratory grow and change.

7.2.2.1 The Flat Format

Most readers are familiar with the flat data model, in which each database is a self-contained table consisting of rows and columns. This is the typical entry format for statistical analysis tools, in which each row is an experimental unit, subject, or record and each column is an attribute or measured variable of that subject. It is critical to most applications that the data in a single column be of a single type—character or numeric. Users of common business spreadsheets have the luxury of entering data of arbitrary types into these cells, rendering import into more sophisticated databases and statistical packages a nightmare! A frequent example is the technician entered comment, e.g., "3.45?," "52+," "6.3–9.6," or "43 gm." Using a rigorous data storage technique, users can create a series of comment codes that can be later used as filtering or analytic criteria. Other common violations of the flat file format, such as entering data from multiple groups in separate areas of the spreadsheet rather than creating a separate column to indicate the grouping variable, should be avoided.

7.2.2.2 The Hierarchical Format

The hierarchical format is a method for data storage in which each object is a subset of a larger class of objects. Most users are familiar with this in the form of file storage systems on Windows®, Mac®, and Unix® operating systems, or even a simple physical file cabinet. The limitation of the utility of such a tree-based system for managing diverse, interrelated sets of data becomes readily obvious. Two folders labeled "QTL analysis," and "microarray analysis" do not provide a unique location for a new paper on quantitative trait loci (QTL) analysis of microarray data. Most individual flat files that comprise the various studies conducted in a single laboratory each have different data structures and often get filed in such a fashion. As a result, it is difficult to integrate them and analyze a novel, synthetic view of the data.

7.2.2.3 The Relational Database Management System (RDBMS)

Relational databases are the most versatile system for storing large complex sets of data. In a relational database, data are stored in multiple tables and operations are performed on tables to result in new tables. The database can be assembled from a set of flat-format tables through a design process called "normalization." Each level of normalization represents a dramatic reduction in the redundancy of data storage. The purpose of this type of data model is to avoid a common problem in data storage systems, that "if there are two copies of a datum, one of them will be wrong." This may seem paradoxical at first, but is a problem that most readers have experienced

TABLE 7.1
Transitioning from Flat Tables to First Normal Form

	Flat Format	
MOUSE_ID	Latency_Trial_1	Latency_Trial_2
A9483	3.42	2.59
A7842	5.43	5.52
	First-Normal Form	
MOUSE_ID	Trial	Latency
A9483	1	3.42
A9483	2	2.59
A7842	1	5.43
A7842	2	5.52

firsthand. Using a relational database eliminates this problem because changes made to one datum will automatically be propagated throughout the entire database. In the first normal form, each field contains different information. For an example, see Table 7.1. If latency measures on a maze task were obtained twice for each individual mouse, the record can be entered in columns with field names "Mouse_ID, latency_1, latency_2." This is a common data entry format for many statistical packages. However, in the first normal form this would correspond to two records each with column field names "Mouse_ID, latency, score." There are four other levels of normalization. The end result is a set of tables to which each possesses a primary key (the unique value for each record in the table) and a set of foreign keys that refer to records in other tables. Accession numbers are a common example of primary keys. To create a purely normalized set of tables is a challenging effort that may exceed the needs of most users, and ideally a balance is struck between efficiency (obtained by a lack of redundant data) and complexity (obtained by a proliferation of tables).

7.3 "NATURAL" KEYS IN BIOLOGICAL DATABASES

The concept of the key is critical to understanding the use and construction of biological databases. Many of the challenges to data integration emerge from the instability and lack of uniqueness of keys. Biological data are often collected in many parallel efforts, in arbitrary forms devised by individual labs using diverse experimental paradigms. There is an often-expressed hope that somehow these data will be magically integrated so that biological hypotheses can be meaningfully extracted and analyzed. To quote J. R. Platt, "We speak piously of taking measurements and making small studies that will add another brick to the temple of science. Most such bricks just lie around in the brickyard."[10] Computational integration of biological data requires that normalized data structures can be created. When data have a great degree of integration, they can be brought together into a single-flat table. This is rarely possible from the many diverse parallel experimental efforts that occur. But genome sequence location provides a natural key. Historically, the assumption that biological data could be brought together *post hoc* in such a fashion probably emerged because of the fortuitous properties of sequence data—it "anneals" spontaneously. Numerous alignment algorithms have been developed to pull this data together optimally, but in general, any sequence data can be combined with other overlapping sequence data to identify and compare individuals, strains, and species. This is not the case with typical experimental observations, for which additional data often translate into additional data structure complexity. The unaided human brain must be used to perform the only integration possible. Other "natural" keys are present in much of the data collected by mouse geneticists, the gene and the mouse strain. To the extent that these natural keys can be well defined and remain stable, they act as references through which data can be aggregated indefinitely and integrated analytically.

7.3.1 GENOME LOCATION AS REFERENCE

Several browsers allow the examination of and comparison of genomic features by location, including Ensembl, www.ensembl.org, (Figure 7.1), and the University of California–Santa Cruz Genome Browser, www.genome.ucsc.org. They are excellent resources that allow efficient viewing of information about genes within a region, of particular interest to those who have identified a QTL. These displays overlay multiple aligned data types. They offer highly customizable views to reveal positional genes, haplotype variation, SNPs, available mutants, gene expression data, conservation and synteny across species, to name a few examples. Most of them can be queried by BLAST search of DNA or protein sequence, feature name (e.g., gene symbol or microsatellite name), or position. The data are displayed in many rows or "tracks," and have tremendous flexibility and customization of the display. Users can identify the location and strain distribution of single nucleotide polymorphisms using the relatively new SNPview http://snp.gnf.org/ database, or Mouse SNP Selector at http://zeon.well.ox.ac.uk/rmott-bin/strains.cgi for examining mouse haplotypes across many strains. Most of these viewers contain links to copious information about genome features, and even allow the comparison across species of syntenic regions of the genome.

7.3.2 GENES AS REFERENCES

Using the gene as a reference one can simultaneously analyze genes, mRNA (especially by transcriptomic assay) and proteins as they vary under various psychological manipulations. This method can be used to ascribe functional roles to genes. Efforts to knock out every gene are proposed.[11] These mice will be screened on a battery of neurological and behavioral assays, allowing further annotation of the genome with higher order biological function. However, there are many challenges to the use of the gene as a natural key. Genes are surprisingly ill defined. They have a complex anatomy, including enhancers, promoters, multiple exons and splice-variants, polymorphisms, post-translational modifications, methylation states and multiple sequence records. The nomenclature of genes, and particularly gene products, has a complex history, with field specific language applying overlapping terms to groups of genes. A few notable efforts to integrate data at the level of the gene are

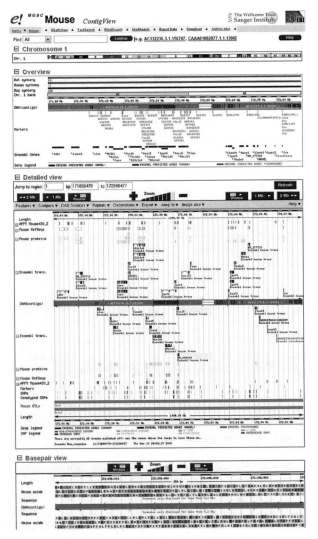

FIGURE 7.1 Genome sequence as reference. A screen image of Ensembl Contig View shows the location of genes, cross-species synteny, microsatellite markers, microarray probes, and polymorphisms in the region of a QTL for seizure susceptibility.

EntrezGene http://www.ncbi.nlm.nih.gov/entrez/query.fcgi?db=gene (a replacement for LocusLink), and MGI[12] http://www.informatics.jax.org/. A database of keys to many gene-centered databases, "GeneKeyDB" http://genereg.ornl.gov/gkdb/, allows users to cross-reference sources readily, using WebGestalt, a Web-based gene set analysis toolkit http://watson.lsd.ornl.gov/genereg/webgestalt/.

7.3.3 GENETIC REFERENCE POPULATIONS

Genetic reference populations (GRPs) including the recombinant inbred strains, consomics and standard inbred strain panels consist of isogenic populations of related

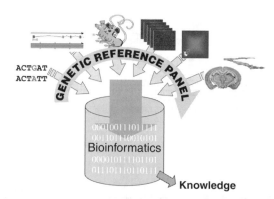

FIGURE 7.2 Using genetic reference populations to meet the challenge of integrating data across levels of biological scale. There is tremendous hope and expectation that compiling biological data using bioinformatics methods will rapidly lead to knowledge. A strong experimental design can greatly facilitate the process. Using a genetic reference population, sequence polymorphisms, gene annotation, strain variation, molecular, behavioral, physiological and neuroanatomical data can be integrated with great facility.

individuals that can be repeatedly sampled through time and space (Figure 7.2). In this approach, the mouse strain becomes the natural key, allowing the analytic integration of many diverse data types. A single matrix can be constructed of attributes by mouse strain, and the data integration is performed using correlational analyses, though various attributes, for example genotype at a QTL, can be ascribed a causal role. Where causality cannot be obtained analytically, a new experiment can be performed to assess the antecedence, necessity and sufficiency of the putative cause. Two resources for the analysis of these populations are WebQTL (www.genenetwork.org)[13–15] and the Mouse Phenome Database.[16]

7.3.3.1 Genetic Correlations and Reference Populations

A major application of the genetic reference population is that diverse trait data can be aggregated and correlated indefinitely, allowing the identification of traits that share pleiotropic regulation. The presence of a genetic correlation implies that the traits share a common heritable source of variance, which is often the result of multiple effects of a single polymorphism or mutation; however, numerous other sources of apparent pleiotropy exist.[17] While individual studies allow the determination of heritability of a phenotype and possible sources of genetic variation, i.e., QTLs, the aggregation of multiple studies allows for analysis of the stability of the phenotype under different environmental conditions, and the interpretation of historical data in the context of novel technologies. For example, in WebQTL, users can examine relationships among the earliest behavioral phenotypes collected in the BXD RI strains in the early 1980s to microarray based measures of gene expression,[14,15] a technology that had not even been invented at the time of the initial data collection! The tremendous advantage of reference populations is that results obtained can be integrated analytically. WebQTL features a variety of analytic tools, ranging from simple descriptive statistics, to complex and highly flexible multivariate data analysis, with results referenced to external genomic resources.[15]

7.3.3.2 GRPs for Mapping QTLs

Some reference populations have the additional advantage that they can be used for genetic mapping. These isogenic lines have some known genetic variation that can be assembled into high precision genome-wide databases and associated with phenotypic data.

The recombinant inbred strains are the first and arguably still the best genetic reference population for mapping. The largest set of these strains, the BXD RIs were developed in the late 1970s and expanded in the 1990s by B. A. Taylor[18] and recently expanded again to a size of nearly 80 strains.[19] The high density of recombinations obtained results in a dense genotypic map that has been refined and incorporated into WebQTL.[13] These genotypes arise from just two progenitors that were allowed to randomly segregate. This is an important contrast to the standard inbred strains, which experienced numerous bottlenecks in their breeding history.[20]

The standard inbred strains are a tantalizing population for use as a genetic mapping panel. Early approaches used limited single nucleotide polymorphism data from a few strains to identify QTLs using these lines of mice,[21] but were challenged due to low statistical power, high error rates, and concerns about the genetic architecture of the population.[22,23] As additional mouse SNP data became available[3,4,24] this controversial approach has been revisited. However, the genetic architecture of these strains is highly complex[2] giving rise to concerns about the effects of population admixture due to the breeding history of the strains that remain.[25,26] This has resulted in the need to restrict these analyses to a small fraction of the existing standard inbred strains—fewer than that available in the larger RI sets. While the use of mouse single nucleotide polymorphisms is an excellent approach for refinement of the QTL interval and selection of additional informative strains for multiple cross mapping,[27,28] the use of these strains for genome wide scans is not yet sufficiently accurate. The interest in this approach has led to the construction of a variety of large SNP data sets. Gradually, this data is being incorporated into existing public genome browsers, and will be a major boon to efforts at QTL candidate gene identification. Construction of a large, powerful, randomly bred and genetically diverse recombinant inbred strain panel is underway.[29]

The consomic lines[30] are also a genetic reference population, but they have low resolution and other limitations. They can be used to map sources of genetic variance to a single chromosome, allowing for determination of the location of genes with main effects on phenotypic variation. However, a significant problem with the consomic lines is that most complex traits are under regulation of multiple genetic loci, and phenotypes may be a consequence of epistasis, a situation that is not modeled well in these lines. Occasional evidence of coadaptive alleles is observed, such that having a pair of genes from common progenitors at two homozygous loci results in a different phenotype than having either recombinant pair of alleles. In consomic analysis, it would appear that these two loci have main effects on phenotype such that the introgressed chromosome has a main effect on the phenotype relative to the background chromosome. The chief advantage of the consomic lines is that they are closer to congenic status than other cross-progeny.

7.3.4   GENE SETS AS REFERENCES

Increasingly, researchers are noting the limitations of gene-at-a-time data presentation of analysis. This is particularly true as large gene sets are identified and need to be interpreted as in microarray analysis. Furthermore, it is becoming clear that it is necessary to examine and catalog the interactions between genes and gene products, in addition to indexing the genes themselves. This shift in emphasis from the nodes to the edges between them is evident in the proliferation of "interactome" databases.

7.3.4.1   Gene Ontology

A major effort is being made to identify meaningful sets of genes that share a common biological structure, function or compartment.[31] Numerous tools exist for the examination of representation of category members among a gene set identified through other biological analysis. These tools also allow users to compare submitted lists of genes with known biological pathways databases KEGG,[32] PubMed's GRIF, and other categorization schemes. Some of these are Web-based, including DAVID[33] and the Gene Ontology Tree Machine[34] http://genereg.ornl.gov/gotm/ (Figure 7.3), whereas others are standalone executables including EASE.[35] A major advantage of the standalone tools is that they allow users to store and compare gene sets generated locally. For gene sets associated with very subtle behavioral phenotypes, the custom development of gene sets will be necessary because few categorization schemes reflect in-depth characterizations of behavior at the present time.

7.3.4.2   The Interactome

Multiple databases have been developed to identify genes that physically interact. This includes DIP,[36,37] BIND,[38,39] and MINT.[40] While the categorization of the physical interactions between gene products is interesting, it is important to note that in the case of behavioral analysis, the interactions of gene products are often indirect, as in the interaction between neurotransmitter synthesis enzymes and receptors. While these gene products may be highly correlated and coregulated, and may co-occur in various tissues, the interaction between them is not physical and would not be encapsulated in the interactome databases.

7.4   APPLICATIONS

A few applications and examples merit further elaboration, though it is impossible for any text to remain current on this topic. As with any biological result, the user should make use of documentation to know and understand the source of the data and the analytic criteria by which results are presented. Regardless of the professional appearance of the interface, the quality of the underlying data and the validity of

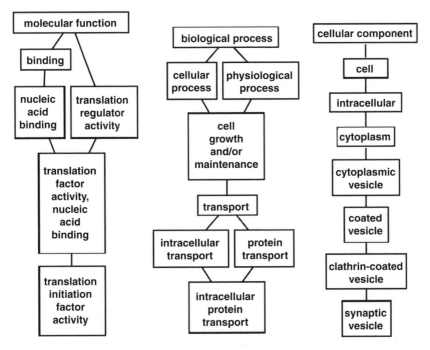

FIGURE 7.3 Directed acyclic graphs (DAGs) from WebGestalt's GO analysis of genes that are co-expressed in the brains of isogenic mice indicate covariance among translation initiation factors, transport mechanisms and synaptic vesicle related genes.

the analytical approach should be evaluated critically. Biological databases are not perfect animals, and are known to contain errors and inaccuracies. Often, the research community is able to deposit data with little verification or review. The more frequently used data are going to be more accurate. This does not diminish the value of biological databases, but is a caveat to their use as one of many biological tools.

7.4.1 Sequence Analysis Scoring Matrices and Phylogeny

Many of the tools for sequence analysis and phylogeny, comparison of sets of sequence data to examine evolutionary questions, have been integrated in a single interface at University of California San Diego's Biology Workbench[41] http://workbench.sdsc.edu/. This is an excellent tool for rapid simultaneous retrieval of sequence data from multiple sources performing multiple sequence alignments, identifying sets of related sequences and simultaneously manipulating them. Biology Workbench has many applications; for example, it can be used to examine sequence variation between strains in relation to probes and primers and use a variety of other tools for sequence analysis. The increasing sophistication of genome browsers has reduced the need for these tools, but significant advantages of Biology Workbench are that many parameters can be adjusted, and results need not be organized with respect to a single feature in the genome. Numerous example applications and tutorials are available at the University of Illinois' Biology Student Workbench http://peptide.ncsa.uiuc.edu/.

7.4.1.1 Scoring Matrices

Sequence comparison, or "alignment" has much in common with its biological counterpart, the annealing of DNA. Just as one could regulate hybridization stringency depending on the purpose of the experiment, one could adjust the sensitivity and specificity of a sequence alignment analysis. Querying sequence data with other sequence data requires a "scoring" matrix and penalties for gaps and mismatches of DNA sequence. Different matrices exist for DNA and amino acid sequence. Low "stringency" matrices are used to identify homologs, orthologs and paralogs. This can be very useful for identification of ESTs. This type of search can be used to identify novel genes with similarity to known genes. For example, one can retrieve the protein sequence of an unknown gene identified as a QTL candidate or upregulated expressed sequence and other databases for sequences using a low-penalty scoring matrix to find similar, but not exact matches.

7.4.1.2 Motif Search and Alignment

A major application for multiple sequence analysis methods is the identification and search for common motifs. A popular system for this is MEME, which can be used to compare several sequences for the identification of motifs, and MAST, which can then be used to search for the conserved motif throughout the entire genome (http://meme.sdsc.edu/meme/website/intro.html). These tools can be powerfully applied to gene sets in which one suspects common functional motifs, and have been applied to upstream DNA to identify novel transcription factor binding sites that may be shared by a coregulated gene set. Another application of the search for motifs is to examine biologically important and conserved sequences.

7.4.1.3 Structure Analysis

Protein structure data can be obtained for many gene products and visualized using NCBIs Cn3D tool, obtained at http://www.ncbi.nlm.nih.gov/Structure/CN3D/cn3d.shtml. This user-friendly tool allows users to examine structure and amino acid sequence relations from structural data retrieved from the Molecular Modeling DataBase, which contains data from the Protein Data Bank (http://www.rcsb.org/pdb/).[42–44] For example, Figure 7.4 shows the ovine serotonin acetyltransferase, highlighted to reveal residues that are different in the mouse. Trait relevant mis-sense polymorphisms are likely to influence protein structure or function. In this example, the ligand binding site is relatively conserved between the two species, but other regions are much more variable. Using actual structure or results from prediction, in concert with knowledge of sequence polymorphisms, one can rapidly predict the molecular consequences of allelic variation and prioritize analysis of polymorphic loci. The rapid development of structure prediction algorithms will

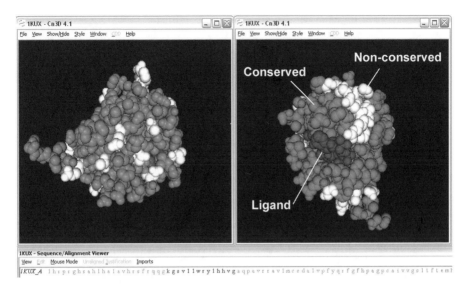

FIGURE 7.4 Structural analysis of *Aanat*, the serotonin *n*-acetyl transferase enzyme. conserved amino acid sequence between mouse and sheep using Cn3D. Protein structure was retrieved from Entrez Structure, and mouse amino acid sequence was retrieved via Ensembl. Sequences were aligned and differences were highlighted using Cn3D.

greatly increase the number of structures available and will soon make this approach readily applicable.[45]

7.4.2 QTL CANDIDATE GENE SELECTION

Identifying the naturally occurring polymorphisms responsible for trait variation, i.e., "genes for" a trait is as relevant today as ever. Bioinformatics has allowed the efficient aggregation of data about the genomic features within a QTL interval, and by including knowledge of candidate polymorphisms alongside information about the genes, the challenge of going from QTL to QTG is reduced. As mentioned before, a wealth of information is incorporated in sequence-based browsers, and these are now becoming a routine first stop in the search for QTL candidate genes. This process is now tremendously enhanced by viewers that incorporate additional data sources and overlay them with historic sequence annotation. For example, the Genome Institute of Novartis has incorporated a large set of tissue specific expression data that can be examined directly within the genome browser. New tools specifically devised for QTL candidate analysis incluing WebQTLs interval map viewer (Figure 7.5) and the protypical PLAD http://proto.informatics.jax.org/prototypes/plad/ bring together data on the genes and polymorphisms within a QTL interval. This organized framework for the analysis of candidate genes allows users to integrate bioinformatic resources with trait-specific information to identify candidates for refined positional analysis, rather than resorting to purely genetic recombination based approaches to refine QTL.

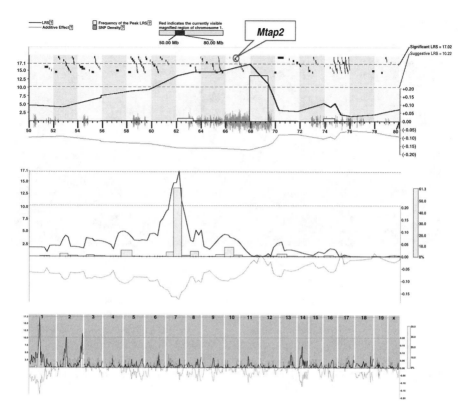

FIGURE 7.5 From QTL to candidate gene. QTL analysis of motor protein *Kif5a* mRNA abundance. *Mtap2* lies in the significant regulatory locus on Chr. 1, *Creb* is near the peak regulatory locus. The bottom panel is a whole genome scan from WebQTL. The middle panel is a view of Chr. 1, and the top is WebQTL analyzer view, which shows the physical map position of the QTL, the location of genes in the region, and the SNP density within genes and across the region. The heavy curve indicates the "likelihood ratio statistic" across the genome. The bars indicate the frequency of the QTL peak at a given location over 2000 bootstraps. Horizonal lines indicate suggestive and significant linkage thresholds based on the permutation analysis. The light line indicates the additive effect of the DBA/2J allele. In this example, the DBA/2J allele decreases expression of the kinesin 5a transcript.

7.4.3 MICROARRAY, PROTEOMIC, AND OTHER HIGH-THROUGHPUT GENE SET ANALYSIS

The accumulation, interpretation and comparison of results from the numerous high-throughput genomic assays can be daunting, but multiple parallel approaches can be used to understand and analyze the result. It is always critical to keep in mind the experimental context of the data set and the goals and interpretation of the analytic method. Without careful attention to these issues, one can easily become mired in a whirlpool of data. From a list of annotated genes, many questions can be answered using external annotation ranging from the simple gene set analysis tools, to more complex approaches aimed at transcriptional (e.g., MEME, MAST) and genetic

(WebQTL) coregulation. Identifying genes that are hypothesized to have these specific relationships in common is an important first step to defining a meaningful result. Not all transcripts that are co-expressed are regulated by the same transcription regulatory pathways. New tools, including MOTIF http://motif.genome.jp/ and PAINT http://www.dbi.tju.edu/dbi/tools/paint/index.php?op=FnetBuilder, can be used to search for known transcription factor motifs among sets of coregulated transcripts. Microarray analysts often then identify known pathways and systems in which these genes reside. Numerous tools are available for this type of annotation, and several incorporate the actual expression results. These include GenMAPP,[46,47] DAVID/EASE,[33,35] and several commercial tools.

7.4.4 TEXT MINING

Often, data relations are not established well enough for direct analytic integration. A whole family of tools has been developed to mine this type of free-text information. The simple PubMed query interface to the Medline database presents lists of related documents. More advanced tools examine sets of literature for network relations extracted via natural language processing (Chilibot.net),[48] latent semantic relationships in Semantic Gene Organize (SGO),[49] and other implicit literature relationships (Arrowsmith).[50] Commercial tools including Ref Viz (ISI ResearchSoft) and PathwayAssist (Ariadne Genomics) are also available for text analysis. Tools such as PubGene allow users to examine the network of literature co-occurences around a gene. These tools allow users to develop hypotheses concerning novel gene–gene or gene–trait relationships that have been observed using other bioinformatics or genomic approaches. All of these tools are subject to the problems of gene annotation, the redundant and high-entropy set of symbols used to refer to genes and gene products. A heroic effort is being undertaken by the National Library of Medicine to annotate the existing literature with standard gene nomenclature. This effort, called Gene Reference Into Function (GRIF) attempts to connect higher-order biological literature with gene products that it refers to. To date, no automated process has been able to match this manual curation effort, so while the data are highly accurate, the database itself will take many years to fully populate.

7.4.5 INTEGRATING THE GENOME AND THE PHENOME FOR SYSTEMS-LEVEL BIOINFORMATICS

For applications in behavioral genetics, the most exciting developments in bioinformatics have been those that allow low-level genomic information to be integrated with higher order phenotypic data. This integration can occur by collecting all data in a reference population, or using a reference gene set or microarray platform. Exploring the phenome space will provide insight into the relationships among biological traits. WebQTL[13,15] and the Mouse Phenome Database[51] are two tools that allow users to examine genetic correlations of traits across levels of biological scale, and organ systems. These tools will allow users to explore relationships between seemingly diverse constructs, such as memory formation and addiction, both related through some form of neural plasticity. As data are accumulated in the recombinant or standard inbred strains, these databases will expand dramatically in their utility.

7.5 TOWARD A BIOINFORMATICS OF BEHAVIOR

Three of the major neurobehavioral genetics questions that can be addressed using bioinformatics are the forward genetics question, "Is there a naturally occurring polymorphism present in this population that influences this behavioral phenotype," the reverse genetics question, "Does manipulation of this gene or gene product influence a phenotype," and a third question, which makes use of genetic approaches to answer neurobehavioral questions, "Are these traits related to one another through a shared biological substrate?" From this latter question, we can ask very high-level questions about the relatedness of psychological processes through the black box of genes and gene products. In this approach, the genetics itself becomes a tool for understanding the relationships among phenomenologically defined psychobiological constructs, i.e., those observed behavioral and personality characteristics that we attempt to explore by careful examination of operationally defined phenotype models. By ascribing biological substrates—e.g., gene sets—to these constructs using bioinformatics approaches, one can discover the naturally occurring categories or "ontology" of central nervous system processes.

This type of analysis can be performed using genetic correlations, which require all data to be obtained in a single reference population, or, the approach can use genes as references. These genes can be identified across species and across experimental paradigms, in many lines of mice. By identifying the molecular components of behavioral processes and examining their overlap, we can define the shared and unique molecular substrates of these processes. This approach can be used to answer fundamental questions about behavior and mouse models for behavioral disorders. For example, one can use these genes and gene networks to define mouse behavioral correlates of human behavioral processes and address challenges of translational genomics. Do two processes share a common molecular network? Is gene expression in the molecular network for spatial water maze learning induced by pretraining exposure to the maze apparatus and other stress-related paradigms? Which genes are also expressed in less stressful maze learning tasks and thus more specific to the processes that subserve memory? Are any of them related to cognitive decline in humans? Identification of the common processes will form an empirically derived framework of nervous system function.

Building the network from genetic polymorphism to genes to phenotypes, and then translating that information to humans have been a challenging process, but compelling successes have been demonstrated. The powerful tools of mouse genetics allow us to perform thorough molecular dissection of phenotypes. Effectively harnessing human–mouse genome relations will allow us to translate these polymorphisms, gene and phenotypes to other species, including our own. As in the past, we can gain much from the experimental precision and control in the mouse, generating novel and highly informative hypotheses of genetic mediation of human behavior.

ACKNOWLEDGMENTS

Jeremy L. Peirce and Susan E. Bergeson provided helpful comments on this chapter.

REFERENCES

1. Waterston, R.H. et al., *Initial sequencing and comparative analysis of the mouse genome*. Nature, 2002. **420**(6915): p. 520–62.
2. Yalcin, B. et al., *Unexpected complexity in the haplotypes of commonly used inbred strains of laboratory mice*. Proc Natl Acad Sci U S A, 2004. **101**(26): p. 9734–9.
3. Wiltshire, T. et al., *Genome-wide single-nucleotide polymorphism analysis defines haplotype patterns in mouse*. Proc Natl Acad Sci U S A, 2003. 100(6): p. 3380–5.
4. Wade, C.M. et al., *The mosaic structure of variation in the laboratory mouse genome*. Nature, 2002. **420**(6915): p. 574–8.
5. Lagos-Quintana, M. et al., *Identification of novel genes coding for small expressed RNAs*. Science, 2001. **294**(5543): p. 853–8.
6. Ambros, V., *microRNAs: tiny regulators with great potential*. Cell, 2001. **107**(7): p. 823–6.
7. Smaglik, P., *Internet gateway planned for neuroinformatics data*. Nature, 2000. **405**(6787): p. 603.
8. McCullough, B.D. and B. Wilson, *On the accuracy of statistical procedures in Microsoft Excel 2000 and Excel XP*. Computational Statistics & Data Analysis, 2002. **40**: p. 713–721.
9. Chesler, E.J. et al., *Identification and ranking of genetic and laboratory environment factors influencing a behavioral trait, thermal nociception, via computational analysis of a large data archive*. Neurosci Biobehav Rev, 2002. **26**(8): p. 907–23.
10. Platt, J.R., *Strong inference. Certain systematic methods of scientific thinking may produce more rapid progress than others*. Science, 1964. **146**(3642): p. 347–353.
11. Austin, C.P. et al., *The knockout mouse project*. Nat Genet, 2004. **36**(9): p. 921–4.
12. Bult, C.J. et al., *The Mouse Genome Database (MGD): integrating biology with the genome*. Nucleic Acids Res, 2004. **32**(Database issue): p. D476–81.
13. Wang, J., R.W. Williams, and K.F. Manly, *WebQTL: web-based complex trait analysis*. Neuroinformatics, 2003. **1**(4): p. 299–308.
14. Chesler, E.J. et al., *Genetic correlates of gene expression in recombinant inbred strains: a relational model system to explore neurobehavioral phenotypes*. Neuroinformatics, 2003. **1**(4): p. 343–57.
15. Chesler, E.J. et al., *WebQTL: rapid exploratory analysis of gene expression and genetic networks for brain and behavior*. Nat Neurosci, 2004. **7**(5): p. 485–6.
16. Bogue, M.A. and S.C. Grubb, *The Mouse Phenome Project*. Genetica, 2004. **122**(1): p. 71–4.
17. Hodgkin, J., *Seven types of pleiotropy*. Int J Dev Biol, 1998. **42**(3): p. 501–5.
18. Taylor, B.A. et al., *Genotyping new BXD recombinant inbred mouse strains and comparison of BXD and consensus maps*. Mamm Genome, 1999. **10**(4): p. 335–48.
19. Peirce, J.L. et al., *Genetic architecture of the mouse hippocampus: identification of gene loci with selective regional effects*. Genes Brain Behav, 2003. **2**(4): p. 238–52.
20. Beck, J.A. et al., *Genealogies of mouse inbred strains*. Nat Genet, 2000. **24**(1): p. 23–5.
21. Grupe, A. et al., *In silico mapping of complex disease-related traits in mice*. Science, 2001. **292**(5523): p. 1915–8.

22. Darvasi, A., *In silico mapping of mouse quantitative trait loci.* Science, 2001. **294**(5551): p. 2423.

23. Chesler, E.J., S.L. Rodriguez-Zas, and J.S. Mogil, *In silico mapping of mouse quantitative trait loci.* Science, 2001. **294**(5551): p. 2423.

24. Pletcher, M.T. et al., *Use of a dense single nucleotide polymorphism map for in silico mapping in the mouse.* PLoS Biol, 2004. **2**(12): p. e393.

25. Mhyre, T. et al., *Heritability, correlations, and in silico mapping of locomotor behavior and neurochemistry in inbred strains of mice.* Genes, Brain, Behavior, 2005. **4**: 209–28.

26. Churchill, G.A., *Researchers didn't scurry around fields looking for mice, says Jackson Lab scientist*, BioMed Central, July 7, 2004.

27. Hitzemann, R. et al., *Multiple cross mapping (MCM) markedly improves the localization of a QTL for ethanol-induced activation.* Genes Brain Behav, 2002. **1**(4): p. 214–22.

28. Shifman, S. and A. Darvasi, *Mouse inbred strain sequence information and yin-yang crosses for quantitative trait locus fine mapping.* Genetics, 2005. **169**(2): p. 849–54.

29. Churchill, G.A. et al., *The Collaborative Cross, a community resource for the genetic analysis of complex traits.* Nat Genet, 2004. **36**(11): p. 1133–7.

30. Singer, J.B. et al., *Genetic dissection of complex traits with chromosome substitution strains of mice.* Science, 2004. **304**(5669): p. 445–8.

31. Ashburner, M. et al., *Gene ontology: tool for the unification of biology.* Nat Genet, 2000. **25**: p. 25–29.

32. Ogata, H. et al., *KEGG: Kyoto Encyclopedia of Genes and Genomes.* Nucleic Acids Res, 1999. **27**(1): p. 29–34.

33. Dennis, G., Jr. et al., *DAVID: Database for Annotation, Visualization, and Integrated Discovery.* Genome Biol, 2003. **4**(5): p. P3.

34. Zhang, B. et al., *GOTree Machine (GOTM): a web-based platform for interpreting sets of interesting genes using gene ontology hierarchies.* BMC Bioinformatics, 2004. **5**(1): p. 16.

35. Hosack, D.A. et al., *Identifying biological themes within lists of genes with EASE.* Genome Biol, 2003. **4**(10): p. R70.

36. Xenarios, I. et al., *DIP, the Database of Interacting Proteins: a research tool for studying cellular networks of protein interactions.* Nucleic Acids Res, 2002. **30**(1): p. 303–5.

37. Xenarios, I. et al., *DIP: the database of interacting proteins.* Nucleic Acids Res, 2000. **28**(1): p. 289–91.

38. Bader, G.D. et al., *BIND—The Biomolecular Interaction Network Database.* Nucleic Acids Res, 2001. **29**(1): p. 242–5.

39. Alfarano, C. et al., *The Biomolecular Interaction Network Database and related tools 2005 update.* Nucleic Acids Res, 2005. **33 Database Issue**: p. D418–24.

40. Zanzoni, A. et al., *MINT: a Molecular INTeraction database.* FEBS Lett, 2002. **513**(1): p. 135–40.

41. Subramaniam, S., *The Biology Workbench—a seamless database and analysis environment for the biologist.* Proteins, 1998. **32**(1): p. 1–2.

42. Berman, H.M. et al., *The Protein Data Bank.* Nucleic Acids Res, 2000. **28**(1): p. 235–42.

43. Berman, H.M. et al., *The Protein Data Bank.* Acta Crystallogr D Biol Crystallogr, 2002. **58**(Pt 6 No 1): p. 899–907.

44. Bhat, T.N. et al., *The PDB data uniformity project.* Nucleic Acids Res, 2001. **29**(1): p. 214–8.

45. Berman, H.M. and J.D. Westbrook, *The impact of structural genomics on the protein data bank.* Am J Pharmacogenomics, 2004. **4**(4): p. 247–52.

46. Dahlquist, K.D. et al., *GenMAPP, a new tool for viewing and analyzing microarray data on biological pathways.* Nat Genet, 2002. **31**(1): p. 19–20.

47. Doniger, S.W. et al., *MAPPFinder: using Gene Ontology and GenMAPP to create a global gene-expression profile from microarray data.* Genome Biol, 2003. **4**(1): p. R7.

48. Chen, H. and B.M. Sharp, *Content-rich biological network constructed by mining PubMed abstracts.* BMC Bioinformatics, 2004. **5**(1): p. 147.

49. Homayouni, R. et al., *Gene clustering by latent semantic indexing of MEDLINE abstracts.* Bioinformatics, 2005. **21**(1): p. 104–15.

50. Smalheiser, N.R. and D.R. Swanson, *Using ARROWSMITH: a computer-assisted approach to formulating and assessing scientific hypotheses.* Comput Methods Programs Biomed, 1998. **57**(3): p. 149–53.

51. Grubb, S.C., G.A. Churchill, and M.A. Bogue, *A collaborative database of inbred mouse strain characteristics.* Bioinformatics, 2004. **20**(16): p. 2857–9.

8 Congenic and Consomic Strains

Lorraine Flaherty and Valerie Bolivar

CONTENTS

INTRODUCTION

Genetically defined mouse strains have become invaluable in biomedical research because of their ability to clarify genetic vs. environmental influences on a wide variety of traits including behavior. They have provided the means to test multiple genetically identical or nearly identical animals at one time, thus reducing the effects of genetic variation to a minimum. Some of the most useful types include inbred, co-isogenic, congenic, consomic, recombinant inbred, knockout, and transgenic strains.

The most commonly used strain of mouse is the inbred strain. Since these strains are made by brother–sister mating mice for more than 20 generations, any two mice within an inbred strain will be genetically identical at 99.99% of their genetic loci. A co-isogenic strain is similar to an inbred strain except that it differs from its inbred partner by only one locus.[1] These co-isogenic strains are usually the result of spontaneous mutations that occur within an inbred stock and have been very useful in analyzing the effects of single genes on a particular behavior or trait. However, the investigator must either wait for mutations to occur spontaneously or induce them by use of mutagenesis techniques, both requiring much time and effort. Because of the usefulness of these co-isogenic strains, in the 1940s, George Snell constructed a breeding scheme to develop a new type of strain, called the congenic strain, which approximates a co-isogenic strain. These congenic strains have become extremely valuable resources in analyzing single gene effects. Below we describe how these strains are made and some of their properties. We also describe a different use for congenic strains in mapping genes that influence quantitative traits such as behavior.

115

Finally we describe a variation of the congenic strain, called the consomic strain, which selects for genetic differences covering an entire chromosome instead of a single genetic locus.

8.1 CONGENIC STRAINS

A congenic strain is made by repeatedly backcrossing one strain to another and selecting for an allele at a particular locus, called the differential locus, at every generation (see description below for details on construction of such strains). This process is called introgression where the donor genetic material is said to be "introgressed" into the recipient genome. There are several different breeding schemes that can be used to construct a congenic strain.[2] The most common is illustrated in Figure 8.1. Two inbred strains are chosen to begin this process—a donor strain and a recipient strain. The donor strain contains the donor allele at the differential locus. The recipient strain is usually chosen based on its breeding ability and characteristics. Usually it is a strain that is readily available and used as a standard in a particular aspect of biology, like behavior. In most cases, it is the C57BL/6 (B6) strain although other strains such as DBA/2 or BALB/c are also used. The first cross is a simple cross between two inbred strains, leading to an F1 (standing for first filial generation), also indicated as N1 (standing for the first backcross generation). This F1 is then crossed to the recipient inbred strain to yield an N2 generation. At the N2 generation and each subsequent generation, the mice are typed at the differential locus and mice are selected which possess the donor allele, symbolized by "a" in Figure 8.1. At the end of 10 or more generations of selection and backcrossing, the mice are intercrossed and a strain that is homozygous at the differential locus is maintained by brother–sister matings. In the Figure 8.1 example, the strain would be called B.G-a. This would stand for the recipient strain and donor strain separated by a period and followed by a hyphen and the selected allele. In some cases, less than 10 generations of backcrossing is used. However, here, the genetic background will contain more contaminant or "passenger" genetic material from the donor strain (Tables 8.1 and 8.2).

Making congenic strains that isolate quantitative trait loci (QTLs) is more difficult, especially when behavioral phenotypes are concerned. Since there is often a large variation in the behavioral phenotypes of animals even with the same genotype, the scoring of an individual animal often does not reflect its exact genotype at the locus in question. Thus, relying on phenotype alone for selection of individuals within the congenic strain breeding scheme will often be misleading. Progeny testing is often necessary to confirm this genotype–phenotype relationship. Thus, many geneticists studying complex traits have chosen to make congenic strains by the alternative way— selection for flanking DNA markers framing their locus of interest.[3,4,5] Since most of the mappings of these QTLs give a log of odds (LOD) probability map for the location, selection of flanking markers is usually based on a confidence interval for the QTL (see Abiola[6] for a discussion of QTL mapping). The distance between these flanking markers varies considerably and depends on the sharpness of the LOD score peak as well as the judgment of the investigator making the strain. In general, the rule of thumb is to pick a wide enough interval (>20 cM on each side of the peak LOD score) so that the investigator is assured that the QTL will not fall outside

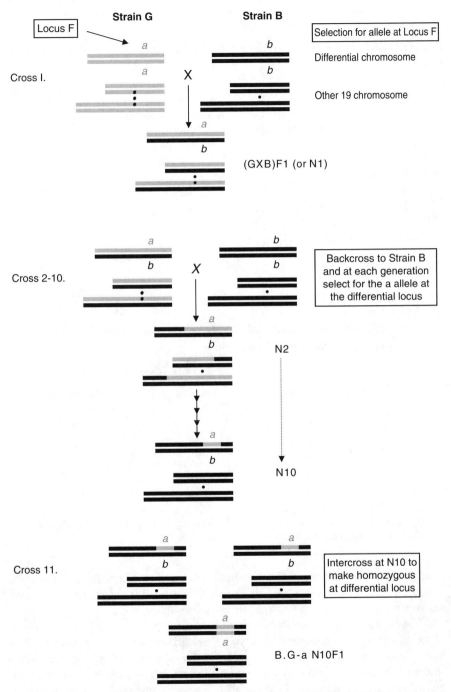

FIGURE 8.1 Derivation of a congenic strain of mice. Selection for the "a" allele at locus F of the G strain (donor) at every generation.

TABLE 8.1
Congenic and Consomic Strains of Mice and Amount of Contaminant Genetic Material*

Strain	Strain: Recipient (donor)	Differential Genetic Region (average) (cM)	Average Amount of Unlinked Contaminant Genetic Region (cM)
Consomics	B6 (A) B6 (Spretus)	160 cM	3 cM
Congenics (>N10)	B6 (various)	20 cM	3 cM
Speed congenics (N5)	B6 (various)	40 cM	16 cM
Knockout/congenics (>N10)	B6 (129)	20 cM	3 cM
Genome tagged mice (N4-N5)	B6 (DBA/2, CAST)	40–50 cM	48 cM

*References for calculations are 1, 2, 8, 17, 22.

TABLE 8.2
Amount of Genomic Material That Is Retained as a Function of the N Number of Backcross Generations

N	Consomic (other 19 chrs)	Congenic	
		Unlinked	Linked
2	1282 Mb*	1282 Mb	140.6 Mb
4	321	321	82.0
8	20	20	46.7
10	5.0	5.0	37.5
16	0.1	0.1	23.4
20	<0.01	<0.01	18.7

*Numbers assume a genome size of 2700 Mb, a recombinant map size of 1600 cM and an average chromosome size of 135 Mb. References are 1, 2.

of it. Additional recombinational events in the interval can then be used to divide the region for further analysis. Congenic strains where selection is based on flanking markers are often called interval-specific congenic strains.[7]

Another breeding scheme that has become more popular in recent years, because of the advent of high-throughput genotyping, is marker-assisted selection and results in quasi-congenic strains, commonly called a "speed congenic."[8,9] Speed congenic strains start the same way as conventional congenics. An F1 is made between two inbred strains of mice and the donor allele is selected in each generation. However, at every subsequent generation, founder mice are selected that (1) retain the differential allele and (2) have a minimum of unlinked genetic material from the donor strain. Here, the unlinked contaminant genetic material is detected by use of a genome scan of DNA markers spaced at regular intervals on the 19 nonsyntenic chromosomes. The optimal spacing of these markers should be at approximately 20 cM intervals.[10,11] At least 20 mice from each generation must be typed in a genome scan to obtain mice with a minimal number of unlinked genes. With this technique, one can make the equivalent of an N10 congenic in 4 to 5 generations, thus, reducing the time for completion by at least one half (Table 8.1). However, the creation of these speed congenics also requires more technician time and genotyping costs because of the number of mice and loci that must be typed. For a more thorough discussion of this technique, see Markel et al.[8]

8.2 CONSOMIC STRAINS

More recently, consomic strains of mice have been made by similar techniques. These strains are also called chromosome substitution strains. Here, whole chromosomes, instead of single genes have been used for selection. These are made in a similar fashion to congenic strains except that markers spread evenly along the selected or differential chromosome are used for selection (Figure 8.2). These consomic strains are useful for preliminary mapping studies of quantitative traits since only 20 of them will be necessary to map a trait to a particular chromosome.

The idea of studying the influence of one chromosome at a time and the construction of consomic strains began with studies of the Y chromosome. Since there is no recombination (or extremely little) between the Y and X chromosomes, a Y consomic is simply made by mating two strains together and selecting for the male at every generation.[12] However, with the advent of a high density DNA marker map, consomics of other chromosomes became possible, including a set of consomics made by Jean-Louis Guenet and his group.[13] These consomics involved B6 introgressed with *Mus spretus*. A full set of consomics involving the B6 and A strains has recently been made in the laboratory of Joseph Nadeau and are now commercially available through the Jackson laboratory.[14] This new consomic panel has already been analyzed in two behavioral paradigms[15,16] and it promises to be a very valuable tool to the behavioral geneticist. A consomic strain is designated by first giving the name of the recipient strain, followed by a hyphen, followed by a number, indicating the transferred chromosome, and then by the name of the donor strain in a superscript. Thus, a B6 strain with an introgressed Chromosome 3 from the A strain would be designated B6-3[A] or B6-3<A>, if a superscript font is not available.

8.3 KNOCKOUT/CONGENIC STRAINS

When homologous recombination is used to inactivate a gene, the process is called gene ablation and the strain is commonly referred to as a knockout strain. This process involves the elimination of a functional gene in embryonic stem cells either by replacement with a nonfunctional one or by its deletion. Since most embryonic stem cells are derived from the 129 strain, the knockout mice are generally on a 129 background. However, most investigators prefer a B6 strain background and cross

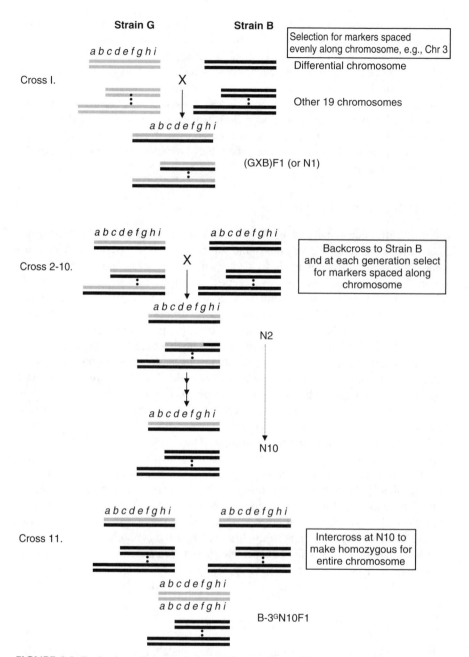

FIGURE 8.2 Derivation of a consomic or chromosome substitution strain. Selection for Chr. 3 of the G strain at every generation.

these 129 knockout strains to B6. This leads to a congenic strain with a section of 129 introgressed into the B6 background. Thus, when these knockout/congenic strains are tested for a phenotype that differs from the B6 strain, this phenotype

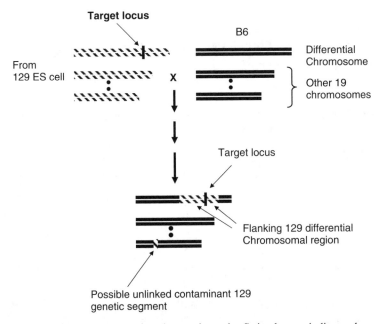

FIGURE 8.3 Derivation of a knockout/congenic strain. Striped areas indicate chromosomal regions from 129. Ablated gene indicated with an arrow.

could either be due to the gene that is ablated or to the flanking 129 chromosomal segment (Figure 8.3).[17] This consideration is important for two reasons. First, one should be cautioned when using knockout/congenic strains to understand that the observed phenotype is not always due to the targeted locus. There have been several reviews, covering the cautions attached to the use of knockout/congenics.[18–21] This situation is particularly poignant when considering phenotypes that are quantitatively different in B6 vs. 129 and when the phenotype is not predicted from the tissue distribution of gene expression and/or the known function of the ablated gene. Second, the knockout/congenic strains can be used, in a positive way, to scan the genome for genes that influence quantitative traits such as behavior.[17] Here, the investigator can use these strains as a panel of congenic strains with different segments of 129 from a wide variety of chromosomes introgressed into the B6 background. Thus, by testing these strains, the investigator will actually be testing for the QTL location somewhere in the flanking regions surrounding the targeted loci (see below). In this respect, there are more than 50 knockout/congenic strains available through commercial sources. Since the average length of the differential 129 chromosomal segment in these strains is about 20 to 30 cM (see Table 8.3 for a representative list), this assortment of knockout/congenic strains offers the researcher a set of strains that will cover most of the mouse genome. Moreover, often these strains provide a set of overlapping 129 chromosomal segments that will map a particular trait. Here the region of overlap is often small and may define the boundaries of the genetic region determining the trait. For example, in Figure 8.4, the QTL could be narrowed to a short segment by the overlap between B and C.

TABLE 8.3
Length of Differential Chromosomal Segments in a Sampling of Knockout/Congenic Strains*

Targeted Gene	N Number	Chr.	Position in cM of Target Gene	Length of 129 Chromosomal Stretch**
Apoe	10	7	4	16–22
Cd3z	8	1	87	9–10
Fcgr3	6	1	92	25–35
Igl-5	6	16	10	16–32
Il6	11	5	17	9–16
Il10	10	1	70	33–41
Lck	12	4	59	8–14
Selp	10	1	86	29–35
Tap1	10	17	19	3–5

*Information on position of targeted gene and number of backcross generations was taken from www.jax.org.

**Numbers indicate minimum to maximum lengths of chromosomal segment as defined by typed flanking DNA markers.

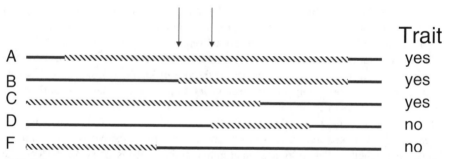

FIGURE 8.4 Mapping QTLs with overlapping differential chromosomes. Striped areas indicate the chromosomal stretch on the differential chromosome that is derived from 129. The arrows indicate the region that maps the trait.

Along this same line of reasoning, Iakoubova et al.[22] constructed a similar assortment of quasi-congenic strains of mice which they have called genome-tagged mice. There are two sets of these mice described by Iakoubova and colleagues—one with small sections of DBA/2 introgressed into the recipient B6 strain and one with small sections of CAST/Ei introgressed into B6. Because of the way that these strains were derived, they often have more unlinked contaminant passenger genetic material than a standard congenic strain (Table 8.1). So far, these are not available commercially; however, they have been used to map several genes affecting behavior.[23,24]

8.4 MAPPING QTLS WITH KNOCKOUT/CONGENIC MICE

The major use of knockout/congenic strains is to study a given gene on a defined genetic background. However, these same strains can be used to map genes affecting quantitative traits such as behavior.[17] It must be remembered that these strains differ not only at the target locus, but also at a short chromosomal stretch immediately surrounding the target locus. Because these strains are readily available, they serve as an extensive series of congenic strains that can be used to study short chromosomal regions that surround the indicated ablated target loci.

Thus, by comparing a series of the knockout/congenic strains with their inbred partner, B6, we were able to map a number of behavioral traits to precise chromosomal locations in or around the target locus. By further matings, we were able to prove that at least some of these traits were not due to the ablated target but rather due to the flanking genetic region.[17]

The data in Figure 8.5 illustrate this point. We screened 15 knockout/congenic mouse strains in our exploratory activity assay. This assay consisted of a 5-minute test in a darkened activity monitor.[25] The strains were selected based on their chromosomal position and the likelihood that they may affect behavior. We also tested these same mice for habituation to the open field by continuing this test system for 2 more consecutive days and measuring the change in activity from day 1 to day 3.[25] We have successfully used these assays in the past to establish exploratory activity and habituation differences among inbred stains/substrains of mice[25–28] and between neurological mutant mice and controls.[29–31]

As shown in Figure 8.5A, we found that a number of these strains behaved differently from B6 in first day exploratory activity. Based on the performance of these knockout/congenic strains, there are genes affecting this activity on Chrs. 1, 9, 14, and 15. By use of additional backcross tests, we have now confirmed some of these map locations on Chrs. 1 and 15. Interestingly, genes affecting habituation scores also seem to be affected by some of these same regions (Figure 8.5B).

Some strains were significantly less active than B6, i.e., behaved more like 129, including B6-Il10. This intermediate activity of the knockout/congenic is explainable by an additive mode of inheritance where the knockout/congenic strain should have an activity between B6 and 129. If the mode of inheritance is either dominant or recessive, the knockout/strain should have an activity approximately equal to one of its parents. This may be the case for B6-Il7r in first day exploratory activity.

The high activity of the two strains, B6-Lipc and B6-Tcrd, is more difficult to explain. Their activity does not correspond to either a dominant, recessive or additive mode of inheritance. There are three possible explanations. First, the increased activity may be due to the effects of the ablated gene itself. In the case of Tcrd (T cell receptor d), this explanation seems unlikely since this gene has no known influence on behavior or brain function. A second explanation could be that a mutation, either linked or unlinked to the differential locus, has occurred in the production of this B6 knockout/congenic strain in a gene that is important to this exploratory behavior. Again, this seems like an unlikely explanation because of the high incidence of this phenomenon with these knockout/congenics (Bolivar and

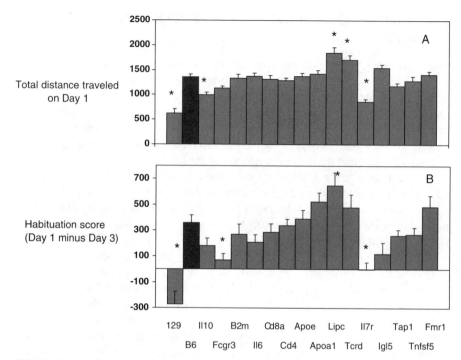

FIGURE 8.5 Exploratory activity and habituation of inbred and knockout/congenic mouse strains. The knockout mouse strains tested included B6.129P2-Il10, B6.129P2-Fcgr3, B6.129P2-B2m, B6.129S2-Il6, B6.129S2-Cd8a, B6.129S2-Cd4, B6.129P2-Apoe, B6.129P2-Apoa1, B6.129P2-Lipc, B6.129P2-Tcrd, B6.129S7-Il7r, B6.129S2-Igl-5, B6.129S2-Tap1, B6.129S2-Tnfsf5 and B6.129P2-Fmr1. Knockout mice are all on a B6 background and are indicated by the name of the ablated target gene. Two inbred strains were also tested: B6/J and 129P2 (129). At least 24 male mice of each strain were tested. Results are shown ± s.e.m. There was a significant effect of strain for first day activity ($F_{16,491} = 12.206$, $p < .0001$; Panel A) and habituation ($F_{16,491} = 8.885$, $p < .0001$; Panel B). Tukey's posthoc tests were then used to make pairwise comparisons. An * indicates that a knockout/congenic strain is significantly different from B6.

Flaherty, unpublished results). Finally, and most likely, there could be B6 modifier genes that are interacting with 129 flanking genes to produce a heightened effect. This explanation seems the most reasonable based on the high frequency of genetic background effects on behavioral traits.

For intersession habituation (Figure 8.5B), we have also found the B6-Il7r has an abnormally low habituation score. Since the Il7r gene is located on Chr. 15, this agrees with our previous data showing a strong gene affecting habituation on this same chromosome.[26] Moreover, we have subsequently confirmed the location of a gene influencing habituation on Chr. 15 by testing a Chr. 15 consomic strain, B6-15A.[16] This strain was significantly less active in the open field (Figure 8.6A), and habituated less (Figure 8.6B) than B6 mice.

Thus, both congenic and consomic strains have been successful tools for the behavioral geneticist in investigating single gene effects as well as in mapping genes

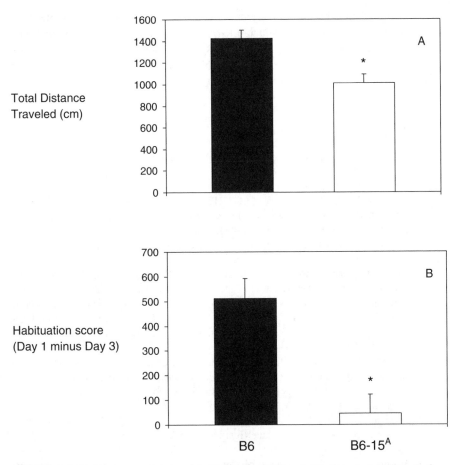

FIGURE 8.6 Exploratory activity and habituation of B6 and B6-15A strains. Twenty-four male mice of each strain were tested. Results are shown + s.e.m. B6 mice were significantly more active on the first day of testing (T46 = 3.747, p = .0005; Panel A) and displayed more habituation (T46 = 4.203, p = .0001; Panel B) than B6-15A mice. An * indicates a significant difference between the two groups.

that influence behavioral traits. The advantage of using these strains for mapping is that they can provide a convenient starting place for fine mapping studies. Here, the creation of subcongenic strains that are derivatives of congenic strains and made by further recombinational events inside the introgressed differential chromosomal region can isolate smaller chromosomal segments that are small enough to be convenient for positional cloning efforts. These new subcongenics can then be used to make an overlapping map that isolates the behavioral QTL to a shorter segment of chromosome (see Figure 8.3 for an example of mapping using overlapping differential chromosomal regions). Since multiple and genetically identical subcongenic mice can be made, the testing of these subcongenic mice will help to resolve weak effects that cannot be detected with more conventional intercross techniques.

Congenics, consomics, and knockout/congenics are readily available from commercial sources. They are valuable tools to the behavioral geneticist and have been successfully used to study genetic background effects as well as map QTLs. However, to use these strains correctly, the investigator should have a proper understanding of their construction and genetic origin.

REFERENCES

1. Silver, L.M. *Mouse Genetics Concepts and Applications*, Oxford University, New York, 1995.
2. Flaherty, L. Congenic strains. in *The Mouse in Biomedical Research*, Vol. 1, eds. Foster, H.L., Small, J.D. & Fox, J.G., 215–22, Academic Press, New York, 1981.
3. Bennett, B. Congenic strains developed for alcohol- and drug-related phenotypes. *Pharmacol Biochem Behav,* 67: 671–81, 2000.
4. Radcliffe, R.A., Bohl, M.L., Lowe, M.V., Cycowski, C.S. & Wehner, J.M. Mapping of quantitative trait loci for hypnotic sensitivity to ethanol in crosses derived from the C57BL/6 and DBA/2 mouse strains. *Alcohol Clin Exp Res,* 24: 1335–42, 2000.
5. Morel, L., Yu, Y., Blenman, K.R., Caldwell, R.A. & Wakeland, E.K. Production of congenic mouse strains carrying genomic intervals containing SLE-susceptibility genes derived from the SLE-prone NZM2410 strain. *Mamm Genome,* 7: 335–9, 1996.
6. Abiola, O. et al. The nature and identification of quantitative trait loci: a community's view. *Nat Rev Genet,* 4: 911–6, 2003.
7. Darvasi, A. Interval-specific congenic strains (ISCS): an experimental design for mapping a QTL into a 1-centimorgan interval. *Mamm Genome,* 8: 163–7, 1997.
8. Markel, P. et al. Theoretical and empirical issues for marker-assisted breeding of congenic mouse strains. *Nat Genet,* 17: 280–4, 1997.
9. Wakeland, E., Morel, L., Achey, K., Yui, M. & Longmate, J. Speed congenics: a classic technique in the fast lane (relatively speaking). *Immunol Today,* 18: 472–7, 1997.
10. Visscher, P.M. Speed congenics: accelerated genome recovery using genetic markers. *Genet Res,* 74: 81–5, 1999.
11. Servin, B. & Hospital, F. Optimal positioning of markers to control genetic background in marker-assisted backcrossing. *J Hered,* 93: 214–7, 2002.
12. Hudgins, C.C., Steinberg, R.T., Klinman, D.M., Reeves, M.J. & Steinberg, A.D. Studies of consomic mice bearing the Y chromosome of the BXSB mouse. *J Immunol,* 134: 3849–54, 1985.
13. Santos, J. et al. A new locus for resistance to gamma-radiation-induced thymic lymphoma identified using inter-specific consomic and inter-specific recombinant congenic strains of mice. *Oncogene,* 21: 6680–3, 2002.
14. Nadeau, J.H., Singer, J.B., Matin, A. & Lander, E.S. Analysing complex genetic traits with chromosome substitution strains. *Nat Genet,* 24: 221–5, 2000.
15. Singer, J.B., Hill, A.E., Nadeau, J.H. & Lander, E.S. Mapping quantitative trait loci for anxiety in chromosome substitution strains of mice. *Genetics,* 169: 855–62, 2005.
16. Singer, J.B. et al. Genetic dissection of complex traits with chromosome substitution strains of mice. *Science,* 304: 445–8, 2004.
17. Bolivar, V.J., Cook, M.N. & Flaherty, L. Mapping of quantitative trait loci with knockout/congenic strains. *Genome Res,* 11: 1549–52, 2001.

18. Crusio, W.E. Flanking gene and genetic background problems in genetically manipulated mice. *Biol Psychiatry,* 56: 381–5, 2004.
19. Wolfer, D.P., Crusio, W.E. & Lipp, H.P. Knockout mice: simple solutions to the problems of genetic background and flanking genes. *Trends Neurosci,* 25: 336–40, 2002.
20. Gerlai, R. Gene targeting: technical confounds and potential solutions in behavioral brain research. *Behav Brain Res,* 125: 13–21, 2001.
21. Gerlai, R. Afraid of complications in gene targeting? Anxiety plays a role! *Trends Neurosci,* 25: 136, 2002.
22. Iakoubova, O.A. et al. Genome-tagged mice (GTM): two sets of genome-wide congenic strains. *Genomics,* 74: 89–104, 2001.
23. Liu, D. et al. Mapping behavioral traits by use of genome-tagged mice. *Am J Geriatr Psychiatry,* 12: 158–65, 2004.
24. Liu, D. et al. Identifying loci for behavioral traits using genome-tagged mice. *J Neurosci Res,* 74: 562–9, 2003.
25. Bolivar, V.J., Caldarone, B.J., Reilly, A.A. & Flaherty, L. Habituation of activity in an open field: A survey of inbred strains and F1 hybrids. *Behav Genet,* 30: 285–93, 2000.
26. Bolivar, V. & Flaherty, L. A region on chromosome 15 controls intersession habituation in mice. *J Neurosci,* 23: 9435–8, 2003.
27. Cook, M.N., Bolivar, V.J., McFadyen, M.P. & Flaherty, L. Behavioral differences among 129 substrains: implications for knockout and transgenic mice. *Behav Neurosci,* 116: 600–11, 2002.
28. Bothe, G.W.M., Bolivar, V.J., Vedder, M.J. & Geistfeld, J.G. Genetic and behavioral differences among five inbred mouse strains commonly used in the production of transgenic and knockout mice. *Genes Brain Behav,* 3: 149–57, 2004.
29. Bolivar, V.J., Ganus, J.S. & Messer, A. The development of behavioral abnormalities in the motor neuron degeneration (*mnd*) mouse. *Brain Res,* 937: 74–82, 2002.
30. Bolivar, V.J., Manley, K. & Messer, A. Exploratory activity and fear conditioning abnormalities develop early in R6/2 Huntington's disease trangenic mice. *Behav Neurosci,* 117: 1233–42, 2003.
31. Bolivar, V.J., Manley, K. & Messer, A. Early exploratory behavior abnormalities in R6/1 Huntington's disease transgenic mice. *Brain Res,* 1005: 29–35, 2004.

9 Animal Resources in Behavioral Neurogenetics

Jean-Michel Lassalle

CONTENTS

9.1 INTRODUCTION

Genetics is a differential approach of the intra-specific variation, either normal or pathologic, based on several types of experimental methods. Most of the behavioral and neural phenotypes are complex traits that require the development of specific tools. Several animal species are currently used in the field of behavioral neurogenetics, from humans to rotifers, through dogs, rodents, birds, fishes and flies.

    Whereas in animals it is possible to create strains according to some specific criteria, and to use them as genuine tools to devise genetic experiments for obvious ethical reasons, in humans experimental approaches are restricted to "natural" experiments. These basically include family studies, twin studies, and adoption studies. On the other hand, in animals, various true experimental tools were developed. They are based on inbreeding, selection experiments, crossing experiments, spontaneous mutations, experimental mutagenesis or, more recently, on genetic engineering.

9.1.1 REQUIREMENTS FOR A SUITABLE ANIMAL MODEL

Genetic approaches have needed the development of specific animal models that constitute genuine experimental tools and that must therefore respond to some criteria.

    The requirements for a suitable animal model in the field of behavioral neurogenetics are particularly demanding. The candidate must be a short-lived species since a high number of generations are needed. It must also have a high reproduction rate and yield abundant progeny to ensure large samples of subjects in a restricted period of time. As far as psychological and cognitive abilities are concerned, a large behavioral repertoire and advanced cognitive abilities are required. The size and the accessibility of its central nervous system must allow the use of the most powerful and specialized techniques of modern neurobiology. Although not mandatory, it is also suitable for its genetic code to have been completely deciphered. Finally, the biology of the species must allow that experimental approaches such as selective breeding, inbreeding, or genetic engineering to be implemented.

9.1.2 THE "GENOME CLUB"

Actually, not one species meets all these requirements, so different species are used—chosen at different levels of the animal scale. After humans and mice, the rat is the third mammal species (quite recently) to join this elite club in biology that almost exclusively admits species the genome of which has been entirely deciphered. The sequence of the genome of the chimpanzee is expected to be completed by the end of the year 2004, whereas the genome analysis of the rhesus macaque monkey and other species of interest, such as the dog, the frog, the chicken, and the cow, are making rapid progress. The genome of the nematode worm, *Caenorabditis elegans,* a species of growing interest in the field, has already been deciphered. Nevertheless, for technical reasons, the mouse remains the most widely used species, particularly when knockout and transgenic approaches are needed.

9.1.3 THE BASIC PRINCIPLE OF ANIMAL TOOLS IS INBREEDING

Inbreeding is the production of offspring by closely related parents (whereas outbreeding is the production of offspring by unrelated parents). A strain is considered inbred after 20 generations of brother × sister mating. At this stage, it remains only 1% residual heterozygosity (excluding any genetic drift), so that all loci of an inbred animal can be considered homozygous and that all animals from the same inbred strain are isogenic, i.e., they carry the same alleles at each locus. Although the system by which homozygosis can be approached more rapidly and conveniently in experimental animals such as mice is that of brother × sister mating, other breeding schemes are also acceptable. Self-fertilization occurring only in a limited group of plants, backcrosses between progeny and one parent, double first cousin, first cousin, and so on can be used. Figure 9.1 represents the percentage of homozygosis reached in successive

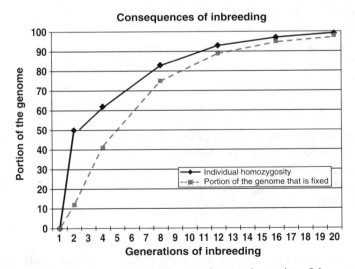

FIGURE 9.1 Effect of inbreeding during 20 generations on the portion of the genome of an animal that is homozygous at each generation (solid line) and the portion that is fixed identically in two animals chosen as parents for the next generation. (Adapted from Silver.[2])

generations under brother × sister mating. Most inbred strains of mice and rats have greatly surpassed 20 generations of inbreeding, which ensures homozygosis at nearly all loci. Nevertheless, not all species support inbreeding; birds generally become sterile after two generations of inbreeding and it seems that *Drosophila* "resists" inbreeding.[1]

9.2   INBRED STRAINS

Numerous inbred strains are available in rodents. Among the 478 strains of mice listed in the Mouse Genome Informatics database by M.F.W. Festing,[3] 100 are commercially available, the others are either extinct or preserved as frozen embryos. A rat strain list can be obtained from the Ratmap database. It is clear from their listings that there are also lots of inbred strains of rats with many substrains, although little is widely known about most of them.

BALB/c (C), C57BL/6 (B6), DBA/2 (D2), for the mouse and Wistar kyoto (WKY), Lewis (LEW), Fisher (FISH), Spontaneous Hypertensive Rat (SHR) are examples of widely used inbred strains.

The characteristic features of inbred strains are their genetic and phenotypic uniformity, although the latter is questionable, due to inbreeding depression that often results in less buffered homeostasis of physiological or behavioral traits.

Their main applications involve inbred strains comparisons, the production of F1 and F2 hybrid generations or of even more complete crossing schedules (Mendelian, diallel). They can also be used to generate controlled and reproducible genetically heterogeneous populations (e.g., four-way, or eight-way crosses).

Among the strengths and weaknesses of inbred strains, it is worth pointing out that they are good for mapping and gene identification. They provide excellent genetic background for studying epistasis and constitute a powerful method for detecting pleiotropism. They are also good to give evidence of gene × environment interactions. Overall, their genotype is invariant over time, so the database of inbred strains is cumulative and rapidly expanding. But the other side of the picture is that inbreeding often leads to a reduction in viability and fertility and also in developmental homeostasis, which makes them more sensitive to environmental short-term variations and increases the phenotypic variance.

9.2.1   Hybrid Crosses

Hybrid crosses result from crosses between two (and in some cases more than two) inbred strains. The aim of hybrid crosses is to restore hybrid vigor (F1), to maintain lethal (targeted) alleles at the heterozygous state (F1), to increase the number of recombination events (F2, F3, etc.) and to generate reproducible, highly heterozygous, segregating populations.

9.2.1.1   F1 Hybrids

Crossing two inbred strains produces F1 hybrids. For instance, B6D2 F1 results from a cross between a C57BL/6 female (B6) and a DBA/2 male (D2), whereas D2B6 F1 is the reciprocal F1 cross.

The characteristic features of F1 hybrids are their genetic and phenotypic uniformity: although partly heterozygous and partly homozygous, depending on the loci where the parents carry different or identical alleles, they are all isogenic. F1 hybrids present advantages that predispose them to specific applications. First of all, F1 hybrids between two inbred strains often display hybrid vigor or heterosis that results in improved development and longevity, better resistance to diseases and stress, larger size and weight, increased prolificacy, etc.). They often also display better behavioral responses to cognitive tests.[4] Secondly, they are useful as hosts for tissue transplants (tumors, skin, ovaries) from either parental strain and as recipients for some deleterious mutations (transgenic mice). The comparison of reciprocal crosses between the two F1s allows detection of maternal effects resulting from cytoplasmic and/or pre- and postnatal maternal environmental influences. Last, they can be repeatedly produced, as long as the inbred parental strains remain available (samples of parental strains can be preserved for years as frozen embryos).

9.2.1.2 F2 Hybrids

Mating of F1 × F1 mice produces an F2 generation. B6129SF2/J is the F2 of a C57BL/6 × 129S/SvImJ strain.

The F2 is highly heterozygous and shows allelic assortment among all loci for which the parents differ, due to allelic segregation and recombination events during meiosis. Applications of F2 hybrids include their use as heterogeneous population for selection experiments, for genetic mapping or to originate a recombinant inbred strains series (RIs, see below). They are also often used as physiological controls for mice carrying targeted mutations maintained on a mixed, e.g., C57BL/6 by 129 F2, background (B6; 129).

9.3 ADVANCED INTERCROSS LINES (AILS)

AILs are made by producing an F2 generation between two inbred strains and then intercrossing in each subsequent generation, avoiding siblings mating.[5] The purpose is to increase the possibility of recombination among tightly linked genes. They can be used to create RIs with large numbers of recombination events, starting inbreeding with breeding pairs from the F8, for instance. Pri:B6,D2-G# is an AIL stock created at Princeton from the inbred strains C57BL/6 × DBA/2. The G number will increase with generations.

AILs constitute a good resource for fine QTL mapping with increased precision.

9.4 STRAINS THAT DIFFER ONLY BY A KNOWN PART OF THEIR GENOME

There are four categories of strains: co-isogenic strains, congenic strains, consomic strains, and conplastic strains that correspond to different kinds of inbred strains differing at only a small part of their genome. Selected strains can enter this category, provided that they have been submitted to an inbreeding procedure after the selection plateau has been reached.

9.4.1 CO-ISOGENIC STRAINS

They are two inbred strains that differ only at a single locus. This condition can be due to spontaneous mutation occurring in a strain, but it can also result from targeted mutation in embryonic stem (ES) cells maintained on the same substrain from which ES cells were derived, or from chemically or radiation induced mutants on an inbred background. C57BL/6 C²J, for instance, is an albino mutation that occurred spontaneously in the C57BL/6J strain.

Co-isogenic strains allow studying the effects of an individual mutation. Unfortunately, the probability that a mutation of interest for a particular research occurs spontaneously is extremely low.

9.4.2 CONGENIC STRAINS

As shown by Figure 9.2, congenic strains are two inbred strains produced by repeated backcrosses to an inbred or (background) strain, with selection for a particular marker (or mutation) from the donor strain.[6,7] After a minimum of 10 backcross generations, the strain can be regarded as congenic. This is a practical way to drive experimentally the construction of strains that are close to the co-isogenic condition. B6.AKR-H2k is a mouse strain with the genetic background of C57BL/6 but differing from that strain by the introduction of a differential allele (H2k) derived from the AKR/J strain. The two congenic B6.AKR-H2k and C57BL/6 strains are expected to be identical at all loci except for a segment of chromosome carrying the transferred locus. The size of this segment decreases with successive backcrosses.

There are many advantages of maintaining a specific mutation on a defined inbred background. First, the increase of genetic and phenotypic homogeneity due to inbreeding results in reduced variability. Secondly, it allows identification of modifier genes present in a specific strain. Finally, the background inbred strain provides an accurate control.

Congenic strains are useful for avoiding genetic background effects, allowing more precise analysis of single single-gene effects, and also for confirming the position of a QTL, for mapping studies and for identifying a QTL.

One pitfall of congenic strains is contamination on another chromosome.

9.4.3 CONSOMIC STRAINS OR CHROMOSOME SUBSTITUTION
STRAINS[8]

They are produced by repeated backcrossing of a whole chromosome of a donor strain onto the genetic background of a host inbred strain for a minimum of 10 generations. For example, in the C57BL/6J-Chr 19 SPR consomic strain, a M. *Spretus* chromosome 19 has been backcrossed onto C57BL/6.

Consomic strains can be used directly to initiate mapping of a single locus responsible for a strain characteristic, or of multiple linked or unlinked loci that contribute in an additive fashion to the phenotype. They can be also useful to confirm weak QTLs identified in linkage crosses.

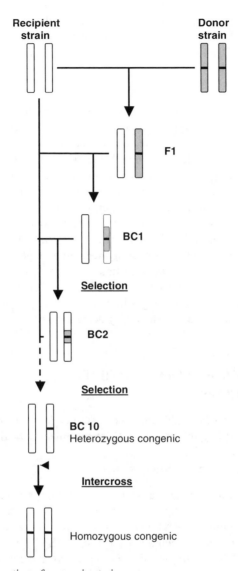

FIGURE 9.2 Construction of congenic strains.

9.4.4 CONPLASTIC STRAINS

These are strains in which the nuclear genome from one strain has been crossed onto the cytoplasm of another (the mitochondrial donor always is the female parent during backcrossing). For instance, crossing C57BL/6 male mice with BALB/c females, followed by repeated backcrossing of female offspring to male C57BL/6, results in a conplastic strain. C57BL/6-mtBALB/c is a conplastic strain with the nuclear genome of C57BL/6 and the cytoplasmic (mitochondrial) genome of BALB/c.

Conplastic strains are used to study cytoplasmic inheritance and to detect point mtDNA mutations. Inherited mtDNA is 99.9% maternal and represents less than 1% of the total cellular DNA in humans and 0.2% in the mouse.[9]

9.4.5 SELECTED STRAINS

Starting with a polymorphic population, bidirectional selection proceeds by selective breeding for a given phenotypic trait in two opposite directions over successive generations, until the selection plateau has been reached. Selection accumulates increasing and decreasing alleles in the high and the low line, respectively. Generally followed by inbreeding, it results in two well-contrasted strains with stable differences. Long sleep (LS) and short sleep (SS) strains result from bidirectional selection for "sleep time" in response to alcohol injection in mice.[10] After 18 generations of selective mating, LS mice slept for an average of two hours, whereas most SS mice were not even "knocked-out" (10 min average "sleep time").

Selected strains are good for estimating the number of genes and mapping QTLs that control the selected trait and selection-correlated traits. They are powerful for detecting pleiotropy in selected genes. They are also good to detect gene × environment interactions when changes in line differences occur in association with environmental changes but are ill-suited for detecting epistasis because the selected genes are unknown.

9.5 STRAINS MADE FROM (MULTIPLE) INBRED STRAINS

This category includes recombinant inbred strains (RIs), recombinant congenic strains (RCs), segregating inbred strains (SIs) and mixed inbred strains (MIs).

9.5.1 RECOMBINANT INBRED STRAINS (RIs)

RIs are issued from *n* breeding pairs chosen at random in the F2 of two inbred strains and followed by 20 or more generations of brother × sister mating.[11,12]

As shown in Figure 9.3, BXD1, BXD2, ... BXD32 are RI strains of a series originating from a cross between a C57BL/6 male (B) by a DBA/2 female (D2). This series includes the 26 strains out of 32 that survived the inbreeding procedure. Mice from the *n* inbred strains of the series, each of them carrying a random sample of the genes of the two parental strains, are homozygous and isogenic within a strain.

9.5.1.1 Seven Major RI Sets Are Currently Available

Originally designed to detect major gene effects, they are more and more used to detect and locate quantitative trait loci (QTLs) on chromosomes, with increased precision as the number of strains in a series increases. Thirty-six BXD RI strains are now available from the Jackson laboratory and the number of strains in the BXD series will soon be greatly increased. The University of Tennessee Health Science

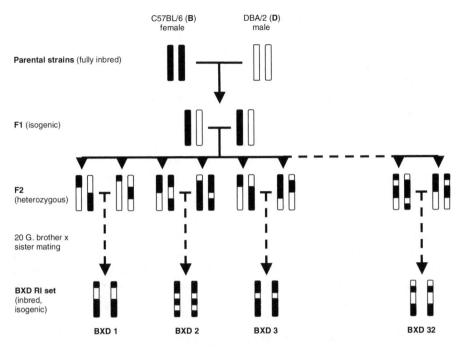

FIGURE 9.3 Construction of the BXD RI series.

Center in Memphis is actively breeding 50 lines of new BXD strains. Forty-six of these lines are likely to be inbred fully in the next 2 years. In order to increase the power and precision of mapping, these 46 newly generated BXD RI strains originated from an advanced intercross (AI: 9 to 14 generations of intercrossing before inbreeding) and thus constitute an advanced RI set.[13] The Complex Trait Consortium is actively pushing to provide the community with 1,000 ARIs derived from an advanced intercross between 5 inbred strains and 3 wild derived strains.[14]

QTL mapping is based on analyzing the pattern of distribution of DNA markers (SSLPs and SNPs) in order to determine whether any of these markers is linked to a given phenotypic trait.

The advantages of the RI strains are the same as those of inbred strains. There are thousands of DNA markers mapped in the RI series for which each strain needs to be genotyped only once. Therefore, genetic correlations can be established across measures, across studies and across laboratories thanks to cooperative data banks.

Their main limitation used to be the reduced number of strains in most series that decreases the power of mapping and allows detecting only QTLs having the largest effects.

9.5.2 RECOMBINANT CONGENIC STRAINS (RCS)

RCs are formed by crossing two inbred strains, followed by a few (generally 2) backcrosses of the hybrids to the parental (recipient) strain, with subsequent inbreeding for 20 generations, without selection for any specific marker.[15] These RCs consist of the background of the recipient strain interspersed with homozygous segments of the donor, the size of which depends upon the number of backcrosses. Two backcrosses result on average 12.5% of the donor strain genome. For example, CcS1, CcS2, CcSX are multiple recombinant congenic strains between a BALB/c (C) recipient and a STS (S) donor.

RCs allow analyzing the individual genes composing a system one by one. A series of RCs is a powerful tool for genetic and functional dissection of quantitative traits controlled by up to 5 or 6 nonlinked genes. They can be employed to study genes–gene interactions in various background combinations.

However, as most existing RC strains have been developed for the genetic analysis of specific traits, they may be of limited use in other applications.

9.5.3 SEGREGATING INBRED STRAINS (SIs)

SIs are inbred strains in which a particular allele or a particular mutation is maintained in the heterozygous state. They are developed by brother × sister mating but with heterozygosity of the particular allele selected at each generation. The 129P3/J mouse strain, for example, segregates for the albino (Tyr^c) and chinchilla (Tyr^{c-ch}) tyrosinase alleles.

9.5.4 MIXED INBRED STRAINS (MIs)

MIs are incipient inbred stocks or mixed strains derived from only two parental strains and carrying a genetic particularity. B6;129-Acvr2tm/Zuk is a mixed strain derived from C57BL/6 and a 129 ES cell line, carrying a targeted knockout of the Acvr2 gene.

9.6 GENETICALLY ENGINEERED MUTANT MICE

This category includes mice carrying a transgene, mice with targeted mutations (knockouts), and mice with retroviral or chemically induced mutations.

9.6.1 TRANSGENIC MICE

Transgenic mice carry a segment of foreign DNA incorporated into the genome via pronuclear microinjection, insertion via infection with a retroviral vector or sometimes via homologous insertion. For instance, Tg(huAPP695.K670N-M671L)2576 is a mouse carrying the human APP695 gene with the double mutation Lys→670 ASN, MET671→Leu, found in a Swedish family with heritable Alzheimer's disease.[16] The βAPP695s we gene has been inserted in the cosmid vector of the hamster's prion protein in the hybrid (C57BL/6xSJL) mouse.

Transgenic mice are generally used as models of genetic human diseases.

One caveat of transgenic lines is that neither the number of transcripts nor the insertion sites of the foreign gene in the genome are known.

9.6.2 MICE WITH TARGETED MUTATIONS (KNOCKOUTS)

They are created by first introducing gene disruptions, replacements or duplications into ES cells by homologous recombination between the exogenous (targeting) DNA and the endogenous (targeted) gene. Genetically modified ES cells are then micro-injected into host embryos at the 8-cell blastocyst stage. These microinjected embryos are in turn transferred into pseudopregnant host females that bear chimeric progeny. Chimeric progeny carrying the targeted mutation in their germ line are then bred to establish the line.

The knockout can be controlled by adding bits to a gene that act as a switch, turning the gene on and off (inducible knockout) when a particular hormone or antibiotic is added or even when the temperature changes (heat shock). The use of appropriate promoters will allow expression of targeting in selected brain areas.

The knockout technology allows analyzing the function of a gene by performing a form of experimental blockage (neurogenetic lesioning).

However, this approach encounters important limitations because numerous genes being also involved in development, the knockout may be nonviable. Compensatory mechanisms may also counterbalance the genetic deficit so that numerous knockouts display "no phenotype."

9.6.3 MICE WITH CHEMICALLY INDUCED MUTATIONS

They are created by exposing male mice to chemical agents, then breeding these treated males with untreated females. The progeny are screened for phenotypes of interest.

A variety of chemicals can be used that result in different kinds of mutations:[17]

Ethylnitrosourea (ENU) is used to induce point mutations.
Chlorembucil, an antibiotic, induces deletion mutations.
Rays exposure results in translocations.
Mutation rate is of 1/1,000 per locus and mutation detection rate of 90%.

These mutant mice are useful tools to elucidate basic biological processes and to study relationships between gene mutations and disease phenotypes. Although only recently used in mice, mutagenesis has been successfully employed in *Drosophila* to study learning and memory mutants early in the 1970s. A neuroscience mutagenesis facility has been created for mice at the Jackson laboratory that is available online.

9.6.4 CHROMOSOMAL ABERRATION STRAINS

Chromosomal aberration strains include various kinds of chromosomal anomalies such as rearrangements of DNA segments within chromosomes (inversions), exchanges of DNA segments between chromosomes (translocations) and deviations from the normal number of diploid chromosomes in somatic cells (aneuploidy) that include both Robertsonian chromosomes, i.e., bi-armed (metacentric) chromosomes formed by the joining of two single-armed (acrocentric or telocentric) chromosomes at the centromere and whole or partial chromosome trisomies.

Chromosomal aberration strains and Robertsonian F1 hybrids stocks are available in mice, e.g.:

- In(7)13Rk in C.Cg-In(7)13Rk	(In = Inversion)
- Is(7;1)40H in C3H/HeJ-p^{j}Is(7;1)40H	(Is = Insertion)
- Rb(6.16)24Lub	(Rb = Robertsonian chromosome)
- Tp(Y)1Ct(y^{Sxrb}) in B6-A$^{w\text{-}J}$-Eda$^{Ta\text{-}6J}$.Cg-Sxr^b Hya$^-$	(Tp = Transposition)
- Ts(17^{16})65Dn in B6EiC3Sn-a/A-Ts(17^{16})65Dn	(Ts = chromosome 16 trisomy)

They constitute useful animal models for studying the effects of chromosomal aberrations in humans. Moreover, Robertsonian chromosomes in combination can be used to produce whole chromosome trisomy for specific mouse chromosomes. For example [Rb(6.16)24Lub × Rb(16.17)7Bnr]F1 mice, produce trisomy for chromosome 16.

9.7 OTHER STRAINS

9.7.1 OUTBRED STRAINS OR OUTBRED COLONIES

An outbred strain is obtained by random mating within a closed colony. NMRI, Swiss Webster, ICR mice, and Wistar, Long-Evans, Zucker, and Sasco-Sprague-Dawley rats are widely used outbred strains.

They are less expensive than inbred strains, but are genetically unstable across generations.

9.7.2 WILD- DERIVED INBRED MICE

They are descendants of a pair or a trio of wild-captured mice.

Much more polymorphic than common laboratory mice, they are a valuable tool for evolution and systematics research. Also, the progeny from interspecific crosses is especially useful for genetic mapping and they often carry Robertsonian chromosomes.

9.8 ORIGINS OF THE OLDEST INBRED STRAINS OF MICE

As can be seen from Figure 9.4, many of the laboratory inbred strains come from the same outbred colony, or share common ancestry. Other strains are derived directly from wild mice. Whereas laboratory rat strains all derive from the *Rattus norvegicus* species, there is increasing evidence that laboratory mice have been developed with contributions from more than one species/subspecies of wild mouse. For example, some strains carry the *Mus musculus domesticus* Y-chromosome (e.g., A/J, BALB/cJ,

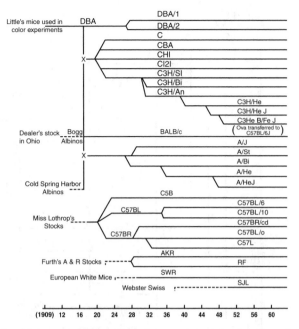

FIGURE 9.4 The origins and relationships of some of the inbred strains of mice.[18]

CBA/J, C3H/HeJ, C57BL/6J, DBA/2J, ...), while others have the *M.m. musculus* type (e.g., AKR/J, SJL/J, SWR/J, etc.).

Most laboratory mice have contributions from both *Mus musculus musculus* and *Mus musculus domesticus*. There is evidence that smaller contributions also may have come from *Mus musculus molossinus* and *Mus musculus castaneus*.[3]

9.9 STRAINS AND SUBSTRAINS OF LABORATORY MICE

Established inbred strains may diverge with time into substrains among which there are discernable genetic differences. This could occur through three different conditions:

If two branches are separated after 20 but before 40 generations of inbreeding, there is still enough heterogeneity for two genetically different substrains to result.[20]

If branches are separated for more than 20 generations from a common ancestor, genetic variation may have occurred by mutation and genetic drift.

If genetic differences are proven by genetic analysis to have occurred.

For example, IS/Kyo is a substrain of IS rat strain originating at Kyoto University. Figure 9.5 gives an example of the various substrains that occurred in the C57BL strain.[19]

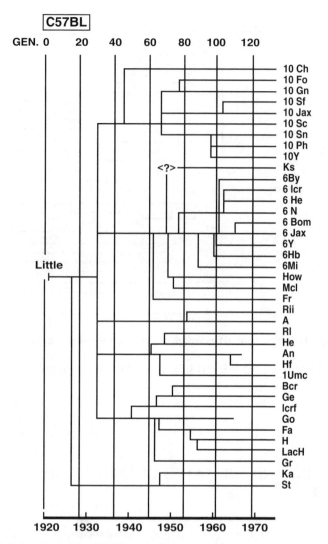

FIGURE 9.5 Substrains of the C57BL strain.[19]

9.10 INTERNATIONAL SUPPLY OF MOUSE AND RAT RESEARCH MODELS

9.10.1 THE JACKSON LABORATORY AND CHARLES RIVER LABORATORIES

The Jackson Laboratory (Bar Harbor, Maine) acts as a conservatory of most inbred, mutant and genetically engineered mice used in biomedical research around the world. These strains are either in production, or preserved as frozen embryos.

In July 2001, The Jackson Laboratory and Charles River Laboratories International, Inc., started to cooperate to supply JAX® mice to researchers located in European and Pacific Rim countries.

As detailed in the JAX Bulletin, (N°11, Sept. 2001), their distribution agreement stipulates that the Jackson laboratory ensures that JAX® mice raised in Charles River's European and Japanese facilities meet the highest standards for health and genetic quality. The main points of this agreement are summarized below.

9.10.1.1 Genetic Quality

JAX animals bred in the United States, Europe and Japan originate from pedigreed identified pairs shipped from The Jackson Laboratory. Authorized breeding colonies are reinfused with new pedigreed stock from The Jackson Laboratory on a routine basis to minimize spontaneously occurring genetic drift. Routine monitoring for genetic quality is performed to ensure both genetic identity and purity.

9.10.1.2 Animal Health Quality

JAX rodents bred in Europe and Japan are monitored for the same pathogenic and opportunistic agents as are the JAX rodents raised at The Jackson Laboratory, on the same basis (sampling plan, frequency of testing, diagnostic test methods).

Such genetic and health quality procedures ensure that all mice used in various countries at different times match the highest standards for genetic uniformity and makes particularly meaningful comparisons of results.

9.10.2 THE COMPLEX TRAIT CONSORTIUM (CTC)

Another important breeding facility is developing at the University of Memphis, which will revolutionize QTL mapping research in RI mice. "The goal of the CTC is to promote the development of genetic resources that can be used to understand, treat and ultimately prevent pervasive human diseases."[14] The CTC is actively trying to generate a collaborative genetic resource called the "Collaborative Cross." Starting with a full diallel cross of eight strains, it should result in a series of 1,000 RIs available to scientists in the next few years.

9.11 OTHER SPECIES OF INTEREST FOR BEHAVIORAL NEUROGENETICS RESEARCH

Two more species, the fly *Drosophila,* and the nematode worm *Cenorhabditis elegans* are more and more widely used in behavioral neurogenetics research. They also gather most of the typical features of a suitable animal resource.

9.11.1 DROSOPHILA MELANOGASTER

D. melanogaster is a very handy tool, highly amenable for behavioral and neurogenetic analysis. Its reproductive cycle is 10 to 14 days at 25°C. Hundreds of eggs are laid on the food medium that hatch after 24 h. The development includes the larval stage (5 days) and the pupa stage (4 days). The adult fly lives about 30 days in the laboratory. *Drosophila* has a compact genome (The X/Y sex chromosomes and 3 pairs of chromosomes, the fourth of which being very tiny) that has been entirely sequenced. The size of the genome is about 165 million bases and contains an estimated 14,000 genes. *Drosophila* displays a rather large behavioral repertoire: learning and memory, courtship, circadian rhythms, food search, olfaction, locomotion, aggression and sleep-like processes can be studied in *Drosophila*, some of them at both the larval and adult stages. Most methods used in forward and backward genetics can be implemented in that species, including polygenic and single-gene analysis, selective breeding, QTL analysis, mutational analysis (EMS and screening), namely for memory and biological clocks, visible markers, transposons, genetic engineering (inducible knockout, reporter genes such as β-Galactosidase, GFP).

One limit of *Drosophila* comes from its resistance to inbreeding so that there are no truly inbred strains in this species.

9.11.2 CAENORHABDITIS ELEGANS

The terrestrial nematode worm *C. elegans* also presents most of the characteristics of a model organism for behavioral neurogenetics. This tiny worm has a very fast reproductive cycle (3 days) at 20°C. Males (XO) are found rarely in laboratory populations (0.05%). Most individuals are hermaphrodites (XX) that self-fertilize, so that it is easy to generate homozygous mutant stocks. *C. elegans* is diploid and has a small genome with 5 pairs of autosomal chromosomes and a pair of sex chromosomes. It has been the first animal genome completely sequenced by the end of 1988. The genome size is 97 Megabases (Mb) and there are about 20,000 protein coding genes.

The nervous system of C. *elegans* is very simple with 302 neurons in the hermaphrodite and 310 in the male that has a more complex phenotype with sexual behavior. It has also an almost invariant connectivity and neurotransmitters (ACh, Glu, GABA, DA, 5-HT) that are common with those of mammals. The behavioral repertoire is reasonably endowed with basic behaviors such as locomotion, feeding, defecation, egg laying, male mating with the hermaphrodite, sensing its environment and responding appropriately. Basic learning processes as habituation and sensitization, some forms of associative memory and internal clocks can also be studied. Reverse genetics and transgenesis can be used in *C. elegans*. Hundreds of genetic markers have been identified. Mutagenesis (EMS and screening) and balancer chromosomes are useful tools in that species.

9.12 USEFUL WEB SITES

Most of the information presented in this chapter can be completed or retrieved from all purpose and more specialized Web sites, a nonexhaustive list of which is presented below.

A general purpose virtual library on model organisms can be found at http://www.ceolas.org. This site is a catalog of Internet resources relating to biological model organisms.

9.12.1 WEB SITES FOR THE MOUSE AND THE RAT

- The Whole Mouse catalog (this Web site is for mice and rats, formerly the Mouse and Rat Research Home Page located at CalTech): http://www.rodentia.com/wmc/.
- Mouse Genome Informatics (MGI): http://www.informatics.jax.org.
- Neuroscience mutagenesis facility: http://nmf.jax.org.
- Complex Trait Consortium (CTC): http://www.complextrait.org.
- The Rat Genome Database (Ratmap group, Sweden)
  - Ratmap: http://ratmap.gen.gu.se/.
  - Rat Resources: http://ratmap.org/ratres.html.
- International Committee on Standardized Genetic Nomenclature for Mice Chairperson: Dr. Janan T. Eppig (jte@informatics.jax.org).
- Rat Genome and Nomenclature Committee, Chairperson: Dr. Eberhard Guenther (eguenth@gwdg.de).
- Current nomenclature rules for naming genes
  - For mouse: http://www.informatics.jax.org/mgihome/nomen/gene.shtml#genenom.
  - For rat: http://rgd.mcw.edu/nomen_rules.html.
- Strain names registration
  - For mouse: http://www.informatics.jax.org/mgihome/submissions/submissions_menu.shtml.
  - For rat: http://rgd.mcw.edu.

9.12.2 RODENTS' SUPPLIERS

- The Jackson Laboratory: http://www.jax.org.
- Charles River laboratories: http://www.criver.com.
- Taconic: http://www.taconic.com.
- Taconic M&B (Europe): http://www.m-b.dk.
- RRRC (Rat Resource and Research Center): http://www.nrrrc.missouri.edu.
- Harlan: http://www.harlan.com.

9.12.3 WEB SITES FOR *DROSOPHILA*

- A quick and simple introduction to *Drosophila melanogaster* can be found at http://www.ceolas.org/fly/introd.html.
- The dedicated database to *Drosophila* is Flybase at http://flybase.net.

9.12.4 WEB SITES FOR *CAENORHABDITIS ELEGANS*

- An introduction to the genetics of *C. elegans* can be found at http://www.nematodes.org.
- Wormbase is a large-scale database covering information on all worm genes at http://www.wormbase.org.

- The *C. elegans* www server at UTSMC is the best on line reference for *C. elegans*: (http://elegans.swmed.edu/).
- *elegans* Net: (http://members.tripod.com/C.elegans/index.html links on *C. elegans* and includes extensive introductory links (ACekit at http://winw.nbr.wisc.edu/outreach/test/celegans.html).

REFERENCES

1. Petit, C., L'influence du mode de croisement sur la structure génétique des populations; la stabilité des populations expérimentales de faible effectif. *Annales de génétique*, 1963, 6, 29–35.
2. Silver, L.E., *Mouse genetics, concepts and applications*. 1995 Oxford University Press. Available online at the Jackson laboratory Mouse Genome Informatics facility: http://www.informatics.jax.org/silver/frame1.1.shtml and also the CTC Web site: http://www.complextrait.org.
3. Festing, M.F.W., Inbred strains of mice and their characteristics In: *Mouse Genome Informatics* (MGI) 1998 at http://www.informatics.jax.org.
4. Lassalle J.M., Le Pape G., and Médioni J., A case of behavioral heterosis in mice: quantitative and qualitative aspects of performance in a water-escape test. *J. Comp. Physiol. Psychol.*, 1976, 93, 116–123.
5. Darvasi, A. and Soller, M., Advanced intercross lines, an experimental population for fine genetic mapping. *Genetics*, 1995 141: 1199–1207.
6. Snell, G.D., Congenic resistant strains of mice. In: *Origins of inbred mice*, Morse, H.C., Ed., Academic Press, New York, 1978 pp 1–31.
7. Flaherty, L., Congenic strains. In: *The mouse in biomedical research, Vol. 1*, Foster, H.L., Small, J.D., Fox, J.G. Eds, Academic Press, New York, 1981 pp. 215–222.
8. Nadeau, J.H., Singer, J.B., Martin, A., and Lander, E.S., Analysing complex genetic traits with chromosome substitution strains. *Nature Genetics,* 2000, 24: 221–225.
9. Lewin, B., 1987. *Genes*, Wiley, New York.
10. Mc Clearn, G.E. and Kakihana, R., Selective breeding for ethanol sensitivity: short sleep and long sleep mice, in: Development of animal models as pharmacogenetic tools. G.E. McClearn, R.A. Dietrich and V.G. Ervin, Eds., Research monograph No. 6, Bethesda, MD, US. 1978 Department of Health and Human Services, 147–159.
11. Bailey, D.W., 1971. Recombinant inbred strains, an aid to finding identity, linkage, and function of histocompatibility and other genes. *Transplantation*, 1971 11: 325–327.
12. Taylor, B.A., Recombinant inbred strains: use in gene mapping. In: *Origins of inbred mice,* Morse, H.C., Ed., Academic Press, New York, 1978 pp 423–438.
13. Peirce, J.L., Lu, L. GU, J., Silver, L.M. and Williams, R., A new set of recombinant inbred lines from advanced intercross populations in mice. *BMC Genetics*, 2004 5:7, 1–17. Article available online from: http://www.biomedcentral.com/1471–2156/5/7.
14. Williams, R.W. and Churchill, G.A., *The collaborative cross: rationale, implementation and costs*. National Institute of General Medical Sciences, Complex trait workshop report. 1998 www.nigms.nih.gov/newx/reports/genetic_arch.html.
15. Demant, P. and Hart, A.A.M., Recombinant congenic strains – a new tool for analyzing genetic traits determined by more than one gene. *Immunogenetics* 1986 24: 416–422.

16. Hsiao, K.A., Chapman, P., Nilsen, S., Eckman, C., Harigaya, Y., Youkin, S., Yang, F., and Cole, G., 1996. Correlative memory deficits, An elevation, and amyloid plaques in transgenic mice. *Science*, 1996, 274: 99–102.

17. Greenspan, R.J., The induction, detection and isolation of mutations. In D. Goldowitz, D. Wahlsten and R.E. Wimer Eds: *Techniques for the Genetic Analysis of Brain and Behavior*. Elsevier Science Publishers BV, 1992, 93–110.

18. Staats, J., Origins of the older inbred strains. In: *Biology of the Laboratory Mouse*. Green, E.L., Ed., 1966, McGraw Hill.

19. Bailey, D.W., Definitions of inbred strains, substrains, sublines, and F1 Hybrids. In *Handbook of Genetically Standardized Jax Mice*. R.R. Fox and B.A. Witham, Eds. 1997 Fifth Edition. The Jackson laboratory, Bar Harbor, Maine.

20. Green, E.L., *Genetics and Probability in Animal Breeding Experiments*. 1981 Oxford University Press, New York.

10 Sample Size Requirements for Experiments on Laboratory Animals

Douglas Wahlsten

CONTENTS

INTRODUCTION

Planning any experiment involves choice of an experimental design and sample size for each of the groups in the study. The design of a study is dictated by the kind of question one seeks to answer, and the choice of appropriate control groups is determined by logic and established genetic principles. Approaches to design are discussed in numerous articles and chapters, including many in the present volume. Once the design is chosen, the researcher must decide on the sample size. This will commonly be done when preparing a grant or thesis proposal. Increasingly, animal ethics committees also request that the investigator justify the number of animals to be used. Ethical and financial considerations, alike, demand that the minimum effective number be employed. At the same time, if too few animals are studied, the experiment may be unable to detect the very effects it was designed to investigate, thereby wasting the animals as well as the researchers' time and the funding agency's money.

The proper sample size is determined by the experimenter's choice of three values. Two are probabilities of making either of two kinds of errors of statistical inference, and the third is the size of an effect that one seeks to detect.[17]

10.1 EFFECT SIZE

Effect size may be expressed in terms of the units of measurement for the phenotype of interest. Suppose one plans to examine brain weight in milligrams. If control mice have an average brain weight of 450 mg and closely related mutants have an average brain weight of 430 mg, then the effect size of the mutant genotype would be –20 mg. When examining learning of a maze, one might find that controls acquire a task in 11 trials while mutants require 15 trials, and the effect size would be +4 trials. Whether sizes of the mutation's effects on brain weight and learning are similar would be difficult to say, because there is no way of equating milligrams of brain tissue with trials in a maze.

Any interesting feature of an animal that we might care to measure is likely to show individual variation within a group having the same genotype. The extent of this variation is commonly expressed as the standard deviation. For normally distributed data, most (about 98%) individuals will score within two standard deviations of the group mean (see Figure 10.1). For the above example, suppose the standard deviation of brain weight is 15 mg and of maze learning is 5 trials. The relative sizes

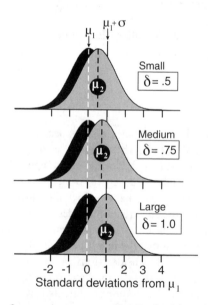

FIGURE 10.1 Overlap of scores in two normally distributed groups for three effect sizes ranging from small to large. Note that most scores in a group are within two standard deviations of its mean.

of the effects of the mutation can be assessed by comparing the difference between group means to the standard deviation within a group. For brain weight, effect size would be -20 mg/15 mg $= -1.33$, whereas for learning it would be 4 trials/5 trials $= 0.80$. Relative to variation within a group, the mutation would appear to have a relatively greater effect on brain weight than on maze learning.

The ratio (δ) of group difference in mean ($\mu_1 - \mu_2$) to standard deviation within a group (σ) is widely used as an index of effect size when two groups are being compared, and this is the value that is commonly used for a sample size estimation. Generally speaking, the smaller the effect that one would like to detect, the larger must be the sample size.

Because sample size calculations are done before any data are collected, these involve population values of parameters rather than estimates from sample data. Symbols used for population parameters and sample statistics are given in Table 10.1. For a study involving comparisons of two groups, the indicator of effect size is the hypothesized true value of the population parameter $\delta = (\mu_1 - \mu_2)/\sigma$, whereas its estimate from sample statistics is $d = (M_1 - M_2)/S$. The value of d is slightly biased and can be adjusted when precision is deemed important.[14,26] In this presentation, the true, population parameter is either a Greek letter or boldface, while the estimated sample statistic is either italicized or the Greek letter with caret ^ to indicate an estimate from data.

On the basis of my review of many studies of a wide variety of phenotypes, guidelines are offered for values of δ that correspond to what are generally regarded as very small, small, medium, large, and very large effects in neurobehavioral genetics: $\delta = 0.25, 0.50, 0.75, 1.0,$ and 1.5, respectively (Table 10.2). These are somewhat larger than values assigned to the same descriptors in psychological research with humans by Cohen[8] and Borenstein et al.[5]

When more than two groups are involved, as in a study of several inbred strains, the difference between any two groups will be unique to that pair. Hence, an indicator that is more broadly representative of the data is needed. Cohen[8] has suggested a convenient approach that compares the population standard deviation among group means (σ_M) to the standard deviation within groups (σ): $f = \sigma_M/\sigma$. When an analysis of variance (ANOVA) partitions total variance into components among and within

TABLE 10.1
Symbols for True Values of Parameters and Their Estimates From Sample Statistics

Characteristic	Population Parameter	Sample Statistic
Mean	μ	M
Standard deviation	σ	S
Effect size for 2 groups	δ	d
Effect size for J groups	ω^2	est. ω^2
Cohen's effect size for J groups	f	—
Linear contrast of J groups	ψ	est. ψ

TABLE 10.2
Values Corresponding to Magnitudes of Effect Size for Animal Research in Neurobehavioral Genetics

Size of Effect	2 Groups		J Groups	
	δ	ω^2	Cohen's f	ω^2
Very small	0.25	1.5%	0.1	1%
Small	0.5	6%	0.2	4%
Medium	0.75	12%	0.35	11%
Large	1.0	20%	0.5	20%
Very large	1.5	36%	0.7	33%

groups, a good descriptor of effect size is ω^2, the proportion of total variance that is attributable to differences among group means in a fixed effects analysis.[13] The f index has a simple relation to this ratio of variances: $\omega^2 = f^2/(f^2 + 1)$. Values of f and ω^2 for effects of various sizes are compared in Table 10.2. When there are only two groups in the study, $\omega^2 = \delta^2/(\delta^2 + 4)$.

Cohen[8] proposed that criteria for small, medium and large effects in a one-way ANOVA design in psychological research be $f = 0.1, 0.25$ and 0.4, respectively, which correspond to $\omega^2 = 0.01, 0.06$ and 0.14. These values appear to be somewhat lower than what is commonly observed in neurobehavioral genetics, for example in comparisons of several inbred strains. A multigroup difference that accounts for only 1% of total variance would be seen as not merely small but almost trivially small. Many inbred-strain studies have found effects exceeding 20% of total variance—and this would be a reasonable criterion for a large effect in our field. Standards are proposed for both f and ω^2 in Table 10.2 for a study with J groups. With these conventions, the verbal labels of size of the effect correspond to similar values of ω^2 for a study with two groups as well as for a large experiment with J groups.

10.2 PROBABILITY OF A FALSE POSITIVE

The size of an effect seen in an experiment is sometimes confused with the "significance" of the statistical test. We commonly read reports that describe "highly significant" effects of a genetic mutation as though this implies the effect was very large. It is easy to understand how people new to data analysis might perceive meaning in this way. *The Oxford English Reference Dictionary*[1] lists the most common use of the word "significance" as "importance; noteworthiness." It lists the technical sense employed by statisticians as "the extent to which a result deviates from that expected on the basis of the null hypothesis." Use of the word "extent" implies size or magnitude, which is not true for statistical significance. If very large samples are studied, very small effect sizes may be judged "significant" by statistical tests.

A null hypothesis usually contends that a treatment had no effect ($\delta = 0$) or several samples observed in the study were drawn from populations that did not differ ($\omega^2 = 0$). After collecting the data, the investigator conducts a statistical test, such as a *t* test when comparing two groups or the *F* test in the ANOVA comparing several groups, in order to assess whether the evidence is sufficient to justify rejection of the null hypothesis. If the null that there was no real effect can be refuted, then we may conclude, tentatively, that there was indeed an effect. This arcane logic, clothed in a most unfortunate choice of words such as "significance," often makes it difficult to keep the characters in the drama of science in their proper roles.

The null may in fact be true or it may be false. If the null is true but we decide to reject it on the basis of the results our study, then the outcome of our study is a *false positive* result and, in statistical parlance, we commit a Type I error of inference. By convention, we agree to set a criterion for rejecting the null hypothesis based on the probability of committing a Type I error. Most commonly this criterion, symbolized α, is set at one chance in 20. If the data suggest that the probability of committing a Type I error is less than 5%, the null may be rejected, but this does not reveal the scientific significance of the result. Instead, it would be better thought of as the false positive rate (FPR). If we are comparing two independent groups of animals, perhaps one homozygous for a knockout (–/–) and the other the wild-type littermate controls (+/+), and the *t* test indicates the FPR is .001 or one chance in 1,000, then it is a fairly safe bet that the knockout really did affect the phenotype we measured.

In those situations where the null is indeed true, the probability of obtaining a false positive result is not influenced in any way by the sample size. That is, even an elegant study of large numbers of animals will lead us to commit a Type I error on about 5% of the tests when the null is true, the same FPR as when only a few animals are studied.

The choice of the Type I error criterion influences the calculated sample size. Generally speaking, the smaller the value of α, the larger must be the size of the observed effect in order to warrant rejection of the null, and the larger the sample size that must be employed. When the results of a simple study depend on just one statistical test of the null hypothesis, then $\alpha = .05$ may be appropriate. If the study involves K independent tests, however, or if the study is one of K similar studies being done independently in different labs, then the Bonferroni corrected Type I error probability α/K is more appropriate. A more sophisticated approach to adjusting for multiple tests has been proposed by Benjamini et al.[4]

If the researcher has a good idea of the direction of the likely effect of a mutation, then it would be appropriate to conduct a one-tailed test of the null hypothesis that two groups do not differ, whereas a two-tailed test is preferred if the direction of the effect is uncertain. In an ANOVA with multiple groups, this distinction is not pertinent.

The overbearing emphasis on null hypothesis testing in psychology has been the subject of debate.[7,12] One of the central issues is the credibility of the null in any particular study. In a genetic linkage study, for example, the hypothesis that alleles at a marker locus are not related to phenotypic variation is highly credible, and we expect that most markers will not show evidence of linkage. In such a case, the proper choice α is critically important. Lander and Kruglyak[16] proposed the use of $\alpha = .0001$ for a genome-wide scan in linkage analysis, while Belknap et al.[3] argued for a lower level of α as a foil against the demise of a study owing to insensitivity to real linkage relationships. Given the surprising but common finding that a gene knockout has little or no effect on many phenotypes, the null may also be credible in this kind of research. For a multiple inbred strain study, on the other hand, it would be astounding if no strain main effect were found. Instead, the more interesting question in strain studies is usually the presence or absence of strain by treatment interactions or strain correlations among phenotypes.

Whereas the null hypothesis is commonly taken to be no effect at all, methods are available for evaluating the possibility that the observed effect exceeds some nonzero effect size.[17] This approach has not yet been widely applied in neurobehavioral genetics, but it is useful when researchers want to know whether a mutation has, for example, something more than merely a small effect on a phenotype.

10.3 PROBABILITY OF FAILING TO DETECT SOMETHING REAL

A test of significance can be done with just the null ($\delta = 0$ or $\omega^2 = 0$) but no specific alternative hypothesis in mind. In this situation, the probability α of Type I error can be controlled, but the test will be oblivious to the other kind of error of inference. Suppose there really is a genetic effect in a study and the null is false. If the statistical test fails to reject the null and detect the real effect, then a Type II error of inference has occurred. This kind of error is much more frequent than Type I errors in most kinds of research, and it is a serious error indeed, given the resources typically devoted to conducting a study. The probability of a Type II error is symbolized β, but there is no way to calculate this probability in any general sense, i.e. when $\delta \neq 0$ or $\omega^2 > 0$. Instead, we must propose a specific value of effect size that can serve as the alternative hypothesis. This value should be a plausible size of an effect that we would like to be able to detect with an experiment.

Suppose that we are seeking to find a medium-sized effect in a comparison of genotypes +/+ and –/–. Type I error probability α is the probability of rejecting the null when the hypothesis that $\delta = 0$ is true. Type II error probability β is the probability of failing to reject $\delta = 0$ when in fact $\delta = 0.75$ (see Figure 10.1). If δ really does equal 0.75, the possibility of Type I error is entirely moot, and Type II error is the major peril.

If β is the probability of an error, $1 - \beta$ is the probability of doing the right thing — rejecting a false null hypothesis when there truly is a genetic effect. The quantity $1 - \beta$ is termed the *power* of the statistical test, and we would like our test to have high power. Type II error probability and power are strongly dependent on sample size; the larger the sample, the higher the power. Because we choose the values of α, β and sample size before conducting the experiment, Type II error probability is under our control. The choice of power has no bearing at all on the probability of a Type I error because that mistaken conclusion can occur only when the null is true. The choice of α, on the other hand, influences power; a more stringent Type I error probability, being a smaller value of α, reduces power to detect an effect of a given size.

There is no universally accepted criterion for a proper level of power in research. Many studies are planned to have power of 80%, meaning that they are willing to risk a 20% rate of Type II error. It is not clear, however, why an investigator should be more concerned about Type I than Type II errors. If both are of equal concern, it would make sense to use $\alpha = \beta = .05$. When working with a novel knockout, $\alpha = .05$ might be reasonable for the purpose of estimating sample size, but power of 99% might be appropriate if great effort and expense are involved in creating and rearing the mice as well as measuring complex phenotypes.

10.4 ESTIMATING EFFECT SIZE FROM PUBLISHED DATA

Statistical analysis is typically done and then reported in an article after data have been collected, and it is unusual to read in the literature the values of δ or f that may have been used to plan the study. If one wishes to use the literature as a guide to choosing sample size for a future study, then it will be useful to know how to find δ or f from published statistics.

In the case of a study with two groups, the task is easy if the authors published the means, standard deviations and sample sizes of both groups. Presuming the variances (S^2) of the two groups were not exactly the same, a pooled estimate needs to be obtained by converting each variance to the sum of squared deviations (SS) from the group mean. For group 1 with n_1 subjects, $SS_1 = (n_1 - 1)S^2$. The pooled estimate of variance is $(SS_1 + SS_2)/(n_1 + n_2 - 2)$, and the pooled estimate of the standard deviation (S_{pooled}) is the square root of this variance. The sample estimate of effect size is then $d = (M_1 - M_2)/S_{pooled}$. If a table provides means and standard errors, then standard error can be converted to standard deviation by dividing by the square root of sample size for that group.

In some instances the authors do not provide a table of descriptive data and instead focus attention on the significance test. For example, a report may state that a group difference was significant ($t = 2.17$, $df = 18$, $P < .05$). Presuming that sample sizes were similar, it follows that the sample size per group was n = 10. As pointed out for a study with equal sample sizes,[26] the t ratio has a concise relation with effect size: $t^2 = d^2 n/2$, so that $d^2 = 2 t^2/n$. Thus, for our example, $d^2 = 2(4.71)/10 = 0.942$ and $d = 0.97$, a large effect size.

Frustration may set in when a report states sparsely that groups differed (e.g., $P = .007$). Provided one can figure out from the methods section that the sample size was perhaps 15 in each group, all is not lost. Degrees of freedom will be $30 - 2 = 28$, and we can then derive the approximate value of what the obtained t statistic must have been from the P value and degrees of freedom: $t = 2.62$. This leads to $d = 0.96$, a large effect size.

If the study noted with disappointment that the genetic difference was not significant at $\alpha = .05$ or stated tersely N.S, there is no way to arrive at their value of d. At best, one could compute an upper limit to the possible value of d. Suppose it is clear that n = 25, making degrees of freedom 48 for the t test. The critical value of t for a two-tailed test at $\alpha = .05$ is $t = 2.01$, and this ratio corresponds to $d = 0.57$, a rather small effect if there was any effect at all.

When an ANOVA is done on J groups, it may be possible to estimate σ_M, σ_{within}, and thereby f from a table of means and standard deviations or standard errors, as described for the case of two groups. If the reader finds only the significance test result ($F = 7.3$, df = 5/64, $P < .05$), it follows that there were six groups with about 12 subjects each. The value of the ω^2 can be found from the relation est. $\omega^2 = (F - 1)/[F + (df_{within} + 1)/df_{between}]$, which for the example is est. $\omega^2 = 0.31$. From the relation $\omega^2 = f^2/(f^2 + 1)$, we then estimate that $f = 0.67$, a very large effect. In the frustrating happenstance when we read only that $P = .034$ and that sample sizes were 20, 25, 22 and 21 in four groups, we know that $df_{between} = 3$ and $df_{within} = 84$. To give $P = .034$, the value of F must have been $F = 3.03$, and from this we find $\omega^2 = 0.065$ and $f = 0.26$, a small effect.

How many previous studies should be consulted in order to arrive at a reasonable guess at the value of δ or f depends on how large the literature on the subject may be. An estimate of δ or f is pertinent to the task at hand only if the methods in a previous study were similar to those to be used by the researcher. If the literature offers a wide range of effect sizes, it may well be that there were important methodological differences among the published studies, but this could also be ascribed to sampling error if samples used in those studies were generally small. Only a formal meta-analysis can assess these alternatives[11,14,20] (see http://www.edres.org/meta). If a meta-analysis is available, then effect size can be taken from the center of the confidence interval for δ in the case of two groups. If the most relevant literature consists of perhaps fewer than 10 studies, it is not difficult to conduct a quick meta-analysis of one's own to find the best estimate of effect size.

10.5 METHODS FOR ESTIMATING SAMPLE SIZE

While there is a widespread and well-founded belief that power and sample size calculations are poorly understood and rarely attempted in our field, there is sufficient interest to inspire the creation of no fewer than 29 computer programs that do most of the work for the informed investigator.[22] These programs entail a wide range of capabilities and ease of use, and there currently is no one implementation that dominates the market. Some are free, while others cost hundreds of dollars. Printed dissertations on the topic, on the other hand, can be purchased on the used book market for $20 in the case of Cohen[8] and implemented with an electronic calculator. Several general purpose statistical data analysis programs provide features for estimating power of a test, but in most cases these presume a considerable degree of expertise from the user.

A mathematically sophisticated approach is available from Kraemer and Thiemann,[15] where the population value of intraclass correlation is adapted to several situations and a master table gives degrees of freedom required to achieve various levels of power under two values of α (.05, .01) for both one- and two-tailed tests. A similar approach based on the F distribution that can be applied to a wide range of statistical tests is offered by Murphy and Myors[17] but provides a master sample size table only for $\alpha = .05$ and power of 80%. A less accurate but very convenient set of sample size tables is provided by Cohen[8] also for two values of α (.05, .01) for both one- and two-tailed tests. The method of Cohen has been implemented as the computer programs *SamplePower* by Borenstein et al.,[6] available from SPSS Inc., and *Power and Precision* by Borenstein et al.,[5] available from Biostat Inc. Bausell and Li[2] rely on a method that is very similar to the Cohen/Borenstein approach, and they adapt this to analysis of covariance as well as the conventional ANOVA. A normal approximation to the noncentral t distribution is described by Wahlsten[25,26] for continuous variables and tests of linear contrasts involving one degree of freedom. Unlike methods that rely on tabled values, it can be applied using only an electronic calculator with any reasonable choice of α, number of tails, desired power, and effect size.

Four methods are compared in Table 10.3 for the case of two independent groups when the effect size is large. The values from Bausell and Li[2] are the same as for the Borenstein et al. program. All methods yield reasonably similar values, with Kraemer and Thiemann[15] consistently being a larger number than Cohen.[8] Wahlsten[26] is generally between those two but is closer to Kraemer and Thiemann[15] for lower levels of power and to Cohen[8] for the higher levels of power. *SamplePower* by Borenstein et al.[6] yields values between those of Cohen and Wahlsten. Considering that the methods never yield values differing by more than three or four subjects at the higher levels of power and two subjects at the lower levels, all four are considered acceptable. The

TABLE 10.3
Four Estimates of Sample Size Needed to Detect a Large Effect Size of $\delta = 1.0$

				Sample Size in Each of Two Groups According to			
α	Tails	Power	$C_{\alpha,\beta}$	Kraemer and Thiemann[14]	Cohen[8]	Wahlsten[26]	Borenstein et al.[6]
.05	1	80%	6.18	15	13	15	14
.05	1	90%	8.56	20	18	20	18
.05	1	95%	10.82	25	22	24	23
.05	1	99%	15.77	36	32	34	33
.05	2	80%	7.85	19	17	18	17
.05	2	90%	10.51	24	22	24	23
.05	2	95%	12.99	30	27	28	27
.05	2	99%	18.37	41	38	39	38
.01	1	80%	10.04	23	22	23	22
.01	1	90%	13.02	30	27	29	28
.01	1	95%	15.77	36	33	34	33
.01	1	99%	21.65	48	45	46	45
.01	2	80%	11.68	27	25	26	26
.01	2	90%	14.88	34	31	32	32
.01	2	95%	17.81	40	37	38	38
.01	2	99%	24.03	53	50	51	50
	Mean of all 16 values:			31.3	28.7	30.1	29.2

Note: In the method of Wahlsten,[25,26] the normal approximation to the noncentral t distribution involves a constant $C_{\alpha,\beta}$ that depends on the standard normal deviates for Type I error (z_α) and Type II error (z_β), such that $C_{\alpha,\beta} = (z_\alpha + z_\beta)^2$. The sample size per group is then $n = C_{\alpha,\beta}/\delta^2 + 2$. The resulting value of n is rounded up to the nearest integer.

accuracy of any estimate of sample size will be undermined by small but inevitable subject attrition during any large study, and the researcher is well advised to plan on using a few more animals than the minimum indicated by a sample size calculation in order to be sure that the final number of good data points is adequate.

10.6 POWER AND SAMPLE SIZE FOR TWO INDEPENDENT GROUPS

The variable having the greatest impact on power and sample size is effect size. For an experiment with two groups (e.g., +/+ versus –/– genotypes), power curves for several values of effect size are shown in Figure 10.2. A family of curves for the case with $\delta = 0.5$ is shown in Figure 10.3. Using *SamplePower* or similar programs, it is possible to generate a series of power curves, print them, and then use the figure to determine sample size simply by drawing a line to the X axis from the point with the appropriate level of power. For effect sizes between those shown in Figure 10.2, interpolation between given values will be reasonably close to the desired value, although a quick calculation would be superior to interpolation.

Perhaps the most difficult thing for someone new to the field to decide is what effect size should be used to choose sample size. This will depend on two things. First, we want to know what effect sizes are commonly reported in published articles in this area of research, for example, in knockout studies. It would be reasonable to expect an effect size in the range of what other labs have observed; grant review and ethics committees are not likely to dispute such a figure.

Second, one must consider the stage of the research. If one is about to collect phenotypes on an entirely new knockout and really has no idea about what the effects may be or whether there may be noteworthy developmental compensation for the missing gene products, there would be no firm basis for expecting either a very large

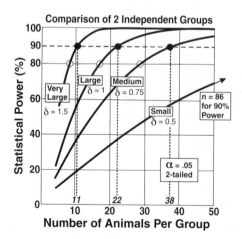

FIGURE 10.2 Power vs. sample size for a two-tailed *t* test of the difference between two independent groups when the null hypothesis is that the true group difference is $\delta = 0$. The sample size needed to yield a power of 90% is shown as an italicized numeral along the x axis.

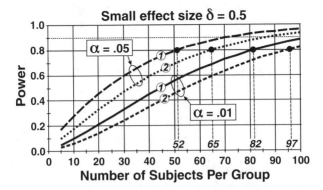

FIGURE 10.3 Power vs. sample size for a *t* test comparing two groups when the null hypothesis is $\delta = 0$ and the alternative is $\delta = 0.5$. Number of tails (1 or 2) for the test is indicated in the circle on each line. The sample size needed to yield a power of 80% is shown as an italicized numeral along the x axis.

or a small effect, and it would seem reasonable to choose sample size in order to have perhaps 90% power to detect a medium effect of $\delta = 0.75$. Usually we would expect a knockout to change a phenotype in a direction that makes the animal less viable, vigorous or competent, and a one-tailed test would be appropriate. If only one major phenotype is to be assessed, a choice of $\alpha = .05$ would make sense, but $\alpha = .01$ would be a better choice if the study will involve measures of several phenotypes, each to be assessed with a separate test of significance.

In the event that the investigator has only 10 or even fewer mutant animals for the first study, it is still important to attend to the power curve. For example, if the test will be done one-tailed with $\alpha = .05$, Figure 10.2 shows that power will be 90% only for a very large effect size, whereas power will be about 35% if the real effect size is $\delta = 0.75$. Thus, by considering the question of power, the investigator will be forewarned that failure to reject the null that $\delta = 0$ in no way proves that the knockout has no effect at all, because the chance of a Type II error occuring would be quite high for a medium or even large effect. The purpose of doing an initial study with fewer than 10 mutant animals and a similar number of normal sibs would be to evaluate the possibility of a severely abnormal phenotype. If such a dramatic deficiency were not found, it would be wise to use larger samples in future research with that particular knockout.

10.7 COMPARISON OF SEVERAL INBRED STRAINS

For many phenotypes, the variation among the more common inbred strains is very large and often exceeds $\omega^2 = 0.2$ or $f = 0.5$. The Mouse Phenome Project[19] (www.jax.org/phenome) has designated 10 strains on the highest priority A list for phenotyping, and these strains are chosen largely because of well-documented differences among them in genotype. Thus, large strain differences are not at all surprising. The Mouse Phenome Database (MPD) now contains a wide range of phenotypic data on these strains, and the investigator can rely on these values to guide an informed choice of f or ω^2 when estimating sample size to achieve a desired level of power. It would make good sense to base the hypothesized value on a phenotype that is in the same neural or behavioral domain as the planned study.

Once the value of ω^2 and then f have been decided and the number of strains is also stipulated, the sample size can be estimated from the tables of Cohen[8] or the program of Borenstein et al.[6] Figure 10.4 shows the power curves for an ANOVA on 8 inbred strains when their means are uniformly distributed from low to high. For example, if mice are tested on an elevated plus maze to assess anxiety levels, the average percent time in the open arms might range from 20 to 55% in steps of 5%. The population standard deviation of strain means would then be 11.5%, and when the standard deviation within strains is about 20%, $f = 11.5/20 = 0.57$, a very large effect. If the null hypothesis of no strain difference is to be evaluated with $\alpha = .05$, then only 10 mice per strain would be sufficient to achieve power of 90%. As the anticipated effect size declines, the necessary sample size becomes much larger.

The sample size is not solely a consequence of α, β and f. The number of strains also must be considered, as shown in Figure 10.5 where the effect size is $f = 0.35$

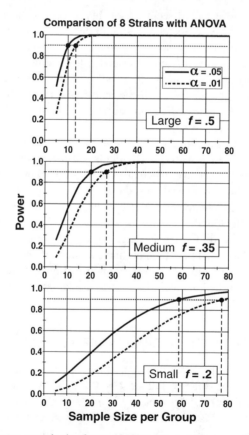

FIGURE 10.4 Power vs. sample size for an ANOVA comparing means of 8 strains of animals for three values of Cohen's f index of effect size. Black dots indicate the sample size for power of 90%.

in each case. For 4 strains, 31 mice per strain would suffice to yield power of 90%, whereas for 8 strains the sample size per group would be considerably less. When power is at issue, it is apparent that studying a smaller number of strains does not reduce the total number of animals greatly. For 10, 8, 6, and 4 strains, the total sample would need to be 170, 160, 144, and 124, respectively. For 2 strains, the sample would be considerably smaller, but the generality of the results would be severely restricted. It does not necessarily follow that a study of 10 strains will cost about the same amount as with 6 strains, because we usually begin the process of choosing strains with the most common ones that are also least expensive and then move to less commonly studied strains that cost more per animal.

10.8 COMPARING SPECIFIC GROUPS IN A ONE-WAY DESIGN

Sometimes the investigator employs a design with multiple groups, but cannot answer the central questions addressed by the design using the omnibus F test of statistical significance in the one-way ANOVA. Instead, the intent is to compare specific groups that differ in biologically interesting ways. Consider the example in Table 10.4 involving four groups of male mice from a backcross of the $A \times B$ F_1 hybrid to strain A. By convention, the first genotype in a cross is the mother, so that the $A \times B$ hybrid is from an A strain female bred with a B strain male. It is of some interest to know whether the four groups differ significantly, but there is deeper meaning in specific group comparisons. All four groups have the same complement of autosomal genes. The backcrosses $[(A \times B) \times A]$ and $[(B \times A) \times A]$ both have the Y chromosome from strain A and a F_1 hybrid maternal environment, but one has the mitochondrial DNA from strain A and the other from strain B. The backcrosses $[A \times (A \times B)]$ and $[A \times (B \times A)]$ differ in the strain contributing the Y chromosome. Groups $[(A \times B) \times A]$ and $[A \times (B \times A)]$ both have strain A mtDNA and strain A Y chromosome, but one has an inbred and the other a hybrid maternal environment. Table 10.4 proposes plausible results of a study in which all three non-Mendelian factors are important but to different degrees.

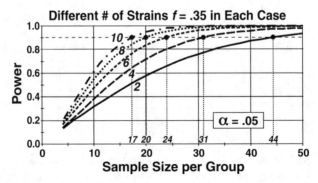

FIGURE 10.5 Power vs. sample size for an ANOVA on different numbers of strains when $f = 0.35$. The sample size needed to yield a power of 90% is shown as an italicized numeral along the x axis.

The method of contrasts[13] can be used to compare specific groups in a logical manner, each comparison being tested with a one degree of freedom t test. For a contrast among J independent groups with true means μ_j, each group is assigned a contrast coefficient c_j, such that $\Sigma c_j = 0$. The contrast value is then the sum of weighted means across all groups: $\psi = \Sigma c_j \mu_j$. For a comparison of any two groups, the coefficients could be +1 and −1 or +0.5 and −0.5; results will be the same. In Table 10.4 it is seen that the maternal environment contrast is twice that of the mtDNA, while the Y effect is midway between them. The method of Wahlsten[25,26] can be used to find the required sample sizes. For $\alpha = .05$ one-tailed and power of 95% ($\beta = .05$), the sample sizes per each of four groups need to be 51, 24, and 15 to detect the mtDNA, Y, and maternal environment effects, respectively. Presuming that the investigator is equally interested in the possible presence of all three kinds of effects, the sample size for the study should be 51 per group.

This logical approach to the dissection of variation among groups can be applied to large and complex genetic experiments with animals.[21,23]

10.9 STRAIN × TREATMENT EXPERIMENTS

Rarely does a researcher adopt a one-way design simply to compare several inbred strains. Often the strain or genotype factor is combined with some environmental treatment to create a two-way factorial design that will be analysed with an ANOVA. The ANOVA will partition the resulting variance among three sources: main effect for strain, main effect for environment, and strain × environment interaction. Each of these three effects will have a certain effect size that can be expressed in terms of Cohen's f or ω^2. In the case of ω^2, the published value will represent a partial ω^2 that compares variance for one factor to variance within groups while excluding from the calculation the variances arising from the other two factors.[9,13]

TABLE 10.4
Contrasts among Males of Reciprocal Backcrosses to Strain A; Agonistic Acts in the Home Cage Intruder Test

| Mother | Father | Mean | Contrast Coefficients to Detect Effect | | |
			mtDNA	Y Chromosome	Maternal Environment
A×B	A	17	1	0	1
B×A	A	15	−1	0	0
A	A×B	10	0	−1	0
A	B×A	13	0	1	−1
		ψ	2	3	4
		n per group	51	24	15

Note: Sample size is found using the method of Wahlsten[25,26] with a one-tailed test at $\alpha = .05$ and power of 95% when $\sigma_{within} = 3.0$. As in the case of two groups (Table 10.3), $C_{\alpha,\beta} = (z_\alpha + z_\beta)^2$. For a one degree of freedom contrast among J groups, $n = C_{\alpha\beta} \Sigma c_j^2 / (\psi/\sigma)^2 + 2$.

In order to estimate sample size for a two-way design when each group is to have the same sample size, one begins by positing cell means that would constitute an effect of interest to the investigator. The values should be the kinds of means that could realistically be observed for the strains, treatment, and phenotype in question, as suggested by preliminary data or the published literature. If the investigator is seriously interested in the possibility of a strain × treatment interaction, then the numerical model must express such an interaction. A computation is done to find the sample size needed to detect the interaction as well as the main effects. In almost all situations, the sample sizes needed to detect the three effects will be different, but the study can actually be done with just one sample size, so the best option is to choose the largest sample size needed to detect the smallest effect of interest. This will ensure that power to detect the other two effects will be even larger than the nominal level of power used in the calculation.

Consider the example shown in Table 10.5 where 4 strains reared in either the usual lab environment or an enriched environment are trained to locate the submerged platform in a water maze. Numbers represent mean latency to escape the water after 4 days of training. It is expected that the standard deviation of latencies within a strain will be about $\sigma = 8.0$ sec. In this case, there are major differences among strains, a substantial effect of environment, and a noteworthy strain × environment interaction. For strain S1 the environmental treatment has no effect at all, it improves latencies by 5 sec for strains S2 and S3, whereas it has a large effect of 10 sec on strain S4. This interaction appears to the researcher's eye as something important that ought to be detectable as a significant interaction effect by the ANOVA, provided enough animals are tested to confer adequate power on the test.

TABLE 10.5
Calculation of Effect Sizes from Hypothetical Results of 4 Strain × 2 Environment Design

	S1	S2	S3	S4	Row Mean	Deviation from G
A. Hypothetical Group Means with Interaction						
E1	10	15	20	25	17.5	+2.5
E2	10	10	15	15	12.5	−2.5
Column mean	10	12.5	17.5	20	G = 15	
Dev. from G	−5	−2.5	+2.5	+5		
B. Group Means under Null Hypothesis of Additivity						
	S1	S2	S3	S4	Row Mean	
E1	12.5	15	20	22.5	17.5	
E2	7.5	10	15	17.5	12.5	
Column mean	10	12.5	17.5	20		
C. Interaction Deviations of Cell Means in A from Additive Model in B						
	S1	S2	S3	S4		
E1	−2.5	0	0	+2.5		
E2	+2.5	0	0	−2.5		

The next step in the planning process is to determine the three effect sizes. This is relatively straightforward for the two main effects; find the mean of all values in a row or column and then find the population standard deviation of all means along one margin. For the strain effect, this will be the standard deviation of the four marginal means 10, 12.5, 17.5 and 20, which is $\sigma_M = 3.95$ or about 4. For the environmental treatment we find $\sigma_M = 2.5$. Thus, the effect sizes for strain and treatment main effects are $\mathbf{f} = 4/8 = .5$ and $\mathbf{f} = 2.5/8 = .31$, respectively. Sample sizes can be found from the tables in Cohen[8] or using the *SamplePower* program by Borenstein et al.[6] The tabular approach requires two steps because the power tables are set up for a one-way design. For $\alpha = .05$ and power of 90%, Cohen's[8] table 8.4.4 calls for 15 subjects per group when four groups have means with $\mathbf{f} = 0.5$. The full factorial study actually has $4 \times 2 = 8$ groups, however. Thus, the next step is to divide the 15 subjects per strain between the two treatment conditions, resulting in a final choice of 8 subjects per group. Similarly for the environmental factor, Table 8.4.4 calls for 57 subjects per group, but when these are distributed across the four strains, the choice is 15 per cell. Full details of the method to find cell sample size from tabled n are provided in Cohen.[8]

The *SamplePower* program uses the same shortcut for finding sample size for the main effects, while concealing the internal workings of the calculations from the user. One can enter means for the marginal values of each main effect, but it is not possible to enter means for all 8 cells separately. Consequently, the program will not compute the effect size for the interaction. Instead, the user must have a good understanding of the statistical definition of interaction in order to compute its effect size using either an electronic calculator, a spreadsheet such as Excel®, or a flexible data analysis program. The procedure is shown in Table 10.5.

The hypothesis of interaction is tested against the null hypothesis that strain and environment effects are additive. Statistically, interaction is defined as departures from additivity. For the example in panel A of Table 10.5, the grand mean G of all eight group means is 15 sec, and the deviations of the two environment means from this grand mean are +2.5 and –2.5 sec, these being the average effects of each environment across the 4 strains. For the strains, deviations from 15 range from –5 to +5 sec. To find the deviation of each cell mean that is expected when the main effects are additive, we simply add the separate deviations. For group S1E1, we add its environmental deviation +2.5 sec to its strain deviation –5 sec to obtain –2.5 sec below 15 or 12.5 sec. The results of these additions are shown in panel B of Table 10.5. Note that the marginal means in panel B are identical to those in panel A; the only difference is that panel A expresses strain × environment interaction, whereas panel B does not. Next, we find the deviation of each cell in panel A from the additive model in panel B and enter these interaction effects into panel C. Note that for strains S2 and S3, the means expected under an additive relation (panel B) are the same as proposed in panel A, while strains S1 and S4 deviate from the additive model. Interaction effect size is then the standard deviation of these eight interaction deviations ($\sigma_M = 1.77$) divided by $\sigma = 8$ to yield effect size $\mathbf{f} = 0.22$. This value can then be used to find sample size from either Cohen's table 8.4.4 or the *SamplePower* program where 0.22 must be entered into the appropriate box, yielding an estimated sample size of 38 animals in each of the eight groups.

The power curves for the three effects in the ANOVA are portrayed in Figure 10.6. As has been demonstrated for several plausible kinds of interactions,[24] at a given sample size the power to detect the interaction effect is substantially inferior to the power to detect the main effects. Likewise, to achieve the same level of power as for the main effects, the sample size to detect the interaction effect must be considerably larger. The exact ratio of sample sizes to detect interaction and main effects depends strongly on the specific kind of interaction. To date, no author has provided a satisfactory rule for judging what constitutes a small, medium and large interaction in neurobehavioral genetics. The only rule offered here is that the interaction should be interesting or noteworthy when an experienced investigator examines the pattern of cell means. Unfortunately, when power to detect an interaction is low, the pattern may look interesting to the educated eye, yet fail to achieve "significance" at $\alpha = .05$.

Bausell and Li[2] devote two chapters to power and sample size analysis for interactions in several kinds of factorial designs. They note that power to detect interaction tends to be lower than for main effects and that a formal power analysis is essential when the investigator is interested in the possibility of interaction effects. Murphy and Myors[17] observe: "In general, the study is most likely to have higher power for testing main effects and lower power for testing complex interactions," and they warn that "very large samples may be needed for detecting complex interaction effects."

10.10 BRIDGING THE GAP: THE 2×2 DESIGN

One variant of the strain × environment factorial experiment, the 2×2 design, lends itself to analysis by the method of contrasts, and for this situation a reasonable standard for what constitutes a noteworthy interaction can be established. The 2×2 design is often employed in neurobehavioral genetics to assess whether the effects of a knockout depend on the strain background (epistasis) or whether two genotypes respond similarly to a treatment (heredity–environment interaction). It is the design of choice when comparing the effects of the Y chromosome or the mitochondrial DNA from two strains backcrossed onto the opposite strain background. Using the

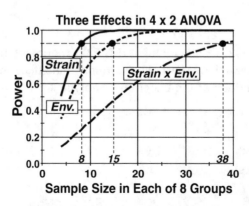

FIGURE 10.6 Power vs. sample size for main and interaction effects in an ANOVA for an experiment with 4 strains reared in 2 environments (Table 10.5). The sample size needed to yield a power of 90% is shown as an italicized numeral along the x axis.

method of Wahlsten,[25,26] it is possible to express the relative sample sizes needed to detect the main effects and the interaction in a most succinct way.

Table 10.6 presents a situation where both genotype and environment have clear effects on a phenotype but the effect size of environment on genotype 2 is *twice* as large as on genotype 1. This is surely an interaction effect that our analysis ought to be able to detect. As noted in Table 10.3, the sample size per group depends on a constant $C_{\alpha,\beta} = (z_\alpha + z_\beta)^2$ that is determined by our choice of Type I and II error probabilities. Provided that we plan to evaluate the main effects and interaction effect in the 2×2 experiment using t tests on contrasts with the same false positive rate α, the value of $C_{\alpha,\beta}$ will be the same for the genotype, environment and interaction effects. The sample size per group to detect a contrast is $n = C_{\alpha,\beta} \Sigma c_j^2 / (\psi/\sigma)^2 + 2$. The magnitude of the contrast ψ is considerably greater for the main effects than the interaction, but the standard deviation within groups σ is the same in each contrast. By expressing the contrast coefficients in Table 10.6 as ± 0.5, the term Σc_j^2 $= 1$ for all three contrasts. Thus, for the situation in Table 10.6 there are only two quantities that vary in the sample size equation, n and ψ. The ratio of the squared contrast terms ψ^2 for main effects and interactions is 9.0 when the treatment effect on one genotype is twice as large as the other in Table 10.6. For this numerical example, the required sample sizes for the genotype and genotype × environment interaction have a very simple relationship: $n_{G\times E} - 2 = 9(n_G - 2)$ or, equivalently, $n_{G\times E} = 9n_G - 16$. The value of n_G will depend on the effect size ψ/σ, being smaller when the within group variance is smaller, as shown in Figure 10.7. For generally large effects where the necessary n_G is quite small, the necessary sample size $n_{G\times E}$ needed to detect the interaction will be about six times larger than n_G, whereas for generally small effects where n_G is itself fairly large, $n_{G\times E}$ will be more than eight times larger than n_G.

The exact ratio of $n_{G\times E}$ to n_G or n_E will depend on the specific model of means for the four groups in the 2×2 design, but for most situations where the interaction is something we might reasonably expect to observe, the need for larger samples to detect the interaction than the main effects is evident in the algebra.[24,25,26]

TABLE 10.6
Genotype (G1, G2) by Environment (E1, E2) Experiment That Expresses a Noteworthy Interaction Where the Environmental Effect on Genotype 2 Is *Twice* the Magnitude of the Effect on Genotype 1

Group	G1E1	G1E2	G2E1	G2E2	Σc_j^2	$\psi = \Sigma c_j \mu_j$	ψ^2
Mean μ_j	10	20	20	40			
G effect c_j	−0.5	−0.5	+0.5	+0.5	1.0	15	225
E effect c_j	−0.5	+0.5	−0.5	+0.5	1.0	15	225
G×E c_j for interaction	−0.5	+0.5	+0.5	−0.5	1.0	−5	25

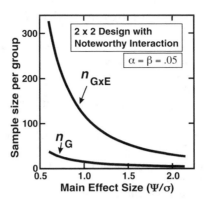

FIGURE 10.7 Sample size needed to yield power of 95% when $\alpha = .05$ for an experiment in which 2 strains are reared in two environments and the environmental effect on one strain is twice the size of the effect on the other strain, as shown in Table 10.6. Sample size needed to detect the interaction is substantially greater than to detect the main effects over a wide range of effect sizes.

10.11 SAMPLE SIZE FOR MORE SPECIALIZED EXPERIMENTS

The methods for determining sample size described here are generic and make no assumptions about the genetics of our subjects. Being generic rather than genetic, they can be applied in a wide variety of experiments, including many not addressed in this chapter. They will not be useful for certain kinds of studies such as detection and mapping of quantitative trait loci, and the reader will want to explore more specialized treatments for determining sample sizes for that purpose.[3,18] Neither do they inform the question of the number of inbred strains to use or the proper number of subjects per strain when assessing genetic correlation.[10]

ACKNOWLEDGMENTS

Preparation of this chapter was supported in part by grant 45825 from Natural Science and Engineering Research Council of Canada and grant 2 R01 AA012714 from the National Institutes of Health. The author is grateful to Elizabeth Munn for assistance.

REFERENCES

1. *The Oxford English Reference Dictionary*, Oxford, Oxford University Press, 1996.
2. Bausell, R.B. and Li, Y.-F., Power analysis for experimental research: a practical guide for the biological, medical, and social sciences, Cambridge, UK, Cambridge University Press, 2002.
3. Belknap, J.K., Mitchell, S.R., O'Toole, L.A., Helms, M.L., and Crabbe, J.C., Type I and type II error rates for quantitative trait loci (QTL) mapping studies using recombinant inbred mouse strains, *Behav. Genet.*, 26: 149–160, 1996.

4. Benjamini, Y., Drai, D., Elmer, G., Kafkafi, N., and Golani, I., Controlling the false discovery rate in behavior genetics research, *Behav. Brain Res.*, 125: 279–284, 2001.

5. Borenstein, M., Cohen, J., Rothstein, H., Schoenfeld, D., Berlin, J., and Lakatos, E., *Power and Precision*, Englewood, NJ, Biostat, 2001.

6. Borenstein, M., Rothstein, H., Cohen, J., Schoenfeld, D., and Berlin, J., *SamplePower 2.0*, Chicago, IL, SPSS, Inc., 2000.

7. Chow, S.L., *Statistical Significance: Rationale, Validity, and Utility*, London, Sage Publications, 1996.

8. Cohen, J., *Statistical Power Analysis for the Behavioral Sciences*, Hillsdale, NJ, Erlbaum, 1988.

9. Crabbe, J.C., Wahlsten, D., and Dudek, B.C., Genetics of mouse behavior: interactions with laboratory environment, *Science*, 284: 1670–1672, 1999.

10. Crusio, W.E., A note on the effect of within-strain sample sizes on QTL mapping in recombinant inbred strain studies, *Genes Brain Behav.*, 3: 249–251, 2004.

11. Egger, M., Smith, G.D., and Phillips, A.N., Meta-analysis. Principles and procedures, *British Medical Journal*, 315: 1533–1537, 1997.

12. Harlow, L.L., Mulaik, S.A., and Steiger, A.H., Eds., *What if There Were No Significance Tests?* Mahwah, NJ, Lawrence Erlbaum Associates, 1997.

13. Hays, W.L., Statistics, 4th edition, New York, Holt, Rinehart, Winston, 1988.

14. Hedges, L.V. and Olkin, I., *Statistical Methods for Meta-Analysis*, Orlando, Academic Press, 1985.

15. Kraemer, H.C. and Thiemann, S., *How Many Subjects? Statistical Power Analysis in Research*, Newbury Park, Sage Publications, 1987.

16. Lander, E. and Kruglyak, L., Genetic dissection of complex traits: Guidelines for interpreting and reporting linkage results, *Nat. Genet.*, 11: 241–247, 1995.

17. Murphy, K.R. and Myors, B., *Statistical Power Analysis: a Simple and General Model for Traditional and Modern Hypothesis Tests*, Mahwah, NJ, Lawrence Erlbaum Associates, 2004.

18. Neumann, P.E., Three-locus linkage analysis using recombinant inbred strains and Bayes' theorem, *Genetics*, 128: 631–638, 1991.

19. Paigen, K. and Eppig, J.T., A mouse phenome project, *Mamm. Genome*, 11: 715–717, 2000.

20. Rosenberg, M.S., Adams, D.C., and Gurevitch, J., *MetaWin Version 2.0*, Sunderland, MA, Sinauer Associates, 1997.

21. Sokolowski, M.B., Genetic analysis of behavior in the fruit fly, *Drosophila melanogaster*, in *Techniques for the Genetic Analysis of Brain and Behavior. Focus on the Mouse*, D. Goldowitz, D. Wahlsten, and R.E. Wimer, Eds. Amsterdam, Elsevier, 497–512, 1992.

22. Thomas, L. and Krebs, C.J., A review of statistical power analysis software, *Bulletin of the Ecological Society of America*, 78: 126–139, 1997.

23. Wahlsten, D., Behavioral genetics and animal learning, in *Psychopharmacology of Aversively Motivated Behaviors*, H. Anisman and G. Bignami, Eds. New York, Plenum, 63–118, 1978.

24. Wahlsten, D., Insensitivity of the analysis of variance to heredity–environment interaction, *Behav. Brain Sci.*, 13: 109–120, 1990.

25. Wahlsten, D., Sample size to detect a planned contrast and a one degree-of-freedom interaction effect, *Psychol. Bull.*, 110: 587–595, 1991.

26. Wahlsten, D., Experimental design and statistical inference, in *Molecular-Genetic Techniques for Behavioral Neuroscience,* W.E. Crusio and R.T. Gerlai, Eds. Amsterdam, Elsevier, 40–57, 1999.

11 The Role of Association Studies in Psychiatric Disorders

Nicolas Ramoz and Philip Gorwood

CONTENTS

INTRODUCTION

In the beginning of the twenty-first century, the fiftieth anniversary of the discovery of DNA was celebrated, and despite the rapid progress in both human genetics and molecular biological technologies since, neurobehavioral genetics has provided us with only its first promising successes. The human genome has a physical size of 3×10^9 base pairs that encode and regulate approximately 32,000 genes. Some of these genes play a role in the aetiology of psychiatric disorders. However, the model of segregation for these disorders does not follow a simple Mendelian trait but follows complex traits where both polygenetic and environmental factors influence the susceptibility to neurobehavioral diseases. The use of association studies to complement linkage analyses provides a powerful tool to identify the genetic component in psychiatric disorders.

11.1 ASSOCIATION STUDIES: DEFINITIONS AND MECHANISMS

Association studies compare the occurrence in a population (case and control) or kindred (family based) of two or more features (usually phenotype vs. genotype) with a frequency greater than that anticipated on the basis of chance alone. Association studies are widely used in genetics to search for the involvement and/or the localization of a gene in the risk of certain disorders.

11.1.1 CASE CONTROL ASSOCIATION STUDY

The case control association study analyzes the distributions of allele frequencies or genotypes in different groups of unrelated patients, with and without a certain phenotype, to those from matched controls represented by unaffected individuals. The case/control study can be performed with one marker or a combination of markers (haplotypes). The χ^2 test determines the existence of a significant association, usually with a fixed 5% risk of type I error. These parameters are explained in Table 11.1.

The measure of association can be both qualitative, with a presence or an absence of a significant association, and quantitative, with a strength of association evaluated by risk difference (RD), relative risk (RR), odds ratio (OR), and attributable risk (AR). The RD computation evaluates the presence of the genetic marker in the affected group, while taking into account the existence of subjects without the phenotype who also carry the marker. The RR calculation is a ratio of these probabilities,

---

TABLE 11.1
Measures of Association

Phenotype	Genetic Marker		
	Present	Absent	Total
Affected	a	b	a+b
Controls	c	d	c+d
Total	a+c	b+d	N

Notes: χ^2 (qualitative approach) = $\Sigma(\text{obs}-\text{calc})^2/\text{calc}$. Where a, b, c, d are the observed (obs) numbers, and (a+b)*(a+c)/N is the calculated (calc) value for a, and so on for b, c, d.

RD (risk difference) = [a/(a+b)]–[c/(c+d)]; RR (relative risk) = [a/(a+c)] / [b/(b+d)]; OR (odds ratio) = ad/bc; AR (attributable risk) = [a/(a+c)]*[(OR–1)/OR].

---

instead of a difference, whereas, the OR value is the ratio of odds for each population. The significance of the association can be computed from the OR.[1] Finally, AR provides the strength of the association and reflects the impact of the allele in explaining the phenotype, since it is dependent on the OR and also the allele frequency in the general population.

The case/control association study is a useful tool that provides a powerful statistical test. However, one major limitation of association analysis can be caused by population stratification, i.e., when cases and controls are not ethnically matched.

11.1.2 FAMILY-BASED ASSOCIATION STUDIES

There are several other approaches that can be used to address the stratification bias, and all use family-based association studies.

Thus, the haplotype relative risk (HRR) test provides OR by comparing allele frequencies of a marker or a haplotype in affected offspring with frequencies in a "virtual control" builds with parental alleles not transmitted to affected offspring (Figure 11.1).[2,3]

The transmission disequilibrium test (TDT) is another approach that considers parents who are heterozygous for an allele associated with the disease, and evaluates the frequency of transmission vs. nontransmission of marker alleles to affected offspring.[4] TDT is also considered as being a test for linkage in presence of association, especially for fine-mapping.[5] Several tests were developed to include multi-allelic markers and haplotypes, such as the extended TDT (ETDT).[6] The major limitation of TDT is the use of complete triads because families with one missing parent could introduce a bias. To counter this problem, the sib-transmission disequilibrium test (S-TDT) was designed by Spielman and Ewens.[7] This test compares the allele frequencies between affected and unaffected siblings in order to measure difference. Since complex disorders may not be characterized by dichotomous traits, without covariates, test of association for quantitative traits (QTDT) was designed for nuclear families of any size, with or without the parental phenotype.[8] One again, statistical power of QTDT is greater than that with the genotyping of parents. The information in pedigrees of any size, in particular the large or "extended" pedigrees, could be taken into account by using the pedigree disequilibrium test (PDT).[9] The PDT provides the statistical significance of linkage disequilibrium for the entire pedigree. Of the tests discussed so far, the PDT appears to be the most powerful.[9] Furthermore, additional statistical tests based on likelihood approaches, and not on χ^2 test, are also available.

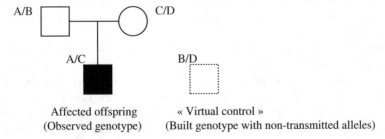

Affected offspring   « Virtual control »
(Observed genotype)   (Built genotype with non-transmitted alleles)

FIGURE 11.1 The haplotype relative risk (HRR) method.

These tests include TRANSMIT program, which can compute association for marker(s) or haplotype(s) to affected subjects or to one randomly selected from families, either in the to absence of parental genotypes or to unknown haplotype phase.[10] These family-based association tests are now available in packages, e.g., FBAT.[11]

11.2 ADVANTAGE OF ASSOCIATION STUDIES

To identify the impact of markers, haplotypes, regions, or candidate genes in the risk for a disorder, the association studies are based on strategy of comparing correlations between phenotype and genotype in case/control or family-based populations.

11.2.1 POWER

Whole genome scans followed by genetic linkage analyses are considered to be the most powerful strategies to identify genes of heritable diseases. Linkage analyses are successfully used for genetic disorders harboring Mendelian inheritance patterns, but may not be applicable for common diseases that have a more complex pattern of inheritance. Linkage analyses present several methodological limitations for determining polygenic disorders when the assumption of a single major gene is incorrect, or genetic heterogeneity and sporadic cases are present, or penetrance of susceptibility genotypes is reduced, or when the genetic aetiology is not well defined.[12] These features all characterize the genetics of animal and human neurobehavior, and therefore association studies may be more relevant.

In fact, linkage analyses are comparisons of the likelihood to observe a link between markers and loci (transmitted together with greater frequency than 50%), while association studies compare discrete traits and alleles (found together with greater frequency than by chance alone). Linkage analyses attempt to detect physical mapping on the chromosome while association studies look for a causal relationship. Linkage analyses also need the screening of multigenerational families, while association studies require only the studying of unrelated individuals within a population, or triad families. According to these characteristics, association studies should be the preferred initial choice rather than linkage studies. Association studies are more appropriate for polygenic (many additive genes, each with a small effect), multiloci (two or more genes involved), multifactorial (epigenetic factors, such as environment, play a significant role in the individual vulnerability), heterogeneous (phenotype contains different entities that are dependent on several factors) disorders or traits.

Furthermore, the power of association studies vs. linkage studies to detect a significant role of a gene involved in complex human disease has been compared recently.[13] When the genotypic RR is low (<2.0), then the number of families needed to detect a significant linkage is so large that it becomes impractical to achieve (12,000 to 300,000 for allele frequencies between 0.01 and 0.50). In comparison, the number of triplets to be investigated with the TDT method is much lower (340 to 5800), for allele frequencies between 0.01 and 0.50. Simulations of the power of association studies can be performed with the recent genetic power calculator program (GPC; http://statgen.iop.kcl.ac.uk/gpc/).[14] These results have shown that association studies are relevant for genetic susceptibility or liability, but not for genetic determinism in complex genetic diseases, such as neurobehavior disorders.

11.2.2 LINKAGE DISEQUILIBRIUM

Association studies are also statistical tools that can be used to test the allele frequencies of some genetic polymorphisms that are potentially close to and in linkage disequilibrium (LD) with the gene involved in the feature analyzed. This analysis is therefore dedicated to localization studies, whether for systematic screening of the whole genome or for a fine-mapping in specific candidate regions of interest.

A certain phenotype can be influenced by a specific allele, and the polymorphisms of a closed genetic marker should be associated with this phenotype if they are in LD. Association studies could be used in conjunction with LD to perform to human genetic analysis. In the absence of selection, population heterogeneity, or other confounding phenomena, LD should by theory be linearly related to distance. The utilization of linkage disequilibrium concepts in association studies has some limitations that could be partially raised, such as the gap between physical and linkage distance.

Single nucleotide polymorphism (SNP) is a recently developed tool in molecular biology and genetics. SNPs are a useful and powerful tool. Genotyping of SNPs is simple, efficient, quick, easily automatizable, and cheap. It is powerful because the resolution with SNPs is 10 times higher than microsatellites, short tandem repeats (STRs), or variable number of tandem repeats (VNTRs). SNPs are present in the human genome once every 600 base pairs. Typically, SNPs segregate as specific combinations or haplotypes, or blocks that harbour a high linkage disequilibrium and a limited diversity.[15] In other words, few SNPs are request and sufficient to screen for association between a disorder trait and a gene. The choice of the SNPs and their information regarding their allelic frequencies and linkage disequilibrium in populations are available in databases, in particular with the international consortium HAPMAP (www.hapmap.org). Looking for a strong and specific excess of an allele, genotype or haplotype in a group of subjects compared to unaffected matched controls according to the presence of linkage disequilibrium allows to the identification of disease associations between genes and human traits. However, it is exposed to certain limitations.

11.3 LIMITATIONS OF ASSOCIATION STUDIES

A major problem in association studies is false positives. They occur when the populations of case controls are not matched ethnically, and when there is a lack of parent data in family-based or with a low frequency of the minor allele of the genotyped marker. The other limitation is the statistical power of the association study.

11.3.1 FALSE POSITIVES: LINKAGE DISEQUILIBRIUM AND CONTROL GROUPS

The genetic heterogeneity of a population is directly linked to the linkage disequilibrium between variants or markers in the genome. Thus, the relation between physical distance and linkage disequilibrium depends on so-called "hot" or "cold" spots of recombination that can be detected with frequent, or low, recombinations on small distance of chromatides.[16] Telomeric regions present a higher frequency of recombinations, particularly to centromeric regions, than the rest of the genome. Furthermore, the decay of linkage disequilibrium in succeeding generations is governed not only by the recombination frequency (i.e., the frequency of crossovers between the relevant loci during meiosis), but also the number of generations since the introduction of the mutation into the population.[17] For example, the "half life" of an association would be 69 generations or about 2,000 years at a recombination fraction of 0.01.[18] Nevertheless, the shared segments extended over large genetic distances (1 megabases) in a Finnish sample established over 2,000 years ago.[19] Thus, linkage disequilibrium may not be used as a constant and uniform concept. In conclusion, linkage disequilibrium studies should be more productive among genetically isolated populations with a well-documented history.[20]

The second limitation of association studies is based on the required comparison of cohorts (affected vs. unaffected). First of all, association studies have to control the stratification effect, i.e., one of the samples, affected or controls, derived from an ethnic group sharing other DNA variants for reasons unrelated to illness. Control and affected cohorts have thus to be matched for ethnic origins. This kind of limitation is now well controlled and measured by genotyping specific markers across the genome and is likely independent of the goal of the study. The control group based on volunteers in human association studies is also subject to bias as volunteer subjects have rates of psychopathology that far exceed the population expectations.[12,21–23] The use of unrelated relatives, spouses, is also a potential artefact because they are more frequently affected (assortative mating).[24] The other way is to perform family-based association study when parental data are available, as we described in Section 11.1.2.

The limitation may come from the proband to study. This question is still conflicting. For some authors, few severely affected probands from numerous familial cases are required; for others, much more representative affected subjects are sufficient. The question is also unresolved for controls cohort: (1) extracted from the general population, regardless of their phenotype (if the mutation analyzed is involved, affected probands should be different from the whole population), (2) extracted from the general population, but excluding affected subjects, and (3) without any potentially linked disease, in the proband interview and also in his first (and sometimes second) degree family. This last cohort of controls (labeled "super-control") has been criticized as favoring the observation of coaggregation of different diseases in the affected group.[25]

Finally, in multifactorial disorders and complex traits, the existence of incomplete penetrance (i.e., subjects with the vulnerability gene that do not express the

TABLE 11.2

Estimation of Each Sample Size Needed to Show a Significant Association According to Allele Frequency in Controls and the Expected Difference between Patients and Controls

Allele Frequency in Control (N)	Expected Difference in Allele Frequency between Patients and Controls							
N	1%	2%	5%	10%	20%	30%	40%	50%
1%	2578	822	207	80	32	19*	13*	10*
2%	4332	1270	284	100	38	22	15*	11*
5%	9312	2515	485	223	68	34	21	14*
10%	16862	4386	778	223	68	34	21	14*
20%	29254	7442	1249	334	92	44	26	17*
30%	39052	9598	1574	407	106	48	27	17*
40%	43259	10854	1754	443	111	48	26	14*
50%	44875	11214	1790	443	106	44	21	8*

*= Sample size lower than 20 subjects requires special test corrections.

phenotype) renders the choice of the control group even more critical. For example, for research on a common disorder such as alcoholism, it may be more relevant to recruit controls that drink alcohol regularly without developing alcohol abuse or dependence, instead of abstainers, who may quickly develop alcoholism if they were exposed to alcohol consumption.

11.3.2 STATISTICAL POWER: ACCEPT OR REJECT AN ASSOCIATION

The statistical power and specificity of the association detected or rejected are other problems. Small sample size and low heterogeneity of allele frequencies are important limitations in the opportunity to show linkage disequilibrium between two closely linked markers.[26] The chances of detecting the association are reduced if the allele conferring susceptibility is common among unaffected individuals.[27] The sample size required to detect a significant association can be evaluated (Table 11.2). These estimations are relevant for testing a single hypothesis (relationship between a marker and a phenotype), without prior notion of the type of association (association or exclusion), and with low type I and type II errors ($\alpha = 5\%$ and $\beta = 10\%$). The number of patients that have to be included are given according to the frequencies of the studied allele in the control population (from 0.01 to 0.50) and the expected difference in allele frequency between patients and controls. These power computations of association studies can now be simulated by investigators prior to promoting any cohort or performing any genetic screening with the genetic power calculator program (GPC; http://statgen.iop.kcl.ac.uk/gpc/).[14]

11.3.3 CONSEQUENCES IN NEUROBEHAVIORAL GENETICS

Statistical corrections may be applied to assess the question of false-positive and the power of association studies to accept or reject the hypothesis.

Thus, the importance of the likelihood of false-positive results in the genetics of behavior and psychiatric disorders, on the basis of the number of candidate genes (namely all genes expressed in the brain) could be raised.[28] Bonferroni's correction could be performed by dividing the 0.05 first risk error by the number of markers screened, and a statistical significance of 0.00001 should be achieved to avoid a false-positive rate of 5%. Using a significant level of 0.0001, 80% of positive findings would be false, and the traditional α level of 0.05 would yield 99.5% false-positive results. According to such a false-positive rate and considering that half of the genes are expressed in the brain, there are potentially 100,000 association studies which could be carried out for each neurobehavioral trait studied. Regarding such a nearly impossible-to-reach level of significance, alternative approaches have been proposed. However, consistent replications, on independent cohorts, would be the best evidence for a true association.[29]

Considering all the advantages and limitations of the methodology previously described for association studies, the role and the impact of these approaches need to be discussed further and compared with the different tests and statistical corrections. One of the frequent criticisms of positive association studies is the lack of reproducibility. Some journals propose to refuse any more positive case-control genetic association studies of complex traits.[30] Association studies have a lower chance to throw light on the biological mechanism underlying the association. A major disease gene at another locus in linkage disequilibrium with the associated marker locus or a minor gene possibly located at the marker locus itself cannot be distinguished. Association studies are nevertheless extremely powerful in detecting small, partial or complex genetic effects, although their high sensibility is associated with low specificity. Often, association studies are in fact specifically required, and different approaches and tests may easily increase their validity. There are family-based association studies described, in Section 11.1.2, that do not request control group but needed parental data. Instead of preferring sensibility to specificity, a step-by-step strategy could be chosen, especially as it is in accordance with the recruitment of affected subjects. Beginning studies with a case-control analysis is strong, but the recruitment of parents may be done later. DNA from parents can be used to see if the allele found in excess in the affected sample is more frequently transmitted to the affected proband than not transmitted from the parents. Affected sibs can be later recruited (sibpair analysis), and once multiplex families are detected, linkage studies are easier to perform (LOD score linkage analysis). Combined with high-throughput genotyping of markers, including SNPs, case-control or family-based association studies allow identification of the genetic component in neurobehavioral disorders.[31-33]

11.4 PROGRESS IN NEUROBEHAVIORAL GENETICS

Apart from the apparent absence of monogenic inheritance, even in some families, there are three main reasons explaining how no definite results were found in psychiatric genetics, although tremendous efforts have been made to depict which genes are involved. The case-control association study approach may be partially helpful in each case.

11.4.1 One Gene ... Many Phenotypes (the Case of Phenotypical Heterogeneity)

International and validated classifications (such as ICD or DSM) of psychiatric disorders are mainly based on clinical intuition, are partly influenced by epidemiological studies, and depend, to a moderate degree, on treatment response. But none of the decisive factors are relying on biological aspects, for an obvious lack of clear results. But classifications based on biological specificities are much easier to use when looking for genes. A good example of this problem in psychiatry is the distinction between mood depressive disorder (MDD) and generalized anxiety disorder (GAD). Indeed, the vulnerability factors involved, and even more specifically for the genes involved, are much more shared than specific. The proposition has even been made that GAD and MDD are two different aspects of the same disorder, an example of genetic pleiotropy.[34]

At the molecular genetic level, the gene which has been the most frequently analyzed is, without a doubt, the gene coding for the serotonin transporter (*5-HTT*). From the year 2000, more than 500 studies analyzed this gene in more than 20 different phenotypes, mainly psychiatric disorders, traits or temperaments. Some of these findings are now convincing, for example regarding the role of the short allele in the risk of suicidal behavior.[35-36] Yet, it is difficult to disentangle at which level the short allele of the *5-HTT* gene is increasing the risk for psychiatric morbidity, as this allele has been found associated at least once in each step of mood disorders, including the temperament: "harm-avoidance,"[37] which is a well-known risk factor for depression; general anxiety disorder,[38] which has a high comorbidity with mood disorders; major depressive disorder and bipolar disorder per se;[39-40] suicidal behavior,[35-36] which is a relatively frequent complication of the latter, and antidepressant treatment response.[41]

An alternative approach is to focus on intermediate phenotype.[42] Such phenotype might be closer to vulnerability allele (as relying on simpler biological and physiological mechanisms) and are directly related to the concept of genetic vulnerability to the disorder (as, for example, being found in excess in nonaffected relatives of the proband). The well-known role of the amygdala in anxiety has been assessed regarding the role of the *5-HTT* gene. An experimental paradigm was thus used, in showing the face of a very anxious person, and looking at which part of the cortex was activated. This study showed that subjects with the short allele are activating the amygdala, contrary to those who do not have this vulnerability allele.[43] As serotonin has a core role in balancing internal vs. external stimuli, and as the amygdala hyperactivity is probably a core mechanism in anxious disorder, such case-control approach might be particularly more powerful to depict the genes involved in such a complex disorder with relatively low heritability.

11.4.2 One Phenotype ... Many Genes (the Case of Genetic Heterogeneity)

There are not many psychiatric disorders for which some evidence exists about the clear involvement of some vulnerability genes. Autism might be one of them. Jamain et al.[44] reported mutations in the X-linked neuroligin 3 and 4 genes in Swedish sibpairs with autism. A frameshift mutation in *NLGN4* appeared *de novo* in the mother, cosegregated with an affected brother with Asperger syndrome and was absent in a normal brother. This frameshift mutation was not present in 600 unrelated control X-chromosomes. A mis-sense mutation in *NLGN3*, was found in the mother and two sibs, one with autism and another with Asperger syndrome, but no other relatives were studied. It was not found in 300 unrelated control X-chromosomes. Laumonnier et al.[45] reported later on a large French family in which 10 males had nonspecific X-linked mental retardation, two had autism, and one had pervasive developmental disorder. All affected patients were found to have the same frameshift mutation (1253delAG) in the *NLGN4* gene. A third study scanned and sequenced 2.5 Mb of the same gene in 150 patients with autism, and revealed an association of *NLGN4* structural variants at highly conserved amino acids with an estimated attributable risk for autism of about 3% in these cohorts.[46]

Thus, important evidence was in favor of the conclusion that the first vulnerability genes in autism were found. On the other hand, screening a large set of individuals affected with autism (N = 196) did not reveal any mutation for these two genes associated with the disorder,[47] showing that many mutations are probably involved in autism. Using case-control or within-family association studies is an important tool to cope with the difficulty of finding rare mutations.

11.4.3 INTERACTION BETWEEN THE ENVIRONMENT AND GENETIC VULNERABILITY

Case-control association studies also offer a unique opportunity to test the interaction between genes and the environment. Archibold Garrod[48] was probably one of the first to raise the importance of the interaction between environment and genetics when studying complex disorders, response to treatment or even pure genetic disorders. Assessing such potential interaction may be particularly important in psychiatric disorders for which various stressors are nearly systematically found before the onset of the disease.

First of all, the estimate of the population's attributable risk for genetic, as well as for environmental, factors might be biased when such interaction is not taken into account.[49] If a gene has an important but specific impact on the risk for a disorder when, and only when, a specific environmental condition is present, then we could expect a large variability in findings, according to the prevalence of the environmental conditions. A good example to highlight this problem is the risk for aggressive behavior at adult age. As antisocial, aggressive and impulsive behaviors belong to the antisocial personality disorder spectrum, a personality disorder that has substantial heritability (around 50%), a cohort study analyzed the role of the gene coding for the MAO-A (the monoamine oxydase-A degrades all monoamines, including serotonin) in the risk of aggressive behavior at adulthood.[50] The rare allele (which codes for a hypoactive enzyme) was indeed associated with an increased risk of aggression at adulthood, but specifically for subjects who were exposed to violence during childhood. In this example, it is the assessment of both these two factors that allowed the possibility to show how genes (in this case the MAO-A gene) and environment (here aggression from the parents during childhood) are significantly involved in the risk of aggression in adults.

A second interesting aspect is the greater chance to discover which environmental factors are interacting with genes, in order to propose a prevention strategy. For example, the role of cannabis in schizophrenia has been the topic of numerous discussions, the difficulty being to disentangle the role of cannabis as a marker (i.e., those who will ultimately develop schizophrenia are more prone to use cannabis) or a risk factor (i.e., those who use cannabis increase their risk for schizophrenia). One study, focusing on gene–environment interactions, showed that cannabis is a risk factor, but mainly for a subgroup of at-risk patients as having familial history of psychosis.[51]

The adequate way to assess the existence of a gene–environment interaction depends on the selected sample. The golden standard would be based on the prospective collection of both environmental and lifestyle data and DNA, if (1) the disease, (2) the vulnerability allele(s), and (3) major environmental risk factors are frequent enough. Collecting DNA at the end of the cohort would expose to stratification bias when DNA is not obtained from all patients and controls. For rare disorders, such as autism, a case-control approach assessing gene–environment interaction retrospectively might be more appropriate, although more exposed to different risks. Such biases could include unnatural selection (for more severe or treated patients), population stratification, survivor bias (suicide being a major problem for bipolar disorder for example) and imperfect recall (age at onset and number of lifetime major depressive episodes are sometimes difficult to remember for old depressed patients).

Statistically, there are two ways to demonstrate a significant gene–environment interaction. Table 11.3 could be used, defining a, b, c and d as the frequency of affected patients, taking into account presence vs. absence of the vulnerability factor and/or the vulnerability allele. Testing the departure from the multiplicative model of interaction is the most frequently used way, comparing the relative risk of the population b (exposed to both factors) from the relative risk of a (only environmental risk) multiplied by the relative risk of d (only genetic risk). Relative risks are created with the reference category c for which the relative risk is, by definition 1 (no exposition), and using the 95% confidence interval of each relative risk to assess the significance of the difference between these two relative risks. The alternative (the additive model) is to use rate differences, the effect of genes (frequency d) plus the effect of environment (frequency a) being smaller than the effect of both factors (b).

TABLE 11.3

Assessing Gene–Environmental Interaction on the Basis of the Percentage of Affected Patients Exposed (vs. Unexposed) to Each Factor

Environmental Exposure	Genotype	
	Wild Type	Variant
Exposed	a	b
Unexposed	c	d

11.5   CONCLUSIONS

Association studies have one of the highest risk of false-positive findings, are exposed to numerous bias because of the selection of probands (frequently variable in different studies, even when having the same diagnosis) and controls [from "too healthy" (super-normal controls) to "to close with the analysed phenotype" (general population)], are extremely sensitive to the problem of mismatching controls with cases for ethnic background (stratification bias), and are particularly exposed to the problem of the nonlinearity of linkage disequilibrium through the genome. Yet, the association study approach is one of the most powerful, i.e., being able to discover the role which has a small attributable risk, and has many advantages that are summed up in this chapter. Thus, this approach should be used cautiously, in conjunction with more specific techniques, and be more appropriate for complex traits, such as polygenic (many additive genes each with a small effect), multiloci (two or more genes involved), multifactorial (epigenetic factors such as environment play a significant role in the individual vulnerability), heterogeneous (phenotype contains different entities which are dependent on several factors) disorders or traits. As all these characteristics may be particularly true for neuropsychiatric disorders, association studies still have a role to play.

REFERENCES

1. Mantel, N. and Haenzel, W., Statistical aspects of the analysis of data from retrospective studies of disease, *J. Natl. Cancer Inst.*, 22, 719, 1959.
2. Falk, C.T. and Rubinstein, P., Haplotype relative risk: an easy reliable way to construct a proper control sample for risk calculation, *Ann. Hum. Gen.*, 51, 227, 1987.
3. Ott, J., Statistical properties of the haplotype relative risk, *Genet. Epidemiol.*, 6, 127, 1989.
4. Spielman, R.S., McInnis, R.E., and Ewens, W.J., Transmission Test for Linkage Disequilibrium: The insulin gene region and insulin-dependent diabetus mellitus (IDDM), *Am. J. Hum. Genet.*, 52, 506, 1993.
5. Spielman, R. and Ewens, W., The TDT and other family-based tests for linkage disequilibrium and association, *Am. J. Hum. Genet.*, 59, 983, 1996.
6. Sham, P.C. and Curtis, D., An extended transmission/disequilibrium test (TDT) for multi-allele marker loci, *Ann. Hum. Genet.*, 59, 323, 1995.
7. Spielman, R.S. and Ewens, W.J., A sibship test for linkage in the presence of association: the sib transmission/disequilibrium test, *Am. J. Hum. Genet.*, 62, 450, 1998.
8. Abecasis, G.R., Cardon, L.R., and Cookson, W.O., A general test of association for quantitative traits in nuclear families, *Am. J. Hum. Genet.*, 66, 279, 2000.
9. Martin, E.R. et al., A test for linkage and association in general pedigrees: the pedigree disequilibrium test, *Am. J. Hum. Genet.*, 67, 146, 2000.
10. Clayton, D., A generalization of the transmission/disequilibrium test for uncertain-haplotype transmission, *Am. J. Hum. Genet.*, 65, 1170, 1999.
11. Laird, N., Horvath, S., and Xu, X., Implementing an untried approach to family based tests of association, *Genet. Epidemiol.*, 19, S36, 2000.
12. Risch, S. et al., Ensuring the normalcy of "normal" volunteers, *Am. J. Psychiatry*, 147, 682, 1990.

13. Risch, N. and Merikangas, K., The future of genetic studies of complex human diseases, *Science*, 273, 1516, 1996.
14. Purcell, S., Sham, P., and Daly, N.J., Parental phenotypes in family-based association analysis, *Am. J. Hum. Genet.*, 76, 249, 2005.
15. Patil, N. et al., Blocks of limited haplotype diversity revealed by high-resolution scanning of human chromosome 21, *Science*, 294, 1719, 2001.
16. Chakravarti, A. and Nei, M., Utility and efficiency of linked marker genes for genetic counseling. II. Identification of linkage phase by offspring phenotypes, *Am. J. Hum. Genet.*, 34, 531, 1982.
17. Jorde, L.B., Linkage disequilibrium as a gene-mapping tool, *Am. J. Hum. Genet.*, 56, 11, 1995.
18. Owen, M. and McGuffin, P., Association and linkage: complementary strategies for complex disorders, *J. Med. Genet.*, 30, 638, 1993.
19. Peterson, A.C., et al., The distribution of linkage disequilibrium over anonymous genome regions, *Hum. Mol. Genet.*, 4, 887, 1995.
20. Nimgaonkar, V., In defense of association studies, *Mol. Psychiatry*, 2, 275, 1997.
21. Gibbons, R., Davis, J., and Hedeker, D., A comment on the selection of healthy controls for psychiatric experiments, *Arch. Gen. Psychiatry*, 47, 785, 1990.
22. Shtasel, D. et al., Volunteers for biomedical research. Recruitment and screening of normal controls, *Arch. Gen. Psychiatry*, 48, 1022, 1991.
23. Buckley, P. et al., Schizophrenia research: the problem of controls, *Biol. Psychiatry*, 32, 215, 1992.
24. Suarez, B.K. and Hampe, C.L., Linkage and association, *Am. J. Hum. Genet.*, 54, 554, 1994.
25. Kendler, K., The super-normal control group in psychiatric genetics: possible evidence for coaggregation, *Psychiatric Genetics*, 1, 45, 1990.
26. Greenberg, D.A., Linkage analysis of 'necessary' disease loci versus 'susceptibility' loci, *Am. J. Hum. Genet.*, 52, 135, 1993.
27. Cox, N.J. and Bell, G.I., Disease associations. Chance, artifact, or susceptibility genes?, *Diabetes*, 38, 947, 1989.
28. Crowe, R.R., Candidate genes in psychiatry: an epidemiological perspective, *Am. J. Med. Genet.*, 48, 74, 1993.
29. Kidd, K.K., Association of disease with genetic markers: deja vu all over again, *Am. J. Med. Genet.*, 48, 71, 1993.
30. Paterson, D., Case control association studies in complex traits-the end of an era?, *Mol. Psychiatry*, 2, 277, 1997.
31. Hodge, S.E., What association analysis can and cannot tell us about the genetics of complex disease, *Am. J. Med. Genet.*, 54, 318, 1994.
32. Baron, M., Association studies in psychiatry: a season of discontent, *Mol. Psychiatry*, 2, 278, 1997.
33. Inoue, K. and Lupski, J.R., Genetics and genomics of behavioural and psychiatric disorders, *Cur. Op. Gen Dev.*, 13, 303, 2003.
34. Gorwood, P., The generalized anxiety disorder and major depressive disorder comorbidity: an example of genetic pleitropy? *Eur. Psychiatry*, 19, 27, 2004.
35. Anguelova, M., Benkelfat, C., and Turecki, G., A systematic review of association studies investigating genes coding for serotonin receptors and the serotonin transporter: II. Suicidal behavior, *Mol. Psychiatry*, 8, 646, 2003.
36. Lin, P.Y., and Tsai, G., Association between serotonin transporter gene promoter polymorphism and suicide: results of a meta-analysis. *Biol. Psychiatry*, 15, 1023, 2004.

37. Munafo, M.R., Clark, T., and Flint, J., Does measurement instrument moderate the association between the serotonin transporter gene and anxiety-related personality traits? A meta-analysis. *Mol. Psychiatry*, 10, 415, 2005.

38. You, J.S., Hu.,S., Chen, B., and Zhang, H.,G., Serotonin transporter and tryptophan hydroxylase gene polymorphisms in Chinese patients with generalized anxiety disorder. *Psychiatr. Genet.*, 15, 7, 2005.

39. Hoefgen, B., Schulze, T.G., Ohlraun, S., von Widdern, O., Hofels, S., Gross, M. et al. The power of sample size and homogeneous sampling: association between the 5-HTTLPR serotonin transporter polymorphism and major depressive disorder. *Biol. Psychiatry*, 57, 247, 2005.

40. Cho, H.J., Meira-Lima, I., Cordeiro, Q., Michelon, L., Sham, P., Vallada, H., and Collier, D.A., Population-based and family-based studies on the serotonin transporter gene polymorphisms and bipolar disorder: a systematic review and meta-analysis. *Mol. Psychiatry*. 2005 Apr. 12 (*online*).

41. Malhotra, A.K., Murphy, G.M., Jr, and Kennedy, J.L., Pharmacogenetics of psychotropic drug response. *Am. J. Psychiatry*, 161, 780, 2004.

42. Leboyer, M., Bellivier, F., Nosten-Bertrand, M., Jouvent, R., Pauls, D., and Mallet, J., Psychiatric genetics: search for phenotypes. *Trends Neurosci.*, 21, 102, 1998.

43. Hariri, A.R., Mattay, V.S., Tessitore, A., Kolachana, B., Fera, F., Goldman, D., Egan, M.F., and Weinberger, D.R., Serotonin transporter genetic variation and the response of the human amygdala. *Science*, 297, 400, 2002.

44. Jamain, S. et al., Mutations of the X-linked genes encoding neuroligins NLGN3 and NLGN4 are associated with autism. *Nat. Genet.*, 34, 27, 2003.

45. Laumonnier, F. et al., X-linked mental retardation and autism are associated with a mutation in the NLGN4 gene, a member of the neuroligin family. *Am. J. Hum. Genet.*, 74, 552, 2004.

46. Yan, J. et al., Analysis of the neuroligin 3 and 4 genes in autism and other neuropsychiatric patients. *Mol. Psychiatry*, 10, 329, 2005.

47. Gauthier, J., Bonnel, A., St-Onge, J., Karemera, L., Laurent, S., Mottron, L., Fombonne, E., Joober, R., and Rouleau, G.A., NLGN3/NLGN4 gene mutations are not responsible for autism in the Quebec population. *Am. J. Med. Genet. B Neuropsychiatr. Genet.*, 132, 74, 2005.

48. Garrod, A., The incidence of alkaptonuria: a study in chemical individuality. *Lancet*, 2, 1616, 1902.

49. Hunter, D., Gene-environment interactions in human diseases. *Nature reviews genetics*, 6, 87, 2005.

50. Caspi, A., McClay, J., Moffitt, T.E., Mill, J., Martin, J., Craig, I.W., Taylor, A., and Poulton, R., Role of genotype in the cycle of violence in maltreated children. *Science*, 297, 851, 2002.

51. Henquet, C., Krabbendam, L., Spauwen, J., Kaplan, C., Lieb, R., Wittchen, H.U., and van Os, J., Prospective cohort study of cannabis use, predisposition for psychosis, and psychotic symptoms in young people. *BMJ*, 330, 11, 2005.

12 Family and Twin Methods

Keith E. Whitfield and Tracy L. Nelson

CONTENTS

12.1 INTRODUCTION

Traditionally, behavioral scientists and epidemiologists have tended to attribute health and behavioral variation to environmental sources. The recent completion of mapping the human genome opens a new era of discovering genetic influences on behavioral and neuroscience phenotypes. Behavioral genetics offers a theoretical and statistical approach for assessing both the genetic and environmental contributions to individual variation in health. This paper provides an introduction to the concepts and statistical techniques used in behavioral genetic research.

12.2 BEHAVIORAL GENETICS AND HEALTH

Behavioral genetics (also known as quantitative genetics) became a field of inquiry in part in opposition to the developmental, or "environmentalist," view that only factors from the environment are involved in interindividual variability in the development of children.[1] Just as erroneous a view is that genes are the most important factors involved in interindividual differences. Recent advances in molecular-genetics techniques have greatly increased our ability to understand the contribution of genes to health and illness. The National Human Genome Research Institute (NHGRI) of the National Institutes of Health announced in June 2000 that it had developed a working draft of the human genome. However, the inclination to believe that genes account for all or none of the differences between people is "genetic determinism" that does not accurately represent the assumptions of behavior genetic methodologies.

Some suggest that disease processes arise from early environment.[2] If the prenatal environment does impact diseases like type II diabetes or hypertension and identical twins are highly concordant for these conditions, genetic explanations are not the only viable interpretations. These conditions may be due to adverse environments in the womb.

One of the potential dangers in emphasizing the presence of genetic influences and ignoring environmental factors is that society and policy makers will invest less in changing environments. One must keep in mind that behavior genetics methods operate under assumptions of the collective contribution of genes *and* environment, *not* nature *vs.* nurture. So the interventions to improve health will be most effective when combinations of environmental manipulations and gene therapies (when appropriate) are developed. In this article, a brief introduction to behavior genetic methodology is presented, followed by a brief review of some recent research on health using this approach from the past 15 years.

12.3 GENERAL BEHAVIORAL GENETIC METHODS

Traditionally, behavioral scientists and epidemiologists have tended to attribute behavioral variation to only environmental sources. During the past decade, however, most researchers have come to accept the notion that genetic factors are, in part, responsible for individual variation on many measures of behaviors. Using a behavioral genetic approach, origins of individual differences are conceptualized to derive from additive genetic and environmental sources of variance[1] and thus, in theory, account for the total behavioral or phenotypic variation.

Additive genetic variation is the sum of the effects from genes influencing a trait. In some disease processes such as phenylketonuria or Huntington's chorea, single genes have been found to be responsible in the formation of the disease states. These examples of one-gene one-disease (OGOD) scenarios are typically not the case for most behavioral phenomena nor the premise of quantitative genetic analyses. The underlying premise is that multiple genes, or pleiotropy,[1] are involved in genetic effects that are observed in a trait.

Although the name "behavioral genetics" (sometimes used in place of quantitative genetics) implies an emphasis on heritable aspects of behavior, the identification of environmental influences on a phenotype is equally important in this type of analysis. Environmental effects are partitioned into those that are common (shared) or unique (nonshared).[1] Common/shared environmental variation is the phenotypic variation due to the subjects' living in the same family, thus sharing the same environment.[1]

Unique/nonshared environmental variation is the component of phenotypic variance that can be attributed to the environmental factors not shared by family members and thus the factor that makes members of the same family different from one another. Plomin and Daniels[3] illustrate how unique environmental variance tracks how individuals from the same family differ from one another on numerous dimensions of behavior.

Behavior genetic designs frequently take the form of twin studies. The classic twin study involves comparisons between identical or monozygotic (MZ) and same sex fraternal or dizygotic (DZ) twin pairs. There are several variations of this design (for examples see Segal[9]) but the classic twin study tends to be the most frequently used design. An assumption of behavioral genetics is that MZ twin pairs share 100%

of their segregating genes and DZ twin pairs share on average 50%. Using data from twin pairs and operating under this assumption, the heritability of a trait can be analyzed using several techniques. Heritability is considered the proportion of phenotypic variance that is due to genetic variation. A popular simple method for calculating heritability is to calculate twice the difference between the intraclass correlations of MZ and DZ twin pairs (for the calculation of intraclass correlations see Plomin et al.[1]). Shared environmental variance can be calculated by doubling the intraclass correlation for the DZ twin pairs and subtracting the MZ intraclass correlation. By subtracting the sum of the heritability and shared environmental estimates from 1.00, an estimate of nonshared environmental variance is obtained.

While twin designs are a useful and interesting method for evaluating genetic and environmental contributions, family studies are a popular alternative. Family studies use the same basic analysis techniques as those in twin studies, but the expected covariances for pairs of relatives are estimated. The following table highlights some of the common associations used in family studies.

Types of Relatives	Additive Genetic Variance	Coefficient for Dominant Genetic Variance	Shared Environmental Variance
Spouse–spouse	0	0	1
Parent–offspring (living together)	0.5	0	1
Full sibs (living together)	0.5	0.25	1
Full sibs (living apart)	0.5	0.25	0
Half sibs (living together)	0.25	0	1
Aunt/uncle–niece/nephew	0.25	0	0
First cousins (living apart)	0.125	0	0

Source: From Khoury et al.[4]

As in other social sciences fields, structural equation modeling (SEM) or biometrical modeling, the fitting of observed data to models of genetic and environmental effects[5] is the preferred analytical method of behavioral geneticists. These SEM techniques provide parameter estimates that represent the relative contribution of genetic and environmental influences. Variances and covariances are used in the calculation of the components of variance (genetic, shared environmental, and unique environmental). The benefit of these techniques over the intraclass correlation method described earlier is that multivariate extensions can be performed to simultaneously assess shared and unique genetic and environmental relationships among two or more variables.

An example of a path diagram of a classic univariate twin model is shown in Figure 12.1 (for examples of other models).[6] Path diagrams are useful in that they provide a visual display of correlational and causal relationships between variables.[7] By employing path analysis, specific hypotheses about relationships between the

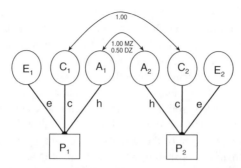

FIGURE 12.1 General structural equation model.

variables are quantified by parameter estimates or path coefficients.[6] In the model shown in Figure 12.1, the overall phenotypic variance is explained using three components: A-additive genetic variation, C-common or shared family environmental variation, and E-unique environmental variation. In Figure 12.1, A, C, and E are latent variables, and h, c, and e are parameter estimates. In this figure, P_1 represents the observed score for twin 1 and P_2 represents the score for twin 2. It should be noted that the example provided here is the general example and that P could represent any phenotype, trait, or variable.

In model fitting, observed data are compared to expected values and the result is a familiar Π^2 statistic. Using these techniques, parameters are dropped from the model to see if a more parsimonious model (one with fewer parameters or latent variables, yet still fitting the data) is available to explain the data. Statistical significance of these parameters is assessed from maximum likelihood-ratio Π^2 comparisons of the models after the parameters have been dropped. Significance of parameters is evaluated by taking the difference between the Π^2s of the full and reduced model and using that difference as a Π^2. The degrees of freedom are calculated by taking the difference between the degrees of freedom for the full and reduced. If there is a significant difference between two models, the parameter that was dropped is significant.[6]

Using this basic model then, if MZ twins are more alike than DZ twins, phenotypic variance can be attributed, in part, to genetic sources. If the genetic variation estimated by the parameter h represents a significant source of variation, this parameter cannot be dropped from the model without causing a significant change in χ^2 between the models. Squaring that parameter, h, provides an estimate of heritability. The contributions of shared and unique environmental factors to the phenotypic variation are also tested in this manner. For ease of comprehension, parameter estimates are typically standardized (procedure outlined in the LISREL manual).[8]

12.4 GENETICS, ENVIRONMENT, AND HEALTH: INTEGRATION OF APPROACHES

The National Human Genome Research Institute (NHGRI) of the National Institutes of Health announced in June of 2000 that they had developed a working draft of the human genome. This historic event places science on the doorstep of limitless possibilities including new insights about diseases and how to treat and prevent them. The trepidation about genetic approaches in the study of health by some scientists may arise from the inherent power in identifying biological mechanisms of disease and illness. Knowing the sequence of the genome, however, is only the beginning. Equally important will be our knowledge of how the environment influences health, disease, and complex behaviors associated with health. This integrative genetic–environmental perspective places behavioral genetic approaches at the forefront of useful research designs and theory that will advance the search for answers about health. Twin studies and other behavioral genetic designs have not been used to their fullest potential to this end.

One of the areas of most interest in the search for answers about health and disease is why disparities exist across economic and ethnic groups. It is a correct assumption that basing investigations of health differentials across ethnic groups solely on the basis of genetic differences will not yield accurate identification of the mechanisms responsible for health disparities. Preconceived notions about genetically based racial inferiority have hindered advances in understanding and reducing health disparities. Attempting to explain the differential health burden ethnic minorities experience by genetic differences goes against probability given there are considerably small genetic differences across racial groups and more variability within each group. The role of genetic influences, however, cannot be completely dismissed. The manner in which genes have the potential of playing a role in creating health differentials requires further explanation. It is not genes defining individuals from different ethnic groups that is key to the elucidation of health differentials per se. Instead, describing health differentials as arising from insults to a complex system represented by the interaction between genes and environments, which creates excess burden of chronic illness and disease within some groups, is a more accurate perspective.

In contrast to simply focusing on genetic explanations, there is ample information that differences in environmental factors between ethnic groups account for disparities in health status. Previous research on the significant impact combinations of socio-demographic and psychosocial factors in disease processes and complex behaviors is perhaps our best indicator that science must avoid a reductionistic view. Genetic reductionism assumes knowing and manipulating the genome will cure all our ills. Rather, we must understand how genetic and environmental influences work in concert to account for health conditions and the psychosocial variables that impact health. Much of previous research has focused on the behaviors and social structures that produce differences in health and disease across ethnic groups. One of the future and formidable challenges to using the information ascertained from adding genetic information to examinations of health differentials is to understand the underlying effect genes have on health and aging within complex environments or contexts. We may find that the polymorphisms that occur in genotypes are deleterious or protective factors related to disease and health that are created, modified, or triggered by cultural and contextual factors.

Complementary, interdisciplinary approaches are desperately needed to harness the important findings that can from the Human Genome Project and continued epidemiological research in the exploration of the underlying causes of health and illness and the related psychosocial behaviors. Continued use of behavior genetic designs (and modified designs such as those mentioned here) will significantly advance our knowledge. Both conceptual and statistical advances in these methods are still required. These methods have the potential to provide the backdrop for exciting new revelations about how genes and environment work in concert to create health and illness.

ACKNOWLEDGMENTS

Keith E. Whitfield is supported by the National Institute on Aging (5-RO1-AG13662-04). Tracy L. Nelson is supported by USPHS Grant K01-DK-64647-01.

REFERENCES

1. Plomin R., DeFries, J. C. and McClearn, G. E. *Behavioral Genetics: A Primer.* W. A. Freeman and Co. New York. 1990, P53 ff.
2. Barker, D. J. A new model for the origins of chronic disease. *Medical Health Care Philosophy,* 4 (1), 31–35, 2001.
3. Plomin, R. and Daniels, D. Why are children in the same family so different from one another? *Behavioral and Brain Sciences,* 10 (1), 1–16, 1987.
4. Khoury, M. J., Beaty, T. H., and Cohen, B. H. *Fundamentals of Genetic Epidemiology*: chapter 7, Oxford University Press, Chap 7, 1993.
5. Jinks, J. L. and Fulker, D. W. Comparison of the biomertical genetical, MAVA, and classical approaches to the analysis of human behavior. *Psychological Bulletin, 73,* 311–349, 1970.
6. Neale, M.C. and Cardon, L. R. (Eds.). *Methodology for Genetic Studies of Twins and Families.* Dordrecht, the Netherlands: Kluwer Academic Press, 1992.
7. Loehlin, J. C. *Latent Variable Models.* Baltimore: Lawrence Erlbaum, 1987.
8. Jöreskog, K. G. and Sörbom, D. *LISREL 7: A Guide to the Program and Applications* (2nd ed.). Chicago: SPSS, Inc., 1989.
9. Segal, N.L. The importance of twin studies for individual differences research. *J. Counsel. Dvlp.,* 68(6), 612–622, 1990.

13 Gene–Environment Interactions

Byron C. Jones and Leslie C. Jones

CONTENTS

13.1 INTRODUCTION: WHAT ARE GENE–ENVIRONMENT INTERACTIONS AND WHY ARE THEY IMPORTANT?

Some of the most important work on neurobehavioral genetics is in the study of psychiatric disorders in twins. As you will recall, monozygotic twins share identical genotypes and dizygotic twins share, on average, 50% of genes (alleles) in common. While we know that there is a genetic component to schizophrenia, most estimated concordance rates among monozygotic twins range from 40 to 50%, and for dizygotic twins, most estimates are below 20%.[1] This tells us that genes are involved in schizophrenia, but the environment also plays a major role. So, those carrying one or more allelic configurations are susceptible to environmental events that produce the disease. Just what the genetic configurations and environmental events are constitutes a subject of intensive study. Environment can be broadly construed and can include internal as well as external events. Thus, exposure to drugs, hormones, toxins, or teratogens *in utero*, nutritional status, viral infections, early rearing conditions, and social situations in adulthood may all be considered as "environment."

13.2 HOW DO GENES AND ENVIRONMENTS COOPERATE?

As discussed in earlier chapters, genes do not work in isolation. They are influenced by the environment and in turn may help select individuals into specific environments.

13.2.1 Gene–Environment Correlation

Gene–environment interaction sometimes is confused with gene–environment *correlation*. Gene–environment correlation may be considered as the degree to which genetically influenced propensities in the individual cause that individual to be associated with a characteristic environment. Plomin, Loehlin and DeFries[2] describe three types of geneotype–environment correlations as described below.

13.2.1.1 Passive Genotype–Environment Correlation

Under this condition, individuals inherit the same relevant alleles and are exposed to the same environment as other family members. For example, some forms of antisocial behaviors show significant heritability. It is also nearly certain that those at genetic risk are exposed to family members who exhibit many of the same behavioral characteristics. The importance of this correlation has obvious implications for developmental and intervention concerns.[2]

13.2.1.2 Reactive Genotype–Environment Correlation

In this situation, genetically influenced behavioral tendencies in the individual elicit characteristic responses from others. Thus children who are impulsive and show an inability to suppress inappropriate behavior are more likely to receive harsh parenting than children with little tendency to behave inappropriately.

13.2.1.3 Active Genotype–Environment Correlation

Children who are verbally and socially gifted tend to seek out social situations that match their tendencies. This is termed "niche-picking." Scarr and McCartney[3] have exploited genotype–environmental correlation to help explain some of the phenomena in child development, parental styles, treatment of behavior disorders, and child–family relations.

13.2.2 Gene–Environment Interaction

Gene–environment interaction is a quite different matter, although there may be co-extant gene–environment correlations. In this situation, individuals with specific genotypes respond differently to the environment than do others of differing genotypes. There are obvious implications for such interactions. For example, given an individual's genotype, we could assess susceptibility to adverse environmental conditions.

13.2.2.1 Definition Vp = Vg + Ve + (Vgxe) + Vε

Phenotypic variability (or more properly, variance) can be partitioned into a number of components. Thus Vg is the genetic part and consists of additive and dominant genetic effects. Ve is the environmental source and Vε is measurement error. Vgxe is the component left over when Vg, Ve, and Vε are accounted for (in the long run, we assume that Vε sums to zero).

Some graphical representations are shown below.

13.2.2.2 Evidence for Gene–Environment

In Cloninger's 1987 seminal paper on alcoholism typologies,[4] two kinds of alcohol-related problems were described. In the type II alcoholism, the afflicted individuals are mostly male, there is an early onset of drinking, and there is a strong association with sensation-seeking and conduct disorder. The estimated heritability is quite high and there seems to be little influence from the environment. Type I alcoholism is also genetically influenced, but its expression is dependent upon the environment— difficult relationships, job-related troubles, etc. So, the individual carries a complement of alleles that interact with environmental situations that shape the phenotype.

13.2.2.2.1 Genotype–Environment Interaction in Humans

The malfunctioning of the raphe serotonin (5HT) system has long been implicated in depression, and polymorphisms in genes involved in the regulation of that system have been considered candidate risk factors. In 1996, a report by Lesch and colleagues[5] described an association between anxiety-related traits and a polymorphism at the promoter region of the gene (SLC6A4) for the serotonin transporter (5HTT). This polymorphism (5HTTLPR) is a 44 base-pair insertion or deletion that results in a long (l) or short (s) allele. The l allele exhibits higher transcriptional

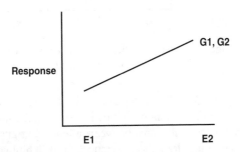

FIGURE 13.1 In this representation, we can see that the environment has an effect on the phenotype, but both genotypes respond equally to the environmental change.

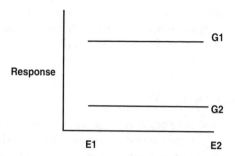

FIGURE 13.2 In this figure, we observe that there is no effect of the environment, but an effect of genotype.

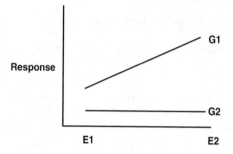

FIGURE 13.3 This figure shows, not only a significant effect of genotype, a possible effect of environment, not only but also a significant genotype–environment interaction.

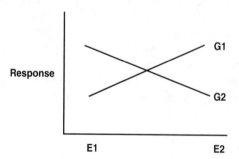

FIGURE 13.4 In this figure, we see a significant interaction between genotype and environment but no main effect of either genotype or environment.

activity on the order of 1.4- to 1.7-fold and cells transfected with the l allele evince increased serotonin uptake compared with cells transfected with the s allele. Lesch then performed an association study between allelic configuration and affective disorders. Despite the rather dramatic differences in 5HT biology, the allelic differences accounted for rather modest variance in anxiety-related traits.

In other studies concerning the relationship between 5HTTLPR polymorphism and affective disorders, either no association was found or, like the Lesch study, the strength of association was modest at best.[6,7]

One reason for conflicting results among studies, suggested by Hoefgen et al.,[7] could be insufficient sample sizes and low power. These researchers reported an association between the 5HTTLPR and depression, in which the s allele mildly increases risk, and they attributed this to the size of their sample (466 patients, 836 controls) and careful adjustment for various stratification factors. Still, the effect that they found was modest. The lack of effect found in many studies and the modest effect found by Hoefgen et al. could also be explained by how these studies isolated the allele as the putative sole factor. In fact, the allele may exert its effects by interacting with environmental factors, and possibly other genetic polymorphisms.

In 2003 Caspi and colleagues[8] studied a cohort of 847 New Zealand native Caucasians who had been studied every 2 years from age 3 to 21. When these individuals were 26 years of age, the subjects were classified based on their 5HTTLPR genotype, l/l, l/s and s/s. They were able to then classify individuals based on history of depression and history of stressful life events (SLEs). Thirty percent of the subjects reported no SLEs in the past 5 years. Twenty-five percent of subjects had experienced one SLE, while 20, 11, and 15% had experienced 2, 3, or more than 4 SLEs, respectfully. Distribution of SLEs showed that the 5HTTLPR gene had no apparent effect on stress exposure, as there were no significant differences in exposure among allele groups. Next, subjects were categorized by symptoms of major depression. Seventeen percent of the subjects met the DSM-IV criteria for having experienced a major depressive episode within the past year. Among subjects who reported SLEs, those with at least one s allele were significantly more likely than subjects homozygous for the l allele to report depression. In fact, subjects who had reported four or more SLEs and carried two s alleles had a 43% likelihood of evincing depression.

Since the results of this study were reported, at least two groups of researchers have replicated their findings.[9,10] One recent study supported the findings of Caspi et al. by showing that individuals who had been subjected to childhood mistreatment were more likely to develop a depressive episode if they contained an s allele of the 5HTTLPR gene. Kaufman and colleagues[11] studied children who had been removed from their homes, for various reasons, by the State of Connecticut Department of Children and Families (DCF) within the past 6 months. They also included a sample of community controls (CCs). More than 80% of the children in the DCF group had experienced more than two or more forms of maltreatment. Most DCF children (81%) had experienced neglect, but other forms of maltreatment included physical abuse (63%), sexual abuse (19%), emotional maltreatment (79%), and exposure to domestic violence (60%). Their results supported a gene–environment interaction, because the s allele of the 5HTTLPR gene increased the likelihood of depression only in the children who had experienced childhood maltreatment. Analysis of the effect of social supports showed that their presence moderates in a compensatory manner the interaction between the 5HTTLPR genotype and childhood mistreatment. Children were grouped into high-risk, moderate-risk, and low-risk categories (s/s without social supports, s/l or low social supports, and l/l plus high social supports, respectively), and it was determined that the depression scores of high-risk children doubled those of children who had the s/s genotype but positive social supports.[10] The authors concluded that social supports further moderated the interaction between the 5HTTLPR gene and childhood maltreatment, reducing the (increased) risk for depression among children with the s allele.[10] This study both supports the gene–environment interaction model and emphasizes further the relative importance of environmental factors.

In summary of the findings of the four studies in agreement, it can be concluded that the s allele of the 5HTTLPR significantly increases susceptibility to depression only in the presence of SLEs, and this increased vulnerability can be at least partially mitigated for by other environmental factors, including social supports.

Hariri and colleagues[12] used an anxiety- and fear-provoking task, involving photos of fearful and angry facial expressions, in conjunction with fMRI, to provide evidence of (right) amygdala hyper-reactivity among subjects carrying the s allele of 5HTTLPR. Indeed, the association between 5HTTLPR and amygdala hyper-reactivity fit nicely with the current neurobiological hypothesis of depression.[13] Of particular interest, however, was that the subjects tested in this study were healthy volunteers with no history of psychiatric illness, including depression. It seems likely, therefore, that the amygdala hyper-reactivity among subjects carrying the s allele indicates that the 5HTTLPR may in fact mediate the intensity of an individual's reaction to negative stimuli, and, hence, increased susceptibility to SLE-induced depression.

Pezawas and colleagues[14] used voxel-based morphometry (VBM) to test the hypothesis that allelic configuration of the 5HTTLPR involves neural circuits that function in emotional regulation. VBM is a highly sensitive method that measures differences in gray matter volume between subjects. Using VBM, Pezawas and colleagues detected reduced gray matter in carriers of the s allele in the right amygdala and the perigenual anterior cingulate cortex (pACC). The volume of pACC and amygdala gray matter were shown to be positively correlated and functionally connected, as shown by fMRI. Two subregions of the pACC are functionally connected to the amygdala, one showing positive and the other showing negative correlated activity. The authors proposed that the three brain regions may form a feedback loop, modulating amygdala reactivity. In carriers of the s allele, the coupling of the amygdala to these regions was impaired, especially the connection between the amygdala and the positively associated subregion of the pACC.

Thus, a body of research findings is accumulating that suggests that the s allele of the 5HTTLPR causes neurobiological differences evident even in "normal" subjects with no history of psychiatric illness. These neurobiological differences do predict personality traits related to depression in healthy s-carriers, but do not necessarily cause depression. When life stressors are not reported, s-carriers do not have an increased risk of developing depression; however, when a sufficient amount of life stress is experienced, s-carriers exhibit an increased risk toward major episodes of depression.

Because the subjects in both the fMRI and VBM studies were healthy individuals, the authors also concluded that the phenotypic effects were developmental.[11] This conclusion is in agreement with recent evidence that the impact of reduced transcription of the 5HTT early in development causes irreversible alterations in the regulation of serotonin in relevant brain regions.[15]

13.2.2.2.2 Gene–Environment Interaction in Animals

For more than half a century, we have known that environmental perturbations during infancy can have long lasting effects on neurobehavioral development in rodents and primates. Thus, brief interruption of maternal contact in rats[16] has dramatic effects on adult stress response, and social isolation in infancy causes severe social behavioral problems in primates.[17] In the case of rodents, one of the observations has been that the brief maternal separation or brief disruptions in preweanling rats tends to alter the hormonal response to stress in adulthood.[18] Moreover, early environmental enrichment has been shown to increase cerebral cortical thickness and increase the functioning of synaptic processes.[19] The story that grew out of such studies was that infantile stimulation had nearly universally beneficial effects, somehow "tempering" the rodent to better face life's challenges in adulthood. In 1970, Henderson[20] published an important paper in which he showed that rearing practices in the laboratory interact with genetically influenced behavioral differences in mice so as to reduce those behavioral differences. The latter findings thus raise the question as to whether early experience might have different effects, depending on genetic background. For providing an answer to this question, mice are better study subjects than rats because there are more genetically defined strains and lines than found in rats (although the gap is narrowing rapidly). The notion that early handling (i.e., brief separation from the nest environment) has universally beneficial effects is called into question by a study by Raymond et al.[21] who showed that early handling caused compromised immune response in adult C57BL/10J but not BALB/cJ mice.

A more comprehensive study designed to address the effect of early handling in genetically defined mice was reported recently by Gariépy and colleagues.[22] The subjects for this study were male mice from two lines selected by Cairns[23] for differential isolation-induced aggression, based on latency to attack and frequency of attacking an intruder. The low aggressive line is NC100 and the high aggression line is NC900. Two experimental treatments were administered. The first treatment was infantile handling. On every other day, the pups were removed from the nest and the entire litter was placed into a plastic beaker for 60 seconds and subsequently gently returned to the nest. At 21 days of age, the animals were weaned into isolate housing or group housing with four animals of same handling condition per group. When the animals were 51 to 60 days of age, they were subjected to a mild stressor, exposure to an open field for 10 minutes. At 20 minutes after this exposure, the animals were sacrificed and blood and brains harvested for measurement of corticosterone—the primary glucocorticoid stress hormone in rodents—and density of dopamine D_1 receptors in mesolimbic and nigro-striatal areas.

Subsequently, a separate group of animals was subjected to infantile handling vs. nonhandled controls to examine the effect of this treatment on aggression in adulthood. Because group-housed animals rarely show aggression, all of the animals were housed in isolation from weaning until testing.

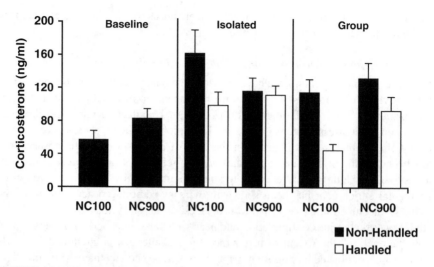

FIGURE 13.5 Handling reduced serum corticosterone response to stress (right two panels) under isolate and group housing conditions in the low aggressive animals (NC100). Moreover, group housing in this line also reduced the response and both treatments had additive effects on the corticosterone response. The high aggressive line presents a different story, as there was no effect of handling under either housing condition. (Reprinted from Gariépy, J. -L. et al., *Pharmacology Biochemistry and Behavior*, 73(1), 7–17. Copyright 2002, with permission from Elsevier.)

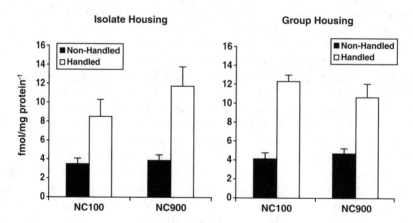

FIGURE 13.6 The effect of infantile handling on the density of dopamine D_1 receptors. Note here that handling dramatically increased the density of dopamine receptors in the nucleus accumbens. There was no effect of differential housing on this measure and there was no effect of either handling or housing on D_1 receptor densities in the caudate-putamen. (Reprinted from Gariépy, J. -L. et al., *Pharmacology Biochemistry and Behavior*, 73(1), 7–17. Copyright 2002, with permission from Elsevier.)

The results from this study showed the hypothalamo-pituitary-adrenal axis to be a good candidate for showing genotype–environment effects (Figure 13.5); however the mesolimbic, dopamine-D_1-related functions are sensitive to handling, regardless of the genetic architecture (Figure 13.6). Figure 13.7 illustrates that

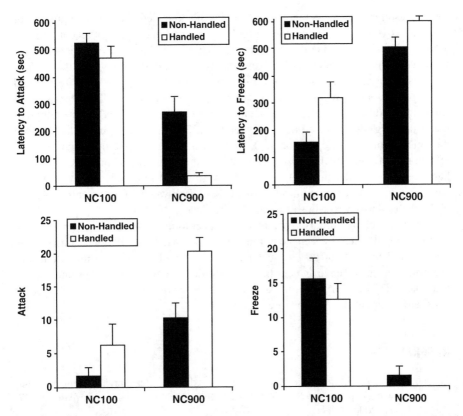

FIGURE 13.7 Handling had no effect on latency to attack in the low aggressive line but significantly shortened this measure in the high aggressive mice. The same was true for a number of attacks, although in the low aggressive line, there was a nonsignificant trend toward an increase in attack frequency in the low aggressive mice. Latency to freeze was also affected by handling, more so in the low line than in the high line, and the number of times that the animal was seen motionless (freezing) was reduced in the low line and abolished in the high line, although there was a clear "floor" effect in these animals.

aggressive and fear-like behaviors are also sensitive to gene–environment interactions. Thus, it is clear that not all neurobehavioral systems are equally susceptible to gene–environment. Is it the case then that systems that are stress-related are more susceptible to gene–environment than other systems? Or, is it the genetic makeup of the individual that determines which system will be more responsive?

13.3 GENERAL CONSIDERATIONS

It is clear that gene–environment-based influences on brain and behavior are common and play a major role in development and function. Learning about the short allele of the 5HTTLPR is important from the standpoint that major depression is the most common psychiatric disorder. Now that we know at least part of the genetic basis for increased risk, there is the possibility that this knowledge can help equip us for more effective means of prevention, prophylaxis, and treatment. In a recent, provocative article, Tienari and colleagues[24] related childhood rearing environments to risk for developing schizophrenia-spectrum disorder (SPD) in young adulthood. The subjects consisted of adoptees whose mothers had been diagnosed with (SPD) and a group of adoptees whose mothers had not been so diagnosed. The rearing environment was evaluated in terms of factors indicating adversity. When the adoptees reached young adulthood, all were assessed for SPD. Among those individuals whose biological mothers were free of SPD, the environment, adverse or benign, had no effect on the frequency diagnosed with SPD. In the group whose biological mothers had been diagnosed with SPD, those who were reared in an adverse environment had a significant increase in SPD diagnoses. The value of intervention—given biologically derived risk in this example—is clear.

The research in animals gives us the prospect to understand what systems seem to be more affected by gene–environment and what kinds of environments are important. The kinds of genetic material that are involved are also important. For example, does gene–environment involve specific structural genes or promoters, for example 5HTTPLR? Are "housekeeping" genes involved? Do genes involved in gene–environment involve specific networks?

REFERENCES

1. Cardno, A. G. and Gottesman, I. I. Twin studies of schizophrenia: from bow-and-arrow concordances to Star Wars Mx and functional genomics. *Am J Med Genet* 97: 12–17, 2000.
2. Plomin, R., Loehlin, J. D. and DeFries, J. C. Genetic and environmental components of "environmental" influences. *Developmental Psychology* 21: 391–402, 1985.
3. Scarr, S. and McCartney, K. How people make their own environments: a theory of genotype–environment effects. *Child Development* 54: 424–435, 1983.
4. Cloninger, C. R. Neurogenetic adaptive mechanisms in alcoholism. *Science* 236: 410–416, 1987.
5. Lesch, K. P. Gene–environment interaction and the genetics of depression. *J. Psychiatry Neurosci* 29: 174–184, 2004.
6. Anguelova, M., Benkelfat, C., Turecki, G. A systematic review of association studies investigating genes coding for serotonin receptors and the serotonin transporter: I. Affective disorders. *Mol Psychiat* 8: 574–591, 2003.
7. Hoefgen, B., Schulze, T. G., Ohlraun, S., Widdern, O. V., Hofels, S., Gross, M., Heidmann, V., et al. The power of sample size and homogeneous sampling: association between the 5-HTTLPR serotonin transporter polymorphism and major depressive disorder. *Biol Psychiat* 57: 247–251, 2005.
8. Caspi, A., Sugden, K., Moffitt, T. E., Taylor, A., Craig, I. W., Harrington, H. L., McClay, J. M., et al. Influence of Life Stress on Depression: Moderation by a polymorphism in the 5-HTT gene. *Science* 301: 386–389, 2003.
9. Kendler, K. S., Kuhn, J. W., Vittum, J., Prescott, C. A., and Riley, B. The interaction of stressful life events and a serotonin transporter polymorphism in the prediction of episodes of major depression. *Arch Gen Psych* 62: 529–535, 2005.

10. Gillespie, N. A., Whitfield, J. B., Williams, B., Heath, A. C., and Martin, N. G. The relationship between stressful life events, the serotonin transporter (5-HTTLPR) genotype, and major depression. *Psychol Med* 35: 101–111, 2005.

11. Kaufman, J., Yang, B-Z., Douglas-Palumberi, H., Houshyar, S., Lipschitz, D., Krystal, J. H., and Gelernter, J. Social supports and serotonin transporter gene moderate depression in maltreated children. *PNAS* 101(49):17316–17321, 2004.

12. Hariri, A. R., Drabant, E. M., Munoz, K. E, Kolachana, B. S., Mattay, B. S., Egan, M. F., and Weinberger, D. R. A susceptibility gene for affective disorders and the response of the human amygdala. *Arch Gen Psych* 62: 146–152, 2005.

13. Hamann, S. Blue Genes: wiring the brain for depression. *Nature neurosc* 8: 701–703, 2005.

14. Pezawas, L., Meyer-Lindenberg, A., Drabant, E. M., Verchinski, B. A., Munoz, K. E., Kolachana, B. S., Egan, M.F., Mattay, V.S., Hariri, A.R., and Weinberger, D.R. 5-HTTLPR polymorphism impacts human cingulate-amygdala interaction: a genetic susceptibility mechanism for depression. *Nature Neurosc* 8(6): 828–834, 2005.

15. Ansorge, M. S., Zhou, M., Lira, A., Hen, R., Gingrich, J. A. Early-life blockade of the 5-HT transporter alters emotional behavior in adult mice. *Science* 306(5697): 879–881, 2004.

16. Treiman, D. M. and Levine, S. Plasma corticosteroid response to stress in four species of wild mice. *Endocrinology* 84: 676–680, 1969.

17. Harlow, H. F., and Soumi, S. Social recovery by isolation-reared monkeys. *Proc Natl Acad Sci USA* 68: 1534–1538, 1971.

18. Levine, S. Developmental determinants of sensitivity and resistance to stress. *Psychoendocrinology* 30: 939–946, 2005.

19. Rosenzweig, M. R. and Bennett, E. L. Psychobiology of plasticity: effects of training and experience on brain and behavior. *Behavioural Brain Research* 78: 57–65, 1996.

20. Henderson, N. D. Genetic influences on the behavior of mice can be obscured by laboratory rearing. *J Comp Physiol Psychol* 72: 505–511, 1970.

21. Raymond, L. N., Tokuda, S., Reyes, E. and Jones, B.C. Differential immune response in two handled inbred strains of mice. *Physiology and Behavior,* 37: 295–297, 1986.

22. Gariépy, J-L., Rodriguiz, R. M. and Jones, B. C. Handling and genetic effects on the stress system, social behavior and dopamine function. *Pharmacol Biochem Behav,* 73(1): 7–17, 2002.

23. Cairns, R. B. Aggression from a developmental perspective: genes, environments and interactions. *Ciba Found Symp* 194: 45–56, 1996.

24. Tienari, P., Wynne, L. C., Sorri, A., Lahti, I., Laksy, K., Moring, J., Naarala, M., Nieminen, P. and Wahlberg, K.-E. Genotype–environment interaction in schizophrenia-spectrum disorder. *Br J Psychiat* 184: 216–222, 2004.

14 And Now It Starts to Get Interesting: Gene–Gene Interactions

Yamima Osher

CONTENTS

14.1 INTRODUCTION

> Genetics was discovered in the nineteenth century by an Austrian monk named Mendel, who spent years observing the reproduction of pea plants (in those days there was no HBO). Mendel noticed that baby pea plants often inherited certain characteristics of the mommy and daddy pea plants, such as height, eye color, and personality. Mendel found that, by mating a certain pea plant with a certain other pea plant, he could cause a third pea plant to go into a violent jealous rage. What can we learn from these experiments? I have no idea.
>
> **—Dave Barry**

In the beginning, there was the OGOD model: one-gene one-disease.[1] Even before the age of molecular genetics, the study of patterns of inheritance made it clear that this model was not applicable to the study of personality traits. When the first replicated reports were published of positive associations between a candidate gene and a personality trait—dopamine D4 receptor gene (D4DR) and novelty seeking

(NS)[2,3]—the authors speculated that that there might be 10 or so different genes, each of small effect size, responsible for the trait.[3] The assumption was that the effects of these genes would be additive, each one responsible for about 10% of the observed genetic variance. However, as Richard Ebstein notes in Chapter 18 of this book, finding these genes has proven frustratingly difficult and results are often inconsistent. More recently, evidence has begun to accumulate, which suggests that part of the answer to this conundrum may be found in gene–gene *interactions*.

14.2 EPISTASIS

Plomin[1] provides an excellent explanation of gene–gene interactions, commonly referred to as epistasis. He explains that while dominance is an *intralocus* interaction—with a given allele interacting in a nonadditive way with the allele at the same locus on the homologous chromosome—epistasis refers to the *interlocus* interaction of alleles at other loci. In Plomin's words, "[C]onsider two loci (A and B) that affect a phenotypic character. Both the additive genetic values and the dominance deviations are summed across the two loci. However, a particular combination of a certain allele at locus A and another at locus B may influence the phenotype in ways not explainable by the additive and dominance effects. Epistasis refers to this sort of effect" (p. 292). It should also be pointed out that epistatic effects are not limited to the interaction of only two genes, but may include any *number* of players.

14.2.1 Two Genes, Interacting

Let us begin with a relatively simple example of an interaction between two candidate genes involved in neurotransmitter systems, and its effect on a personality (temperament) trait. Harm avoidance[4] is an anxiety-related trait which has sometimes[5,6] but not always (for review see Lesch[7]) been found to be associated with the short allele of the 5-HTTLPR (serotonin transporter) gene. Might the impact of this allele be mitigated, or influenced, by another allele which could be related to curiosity, adventurousness, and the desire for stimulation—specifically, the long allele of the D4DR? One could speculate that individuals prone to anxiety might express this trait more if, at the same time, they felt the urge or the need to expose themselves to exciting (but also anxiety-provoking) experiences and stimuli. Interestingly, this seems to be the pattern that emerges when the interaction of the 5-HTTLPR and the D4DR genotypes is examined: whether the dependent variable is anxious behavior in infants[8] or harm avoidance in adults as measured by Cloninger's temperament and character inventory (TCI)[9], the group with BOTH the s/s 5-HTTLPR genotype AND the long allele of the D4DR exon III polymorphism show the highest scores. That this is a true interaction and not an additive effect can be clearly seen in the Szekely[9] data, as the D4DR genotype has no influence on harm avoidance scores in the absence of the s/s 5-HTTLPR genotype.

Findings regarding gene–gene interactions in personality genetics have been noticeably robust (in contrast to the inconsistent findings regarding most single candidate genes). An illustrative example, from the early years of molecular personality genetics: Ebstein et al.[10] looked at the interaction between a serotonin-receptor polymorphism (5-HT$_{2c}$) and the D4DR polymorphism. The 5-HT$_{2c}$ polymorphism involves a cystein to serine substitution. The relatively rare 5-HT$_{2Cser}$ allele was associated with significantly reduced reward dependence (RD) and persistence scores (persistence was originally one of the four subscales of Cloninger's reward dependence, but was later extracted and considered a fourth temperament dimension). This effect was dramatically increased in those subjects who also had the long allele of the D4DR exon III polymorphism, such that having the 5-HT$_{2Cser}$ allele and the long D4DR allele together reduced mean RD scores by more than two standard deviations. Originally, however, this finding was regarded with some skepticism, as the group of subjects with the rarer forms of both alleles was quite small ($n = 6$). Within 2 years, however, an independent group from Germany failed to replicate the main effects (long D4DR/higher NS, 5-HT$_{2Cser}$/lower RD [found only a trend])— but *did* find a highly significant effect of the *interaction* of the two alleles on levels of RD.[11] The lowest mean RD score was again seen in the 5-HT$_{2Cser}$ allele/long D4DR allele group.

14.2.2 THREE CAN PLAY, TOO

In addition to the serotonin and dopamine neurotransmitter systems, a third system, catechol O-methyltransferase (COMT), has also been found to play a role in personality and temperament. COMT and monoamine oxidase (MAO) are the two enzymes that account for the first steps in the elimination of catecholamines.[12] The physiological substrates of COMT include dopamine, norepinephrine and adrenaline but in contrast to MAO, serotonin is not a substrate for this enzyme.[13] A common COMT variant coding for a thermolabile low activity enzyme has been identified.[14,15] A single amino acid substitution (val$_{108}$ → met) in exon IV accounts for differences in thermolability: val/val homozygotes display high levels, val/met intermediate levels and met/met 4 to 5 times lower COMT activity.[16] When the effects on personality of three polymorphisms—D4DR exon III, 5-HTTLPR, and COMT— were examined simultaneously, a complex set of interactions was revealed.[17] The long allele of D4DR was associated with higher NS scores *except* in the presence of the short allele of 5-HTTLPR, and this effect was most powerful in the presence of the high activity (val/val) version of COMT; the effect of D4DR on NS virtually disappeared, however, in those subjects with the val/met version of COMT—regardless of 5-HTTPLR genotype. In this case, as with the 5-HT$_{2c}$ and D4DR interaction noted above, this finding was subsequently replicated by a different group in a separate population.[18]

14.3 NOT ONLY IN HUMANS

Association studies in human populations are not the only possible way to study gene–gene interactions. Research with knockout and other types of genetically manipulated animals has strongly supported the case for complex gene–gene interaction effects, as the effect of many specific manipulations has been found to be profoundly influenced by the genetic background of the animal.[19] Single-gene knockout mice have now been interbred in order to study interactions between alterations in two or more genes. A recent review by Murphy et al.[20] summarizes studies that looked at behavioral, physiological and/or neurochemical phenotypes in crosses between mice with single allele or single gene targeted disruptions in the serotonin transporter (SERT) and five other mouse genes, all of which are known to be polymorphic in humans and all of which are considered candidate genes for neuropsychiatric and other disorders: dopamine transporter (DAT), brain-derived neurotropic factor (BDNF), monoamine oxidase type A (MAOA), serotonin 1B receptor (5-HT$_{1B}$), and norepinephrine transporter (NET). An impressive variety of interaction effects was found. As summarized in the review, some interactions were additive, such as effects on brain serotonin levels of the SERT × BDNF cross. An epistatic effect on striatal serotonin concentrations was found in the SERT × DAT cross, as was an epistatic effect on behavior (reduced cocain-conditioned place preference). Other interactions were synergistic, antagonistic, or even more complex.[20]

14.4 GENE–GENE INTERACTIONS IN COMMON HUMAN DISEASES

As the importance of gene–gene interactions in personality is becoming increasingly clear, it may be said that the role of such interactions is already considered a fact in the realm of many illnesses and disorders. It has even been proposed that "epistasis is a *ubiquitous* component of the genetic architecture of common human diseases, and that complex interactions are *more important* than the independent main effects of any one susceptibility gene" (Moore,[21] emphasis mine). Understanding some of these interactions could have important implications for somatic therapies of neuropsychiatric disorders, as illustrated by the work of Anttila et al.[22] That group has found that those schizophrenia patients who had specific variants of two polymorphisms—the met/met form of COMT, and the c/c version of the gene *NOTCH4* (involved in the regulation of generating neurons and glial cells)—had more than ten times the risk of not responding to conventional antipsychotic drugs, as opposed to patients with other genotypes.

14.5 STATISTICAL METHODS

Many of the gene–gene interactions discussed above have been pinpointed using relatively simple mathematical tools—ANOVA or MANCOVA (multiple analysis of covariance) analyses. This is the analysis of choice in association studies when the dependent variable is continuous (as are personality or temperament trait measures) and the genetic variables are nominal (specific allele/s present or absent). Sex is most appropriately entered as an independent variable (main effect), while other factors known to influence the dependent variable, such as age, are entered as covariates. Note that significant interaction effects can be found even in the absence of main effects for the individual genotype or allele. When study designs include measures of relatedness (twins, sib pairs, linkage studies using parent–offspring triads, and so on), the analysis of choice will be some form of multiple regression.

Increasingly sophisticated analyses are being proposed, which should help carry this work forward. For a good discussion of sample size requirements for finding gene–gene and gene–environment interactions in various models of association studies (case-control, affected sib, and case-parent/ case only designs), see Gauderman.[23] Holmans[24] discusses the conditions under which the use of a modification of logistic regression should be used to examine possible gene–gene interactions in linkage studies. An expanded method for model fitting which uses a conditional logistic approach and allows for the analysis of gene–gene interactions at unlinked loci is presented by Cordell et al.[25]

14.6 SUMMARY

What, as asked above, can we learn from these experiments? One important lesson might be that when we begin to think about genetic engineering for human personality, caveat emptor—let the buyer be very very beware. We already know that many human genes are highly pleiotropic, and that changing a gene could have far-ranging but unintended effects. Now things have become even more complex. Not only the multiple effects of any given gene must be taken into account, but also all possible interactions with any number of other genes. A second lesson, perhaps more immediate than the first, is that meta-analysis may not be the most appropriate way to understand the matrix of replications and non-replications of the effects of individual candidate genes on personality or temperament. It seems increasingly likely that summing studies across different populations will yield false negatives, as differing allele frequencies for other genes interacting with the candidate gene will be impacting upon the dependent variable. Systematic estimation of allele frequencies in different populations, and wide-ranging examination of possible interactions between known candidate genes will be required in order to elucidate the true nature of genetic factors in personality, and no less, in the search for the genetic components of major mental illnesses.

REFERENCES

1. Plomin, R., DeFries, J.C., McClearn, G.E., and Rutter, M. (1997). *Behavioral Genetics*. NY: W.H. Freeman and Company.
2. Ebstein, R.P., Novick, O., Umansky, R., Priel, B., Osher, Y., Blaine, D., Bennett, E.R., Nemanov, L., Katz, M., and Belmaker, R.H., (1996). Dopamine D4 receptor (D4DR) exon III polymorphism associated with the human personality trait of Novelty Seeking. *Nat Genet* 12, 78–80.
3. Benjamin, J., Li, L., Patterson, C., Greenberg, B.D., Murphy, D.L., and Hamer, D.H. (1996). Population and familial association between the D4 dopamine receptor gene and measures of Novelty Seeking. *Nat Genet* 12, 81–84.
4. Cloninger, C.R., Przybeck, T., Svrakic, D., and Wetzel, R. (1994). *The Temperament and Character Inventory: A guide to its development and use*. St. Louis: Center for Psychobiology of Personality, Washington University.

5. Lesch, K.P., Bengel, D., Heils, A., Sabol, S.Z., Greenberg, B.D., Petri, S., Benjamin, J., Muller, C.R., Hamer, D.H., and Murphy, D.L. (1996). Association of anxiety-related traits with a polymorphism in the serotonin transporter gene regulatory region. *Science* 274:1527–1531.

6. Osher, Y., Hamer, D., and Benjamin, J. (2000). Association and linkage of anxiety-related traits with a functional polymorphism of the serotonin transporter gene regulatory region in Israeli sibling pairs. *Mol Psychiat* 5:216–219.

7. Lesch, K.P. (2001). Molecular foundation of anxiety disorders. *J Neural Transm* 108:717–746.

8. Lakatos, K., Nemoda, Z., Birkas, E., Ronai, Z., Kovacs, E., Ney, K., Toth, I., Sasvari-Szekely, M., and Gervai, J. (2003). Association of dopamine receptor gene and serotonin transporter promoter polymorphisms with infants' response to novelty. *Mol Psychiatry* 8:90–97.

9. Szekely, A., Ronai, Z., Nemoda, Z., Kolmann, G., Gervai, J., and Sasvari-Szekely, M. (2004). Human personality dimensions of persistence and harm avoidance associated with D4DR and 5-HTTLPR polymorphisms. *Am J Med Genet* 126B:106–110.

10. Ebstein, R.P., Segman, R., Benjamin, J., Osher, Y., Nemanov, L., and Belmaker, R.H. (1997). 5-HT$_{2C}$ Serotonin receptor gene polymorphism associated with human personality trait of reward dependence: interactins with D4DR and D3DR polymorphisms. *Am J Med Genet* 74:65–72.

11. Kuhn, K.U., Meyer, K., Nothen, M.M., Gansicke, M., Papassotiropoulos, A., and Maier, W. Allelic variants of dopamine receptor D4 (DRD4) and serotonin receptor 5HT2c (HTR2c) and temperament factors: replication tests. *Am J Med Genet* 88(2):168–72.

12. Boulton, A.A. and Eisenhofer, G. (1998). Catecholamoine metabolism: From molecular understanding to clinical diagnosis and treatment. Overview. *Adv Pharmacol* 42:273–292.

13. Mannisto, P.T. (1998). Catechol O-methyltransferase: characterization of the protein, its gene, and the preclinical pharmacology of COMT inhibitors. *Adv Pharmacol* 42:324–328.

14. Weinshilboum, R. and Dunnette, J. (1981). Thermal stability and the biochemical genetics of erythrocyte catechol O-methyl-transferase and plasma dopamine-beta-hydroxylase. *Clin Genet* 19:426–437.

15. Lotta T., Vidgren J., Tilgmann C., Ulmanen I., Melen K., Julkunen I., and Taskinen J. (1995). Kinetics of human soluble and membrane-bound COMT: a revised mechanism and description of the thermolabile variant of the enzyme. *Biochemistry* 34:4202–4210.

16. Lachman, H.M., Papolos, D.F., Saito, T., Yu, Y.M., Szumlanski, C.L., and Weinshilboum, R.M. (1996). Human COMT pharmacogenetics: description of a functional polymorphism and its potential application to neuropsychiatric disorders. *Pharmacogenetics* 6:243–250.

17. Benjamin, J., Osher, Y., Kotler, M., Gritsenko, I., Nemanov, L., Belmaker, R.H., and Ebstein, R.P. (2000). Association between TPQ traits and three functional polymorphisms: D4DR, 5-HTTLPR, and COMT. *Mol Psychiat* 5:96–100.

18. Strobel, A., Lesch, K.P., Jatzke, S., Paetzold, F., and Brocke, B. (2003). Further evidence for a modulation of NS by DR exon III, 5-HTTLPR, and COMT val/met variants. *Mol Psychiatry* 8:371–372.

19. McClearn, G. E. (2004). Nature and nurture: Interaction and coaction. *Am J Med Gen* 124B: 124–130.

20. Murphy, D.L., Uhl, G.R., Holmes, A., Ren-Patterson, R., Hall, F.S., Sora, I., Detera-Wadleigh, S., and Lesch, K.P. (2003). Experimental gene interaction studies with SERT mutant mice as models for human polygenetic and epistatic traits and disorders. *Genes Brain Behav* 2:350–364.
21. Moore, J.H. (2003). The ubiquitous nature of epistasis in determining susceptibility to common human diseases. *Hum Hered* 56:73–82.
22. Anttila, S., Illi, A., Kampman, O., Mattila, K.M., Lehtimaki, T., and Leinonen, E. (2004). Interaction between NOTCH4 and COMT genotypes in schizophrenia patients with poor response to typical neuroleptics. *Pharmacogenetics* 14: 303–307.
23. Gauderman, W.J. (2002). Sample size requirements for association studies of gene–gene interaction. *Am J Epidemiol* 155: 478–484.
24. Holmans, P. (2002). Detecting gene–gene interactions using affected sib-pair analysis with covariates. *Hum Hered* 53:92–102.
25. Cordell, H.J., Barratt, B.J., and Clayton, D.G. (2004). Case/pseudocontrol analysis in genetic association studies: A unified framework for detection of genotype and haplotype associations, gene–gene and gene–environment interactions, and parent-of-origin effects. *Genet Epidemiol* 26:167–185.

15 Schizophrenia: Study of a Genetically Complex Phenotype

Michael F. Pogue-Geile and Irving I. Gottesman

CONTENTS

ABSTRACT

Genetic research on the phenotype of schizophrenia is reviewed to illustrate the strategies, problems, and findings encountered in the study of a genetically complex, relatively common, dysfunctional behavioral phenotype. Schizophrenia provides an excellent case history for these purposes because it has one of the longest histories and most socially costly presence of any psychopathological phenotype. Family, twin, and adoption studies are consistent in indicating that genetic effects on schizophrenia are both important and complex. Genetic "model-fitting" to risk data for different relative classes also suggests that: (1) environmental experiences that are uncorrelated among siblings probably interact with the relevant genotypes to contribute to (or protect from) schizophrenia, (2) action of "abnormal" alleles in multiple genes seems likely for most schizophrenia cases, and (3) phenotypically normal (non-schizophrenic) individuals with "unexpressed" genetic liability exist, confusing molecular genetic studies as false negatives. Genetic linkage and association studies attempting to locate and identify individual genes of probably modest effect have to-date produced numerous positive and negative findings with frequent failures to replicate. However, a second generation of more sophisticated studies suggests increased reasons for cautious optimism that the nature of the genetic effects on schizophrenia will be revealed in the foreseeable future.

15.1 INTRODUCTION

Schizophrenia, which was first defined and initially named "dementia praecox" by the German psychiatrist Emil Kraepelin in 1896,[31] is a diagnosis based upon abnormalities of experience and behavior that include hallucinations, delusions, disorganized speech, affective flattening, and bizarre behavior (for current criteria see DSM IV[1]). Individuals with the diagnosis suffer severe personal and family distress as well as frequent chronic impairments in social and occupational functioning. Not only is the illness usually clinically severe, it is also relatively common and widespread, with approximately a 1% lifetime risk in the general population, worldwide.[16,19] It is therefore a major public health problem warranting concerted effort to understand its etiology as a guide to effective prevention and treatment.

The aim of the present chapter is a circumscribed one. For pedagogical purposes, we will consider schizophrenia as an example of a genetically complex behavioral phenotype and will selectively survey research on genetic effects in order to demonstrate the major questions, strategies, problems, and findings that have been confronted in its study and that may therefore be relevant to other behavioral phenotypes. This is a daunting task, as schizophrenia has both one of the longest and most active histories of genetic research of any psychopathological phenotype, beginning with Rüdin[57] in 1916[12,22] and continuing until today. With these caveats, it should be clear that the following is of necessity not a comprehensive and critical review of all genetic findings on schizophrenia.[8,10,21,29,43]

15.2 DEFINING THE PHENOTYPE

A common issue in the study of many behavioral phenotypes is that they are usually developed and defined based on nongenetic criteria. In the case of schizophrenia, the definition was initially developed based upon clinicians' observations and hypotheses that a number of signs and symptoms clustered together cross-sectionally and longitudinally in a manner that might suggest the existence of some common, albeit unknown, pathophysiology and perhaps etiology. Sometimes the mapping of such phenotypically defined behavioral syndromes on to genetic causes works well and relatively simply, as was the case for the clinically developed definition of Huntington's disease, which is primarily caused by necessary and sufficient mutations of a single gene.[46] However, it is certainly not necessarily the case that one disorder is the result of one gene mutation.[50] If the relationship between a clinical phenotype and genetic causes is a complex one, then it can be argued that the particular phenotype may not be the most useful one for genetic analysis. In which case, it makes sense to begin to define new phenotypes and endophenotypes using genetic validating criteria that may reflect gene effects more simply.[13,61] This "boot strapping" strategy in which the results of genetic studies serve to change the definition of the phenotype studied is an important approach in general and one that appears to have particular relevance to schizophrenia. However, for reasons of space we will focus here primarily on the clinically defined phenotype of schizophrenia itself and not on potentially related phenotypes, such as schizoaffective disorder, schizotypal personality disorder, or various cognitive and biological abnormalities in the long causal chain from gene to endophenotype to a named clinical phenotype, despite their potential interest and importance.[13,52,53,66] For similar reasons, we will also not consider in detail any potential genetic differences among the various operational definitions of clinical schizophrenia itself.

15.3 DOES GENETIC VARIATION CONTRIBUTE TO THE ETIOLOGY OF SCHIZOPHRENIA?

One of the first questions to be asked concerning the etiology of any phenotype is whether variation in genotypes in the population contributes at all to individual differences in the phenotype (it usually does). In this section we will briefly consider several designs that have addressed this question for schizophrenia. Those aspects of the designs most relevant to detecting genetic rather than environmental influences will be stressed.

15.3.1 FAMILY STUDIES

Traditional family studies of first-degree relatives (parents, offspring, and siblings) are popular and often a first step in studying etiology because of the relative ease of subject recruitment. However, the inferences that can be drawn from them are somewhat limited, with the greatest limitation being that genetic and shared environmental causes for any observed phenotypic resemblance among relatives cannot be distinguished because family members share half their genes and many similar environments and experiences. Nevertheless, the presence of some family resemblance for the phenotype under study is a sine qua non for the presence of genetic influence, although phenotypes that are primarily caused by interactions among a large number of gene loci (epistasis) or among a large number of shared environmental experiences may show little or no family resemblance, unless identical (MZ) twins are studied. Detailed comments on the methodology of family studies are provided in Whitfield and Nelson (see Chapter 12 of this volume).

Beginning with Rüdin in 1916,[58] who worked with Kraepelin in Munich, numerous family studies of schizophrenia have been performed and the findings prior to 1991 have been summarized by Gottesman[10] (see Table 15.1). More recent and methodologically stronger individual studies[24,37] echo these earlier aggregated observations. Evidence of family resemblance for a categorical characteristic such as schizophrenia in studies using selected ascertainment via patient probands comes

TABLE 15.1
Risk of Definite or Probable Schizophrenia among Relatives of Schizophrenic Probands Aggregated across Studies

Relationship to Schizophrenic Proband	% Genes Shared with Observed Schizophrenic Proband	Morbid Risk (%)[1]
Monozygotic cotwins	100	48
Dizygotic cotwins	50	17
Siblings	50	9
Offspring (1 parent affected)	50	13
Parents	50	6
Second degree[2]	25	3–4
General population	0	1

[1] Estimate of the probability of receiving a diagnosis of schizophrenia at some time during one's life (all risks age-corrected, except twins).
[2] Includes half-siblings, uncles/aunts, nephews/nieces, and grandchildren.
Source: After Gottesman I.I., *Schizophrenia Genesis: The Origins of Madness*. WH Freeman & Co., New York, 1991. With permission.

primarily from observing an increased age adjusted lifetime risk for schizophrenia among the different relative classes (e.g., parents, offspring, siblings, etc.) compared with the lifetime risk among control samples or the general population, which is usually estimated to be about 1%.[19] As can be seen in Table 15.1, risks for schizophrenia are increased among all classes of relatives, indicating familiality for schizophrenia, which by itself is consistent *either* with the effects of shared genes and/or shared environmental experiences.

If, for the sake of argument at this stage, we were to presume that this observed resemblance were due only to shared genes and not shared environment, then the data suggest several tentative interpretations. First, there is a relative risk (risk among family members/risk in the general population) among individuals sharing 50% of genes with a schizophrenic proband (i.e., offspring and siblings) of approximately 10 times the general population rate. Although substantial, a risk of 13% among offspring of a schizophrenic parent is considerably less than the 50% predicted from a rare, single gene with dominance and complete penetrance (i.e., all individuals with a disease genotype [cf. Huntington's disease] become schizophrenic). The similarity between risks in offspring and siblings argues against genetic recessive effects. Somewhat of an anomaly is the risk among parents, who also share 50% of their genes with a schizophrenic proband, but only have about half the risk of offspring and siblings. This discrepancy is usually interpreted as being a result of the common-sense screening for mental health that occurs in the mating process and the resulting reduced reproductive fitness (number of offspring) among schizophrenic patients. The risk among second-degree relatives (e.g., uncles/aunts, nephews/nieces, and grandchildren), who share 25% of their genes with the proband, is also less than half of the risk among first-degree relatives, the prediction if only a single-gene were involved (see below). Therefore, *if* these family data were due solely to genetic effects, their pattern is inconsistent with a simple, single-gene transmission, but instead tentatively suggests a complex mode of transmission involving perhaps multiple genes and/or incomplete penetrance (i.e., not all individuals with a schizophrenia genotype develop schizophrenia). Of course, based only on family study data, these findings could also all be explained by experiences shared within families. Thus, although the family study data are consistent with genetic influences on schizophrenia, they cannot prove their existence (or rather, fail to rule them out).

15.3.2 TWIN STUDIES

The classical twin study attempts to surmount this constraint of the family study by taking advantage of an experiment of nature (the existence of two kinds of twinning) *and* making an important assumption about environmental effects. Inferences from the classical twin study are based on comparisons of phenotypic resemblance between two twin groups (monozygotic, MZ vs. dizygotic, DZ) that differ in genetic similarity (100% vs. 50% shared genes), but are assumed to be equivalent on phenotype-relevant environmental similarity. Thus, the extent to which MZ twins are observed to resemble each other phenotypically more than DZ twins can be attributed to genetic effects, if the equal environmental similarity assumption is correct. Detailed discussion of twin study methodology is provided in Whitfield and Nelson (see Chapter 12 of this volume).

Beginning with Luxenburger[36] in Munich in 1928, a number of twin studies of schizophrenia have been reported and have again been helpfully aggregated[10] (see Table 15.1). Once again, results from more recent and methodologically sophisticated studies reprise these summaries of earlier work.[3,30,47] Resemblance for a categorical trait such as a diagnosis of schizophrenia from twin studies using selected ascertainment through probands is usually indexed by the probandwise concordance, or rate of schizophrenia among MZ and DZ cotwins of schizophrenic probands.[38] As can be seen in Table 15.1, the MZ concordance rate is considerably greater than that for DZ twins (48% vs. 17%), which suggests the importance of genetic influences on the etiology of schizophrenia, although this interpretation relies on the validity of the assumption that experiences relevant to schizophrenia are shared equally between MZ and DZ twins. For other phenotypes, such as personality traits, evidence generally supports this, but for schizophrenia, it remains an untested, albeit plausible assumption. Again, like the family data discussed above, the twin data point toward genetic complexity. First, the MZ concordance rate is not 100%, which may suggest an important role (in addition to genes) for environmental experiences that are not always shared between twins and therefore the existence of individuals with a genetic liability for schizophrenia who are not clinically schizophrenic (i.e., incomplete penetrance). Furthermore, the MZ concordance rate is considerably more than twice the DZ rate, which would not be predicted if a single gene without recessivity were operating. Thus, twin studies of schizophrenia provide evidence for genetic influences of some sort on schizophrenia and hint at genetic complexities and environmental effects.

A "hybrid" twin-family design has also been employed with schizophrenia to investigate further these results from classical twin studies. Schizophrenic twins and their adult offspring were initially studied by Fischer[7] in Denmark and subsequently followed up by Gottesman and Bertelsen.[11] One of the questions addressed by these studies was the presence of genetic liability in discordant MZ twin pairs.[32] One interpretation of the MZ twin concordance/discordances described above in the classical twin studies involves extreme etiological heterogeneity and posits that there are basically two distinct kinds of schizophrenia—those caused solely by genetic factors and those caused solely by environmental insults ("environmental phenocopies"). In this view, genetic liability would not be necessary for schizophrenia, although it might be sufficient and concordant MZ twin pairs would have "genetic" schizophrenia and discordant pairs "environmental" schizophrenia. An alternative interpretation of the classical twin study results emphasizes relative etiological homogeneity and instead hypothesizes that concordant and discordant MZ twin pairs all have at least some genetic liability. In the discordant pairs, however, the diagnosed twin has been exposed to environmental experiences that interact or co-act with their genetic liability to produce clinical schizophrenia, but the undiagnosed twin has not (or vice versa for protective environmental experiences). Presumably in concordant pairs, both members were exposed to the relevant experiences. Under this model, some genetic liability would probably be necessary, but not sufficient for schizophrenia. In order to address the questions of whether genetic liability may be necessary and/or sufficient to cause schizophrenia, Fischer,[7] Gottesman and Bertelsen identified schizophrenic twins who had become parents by crossing Denmark's National Twin Register with other national registers. The twins' adult offspring were then identified and their rate of schizophrenia was determined using the national psychiatric registry, hospital charts, and interviews. The risk for schizophrenia among the offspring of the non-schizophrenic MZ cotwins of schizophrenic twins was 17%, which was significantly greater than the 2% found among the offspring of non-schizophrenic DZ cotwins of schizophrenic twins. This result suggests that discordant MZ twins have significant genetic liability that can be transmitted to their offspring even though they do not manifest it themselves. Data from this design reinforce the importance of genetic influences, demonstrate the existence of "unexpressed genotypes," and indicate the importance of (unspecified) environmental experiences that are uncorrelated between twins that may interact with genetic liability. These data suggest that genetic effects may be necessary but not sufficient to cause schizophrenia.

15.3.3 ADOPTION STUDIES

Adoption studies are the second general design that attempts to estimate genetic effects and to overcome the limitations of the traditional family study.[51] Adoption studies aim to "uncorrelate" the effects of shared genes and shared experiences within families by identifying individuals (adoptees) who share only half their genes with one set of relatives (biological parents and siblings) and only their family environment with another set (adoptive parents and siblings). If genes and environments are truly uncorrelated in the adoption study, then any observed phenotypic resemblance between adoptees and their biological relatives can be attributed to shared genetic effects and any observed phenotypic resemblance between adoptees and adoptive relatives can be attributed to shared environmental effects. The two inferences, however, depend on the absence of *selective placement* of adoptees in adoptive homes that induces a correlation between their genotypes and rearing environments that are relevant to the phenotype under study. Selective placement is often indexed by an observed resemblance for relevant phenotypes between biological parents (as an estimate of the adoptees genotype) and adoptive parents (as an estimate of the adoptees' rearing environment). For example, it could be that adoptees were *selectively* placed in adoptive homes based on matching the religion or language spoken (e.g., Spanish or English) of the adoptive parents with that of the adoptees' biological parents as part of an adoption agency policy attempting to increase similarity between adoptees and their adoptive families for ethnic/cultural characteristics. In such a situation (selective placement for and environmental transmission of language spoken) it would be obviously incorrect to attribute any observed resemblance for language spoken between adoptees and biological relatives to genetic effects. Instead, adoptees and biological relatives spuriously resemble each other for language spoken due to the effects of selective placement and cultural transmission from adoptive parents to adoptees. Similar examples (e.g., selective placement for hair color) could be developed in which selective placement spuriously inflates estimates of environmental effects based on resemblance between adoptees and their adoptive relatives. Unfortunately, this important assumption of the adoption design is difficult to evaluate for studies of schizophrenia because so little is known about which phenotypes are relevant and should be correlated between biological parents and adoptive parents. More optimistically, if researchers do not know what to measure to detect selective placement, then it seems unlikely that adoption agencies do either.

Adoption studies of relatively rare, categorical phenotypes such as schizophrenia rarely use unselected random samples of adoptees, but typically employ some selective ascertainment strategy to increase statistical power and efficiency. Two complimentary techniques have been used for estimating genetic effects based on the resemblance between adoptees and their biological relatives. The "adoptees method" ascertains index families through a schizophrenic biological parent who has adopted away an offspring and control families through a demographically matched non-schizophrenic biological parent who has adopted away an offspring. The rate of schizophrenia in the adopted away offspring ("adoptees") of the index and control biological parents is then assessed and compared. If genetic influences contribute to schizophrenia, then schizophrenia should be more frequent in the index adoptees compared with the control adoptees. In contrast, the "adoptees' families method" ascertains index probands who are schizophrenic adoptees and control probands who are demographically matched non-schizophrenic adoptees. The rate of schizophrenia in the biological relatives ("adoptees' families") of the index adoptees is then assessed and compared with that among the biological relatives of the control adoptees.

The adoption strategy was pioneered in schizophrenia research by Heston in Oregon in 1966[17] and Rosenthal[56] and Kety[26] from the NIMH who studied adoptees and their families in Denmark. The results across these and other adoption studies[68] have been consistent in indicating significant resemblance for schizophrenia between biological relatives and adoptees using both adoptees and adoptees' families methods, thus suggesting the presence of genetic influences (see Gottesman[16] for a detailed review). For example, a final report from Kety and colleagues of their Danish adoptees families study based on personal interviews found a morbid risk for schizophrenia (using DSM-III criteria) among the first degree biological relatives of schizophrenic adoptees of 8% compared to about 1% in the first degree biological relatives of control adoptees.[23,27] Inclusion of diagnoses of schizoaffective disorder, schizotypal, and paranoid personality disorder as part of a schizophrenia "spectrum" increases the prevalences to 24% vs. 5%, respectively. The convergence of evidence indicating genetic influences from both adoption and twin studies is impressive. Although each kind of study makes assumptions that could be argued separately, they are *different* assumptions, and thus strengthen the case for an important role for genetic effects among the distal causes of schizophrenia.

15.4 WHAT IS THE NATURE OF GENETIC EFFECTS ON SCHIZOPHRENIA?

Decades of work around the world have established indirectly that genes affect risk for schizophrenia. As important as this fact is, however, it is just the beginning of the important questions concerning the *nature* of these genetic effects. One of the first such questions is: Is schizophrenia liability due only to genes or is environmental variation also important?

15.4.1 ONLY GENETIC EFFECTS?

As discussed above, the *dis*cordance rate of about 50% among MZ twins argues strongly for the importance of nongenetic factors that are uncorrelated among family members. Typically, such factors are termed "nonshared" environmental effects, which might include such experiences as accidents, peer relations, and other environmental experiences that are uncorrelated in siblings and MZ twins. However, it may also be that differences between individuals with identical genotypes arise for purely stochastic, "chance" reasons, as suggested by asymmetries in paired bilateral body structures,[71] and epigenetic factors.[70] Although these possibilities cannot be resolved currently, it nevertheless seems clear that some such nonshared factors contribute in an important manner along with genotypes to the liability to schizophrenia. The specific nature of these nonshared nongenetic factors remains one of the many major questions concerning the etiology of schizophrenia. In contrast to the apparent importance of nonshared effects, other results, for example from adoption studies, find little evidence for important effects of experiences that are shared among relatives (often termed shared environmental effects). Thus, to be plausible, models of genetic effects on schizophrenia must include a role for nonshared environmental effects, although apparently shared experiences are not needed. Incorporation of either sort of environmental effect is also often considered as one way of allowing for "incomplete penetrance" of the genotype, where penetrance is the probability of developing schizophrenia given the relevant genotype.

15.4.2 How Many Genes?

Assuming that any genetic model of schizophrenia needs to include nonshared environmental effects, the next important question concerns *how many* genes contribute to schizophrenia liability. The most parsimonious genetic model for the transmission of schizophrenia, the generalized single locus (GSL) model,[20] involves only one gene, although for reasons discussed above, incomplete penetrance due to nonshared environmental effects also needs to be included. More complex alternatives incorporate multiple genes in various ways (in addition to the usual nonshared environmental effects).

"Oligogenic" models posit a small, defined number of genes. One popular oligogenic model specifies that *all* risk-increasing genes are required to produce schizophrenia. For example, if such an oligogenic model hypothesizes the existence of three genes that can contribute to risk for schizophrenia, then only individuals with all three liability alleles are at risk for schizophrenia. Individuals with one or any two alleles will not become schizophrenic. This sort of model involves genetic "epistasis," or interaction among genetic loci, in that the effect of each individual allele depends on the presence of the other alleles.

A more complex multiple gene model is the related "multifactorial threshold" (MFT) model,[6] which was first proposed for schizophrenia by Gottesman and Shields in 1967.[15] This model posits: (1) a large, unspecified number of genes each of relatively small effect (polygenes) and nonshared environmental experiences that are interchangeable and additive in their effect on liability and (2) a categorical threshold on the liability dimension beyond which an individual becomes schizophrenic. For example, if a total of eight different genetic loci and three nonshared experiences are hypothesized to increase risk for schizophrenia and the threshold to produce schizophrenia is four factors, then an individual with *any* four or more of the eight alleles or three experiences will develop schizophrenia. Importantly, although the MFT model assumes additive effects of genetic and environmental factors on quantitative liability, the imposition of a threshold makes for an epistatic or interactive effect on the categorical diagnosis itself as the effect of any one gene depends upon the effects of others in determining whether an individual crosses the threshold and receives a diagnosis. The primary difference therefore between typical MFT and oligogenic–epistatic models is in the number of risk-increasing genes hypothesized. There are also usually differences between the two in where the threshold is placed. The typical MFT posits a greater number of potential risk-increasing genes than is necessary to produce schizophrenia, whereas the oligogenic–epistatic model usually hypothesizes that the number required to produce schizophrenia equals the number of potential risk-increasing genes. This difference has the effect that the MFT (but not the oligogenic–epistatic model) predicts etiological variation among schizophrenic patients because one individual may have risk-increasing alleles for gene numbers 1, 2, 3, and 4, whereas another patient may have alleles for genes 1, 2, 3, and 5.

How is one to decide among these several theoretical possibilities for the number of genes affecting liability to schizophrenia? Currently, the primary data available to evaluate these models are the risks for schizophrenia observed among the different classes of relatives of schizophrenic probands. In addition, segregation analyses have been applied to raw data from individual studies.[67] The former strategy uses one of the genetic models described above to generate predicted risks for schizophrenia among different relative classes that can then be compared with the observed risks (see Table 15.1) to evaluate whether the hypothesized model fails to "fit" the observed data within sampling variation.[39,54] The consistent finding from such analyses, as well as segregation analyses, is that a single gene GSL model even with incomplete penetrance does not fit the observed risks, whereas the multiple gene MFT or oligogenic models cannot be statistically rejected. The essential problem is that the GSL model predicts a linear decrease in risk across the different relative classes because the probability of sharing alleles at *one* genetic locus with a schizophrenic proband decreases by a constant 50% between adjacent relative classes (e.g., from MZ twin to first-degree relatives (including DZ twins, siblings, offspring) 50%/100% = 50%; from first-degree relatives to second-degree relatives 25%/50% = 50%). It is important to note that this linear prediction holds only if there is no recessivity, as seems to be the case, because sibling and offspring risks are generally equal. If penetrance is set at approximately 50% due to nonshared experiences (so that the observed risk of about 50% in MZ cotwins is correctly "predicted"), then the simple GSL model predicts a risk of 25% for first-degree relatives (50% probability of sharing allele with proband × 50% penetrance of allele = 25%) and a risk of 12.5% for second degree relatives (25% probability of sharing allele with proband × 50% penetrance = 12.5%); a linear decrease. Inspection of Table 15.1 shows that the observed risks do not follow such a linear pattern and that this GSL model considerably over-predicts the risks in first- (25% predicted vs. about 12% observed) and second-degree relatives (12.5% predicted vs. 3 to 4% observed). Alternatively, the penetrance may be reduced enough (to about 25%) to fit more closely the observed first (50% × 25% penetrance = 12.5% predicted vs. about 12% observed) and second-degree risks (25% × 25% penetrance = 6% predicted vs. 3% observed), but in this case the GSL considerably under-predicts the MZ twin risk (100% × 25% penetrance = 25% predicted vs. 50% observed). Importantly, these prediction errors of the GSL model hold even in the face of a kind of genetic heterogeneity in which schizophrenia is caused by several *different single* major genes in different families because a linear pattern is predicted for each single major locus and the average across individual families remains linear. An important assumption of all these analyses (in addition to no recessivity) is that there are either minimal effects of environmental experiences shared within families (which seems largely justified from the literature), or that their effects also decrease linearly at a slope of .50 across relative classes. However, if environmental experiences relevant to schizophrenia were shared only among MZ twins, then a more complex GSL model that incorporated this feature would probably be able to fit the observed data and not under-predict the MZ twin risk. Overall, therefore, after making the reasonable assumptions of no special MZ twin shared environment and no recessivity, it is highly improbable that a single gene causes schizophrenia. In contrast to the problems of the GSL models, the more complex multiple-gene hypotheses fit the observed data much more easily because they predict a curvilinear decrease in risk across relatives rather than a linear relationship. As a simplified exercise (ignoring issues of allele frequency effects and phenocopies), let us step through a two-gene epistatic oligogenic model with 50% penetrance of the joint two locus genotype due to nonshared experiences (to "predict" MZ twin risk satisfactorily). The probability that a MZ twin of a schizophrenic proband shares alleles at two loci with the proband is 100% (because MZ twins share all their genes) and given a penetrance of 50% the model "predicts" a risk for MZ twins of 50% (100% × 50% penetrance = 50% predicted vs. about 50% observed). The probability that a first-degree relative shares alleles at *two* loci with a proband is 25% (probability of sharing at one locus, 50%, × probability of sharing at second locus, 50%, = 25%) and thus the model predicts a risk of 12.5% for first-degree relatives (25% × 50% penetrance = 12.5% predicted vs. about 12% observed). Similarly, the probability that second-degree relatives share two loci with a proband is 6.25% (25% × 25% = 6.25%) and their predicted risk would be 3% (6.25% × 50% penetrance = 3.125% predicted vs. 3% observed). This rough example shows how well even a simple multigene model can fit the observed data and MFT models fit similarly well with the heritability of liability estimated at about 80%.[39]

Given such results, it is very likely that schizophrenia's genetic contributions involve two or more gene loci acting together. However, it is difficult to distinguish among epistatic oligogenic models with a few necessary genes, MFT models with many possible genes of quite small effect, or MFT models with some genes of moderate effect along with many of small effect. Attempts at distinguishing some of the multigene alternatives have investigated the properties of integrative "mixed" models, in which there is both one gene of "major" effect *and* a polygene background both contributing to schizophrenia risk.[45] Although fitting mixed models to individual data sets has usually resulted in a failure to detect a single gene of major effect,[67] analyses of aggregated data sets suggest some interesting possibilities. Gottesman and McGue[14] found the following three general scenarios that were statistically consistent with observed familial risks: (1) presence of a highly penetrant, but very rare, single major locus and high polygene heritability, (2) presence of a low penetrance, but relatively common, single locus and high polygene heritability, and, (3) absence of a single locus and high polygene heritability. Risch[55] used the same data and found that several oligogenic models also fit the observed data well. In summary, it appears quite likely that most cases of schizophrenia are caused by two or more gene loci acting along with nonshared environmental influences, although the further details of the number and importance of the loci remain speculative at this time.[9] It is also important to note that the individual genes in such multigene systems need not be specific to schizophrenia liability.

15.5 WHERE ON THE CHROMOSOMES ARE SCHIZOPHRENIA LIABILITY GENES LOCATED?

Studies that attempt to discover the chromosomal *location* of liability genes for schizophrenia usually employ some variant of *linkage* analysis. Although such studies have a long history in research on schizophrenia,[40] considerably antedating the recent "molecular revolution," the first positive findings using DNA markers were reported by Gurling and colleagues in 1988 for a region on chromosome 5.[60] Foreshadowing future developments, a failure to replicate was also published as a companion piece in the same journal issue.[25] Since these initial reports, the history of linkage studies in schizophrenia has been an active one, but until recently with few positive findings and fewer replications. This brief history can be divided into two phases. Early linkage studies often seemed to be predicated on a search for *the* gene for schizophrenia, despite the evidence discussed above suggesting the involvement of multiple genes. Designs often employed large multiplex families, with many affected members in an attempt to identify families segregating a single major gene. Candidate chromosomal regions were usually investigated based on suggestions from other observations. Sample sizes were usually relatively small based on the optimistic assumption of finding a gene of major effect. Methods of analysis usually emphasized "parametric" linkage approaches that depend on specifying a model of transmission that includes values for: dominance/recessivity, frequency of phenocopies, penetrance of genotypes, and frequency of alleles. Incorrect specification of these parameters should decrease power to detect linkage, however, testing of multiple models tends to inflate alpha error rates. Phenotype definitions were multiple, including diagnoses of schizophrenia, "schizophrenia-spectrum," and often even more inclusive diagnostic groupings. Failures to replicate positive findings during this early phase were often optimistically interpreted in terms of genetic heterogeneity across studies.

The frustrations of these initial studies and methodological developments have encouraged a more sophisticated and realistic wave of studies and reports.[29,42,44,49] Increasingly the view is that the aim of linkage studies is to identify multiple genes, probably each of small to modest effect. Studies now typically utilize complete genome scan approaches and, given the multitude of statistical tests applied, failures to replicate are more often interpreted from the perspective of statistical false positives, leading to proposals for more stringent statistical criteria.[33] Designs have been increasingly based on affected sibling pairs and analyzed using "nonparametric" affected pedigree member techniques.[34,69] In contrast to the parametric methods, these analytic approaches do not require specification of a transmission model and include only schizophrenic relatives, thus avoiding the potential problems of incorrect model specification and diagnostic false negatives among non-schizophrenic relatives, although at a cost of statistical power. Sample sizes have also become much larger. Meta-analyses and pooling of data across studies have also become more common (sometimes mandated by grant funders) and have been very useful. Although still very much of a moving target, results from some of these newer studies have generated considerable enthusiasm recently about potential replicated linkages on chromosomes 1, 2, 6, 8, 13, and 22 and others that have been reported by several groups.[43,49,59,63] However, skeptics remain[28] and it will be for future work to resolve these controversies.[2,29,35,41]

15.6 WHICH GENES CONTRIBUTE TO SCHIZOPHRENIA LIABILITY?

Answers to questions about specific contributing genes must rely on genetic association studies in which associations between particular alleles and the schizophrenia phenotype are identified. Valid phenotype–allele associations may arise either because the allele actually contributes causally to the phenotype or because it is in linkage disequilibrium with a locus that does. Linkage disequilibrium occurs when alleles at different loci become correlated. Association studies have the advantage of being much more powerful than linkage studies to detect loci of small effect; however, they are much less sensitive than linkage studies to loci beyond a narrow distance surrounding the marker. Association studies employ either case-control designs or various sorts of within-family designs to avoid spurious findings due to population stratification. Increasingly, association studies utilize multiple polymorphisms (i.e., single nucleotide polymorphisms, SNPs) within a gene, in order to measure haplotypes, which increases power. A haplotype is the sequence of alleles across loci on a particular chromosome. Given their precision, genome-wide association studies have been impractical to date, perhaps requiring well over 100,000 markers to screen the entire genome.[56] New technology using microchips that allows rapid genotyping of individuals for such dense maps of polymorphisms, however, is becoming less expensive and will make this strategy increasingly useful. Techniques, such as "DNA pooling," in which genotyping is performed on "pools" of groups of individuals (e.g., all cases vs. all controls), simultaneously, are also reducing genotyping costs, although the statistical issues surrounding performing thousands of tests remain complicating factors.

For these reasons, association studies to date have employed some sort of "candidate" strategy to narrow the search to a smaller number of markers. Candidate strategies have either been based on hypotheses drawn from models of pathology or on positional information provided by linkage studies. For schizophrenia, there have been numerous such association studies using candidate genes most frequently chosen because they might be relevant to schizophrenia's hypothesized pathophysiologies (e.g., genes coding for characteristics affecting dopamine or glutamate neurotransmission). To date, most such association results have been negative and positive findings have been rarely replicated. One particular variation of this strategy that should hold special promise is exemplified by the history of the RGS4 (Regulator of G protein signaling 4, chr 1) gene. RGS4 was identified as a candidate through a gene expression study of post-mortem brain in schizophrenia, which found it to be under-expressed.[42] Based on this gene-expression finding, gene association studies using RGS4 as a candidate have been performed, with promising results to date.[4] As knowledge of pathology increases and more genes are identified and sequenced, these approaches should become more fruitful.[48,49]

In contrast, positional candidate strategies have generated considerable excitement recently. Based on promising results from linkage studies mentioned above, markers in several candidate chromosomal regions have been investigated using association techniques. Although still controversial, several genes have been proposed as potential contributors to schizophrenia liability based on this strategy. The following genes have had at least some positive replications (along with some negative): Dystrobrevin-binding protein 1 (DTNBP1 dysbindin, chr 6),[64] Neuregulin 1 (NRG1, chr 8),[62] and D-amino acid oxidase (DAO) and D-amino acid oxidase activator (DAOA) (chr 13).[5] Future study will show whether the current enthusiasm for these particular genes is well placed; denser "SNP-otyping" may reveal unexpected candidate genes very close to the genes named here.

15.7 CONCLUSIONS

The aim of this chapter is to provide a "case history" of the primary strategies used to investigate genetic influences on the complex phenotype of schizophrenia, as well as to provide a sketch of current findings. The Internet is your best friend for keeping up with insertions and deletions in the corpus of received knowledge. From the long history of research on schizophrenia, the following points seem certain enough to conclude: (1) genes play a major causal role in schizophrenia, (2) effects of multiple genes, combined with environmental and stressful experiential factors, epigenetic, and stochastic factors are safe bets for explaining most schizophrenia cases, (3) overall effects at any single gene locus are small with unhelpful positive predictive power, (4) some sort of genetic heterogeneity across schizophrenia patients is very likely, (5) environmental experiences that are uncorrelated among siblings probably interact with the relevant genotypes to contribute to (or protect from) schizophrenia liability, and (6) non-schizophrenic individuals with unexpressed genetic liability exist and may become transmitters of liability. It is very likely that most other psychological and psychopathological phenotypes share many of these features.

Although we have covered many aspects of genetic research on schizophrenia, we have also omitted many important areas due to space constraints.[18] Among those that may prove most relevant to future genetic studies are work on environmental influences, gene–environment interaction, etiological heterogeneity, endophenotype and phenotype definition, developmental neurobiology,[65] and evolutionary issues. Despite these omissions, we have tried to communicate at least a hint of the clinical importance, intellectual challenge, and excitement to be found in the etiological study of schizophrenia.

REFERENCES

1. *American Psychiatric Association: Diagnostic and Statistical Manual of Mental Disorders* (4th ed.), American Psychiatric Association Press, Washington, DC, 1994.
2. Badner, J.A. and Gershon, E.S., Meta-analysis of whole-genome linkage scans of bipolar disorder and schizophrenia. *Mol. Psychiatry*, 2002, 7: 405–411.
3. Cardno, A.G. and Gottesman, I.I., Twin studies of schizophrenia: From bow-and-arrow concordances to star wars Mx and functional genomics. *American Journal of Medical Genetics*, 2000, 97, 12–17.
4. Chowdari, K.V., Mirnics K., Semwal, P. et al., Association and linkage analyses of RGS4 polymorphisms in schizophrenia. *Hum. Mol. Genet.*, 2002, 11: 1373–1380.
5. Chumakov, I., Blumenfeld M., Guerassimenkom, O., et al., Genetic and physiological data implicating the new human gene for D-amino acid oxidase in schizophrenia. *Proc. Natl. Acad. Sci. USA*, 2002, 99: 13675–13680.
6. Falconer, D.S., The inheritance of liability to certain diseases estimated from incidence among relatives. *Annals of Human Genetics*, 1965, 29: 51–76.
7. Fischer, M., Psychoses in the offspring of schizophrenic monozygotic twins and their normal co-twins. *British Journal of Psychiatry*, 1971, 115: 981–990.
8. Gershon, E.S. and Cloninger, C.R., *Genetic Approaches to Mental Disorders*. American Psychiatric Association Press, Washington, DC, 1994.
9. Gottesman, I.I., Complications to the complex inheritance of schizophrenia. *Clinical Genetics*, 1994, 46, 116–123.

10. Gottesman, I.I., *Schizophrenia Genesis: The Origins of Madness*. WH Freeman & Co., New York, 1991.
11. Gottesman, I.I. and Bertelsen A., Confirming unexpressed genotypes for schizophrenia: Risks in the offspring of Fischer's Danish identical and fraternal discordant twins. *Archives of General Psychiatry*, 1989, 46: 867–872.
12. Gottesman, I.I. and Bertelsen, A., Editorial: Legacy of German psychiatric genetics: Hindsight is always 20/20. *American Journal of Medical Genetics*, 1996, 67: 317–322.
13. Gottesman, I.I. and Gould, T.D., The endophenotype concept in psychiatry: Etymology and strategic intentions. *Am. J. Psychiatry*. 2003, 160: 636–645.
14. Gottesman, I.I., and McGue, M., Mixed and mixed-up models for the transmission of schizophrenia. In D. Cichetti, W. Grove (Eds.), *Thinking Clearly about Psychology: Essays in Honor of Paul E. Meehl*. University of Minnesota Press, Minneapolis, 1991, pp. 295–312.
15. Gottesman, I.I. and Shields J., A polygenic theory of schizophrenia. *Proceedings of the National Academy of Sciences of the United States of America*, 1967, 58: 199–205.
16. Gottesman, I.I. and Shields, J., *Schizophrenia: The Epigenetic Puzzle*. Cambridge University Press, New York, 1982.
17. Heston, L.L., Psychiatric disorders in foster home reared children of schizophrenic mothers. *British Journal of Psychiatry*, 1966, 112: 819–825.
18. Hirsch, S.R. and Weinberger, D.R., *Schizophrenia*. Blackwell, Oxford, UK, 1995.
19. Jablensky, A., Schizophrenia: Recent epidemiologic issues. *Epidemiologic Reviews*, 1995, 17, 10–20.
20. James, J.W., Frequency in relatives for an all-or-none trait. *Annals of Human Genetics*, 1971, 35: 47–49.
21. Kendler, K.S. and Diehl, S.R., The genetics of schizophrenia: A current, genetic–epidemiologic perspective. *Schizophrenia Bulletin*, 1993, 19: 261–285.
22. Kendler, K.S. and Zerbin-Rüdin, E., Abstract and review of "Studien über Vererbung und Intstehung geistiger Störungen. I. Zur Vererbung und Neuentstehung der Dementia Praecox," *American Journal of Medical Genetics*, 1996, 67: 338–342.
23. Kendler, K.S., Gruenberg, A.M., and Kinney, D.K., Independent diagnoses of adoptees and relatives as defined by DSM-III in the provincial and national samples of the Danish adoption study of schizophrenia. *Archives of General Psychiatry*, 1994, 51: 456–468.
24. Kendler, K.S., McGuire, M., Gruenberg, A.M., O'Hare, A., Spellman, M., and Walsh, D., The Roscommon family study: I. Methods, diagnosis of probands and risk of schizophrenia in relatives. *Archives of General Psychiatry*, 1993, 50: 527–540.
25. Kennedy, J.L., Giuffra, L.A., Moises, H.W., Cavalli-Sforza, L.L., Pakstis, A.J., Kidd, J.R., Castiglione, C.M., Sjogren, B., Wetterberg, L., Kidd, K.K., Evidence against linkage of schizophrenia to markers on chromosome 5 in a northern Swedish pedigree. *Nature*, 1988, 336: 167–170.
26. Kety, S.S., Rosenthal, D., Wender, P.H., Schulsinger, F., The types and prevalence of mental illness in the biological and adoptive families of adopted schizophrenics. In D. Rosenthal, S.S. Kety (Eds.): *The Transmission of Schizophrenia*. Pergamon Press, Oxford, 1968, pp. 345–362.
27. Kety, S.S., Wender, P.H., Jacobsen, B., Ingraham, L.J., Jansson, L., Faber, B., Kinney, D.K., Mental illness in the biological and adoptive relatives of schizophrenic adoptees: Replication of the Copenhagen study in the rest of Denmark. *Archives of General Psychiatry*, 1994, 51: 442–455.
28. Kidd, K.K., Can we find genes for schizophrenia? *American Journal of Medical Genetics (Neuropsychiatric Genetics)*, 1997, 74: 104–111.

29. Kirov G., O'Donovan, M.C., and Owen, M.J., Finding schizophrenia genes. *The Journal of Clinical Investigation*, 2005, 115: 1440–1448.
30. Kläning, U., Schizophrenia in twins: Incidence and risk factors. Department of Psychiatric Demography: University of Aarhus, Aarhus, Denmark. Unpublished doctoral dissertation, 1996.
31. Kraepelin, E., *Psychiatrie*, Barth, Leipzig, 1896, pp. 426–441.
32. Kringlen, E. and Cramer, G., Offspring of monozygotic twins discordant for schizophrenia. *Archives of General Psychiatry*, 1989, 46: 873–877.
33. Lander, E. and Kruglyak, L., Genetic dissection of complex traits: Guidelines for interpreting and reporting linkage results. *Nature Genetics*, 1995, 11: 241–247.
34. Lander, E. and Schork, N.J., Genetic dissection of complex traits. *Science*, 1994, 265, 2037–2048.
35. Lewis, C.M. et al. Genome scan meta-analysis of schizophrenia and bipolar disorder, part II: Schizophrenia. *Am. J. Hum. Genet.* 2003, 73: 34–48.
36. Luxenburger, H., Vorläufiger Bericht über psychiatrische Serienuntersuchungen an Zwillingen. *Zeitscrift für die gesamte Neurologie und Psychiatrie*, 1928, 116: 297–326.
37. Maier, W., Lichtermann, D., Minges, J., Hallmeyer, J., Heun, R., Benkert, O., and Levinson, D.F., Continuity and discontinuity of affective disorders and schizophrenia. *Archives of General Psychiatry*, 1993, 50, 871–883.
38. McGue, M., When assessing twin concordance, use the probandwise not the pairwise rate. *Schizophrenia Bulletin*, 1992, 18: 171–176.
39. McGue, M. and Gottesman, I.I., Genetic linkage in schizophrenia: perspectives from genetic epidemiology. *Schizophrenia Bulletin*, 1989, 15: 453–464.
40. McGuffin, P. and Sturt, E., Genetic markers in schizophrenia. *Human Heredity*, 1986, 36: 65–88.
41. McGuffin, P., Owen, M.J., and Gottesman, I.I., *Psychiatric Genetics and Genomics*. London and New York: Oxford University Press, 2004.
42. Mirnics, K., Middleton, F.A., Stanwood, G.D., Lewis, D.A., and Levitt, P., Disease-specific changes in regulator of G-protein signaling 4 (RGS4) expression in schizophrenia. *Mol. Psychiatry*, 2001, 6: 293–301.
43. Moises, H.W., Yang, L., Kristbjarnarson, H., Wiese, C. et al., An international two-stage genome-wide search for schizophrenia susceptibility genes. *Nature Genetics*, 1995, 11: 321–324.
44. Moldin, S.O. and Gottesman, I.I., Genes, experience, and chance in schizophrenia: Positioning for the 21st century. *Schizophrenia Bulletin*, 1997, 23: 547–561.
45. Morton, N.E. and MacLean, C.J., Analysis of family resemblance: III. Complex segregation analysis of quantitative traits. *American Journal of Human Genetics*, 1974, 26: 489–503.
46. Nance, M.A., Invited editorial—Huntington's disease: Another chapter rewritten. *American Journal of Human Genetics*, 1996, 59, 1–6.
47. Onstad, S., Skre, I., Torgersen, S., and Kringlen, E., Twin concordance for DSM-III-R schizophrenia. *Acta Psychiatrica Scandinavica*, 1991, 83: 395–401.
48. Owen, M.J., O'Donovan, M.C., and Harrison, P.J., Schizophrenia: Disorder of the synapse? *British Medical Journal*, 330: 158–159, 2005.
49. Owen, M.J., Williams, N.M., and O'Donovan, M.C., The molecular genetics of schizophrenia: New findings promise new insights. *Mol. Psychiatry*, 2004, 9: 14–27.
50. Plomin, R., Owen, M.J., and McGuffin P., The genetic basis of complex human behaviors. *Science*, 1994, 264: 1733–1739.

51. Pogue-Geile, M.F. and Rose R.J. Psychopathology: A behavior genetic perspective. In T. Jacob (Ed.), *Family Interaction and Psychopathology*. Plenum Press, New York, 1987, pp. 629–650.

52. Pogue-Geile, M.F. (2003). Schizophrenia spectrum disorders. In *Encyclopedia of the Human Genome*, D. Cooper (Ed.), Vol. 5, pp. 185–189. John Wiley & Sons, Ltd., Chichester, UK.

53. Prescott C.A. and Gottesman, I.I., Genetically mediated vulnerability to schizophrenia. *Psychiatric Clinics of North America*, 1993, 16, 245–267.

54. Risch, N., Genetic linkage and complex diseases, with special reference to psychiatric disorders. *Genetic Epidemiology*, 1990, 7: 3–16.

55. Risch, N., Linkage strategies for genetically complex traits: I Multilocus models. *American Journal of Human Genetics*, 1990, 46: 222–228.

56. Risch, N. and Merikangas, K., The future of genetic studies of complex human diseases. *Science*, 1996, 273, 1516–1517.

57. Rosenthal, D., Wender, P.H., Kety, S.S., Schulsinger, F., Welner, J., Østergaard, L., Schizophrenics' offspring reared in adoptive homes. In D. Rosenthal, S.S. Kety (Eds.), *The Transmission of Schizophrenia*. Pergamon Press, Oxford, 1968, pp. 377–391.

58. Rüdin, E., *Zur Vererbung und Neuentstehung der Dementia Praecox*. Springer, Berlin, 1916.

59. Schizophrenia Linkage Collaborative Group for chromosomes 3,6, and 8: Additional support for schizophrenia linkage on chromosomes 6 and 8: A multicenter study. *American Journal of Medical Genetics (Neuropsychiatric Genetics)*, 1996, 67, 580–594.

60. Sherrington, R., Brynjolfsson, J., Petursson, H., Potter, M., Dudleston, K., Barraclough, B., Wasmuth, J., Dobbs, M., and Gurling, H., Localization of a susceptibility locus for schizophrenia on chromosome 5. *Nature*, 1988, 336, 164–167.

61. Shields J. and Gottesman I.I., Cross-national diagnosis of schizophrenia in twins. *Archives of General Psychiatry*, 1972, 27, 725–730.

62. Stefansson, H. et al., Neuregulin 1 and susceptibility to schizophrenia. *Am. J. Hum. Genet.* 71: 877–892, 2002.

63. Straub, R.E., MacLean, C.J., O'Neill, F.A., Burke, J., Murphy, B., Duke, F., Shinkwin, R., Webb, B.T., Zhang, J., Walsh, D., Kendler, K.S., A potential vulnerability locus for schizophrenia on chromosome 6p24–22: Evidence for genetic heterogeneity. *Nature Genetics*, 1995, 11, 287–293.

64. Straub, R.E. et al. Genetic variation in the 6p22.3 gene DTNBP1, the human ortholog of the mouse dysbindin gene, is associated with schizophrenia. *Am. J. Hum. Genet*, 2002, 71: 337–348.

65. Thompson, J.L., Pogue-Geile, M.F., and Grace, A.A. Developmental pathology, dopamine, and stress: A model for the age of onset of schizophrenia symptoms. *Schizophrenia Bulletin*, 2004, 30, 875–900.

66. Thompson, J.L., Watson, J.R., Steinhauer, S.R., Goldstein, G., and Pogue-Geile, M.F. Indicators of genetic liability to schizophrenia: A sibling study of neuropsychological performance. *Schizophrenia Bulletin*, 2005, 31, 85–96.

67. Vogler, G., Gottesman, I.I., McGue, M.K., Rao, D.C., Mixed-model segregation analysis of schizophrenia in the Lindelius Swedish pedigrees. *Behavior Genetics*, 1990, 20: 461–472.

68. Wahlberg, K-E., Wynne, L.C., Oja, H., Keskitalo, P., Pykäläinen, L., Lahti, I., Moring, J., Naarala, M., Sorri, A., Seitamaa, M., Läksy, K., Kolassa, J., Tienari, P., Gene–environment interaction in vulnerability to schizophrenia: Findings from the Finnish adoptive family study of schizophrenia. *American Journal of Psychiatry*, 1997, 154: 355–362.
69. Weeks, D.E. and Lange, K., The affected-pedigree-member method of linkage analysis. *American Journal of Human Genetics*, 1988, 42: 315–326.
70. Wong, A.H.C., Gottesman, I.I., and Petronis, A., Phenotypic differences in genetically identical organisms: the epigenetic perspective. *Human Molecular Genetics*, 2005, 14, 11–18.
71. Woolf, C.M., Does the genotype for schizophrenia often remain unexpressed because of canalization and stochastic events during development? *Psychological Medicine*, 1997, 27, 659–668.

16 Genetics of Major Affective Disorders

Wade Berrettini

CONTENTS

INTRODUCTION

This chapter reviews some aspects of the genetic epidemiology and molecular genetic research on bipolar disorders (BDs) and recurrent unipolar disorders (RUPs). Genetic concepts of linkage and linkage disequilibrium (LD) are reviewed. Given that the inherited susceptibilities for BD and RUP are explained by multiple genes of small effect, simulations indicate that universal confirmation of vulnerability genes cannot be expected due to power issues, sampling variation, and genetic heterogeneity. With this background, several valid linkages of BD to genomic regions are reviewed, including some that may be shared with schizophrenia. These results suggest that nosology must be changed to reflect the genetic origins of the multiple disorders that are collectively described by the term *BD*. The briefer history of RUP molecular linkage and LD studies is also reviewed.

16.1 BIPOLAR DISORDERS

16.1.1 GENETIC EPIDEMIOLOGY: FAMILY STUDIES

The optimal design for a BD family study is one in which relatives of controls and relatives of BD probands are directly interviewed, and diagnoses are made by individuals who are blind to the identity and family origin of the individual under scrutiny. Family studies of bipolar illness show that a spectrum of mood disorders is found among the first-degree relatives of BD probands: DDBDI, BDII with major depression (hypomania and RUP illness in the same person), schizoaffective disorders and RUP illness.[1-12] The family studies suggest shared liability for BD and RUP disorders.

No BD family study conducted in an optimal manner reports increased risk for schizophrenia (SZ) among relatives of BD probands. Similarly, no SZ family study reports increased risk for BD disorders among relatives of SZ probands; however, several SZ family studies report increased risk for RUP and schizoaffective disorders among relatives of SZ probands.[11-14] These family studies are consistent with some degree of overlap in susceptibility to RUP and schizoaffective disorders for relatives of BP probands and relatives of SZ probands. Kendler et al.[14] specifically note an increase in risk for psychotic affective disorders among the relatives of SZ probands.

Potash et al.[15,16] reported that psychotic affective disorders cluster in families. Risk for psychotic affective disorders was significantly higher among the relatives of psychotic BD probands compared to the risk for relatives of nonpsychotic BD probands. This raises the possibility that the partial overlap in risk BD and SZ nosological categories is due to a subset of BD characterized by psychotic symptoms. This subset of BD is probably quite common, as the majority of bipolar (BP) probands from the Potash et al.[16] study were psychotic.

16.1.2 GENETIC EPIDEMIOLOGY: TWIN STUDIES

Twin studies of BD have been conducted over the past 70 years. In nearly all of these reports, RUP illness in the cotwin of a BP index case was grounds for categorizing the twin pair as concordant. Bertelsen et al.[17] and Allan et al.[18] report that approximately 20 percent of concordant monozygotic twin pairs were comprised of a BP index twin and a RUP cotwin. These older studies (Table 16.1) were conducted prior to the introduction of operationalized diagnostic criteria and semi-structured interviews. However, the older results are quite consistent with the more recent studies,[14,19] reporting significantly higher monozygotic twins concordance rates compared with those for dizygotic twins (Figure 16.1). Recent estimates of heritability are approximately 80%, and approximately 30% of this is shared liability with RUP illness.[19]

Mendlewicz and Rainer[25] reported a controlled adoption study of BD probands, including a control group of probands with poliomyelitis. The biological relatives of the BP probands had a 31% risk for BD or unipolar (UP) disorders, as opposed to 2% in the relatives of the control probands. The risk for affective disorder

TABLE 16.1
Concordance Rates for Affective Illness in Monozygotic and Dizygotic Twins*

Study	Monozygotic Twins Concordant Pairs/Total Pairs (%)		Dizygotic Twins Concordant Pairs/Total Pairs (%)	
Luxenberger[20]	3/4	75.0	0/13	0.0
Rosanoff et al.[21]	16/23	69.6	11/67	16.4
Slater[22]	4/7	57.1	4/17	23.5
Kallman[23]	25/27	92.6	13/55	23.6
Harvald and Hauge[24]	10/15	66.7	2/40	5.0
Allen et al.[18]	5/15	33.3	0/34	0.0
Bertelsen et al.[17]	32/55	58.3	9/52	17.3
Totals	95/146	65.0	39/278	14.0

*Data not corrected for age. Diagnoses include both bipolar and unipolar illness.

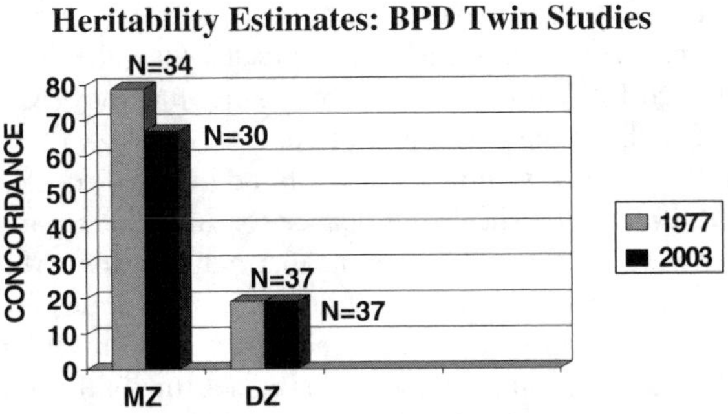

FIGURE 16.1 Two BD twin studies, conducted 25 years apart, are compared for concordance rates in monozygotic (MZ) and dizygotic (DZ) twins. These studies yield very similar results, including estimates of heritability of ~80%. Twins were clinically ascertained through a BPD proband. Co-twins were concordant if BPD or RUP was present. From McGuffin P., et al. *Arch Gen Psychiatry.* 2003;60:497–502, and Bertelsen A, et al. *Br J Psychiatry.* 1977;130:330–351.

in biological relatives of adopted BD patients was similar to the risk in relatives of BD patients who were not adopted (26%). Adoptive relatives do not show increased risk compared with relatives of control probands.

Wender et al.[26] and Cadoret et al.[27] studied UP and BD probands. Although evidence for genetic susceptibility was found, adoptive relatives of affective probands had a tendency to excess affective illness themselves, compared with the adoptive relatives of controls. Von Knorring et al.[28] did not find concordance in psychopathology between adoptees and biological relatives when examining the records of 56 adoptees with UP disorders.

16.1.3 Molecular Linkage Studies

Linkage refers to the observation that two DNA sequences found near each other on the same chromosome tend to be inherited together more often than expected by chance within families. Such DNA sequences are said to be linked. The log of odds (LOD) score refers to the probability that observed cosegregation of alleles at distinct DNA sequences within a family has occurred because the two DNA sequences are linked. An LOD score greater than 3 is evidence, but not proof, that two DNA sequences are linked. The numerical value of the LOD score is dependent on the proposed mode of inheritance (dominant, recessive, sex-linked) and penetrance. Because the LOD score is dependent on these parameters (mode of inheritance and penetrance), it is sometimes termed a parametric statistic. This dependence on inheritance mode and penetrance distinguishes the LOD score from nonparametric statistics (including affected sibling pair and affected pedigree member methods), because such statistics are not dependent on mode of inheritance or penetrance. These nonparametric statistics use kinship coefficients to estimate the randomly expected degree of DNA sharing among affected members of the same family. For example, siblings share 50% of DNA sequences randomly because they have the same parents. If one has DNA samples from 1,000 pairs of ill siblings, one could search through the genomes of those 2,000 persons to find regions where DNA sequences that are shared are significantly greater than the baseline 50% rate. Regions of increased DNA sharing may harbor genes that may explain in part why both members of each sibling pair are affected.

What level of statistical significance should be required for declaring linkage? A recommended level of statistical significance for an initial report ($p < \sim0.00002$) is a stringent criterion, based on simulations that indicate this level of significance would occur less than five times randomly in 100 genome scans for linkage.[29] This statistical criterion assumes that all the genetic information within the pedigrees studied would be extracted, an assumption that is not true in practice. Typically, no more than approximately 80% of the genetic information in a pedigree series is extracted through genotyping; however, as in any area of science, no single report of linkage should be accepted as valid without independent confirmation. The requirement for independent confirmations (at $p < 0.01$) is not waived, no matter the level of statistical significance achieved in a single report. This confirmation requirement should be seen within the context that valid linkages will not be confirmed in some studies. Indeed, nonconfirmations should be expected, intuitively, because of population (ethnic) differences, sampling procedures and genetic heterogeneity.

The probability of confirmation in simulations has been examined by Suarez et al.,[30] who simulated a disorder caused by any one of six loci and determined that a large sample size and substantial time will be required for an initial linkage to be confirmed in a second sample. It is clear from Suarez's simulations that consistent detection of a locus of moderate effect cannot be expected. Nonconfirmatory studies will always occur when an initially detected linkage is valid.

One of the most critical issues in confirmation of reported linkages is power. Attempts at confirmation of a reported susceptibility locus should state what power has been achieved to detect the locus initially described. For example, if a locus increases risk for BP illness by a factor of two, it may be necessary to study approximately 200 affected sibling pairs in order to have adequate (90%) power to detect such a locus.[31]

Unfortunately, few studies address this key issue. If 200 affected sibling pairs are required to achieve adequate (90%) power to detect a previously described locus, then a publication with less than 150 sibling pairs does not address the central issue of confirmation. However, such power-limited publications may have an important role in meta-analyses, in that they identify invaluable sources of additional data. Comprehensive scans of the human genome have been completed with sufficient numbers (e.g., >100) of BP individuals.[32–40] If a major locus (explaining >50% of the risk in >50% of BPD persons) existed, it would have been detected in many of these studies. Thus, no such major locus exists for BPD. There are several confirmed reports of loci of smaller effect, which can be termed susceptibility loci. These loci are neither necessary nor sufficient for disease, but increase risk for the disorder in a non-Mendelian manner.

From these genome scans and from additional, smaller studies, a picture has emerged in which there are approximately 10 confirmed DBD linkage regions across the genome. It is highly probable that additional confirmed BD linkages will be identified through future linkage studies. These BP linkage regions are confirmed by virtue of at least one study with strong statistical significance ($p < 0.0001$) and at least two confirmatory studies ($p < 0.01$). As noted elsewhere,[41] in some cases, these confirmed BP linkage regions overlap with schizophrenia linkage reports, suggesting that the same loci may be involved in some aspects of both disorders. In Figure 16.2, these are mapped onto an ideogram of the human genome.

The studies which support these findings are listed in Table 16.2, with the identified primary report cited as the first study with genome-wide statistical significance.

Two methodological approaches have been used to conduct meta-analyses of BP linkage studies.[73,74] Badner and Gershon[73] analyzed linkage results using a multiple scan probability approach, in which p values are combined across studies, after adjusting for the size of the linkage region. These authors concluded that two genomic regions, 13q32 and 22q,[11–13] were the most promising loci for DBD disorder. Segurado et al.[75] used the method of Levinson et al.[74] which ranks the p values across the genome of each study, then sums the rankings for each genomic "bin." In this approach, no genomic region reached genome-wide significance, although the region that seemed most promising was the pericentromeric region of 18.[75]

16.1.4 LINKAGE DISEQUILIBRIUM STUDIES

The human genome consists of approximately three billion base pairs of DNA.[76] The recent completion of draft genomic sequences of the human genome[76] is consistent with approximately 35,000 to 40,000 genes. Physical distance along the linear sequence of DNA can be expressed in terms of base pairs of DNA. The most common sequence variation in the human genome is a single nucleotide polymorphism (SNP), where at the same position on different chromosomes there

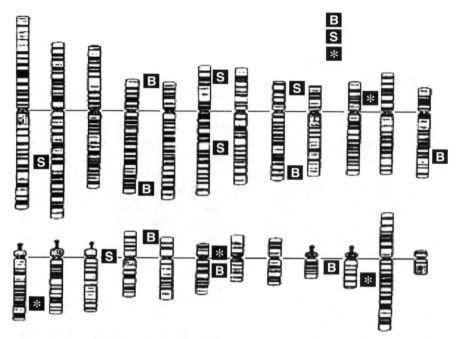

FIGURE 16.2 (SEE COLOR INSERT FOLLOWING PAGE 236) Confirmed linkage loci for bipolar disorder (B), schizophrenia (S), and both disorders (*) on a diagram of the human genome.

are two different nucleotides (from the possible four found among *homo sapiens*: adenine [A], guanine [G], thymidine [T], and cytosine [C]). SNPs with a common minor allele (frequency of approximately 20% or more) occur approximately every 1,000 base pairs of DNA.[76] Analysis of closely spaced SNPs in outbred populations suggests a complex pattern of inheritance in which recombination is inhibited in a small region of DNA, such that blocks of DNA (containing multiple SNPs) tend to be inherited intact over many generations.[77] Thus, blocks of DNA are shared among present-day individuals who may have had a common ancestor 10,000 generations ago. These blocks are variable in length and often contain multiple SNPs, but among outbred human populations the block length rarely exceeds approximately 100,000 base pairs. Alleles of SNPs within a block form a haplotype (a set of alleles) that is usually inherited together across many generations. Such SNPs are said to be in strong LD with one another. LD refers to the fact that two (or more) alleles can be found together in unrelated individuals more often than predicted by chance. The interested reader is referred to primary reports concerning LD.[77]

LD is a useful tool to investigate the relatively small genomic regions that have been implicated in the genetic origination of DBD through linkage studies (Table 16.2). In this process, SNPs spaced across genes in the linkage region are assessed in large groups (ideally at least several hundreds) of ethnically matched

TABLE 16.2
Confirmed Linkage Regions in Bipolar Disorder

Location	Primary Report	Independent Confirmations	Comments
18p11	Berrettini et al., 1994 and 1997	Stine et al., 1995; Nothen et al., 1999; Turecki et al., 1999; Bennett et al., 2002	Paternal parent-of-origin effect; see Schwab et al., 1998; Lin and Bale, 1997
21q22	Straub et al., 1994	Detera-Wadleigh et al., 1996; Smyth et al., 1996; Kwok et al., 1999; Morissette et al., 1999; Aita et al., 1999	
22q11	Kelsoe et al., 2001	Detera-Wadleigh et al., 1999; Lachman et al., 1997	Velocardiofacial syndrome region; also a SZ locus: Gill et al., 1996
18q22	Stine et al., 1995	McInnes et al., 1996; McMahon et al., 1997; De Bruyn et al., 1996; McInnis et al., 2003	See Freimer et al., 1996
12q23	Morissette et al., 1999	Ewald et al., 2002; Maziade et al., 2002; Ekholm et al., 2003; Curtis et al., 2003; Dawson et al., 1995	Primary report in a Canadian isolate; Abkevich et al., 2003
8q24	Cichon et al., 2001	McInnis et al., 2003; Dick et al., 2003	
13q32	Detera-Wadleigh et al., 1999	Kelsoe et al., 2001; Potash et al., 2003; Liu et al., 2004; Badenohop et al., 2001	See Brzustowicz et al., 1999; Blouin et al., 1998; Chumakov et al., 2002
16p12	Ewald et al., 2002	Ekholm et al., 2003; Dick et al., 2003	
4q32	Ekholm et al., 2003	Adams et al., 1998; McInnis et al., 2003; Liu et al., 2004	
4p15	Blackwood et al., 1996	Ewald et al., 2002; Cichon et al., 2001; Morissette et al., 1999; Detera-Wadleigh et al., 1999	See Ginns et al., 1998

cases and controls. Investigators compare allele and genotype frequencies among groups of cases and controls.

There have been a multitude of LD studies in BD over the past decade. In a typical report, allele and genotype frequencies in the BD cases and controls are examined at a single candidate gene variant. If nominally significant differences in allele or genotype frequencies are found between groups, the authors might conclude that the variant influences the risk for BD disorder.

Most often, these studies have assembled a small group of BD patients and unaffected controls from a population. These studies have typically employed one variant in a single candidate gene which is selected based on presumed central nervous system (CNS) function, in relation to BD pathophysiologic theories. Unfortunately, the nearly complete absence of pathophysiologic data in BDBD makes the process of rational candidate gene selection difficult.

Additionally, these studies have typically involved smaller numbers of BD patients than is optimal, given that the effect size of individual alleles on risk must be small. Lastly, these studies have often used gene variants which are not known to confer functional differences in the gene. Despite these difficulties, there are several candidate genes which deserve mention.

One promising candidate gene is the G72 locus on 13q32, the site of a confirmed linkage in BD and SZ (Table 16.2). G72 is a primate-specific brain-expressed gene that activates D-amino acid oxidase.[69] D-amino acid oxidase may control levels of D-serine, which regulates glutamatergic receptors.[78] Chumakov et al.[69] identified a haplotype from G72 SNPs (without obvious functional significance), which were in LD with SZ in a French-Canadian sample. This has been confirmed in distinct SZ populations, including Russian[69] and German[79] populations, although different haplotypes have been associated in different ethnic populations. Similarly, in BD there have been several positive findings with distinct haplotypes in different populations, including American[80,81] and German[79] BD samples. Although the data is promising, no clear functional variants have been defined at this locus.

A second promising candidate gene is brain-derived neurotropic factor (BDNF), for which there are several positive reports at a functional mis-sense variant.[82,83] In the several European-origin populations studied at a G/A SNP (Val/Met), the G (Val) allele was over-transmitted. Egan et al.[84] demonstrated that this SNP confers functional difference.

These reports are promising because they consistently identify a functional variant contributing to genetic risk for DBD. However, there are small studies of Japanese and Chinese BD patients that do not show any evidence of this variant influencing risk for BD disorder.[85,86] The effect may be limited to populations of European origins. Alternatively, the negative studies may have been underpowered. It must be remembered that some variants will be relatively specific to particular ethnic groups. Consider that the protective effect of aldehyde dehydrogenase deficiency on risk for alcoholism is easily demonstrated in Chinese, Korean, and Japanese populations because the deficiency allele has a frequency of approximately 30%.[87,88] Much larger sample sizes are required to detect this influence in European populations because the deficiency allele frequency is lower by an order of magnitude.

There have been numerous independent-association studies of BD and RUP and an MAOA (CA) repeat polymorphism in European[89,94] and Asian[95,96] populations. Those studies reporting a positive association[90,93-95] generally detect an overrepresentation of allele 5 or 6 of the MAOA (CA) n repeat among BD patients compared with controls, which is an observation that may be particularly evident with respect to women. The effect size is small, the odds ratio being 1.49,[94] and the sample size required for adequate power to detect is larger than most of the negative studies.[89,91,92,96,97] There is also an MAOA promoter polymorphism.[98] These studies involve multiple ethnic groups, case-control methods, and family-based designs, with some studies having limited power to detect a small effect size. Thus, it is understandable that conflicting studies are reported.

Another intensively studied candidate gene is the serotonin transporter (5HTT), a functional candidate gene for which multiple BDLD studies have been published. The 5HTT represents a logical candidate gene, as many antidepressants act through binding to the 5HTT protein.[99] There are two variants of the 5HTT that have been studied in BD, and both have functional significance, based on *in vitro* analysis of these noncoding polymorphisms. The first variant is an insertion/deletion polymorphism in the 5HTT promoter region. The shorter allele has much less transcriptional activity than the longer allele.[100] Moreover, the shorter allele has been associated with anxiety-related personality traits in humans.[101] The second variant is a variable number of tandem repeats (VNTRs) polymorphism in intron 2. The two most common alleles are the 10 and 12 repeats, which confer differential transcriptional activity in an embryonic stem cell line.[102] Collier et al.[100] first reported that the 5HTT intron 2 VNTR allele 12 was in LD with BD among patients from the U.K. Collier et al. also reported that the short allele of the 5HTT promoter variant was more common among 454 European BD and RUP patients, compared with 570 European controls, although the statistical significance was marginal ($p = 0.03$), emphasizing the small effect size involved. Analysis by genotype suggested that homozygosity for the short allele was associated with BD ($p < 0.05$) and RUP ($p < 0.01$).

Rees et al.[104] confirmed the observation of Collier et al.[100] that allele 12 of the intron 2 VNTR was in LD with BD among 171 BD probands, compared with 121 controls ($p = 0.031$). Similarly, Rees et al. studied BD patients and controls, reporting an excess of BD patients among individuals homozygous for the shorter promoter allele, implying a recessive mode on inheritance. Note that the sample sizes for Rees et al.[104] and for Collier et al.[103] were in the hundreds.

Vincent et al.[105] studied an initial sample of approximately 100 BD probands from Canada, confirming the observation that the promoter short allele was in LD with BD, compared with approximately 100 controls; however, he then failed to confirm this observation in a second set of approximately 100 BD probands. Sampling variation and the small effect size, coupled with limited power of this sample size, are probable explanations for these results.

Gutierrez et al.[106] studied BD probands and controls from Spanish origin. They reported no evidence for LD with 5HTT alleles. This may be secondary to the ethnic background of patients or to small sample size. Bocchetta et al.[107] studied approximately 55 Sardinian parent–child BD trios, finding no evidence for transmission disequilibrium in the *5HTT* gene, although sample size was a limiting factor in their conclusions. Studying 123 BD parent–child trios of European origin, Mundo et al.[108] reported no evidence for LD with the 5HTT promoter alleles. Mynett-Johnson et al.[109] studied approximately 100 Irish BD parent–child trios from multiplex families, reporting that a haplotype including the shorter promoter allele and a 3′UTR SNP conferred risk for BD.

Kirov et al.[110] studied 122 parent–child trios of British ethnic background, with no nominally significant results at either polymorphism. In a study of 50 Indian BD patients and controls using the VNTR variant, no evidence for LD was reported,[111] this result being limited by the small sample size. From another ethnic perspective,

Mendes de Oliveira et al.[112] studied a small number of Brazilian BD patients, finding no evidence for LD with the 5HTT promoter polymorphism. Kunugi et al.[113] studied these two polymorphisms in a Japanese sample of 191 patients with affective disorders (142 bipolar and 49 unipolar) and 212 controls. They report nominally significant LD between the VNTR and BD, with no evidence for LD with the promoter variant.

Furlong et al.[114] reported results of a meta-analysis for approximately 1,400 individuals of European origin, including 772 controls, 375 bipolar patients and 299 unipolar patients. Although there was no evidence for LD with affective disorders for the VNTR, a marginally significant result was found for the short allele of the 5HTT promoter polymorphism. This result is important because it suggests that samples in the thousands will be necessary to draw firm conclusions, due to the small effect sizes involved.

What clinical characteristics might describe those BD probands with illness secondary to 5HTT alleles? One preliminary answer is derived from two studies on independent populations. Coyle et al.[115] studied BD women with post-partum psychotic episodes, demonstrating a large effect size for the 5HTT promoter short allele, with an attributable risk of 69%. Ospina-Duque et al.[116] studied 100 BD patients from a Columbian population isolate and approximately 100 controls, showing that those patients with psychosis as part of the phenotype showed LD with the shorter 5HTT promoter allele. Their conclusions may be similar to those of Coyle et al.[115] and suggest that this deserves further study.

In another large European study, Mendlewicz et al.[117] examined the genetic contribution of the 5-HTT promoter polymorphism in a case-control sample, including 539 RUP patients, 572 BD patients, and 821 controls. No evidence of LD was found for RUP or BD, and subdividing the sample according to family history, suicidal attempts, or psychotic features did not reveal any role of the promoter variant in the genetic susceptibilities to these disorders.

16.2 RECURRENT UNIPOLAR DISORDERS

16.2.1 GENETIC EPIDEMIOLOGY: FAMILY STUDIES

In considering optimally designed family studies of RUP (with blinded interviews and simultaneous examination of control relatives), there are five reports in the literature.[1,2,10,12,118] These five reports yield remarkably similar results, each study concluding that the first-degree relatives of RUP probands were at increased risk for RUP disorders, compared to first-degree relatives of control probands. Across the five studies, there was a two- to fourfold increased risk for RUP among the first-degree relatives of RUP probands.

Characteristics of RUP disorders that yield a more heritable phenotype include early onset (e.g., before age 30)[27,118–122] and a high degree of recurrence.[13,121,124–126] A third characteristic that may identify a separate group of disorders is the presence of psychosis.[14] Additional genetic subtypes of RUP may be identified through examination of comorbidities with panic disorder and other anxiety disorders, as well as with alcoholism.[127–129]

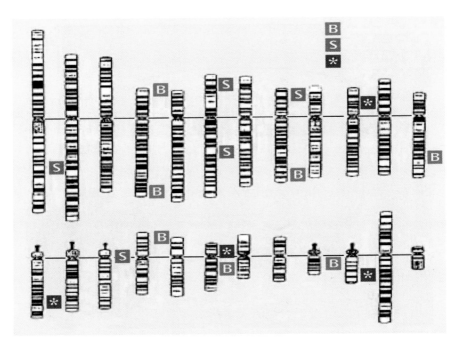

FIGURE 16.2 Confirmed linkage loci for bipolar disorder (B), schizophrenia (S), and both disorders (*) on a diagram of the human genome.

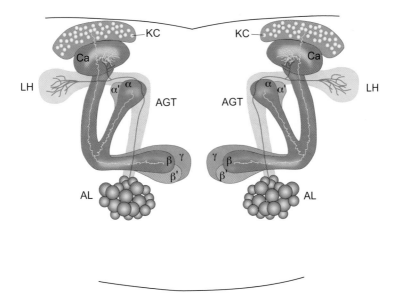

FIGURE 23.1 *Drosophila* olfactory memory system. Olfactory sensory neurons project to the antennal lobes (ALs). From there, projection neurons project through the antenno-glomerular tract (AGT) and connect mushroom body (MB) dendrites localized in the calyx (Ca), as well as the lateral horn (LH). Each MB is composed of about 2,500 neurons, the Kenyon cells (KC). Three types of KC project in five lobes: α/β, α'/β' and γ.

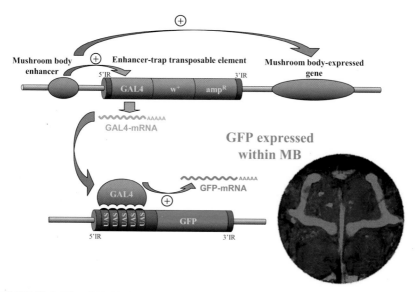

FIGURE 23.2 The GAL4/UAS system. When an enhancer-trap transposable element is inserted near transcription enhancers that control expression in a given structure, the GAL4 gene that is contained in the P-element is expressed in the same structure. If this fly contains also a reporter gene downstream of the UAS sequences, GAL4 will transcribe the reporter.

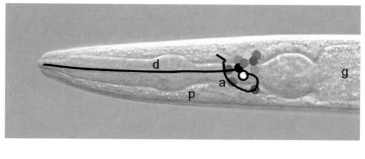

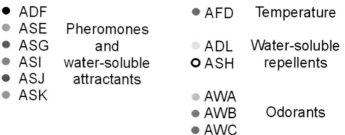

FIGURE 24.3 DIC photograph of the head of a worm. The outline of the pharynx (p) can be seen and the beginning of the gut (g). The approximate positions of the amphid sensory neuron cell bodies are indicated with colored circles. The dendrite (d) and an axon (a) of one of the cells are drawn. Below, the names of the corresponding cells are given, and the compounds perceived by these cells.

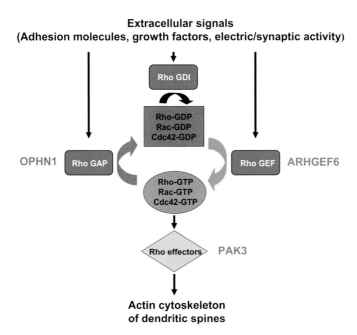

FIGURE 28.1 One cluster of nonsyndromic MR genes relates to the Rho family of small GTPases. Rho family GTPases transduce extracellular signals into adaptive responses of the actin cytoskeleton. Actin-dependent processes at the synapse include the regulation of the morphology and the dynamics of the dendritic spines. The Rho family comprises three members, termed Rho, Rac, and Cdc42. They shuttle between an inactive (red) and an active (green) state under the control of three types of regulatory proteins (blue). GEFs (guanine nucleotide exchange factors) promote the release of GDP and its replacement by GTP and, thereby, mediate the transition of Rho, Rac, and Cdc42 from the inactive into the active state. GAPs (GTPase-activating proteins) activate the endogenous GTPase function and, thus, the self-inactivation of Rho, Rac, and Cdc42. GDIs (GDP dissociation inhibitors) bind to the GDP form of the GTPases and prevent their premature activation as long as the GTPase is not in the correct place and situation for a new round of activation. Further downstream activators (yellow) eventually act directly or indirectly on regulatory components of the actin cytoskeleton. The genes affected in MRs are printed red.

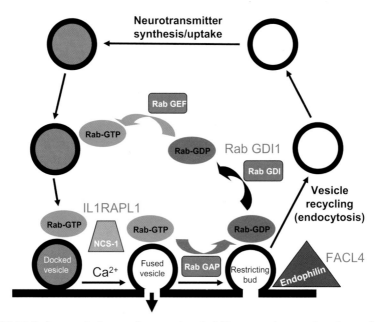

FIGURE 28.2 A second cluster of nonsyndromic MR genes relates to the release of neurotransmitters from the presynaptic nerve terminal and the regeneration of new releasable vesicles by the budding and endocytosis of vesicles from the presynaptic membrane. The release of neurotransmitters from storage vesicles occurs in several consecutive steps. First, the vesicles are docked to the release site, the presynaptic active zone, a process that involves the interaction of several proteins of the vesicle and the active zone. Docked vesicles fuse with the presynaptic membrane in a process regulated by transiently increased intracellular calcium, and release their content into the synaptic cleft. In order to maintain a constant supply of releasable neurotransmitters, vesicles recycle by budding off the presynaptic membrane and endocytosis. The budding occurs in a clathrin-dependent manner. The restriction of the bud neck, a process that prepares the bud for scission, requires endophilin. Endophilin is a lysophosphatidic acid acyltransferase, which introduces arachidonic acid into the cytoplasmic face of the bud neck. It, thereby, facilitates the formation of a strong negative curvature of the membrane required for the restriction of the bud neck in order to allow the scission process to begin. It is conceivable that the supply of arachidonoyl-CoA requires FACL4. The genes affected in MR are printed in red.

16.2.2 GENETIC EPIDEMIOLOGY: TWIN STUDIES

A recent review of twin studies in RUP disorder estimated heritability at 37%, with a substantial component of unique individual environmental risk but little shared environmental risk.[130] These twin studies included four community-ascertained samples[131–134] and two clinically ascertained samples, one from the United Kingdom[135] and one from Sweden.[132] The results are quite consistent in concluding that genetic influence is a significant factor in risk for RUP, independent of ascertainment and country or origin.

16.2.3 MOLECULAR LINKAGE STUDIES

There have been just several RUP genome scans with greater than approximately 100 affected individuals, in contrast to BD disorder. Holmans et al.[136] reported on the first phase of a multisite collaborative effort. The sample consisted of 297 informative multiplex families (containing 685 informative affected relative pairs, 555 sibling pairs, and 130 other pair types). Affected cases had RUP with onset before age 31 for probands or 41 for other affected relatives; the mean age at onset was 18.5 and the mean number of depressive episodes was 7.3, indicating a highly recurrent form of illness. Families were excluded if there was a BD first-degree or second-degree relative. Linkage was observed on chromosome 15q25.3–26.2 (empirical genome-wide $p = 0.023$). The linkage was not sex-specific. This was the sole significant linkage peak observed by this group.

Abkevich et al.[65] reported a genome scan on 110 Utah pedigrees (each with at least four affected individuals), in which there were 784 RUP individuals, 161 persons with single-episode major depressive disorder and 162 BD individuals, who were also considered affected. They observed a highly significant linkage signal at 12q23 ($p = 0.0000007$), confirming a previously identified DBD locus (Table 16.2). There were no other linkage peaks approaching statistical significance. It is probable that this study has detected the same BD 12q23 locus—even though the families were ascertained from a RUP proband because most kindreds probably had at least one BD individual. These results confirm family and twin studies, suggesting genetic overlap between BD and RUP disorders, and this study identifies the 12q23 region as a locus which increases risk for both BD and RUP disorders.

Zubenko et al.[137] reported on a genome scan of 81 families ascertained through a proband with early-onset, nonpsychotic RUP disorder. They describe a highly significant linkage ($p < 0.0001$) of this phenotype to 2q35 near marker D2321, which is near a candidate gene, CREB1 (cyclic AMP response element binding protein 1). Sequence variants in the *CREB* gene were found to segregate with RUP disorder among women in 2 of these 81 extended kindreds,[138] thus nominating CREB as a RUP susceptibility gene. These intriguing results await independent confirmation.

SUMMARY

Family, twin, and adoption studies of BD and RUP disorders are reviewed. They are, in general, consistent with substantial heritable components to risk, with the

BD disorders having higher heritability than the RUP disorders. Multiple regions of the genome, including 18p11, 18q22, 12q24, 21q21, 13q32, 4p15, 4q32, 16p12, 8q24, and 22q11, have been implicated by several independent groups in the genetic origins of BD. It is likely that most of these regions will yield susceptibility genes within the next 5 years through the application of LD mapping methods to large sample sizes. LD approaches to candidate genes have yielded several promising prospects, including G72 and BDNF for DBD. For each of these candidate genes, there are several independent BD populations yielding data consistent with the existence of one or more haplotypes as susceptibility sequences. Only several genome scans for RUP disorders have been published, and confirmations are required.

REFERENCES

1. Gershon, ES, Hamovit, J, Guroff, JJ, et al. A family study of schizoaffective, bipolar I, bipolar II, unipolar, and normal control probands. *Arch Gen Psychiatry* 1982;39:1157–67.
2. Weissman, MM, Gershon, ES, Kidd, KK, et al. Psychiatric disorders in the relatives of probands with affective disorder. *Arch Gen Psychiatry* 1984;41:13–21.
3. Baron, M, Gruen, R, Anis, L, Kane, J. Schizoaffective illness, schizophrenia and affective disorders: morbidity risk and genetic transmission. *Acta Psychiatrica Scand* 1983;65:253–62.
4. Winokur, G, Tsuang, MT, Crowe, RR. The Iowa 500: Affective disorder in relatives of manic and depressed patients. *Am J Psychiatry* 1982;139:209–12.
5. Winokur G, Coryell, Keller, M, et al. A family study of manic-depressive (bipolar I) disease. Is it a distinct illness separable from primary unipolar depression? *Arch Gen Psychiatry* 1995;52:367–73.
6. Helzer, JE and Winokur, G. A family interview study of male manic-depressives. *Arch Gen Psychiatry* 1974;31,73–7.
7. James, NM and Chapman, CJ. A genetic study of bipolar affective disorder. *Br J Psychiatry* 1975;126:449–56.
8. Johnson, GFS and Leeman, MM. Analysis of familial factors in bipolar affective illness. *Arch Gen Psychiatry* 1977;34:1074–83.
9. Angst, J, Frey, R, Lohmeyer, R, and Zerben-Rubin, E. Bipolar manic depressive psychoses: results of a genetic investigation. *Hum Genet* 1980;55:237–54.
10. Tsuang, MT, Winokur, G and Crowe, RR. Morbidity risks of schizophrenia and affective disorders among first-degree relatives of patients with schizophrenia, mania, depression, and surgical conditions. *Br J Psychiatry* 1980;137:497–504.
11. Taylor, MA, Berenbaum, SA, Jampala, VC and Cloninger, CR. Are schizophrenia and affective disorder related? Preliminary data from a family study. *Am J Psychiatry* 1993;150:278–85.
12. Maier, W, Lichtermann, D, Minges, J, et al. Continuity and discontinuity of affective disorders and schizophrenia. Results of a controlled family study. *Arch Gen Psychiatry* 1993;50:871–83.
13. Gershon, ES, DeLisi, LE, Hamovit, J, et al. A controlled family study of chronic psychoses. *Arch Gen Psychiatry* 1988;45:328–36.
14. Kendler, KS, McGuire, M, Gruenberg, AM, et al. The Roscommon family study. *Arch Gen Psychiatry* 1993;50:527–40.

15. Potash, JB, Willour, VL, Chiu, YF, et al. The familial aggregation of psychotic symptoms in bipolar disorder pedigrees. *Am J Psychiatry* 2001;158:1258–64.

16. Potash, JB, Chiu, Y-F, MacKinnon, DF, et al. Familial aggregation of psychosis in a replication set of 69 bipolar pedigrees: *Am J Med Genet* 2003;116B:90–7.

17. Bertelsen, A, Harvald, B, and Hauge, M. A Danish twin study of manic-depressive disorders. *Br J Psychiatry* 1977;130:330–51.

18. Allen, MG, Cohen, S, Pollin, W, Greenspan, SI. Affective illness in veteran twins. A diagnostic review. *Am J Psychiatry* 1974;131:1234–9.

19. McGuffin, P, Rijsdijk, S, Andrew, M, et al. The heritability of bipolar affective disorder and the genetic relationship to unipolar depression. *Arch Gen Psychiatry* 2003;60: 497–502.

20. Luxenberger, H. Psychiatrisch-neurologische Zwillings pathologie. *Zentralblatt fur diagesamte Neurologie and Psychiatrie* 1930;14, 56–7,145–80.

21. Rosanoff, AJ, Handy, L, and Plesset, IR. The etiology of manic-depressive syndromes with special reference to their occurrence in twins. *Am J Psychiatry* 1935;91:725–62.

22. Slater, E. The inheritance of manic-depressive insanity. *Proceedings of the Royal Society of Medicine* 1953;29:981–90.

23. Kallman, F. In: Hoch, PH and Zubin, J(eds). *Depression*. New York:Grune & Stratton;1954:1–24.

24. Harvald, B, Hauge, M. Biopolar disorders, in, *Genetics and the Epidemiology of Chronic Diseases*. (PHS Publ No 1163). Edited by Neal, JV, Shaw, MW, and Shull, WJ. Washington, DC, U.S. Department of Health, Education, and Welfare, 1965 pp 61–76.

25. Mendlewicz, J and Rainer, J.D. Adoption study supporting genetic transmission in manic-depressive illness. *Nature* 1977;268:327–9.

26. Wender, H, Kety, SS, Rosenthal, D, et al. Psychiatric disorders in the biological and adoptive families of adopted individuals with affective disorders. *Arch Gen Psychiatry* 1986;43:923–9.

27. Cadoret, RJ, Woolson, R, and Winokur, G. The relationship of age of onset in unipolar affective illness to risk of alcoholism and depression in parents. *J Psychiatr Res* 1977;13:137–42.

28. Von Knorring, AL, Cloninger, CR, Bohman, M, and Sigvardsson, A. An adoption study of depressive disorders and substance abuse. *Arch Gen Psychiatry* 1983;40:943–50.

29. Lander, E, and Kruglyak, L. Genetic dissection of complex traits: Guidelines for interpreting and reporting linkage results. *Nature Genet* 1995;11:241–7.

30. Suarez, B, Harpe, CL, and Van Eerdewegh, P. Problems of replicating linkage claims in psychiatry. In: Gershon ES, Cloninger CR (eds). *Genetic Approaches to Mental Disorders. Proceedings of the 82nd Annual Meeting of the American Psychopathological Assn.*, Washington (DC): American Psychiatric Press;1994:23–46.

31. Hauser, ER, Boehnke, M, Guo, SW, and Risch, N. Affected sib pair interval mapping and exclusion for complex genetic traits. *Genetic Epidemiology* 1996;13:117–37.

32. Detera-Wadleigh, SD, Badner JA, Berrettini, WH, et al. A highdensity genome scan detects evidence for a bipolar-disorder susceptibility locus on 13q32 and other potential loci on 9q32 and 18p11.2. *Proc Natl Acad Sci USA* 1999;96:5604–56099.

33. Dick, DM, Foroud, T, Flury, L, et al. Genome-wide linkage analyses of bipolar disorder: a new sample of 250 NIMH Genetics Initiative pedigrees. *Am J Hum Genet* 2003;73:107–14.

34. McInnis, MG, Lan, T-H, Willour, VL, et al. Genome-wide scan of bipolar disorder in 65 pedigrees: supportive evidence for linkage on 8q24, 18q22, 4q32, 2p12, and 13q12. *Mol Psychiatry* 2003;8:288–98.

35. Bennett, P, Segurado, R, Jones, I, et al. The Wellcome trust UK-Irish bipolar affective disorder sibling pair genome screen: First stage report. *Mol Psychiatry* 2002;7: 189–200.

36. Rice, JP, Goate, A, Williams, JT, et al: Initial genome scan of the NIMH Genetics Initiative bipolar pedigrees: Chromosomes 1, 6, 8, 10, and 12. *Am J Med Genet* 1997;74:247–53.

37. Ekholm, JM, Kieseppa, T, Hiekkalinna, T, et al. Evidence of susceptibility loci on 4q32 and 16p12 for bipolar disorder. *Hum Mol Genet* 2003;12:1907–15.

38. Cichon, S, Schumacher, J, Muller, DJ, et al. A genome screen for genes predisposing to bipolar affective disorder detects a new susceptibility locus on 8q. *Hum Mol Genet* 2001;10:2933–44.

39. Kelsoe, JR, Spence, MA, Loetscher, E, et al. A genome survey indicates a suscepti-bility locus for bipolar disorder on chromosome 22. *Proc Natl Acad Sci* 2001;98:585–90.

40. Liu, J, Juo, SH, Dewan, A, et al. Evidence for a putative bipolar disorder locus on 2p13-16 and other potential loci on 4q31, 7q34, 8q13, 9q31, 10q21 –24, 13q32, 14q21 and 17q11-12. *Molecular Psychiatry* 2003;8:333–42.

41. Berrettini, WH: Evidence for shared susceptibility in bipolar disorder and schizo-phrenia. *Am J Med Genet* 2003;123C:59–64.

42. Berrettini, W, Ferraro, T, Choi, H, et al. Linkage studies of bipolar illness. *Arch Gen Psych* 1997;54:32–9.

43. Stine, OC, Xu, J, Koskela, R, et al. Evidence for linkage of bipolar disorder to chromosome 18 with a parent-of-origin effect. *Am J Hum Genet* 1995;57:1384–94.

44. Nothen, MM, Cichon, S, Rohleder, H, et al. Evaluation of linkage of bipolar affective disorder to chromosome 18 in a sample of 57 German families. *Mol Psychiatry* 1999;4:76.

45. Turecki, G, Grof, P, Cavazzoni, P, et al. Lithium responsive bipolar disorder, unilin-eality and chromosome 18: A linkage study. *Am J Med Genet* 1999;88:411–15.

46. Schwab, SG, Hallmayer, J, Lerer, B, et al. Support for a chromosome 18p locus conferring susceptibility to functional psychoses in families with schizophrenia, by association and linkage analysis. *Am J Hum Genet* 1998;63:1139.

47. Lin, JP, and Bale, SJ. Parental transmission and D18S37 allele sharing in bipolar affective disorder. *Genetic Epidemiology* 1997;14:665–8.

48. Straub, RE, Lehner, T, Luo, Y, et al. A possible vulnerability locus for bipolar affective disorder on chromosome 21q22.3. *Nature Genet* 1994;8:291–6.

49. Detera-Wadleigh, SD, Badner, JA, Goldin, LR, et al. Affected sib-pair analyses reveal support of prior evidence for a susceptibility locus for bipolar disorder on 21q. *Am J Hum Genet* 1996;58:1279–85.

50. Smyth, C, Kalsi, G, Brynjolfsson, J, et al. Further tests for linkage of bipolar affective disorder to the tyrosine hydroxylase gene locus on chromosome 11p15 in a new series of multiplex British affective disorder pedigrees. *Am J Psychiatry* 1996;153:271–4.

51. Kwok, JB, Adams, LJ, Salmon, JA, et al. Non-parametric simulation-based statistical analyses for bipolar affective disorder locus on chromosome 21q22.3. *Am J Med Genet* 1996;88:99–102.

52. Morissette, J, Villeneuve, A, Bordeleau, L, et al. Genome-wide search for linkage of bipolar affective disorders in a very large pedigree derived from a homogeneous population in Quebec points to a locus of major effect on chromosome 12q23-q24. *Am J Med Genet* 1999;88:567–87.

53. Aita, VM, Liu, J, Knowles, JA, et al. A comprehensive linkage analysis of chromo-some 21q22 supports prior evidence for a putative bipolar affective disorder locus. *Am J Hum Genet* 1999;64:210–17.

54. Detera-Wadleigh, SD, Badner, JA, Berrettini, WH, et al. A high density genome scan detects evidence for a bipolar-disorder susceptibility locus on 13q32 and other potential loci on 1q32 and 18p11.2. *Proc Natl Acad Sci USA* 1999;96:5604–9.

55. Lachman, HM, Kelsoe, JR, Remick, RA, et al. Linkage studies support a possible locus for bipolar disorder in the velocardiofacial syndrome region on chromosome 22. *Am J Med Genet* 1997;74:121–8.

56. Gill, M, Vallada, H, Collier, D, et al. A combined analysis of D22S278 marker alleles in affected sib-pairs: support for a susceptibility locus for schizophrenia at chromosome 22q12. *Am J Med Genet* 1996;67:40–5.

57. McInnes, LA, Escamilla, MA, Service, SK, et al. A complete genome screen for genes predisposing to severe bipolar disorder in two Costa Rican pedigrees. *PNAS* 1996;93:13060–5.

58. McMahon, FJ, Hopkins, PJ, Xu, J, et al. Linkage of bipolar affective disorder to chromosome 18 markers in a new pedigree series. *Am J of Hum Genet* 1997;61: 1397–404.

59. De Bruyn, A, Souery, D, Mendelbaum, K, et al. Linkage analysis of families with bipolar illness and chromosome 18 markers. *Biol Psychiat* 1996;39:679–88.

60. Freimer, NB, Reus, VI, Escamilla, MA, et al. Genetic mapping using haplotype, association and linkage methods suggests a locus for severe bipolar disorder (BPI) at 18q22-q23. *Nature Genet* 1996;12:436–41.

61. Ewald, H, Flint, T, Kruse, TA, et al. A genome-wide scan shows significant linkage between bipolar disorder and chromosome 12q24.3 and suggestive linkage to chromosomes 1p22-21, 4p16, 6q14-22, 10q26 and 16p13.3. *Mol Psychiatry* 2002;7: 734–44.

62. Maziade, M, Roy, MA, Rouillard, E, et al. A search for specific and common susceptibility loci for schizophrenia and bipolar disorder: a linkage study in 13 target chromosomes. *Mol Psychiatry* 2002;6:684–93.

63. Curtis, D, Kalsi, G, Brynjolfsson, J, et al. Genome scan of pedigrees multiply affected with bipolar disorder provides further support for the presence of a susceptibility locus on chromosome 12q23-q24, and suggests the presence of additional loci on 1p and 1q. *Psychiatr Genet* 2003;13:77–84.

64. Dawson, E, Parfitt, E, Roberts, Q, et al. Linkage studies of bipolar disorder in the region of the DarierAEs disease gene on chromosome 12q23-24.1. *Am J Med Genet* 1995;60:94–102.

65. Abkevich, V, Camp, NJ, Hensel, CH, et al. Predisposition locus for major depression at chromosome 12q22-12q23.2. *Am J Hum Genet* 2003;73:1271–81.

66. Badenhop, RF, Moses, MJ, Scimone, A, et al. A genome screen of a large bipolar affective disorder pedigree supports evidence for a susceptibility locus on chromosome 13q. *Mol Psychiatry* 2001;6:396.

67. Brzustowicz, LM, Honer, WG, Chow, EWC, et al. Linkage of familial schizophrenia to chromosome 13q32. *Am J Hum Genet* 1999;65:1096–103.

68. Blouin, JL, Dombroski, B, Nath, SK, et al. Schizophrenia susceptibility loci on chromosomes 13q32 and 8p21. *Nat Genet* 1998;20:70.

69. Chumakov, I, Blumenfeld, M, Guerassimenko, O, et al. Genetic and physiological data implicating the new human gene G72 and the gene for D-amino acid oxidase in schizophrenia. *PNAS* 2002;99:13365–7.

70. Adams, LJ, Mitchell, PB, Fielder, SL, et al. A susceptibility locus for bipolar affective disorder on chromosome 4q35. *Am J Hum Genet* 1998;62:1084–91.

71. Blackwood, DHR, He, L, Morris, SW, et al. A locus for bipolar affective disorder on chromosome 4p. *Nature Genet* 1996;12:427–30.

72. Ginns, EI, St Jean, P, Philibert, RA, et al. A genome search for chromosomal loci linked to mental health wellness in relatives at high risk for bipolar affective disorder among the Old Order Amish. *PNAS* 1998;95:15531.

73. Badner, JA, and Gershon, ES. Meta-analysis of whole-genome linkage scans of bipolar disorder and schizophrenia. *Mol Psychiatry* 2002;7:405–11.

74. Levinson, DF, Levinson, MD, Sequardo, R, and Lewis, CM. Genome scan meta-analysis of schizophrenia and bipolar disorder, part I: Methods and power analysis. *Am J Hum Genet* 2003;73:17–33.

75. Segurado, R, Detera-Wadleigh, SD, Levinson, DF, et al. Genome scan meta-analysis of schizophrenia and bipolar disorder part III: bipolar disorder. *Am J Hum Genet* 2003;3:49–62.

76. Venter, JC, Adams, MD, Myers, EW, et al. The sequence of the human genome. *Science* 2001;291: 1304–51.

77. Gabriel, SB, Schaffner, SF, Nguyen, H, et al. The structure of haplotype blocks in the human genome. *Science* 2002;296:2225–7.

78. Johnson, JW, and Ascher, P. Glycine potentiates the NMDA response in cultured mouse brain neurons. *Nature* 2003;325:529–31.

79. Schumacher, J, Jamra, RA, Freudenberg, J, et al. Examination of G72 and D-amino acid oxidase as genetic risk factors for schizophrenia and bipolar affective disorder. *Mol Psychiatry* 2004;9:203–7.

80. Hattori, E, Liu, C, Badner, JA, et al. Polymorphisms at the G72/G30 gene locus on 13q33 are associated with bipolar disorder in two independent pedigree series. *Am J Hum Genet* 2003;72:1131–40.

81. Chen, YS, Akula, N, Detera-Wadleigh, SD, et al. Findings in an independent sample support an association between bipolar affective disorder and the G72/G30 locus on chromosome 13q33. *Mol Psychiatry* 2004;9:87–92.

82. Sklar, P, Gabriel, SP, McInnis, MG, et al. Family-based association study of 76 candidate genes in bipolar disorder: BDNF is a susceptibility locus. *Mol Psychiatry* 2002;7:579–93.

83. Neves-Pereira, M, Mundo, E, Muglia, P, et al. The brain-derived neurotrophic factor gene confers susceptibility to bipolar disorder: evidence from a family-based association study. *Am J Hum Genet* 2002;71:651–5.

84. Egan, MF, Goldberg, TE, Kolachana, BS, et al. Effect of COMT Val108/158 Met genotype on frontal lobe function and risk for schizophrenia. *PNAS* 2001;98: 6917–22.

85. Nakata, K, Ujike, H, Sakai, A, et al. Association study of the BDNF gene with bipolar disorder. *Neurosci Lett* 2003;337:17–20.

86. Hong, CJ, Huo, SJ, Yen, FC, et al. Association study of a brain-derived neurotrophic factor genetic polymorphism and mood disorders, age of onset and suicidal behavior. *Neuropsychobiology* 2003;48:186–9.

87. Harada, S, Agarwal, DP, and Goedde, HW. Aldehyde dehydrogenase deficiency as cause of the flushing reaction to alcohol in Japanese. *Lancet* ii 1982;8253:982.

88. Thomasson, HR, Edenberg, HJ, Crabb, DW, et al. Alcohol and aldehyde dehydrogenase genotypes and alcoholism in Chinese men. *Am J Hum Genet* 1991;48:677–81.

89. Craddock, N, Daniels, J, Roberts, E, et al. No evidence for allelic association between bipolar disorder and MAOA gene polymorphisms. *Am J Med Genet* 1995;60:322–4.

90. Lim, LC, Powell, J, Sham, P, et al. Evidence for a genetic association between alleles of MAOA gene and bipolar affective disorder. *Am J Med Genet* 1995;60:325–31.

91. Nothen, MM, Eggermann, K, Albus, M, et al. Association analyses of the MAOA gene in bipolar affective disorder by using family based controls. *Am J Human Genet* 1995;57:975–7.

92. Parsian, A, and Todd, RD. Genetic association between monoamine oxidase and manic-depressive illness: comparison of relative risk and haplotype relative risk data. *Am J Med Genet* 1997;74:475–9.
93. Furlong, RA, Ho, L, Rubinszstein, JS, et al. Analysis of the MAOA gene in bipolar affective disorder by association studies, meta-analyses and sequencing of the promoter. *Am J Med Genet* 1999;88:398–406.
94. Preisig, M, Bellivier, F, Fenton, BT, et al. Association between bipolar disorder and MAOA gene polymorphisms: results of a multicenter study. *Am J Psychiatry* 2000;157:948–59.
95. Kawada, Y, Hattori, M, and Dai, XY. Possible association between MAOA gene and bipolar affective disorder. *Am J Hum Genet* 1995;56:335–6.
96. Muramatsu, T, Matsushita, S, Kanba, S, et al. MAO genes polymorphisms and mood disorder. *Am J Med Genet* 1997;19:494–6.
97. Turecki, G, Grof, P, Cavazzoni, P, et al. MAOA: Association and linkage studies with lithium responsive bipolar disorder. *Psychiatric Genet* 1999;9:13–6.
98. Kunugi, H, Ishida, S, Kato, T, et al. A functional polymorphism in the promoter region of MAOA gene and mood disorders. *Mol Psychiatry* 1999;4:393–5.
99. Ramammorthy, S, Bauman, AL, Moore, KR, et al. Antidepressant and cocaine sensitive human transporter: Molecular cloning, expression and chromosomal localization. *Proc Natl Acad Sci* 1993;90:2542–6.
100. Collier, DA, Arranz, MJ, Sham, P, et al. The serotonin transporter is a potential susceptibility factor for bipolar affective disorder. *Neuroreport* 1996;7:1675–9.
101. Lesch, KP, Bengel, D, Heils, A, et al. Association of anxiety related traits with a polymorphism in the serotonin transporter gene regulatory region. *Science* 1996; 274: 1527–31.
102. Fickerstrand, CE, Lovejoy, EA, and Quinn, JP. An intronic polymorphic domain often associated with susceptibility to affective disorders has allele dependent differential enhancer activity in embryonic stem cells. *FEBS Lett* 1999;458:171–4.
103. Collier, DA, Stober, G, Li, T, et al. A novel functional polymorphism within the promoter of the serotonin transporter gene: possible role in susceptibility to affective disorders. *Mol Psychiatry* 1996a;1:453–60.
104. Rees, M, Norton, N, Jones, T, et al. Association studies of bipolar disorder at the human serotonin transporter gene(hSERT;5HTT). *Mol Psychiatry* 1997;2:398–402.
105. Vincent, JB, Masellis, M, Lawrence, J, et al. Genetic association analysis of serotonin system genes in bipolar affective disorder. *Am J Psychiatry* 1999;156:136–8.
106. Gutierrez, B, Arranz, MJ, Collier, DA, et al. Serotonin transporter gene and risk for bipolar affective disorder: An association study in Spanish population. *Biol Psychiatry* 1998;43:843–7.
107. Bocchetta, A, Piccardi, MP, Palmas, MA, et al. Family-based association study between bipolar disorder and DRD2, DRD4, DAT, and SERT in Sardinia. *Am J Med Genet* 1999;88:522–6.
108. Mundo, E, Walker, M, Tims, H, et al. Lack of linkage disequilibrium between serotonin transporter protein gene (SLC6A4) and bipolar disorder. *Am J Med Genet* 2000;96:379–83.
109. Mynett-Johnson, L, Kealey, C, Claffey, E, et al. Multimarker haplotypes within the serotonin transporter gene suggest evidence of an association with bipolar disorder. *Am J Med Genet* 2000;96:845–9.
110. Kirov, G, Rees, M, Jones, I, et al. Bipolar disorder and the serotonin transporter gene: A family-based association study. *Psychol Med* 1999;29:1249–54.
111. Saleem, Q, Ganesh, S, Vijaykumar, M, et al. Association analysis of 5HT transporter gene in bipolar disorder in the Indian population. *Am J Med Genet* 2000;96:170–2.

112. Mendes de Oliveira, JR, Otto, PA, Vallada, H, et al. Analysis of a novel functional polymorphism within the promoter region of the serotonin transporter gene (5-HTT) in Brazilian patients affected by bipolar disorder and schizophrenia. *Am J Med Genet* 1998;81:225–7.

113. Kunugi, H, Hattori, M, Kato, T, et al. Serotonin transporter gene polymorphisms: ethnic difference and possible association with bipolar affective disorder. *Mol Psychiatry* 1997;2:457–62.

114. Furlong, RA, Ho, L, Walsh, C, et al. Analysis and meta-analysis of two serotonin transporter gene polymorphisms in bipolar and unipolar affective disorders. *Am J Med Genet* 1998;81:58–63.

115. Coyle, N, Jones, I, Robertson, E, et al. Variation at the serotonin transporter gene influences susceptibility to bipolar affective puerperal psychosis. *Lancet* 2000;356:1490–1.

116. Ospina-Duque, J, Duque, C, Carvajal-Carmona, L, et al. An association study of bipolar mood disorder (type I) with the 5-HTTLPR serotonin transporter polymorphism in a human population isolate from Colombia. *Neurosci Lett* 2000;292:199–202.

117. Mendlewicz, J, Massat, I, Souery, D, et al. Serotonin transporter 5HTTLPR polymorphism and affective disorders: No evidence of association in a large European multicenter study. *Eur J Hum Genet* 2004;12:377–82.

118. Weissman, MM, Wickramaratne, P, Adams, PB, et al: The relationship between panic disorder and major depression: a new family study. *Arch Gen Psychiatry* 1993; 50:767–80.

119. Mendlewicz, J and Baron, M. Morbidity risks in subtypes of unipolar depressive illness: Differences between early and late onset forms. *Br J Psychiatry* 1981;139:463–6.

120. Weissman, MM, Merikangas, KR, Wickramaratne, P, et al. Understanding the clinical heterogeneity of major depression using family data. *Arch Gen Psychiatry* 1986; 43:430–4.

121. Bland, RC, Newman, SC, and Orn, H. Recurrent and nonrecurrent depression. *Arch Gen Psychiatry* 1986;43:1085–9.

122. Stancer, HC, Persad, E, Wagener, DK, and Jorna, T. Evidence for homogeneity of major depression and bipolar affective disorder. *J Psychiatr Res* 1987;21:37–53.

123. Ekholm, JM, Kieseppa, T, Hiekkalinna, T, et al. Evidence of susceptibility loci on 4q32 and 16p12 for bipolar disorder. *Hum Mol Genet* 2000;12:1907–15.

124. Reich, T, Van Eerdewegh, P, Rice, J, et al. The familial transmission of primary major depressive disorder. *J Psychiatr Res* 1987;21:613–24.

125. Kendler, KS, Neale, MC, Kessler, RC, et al. The clinical characteristics of major depression as indices of the familial risk to illness. *Br J Psychiatry* 1994;165:66–72.

126. Kendler, KS, Gardner, CO, and Prescott, CA. Clinical characteristics of major depression that predict risk of depression in relatives. *Arch Gen Psychiatry* 1999;56:322.

127. Merikangas, KR, Risch, NJ, and Weissman, MM. Co-morbidity and co-transmission of alcoholism, anxiety and depression. *Psychol Med* 1994;24:69–80.

128. Winokur, G, Cadoret, R, Dorsab, J, et al. Depressive disease: a genetic study. *Arch Gen Psychiatry* 1971;24:135–44.

129. Nurnberger, JI, Foroud, T, Flury, L, et al. Evidence for a locus on chromosome 1 that influences vulnerability to alcoholism and affective disorder. *Am J Psychiatry* 2001;158:718–24.

130. Sullivan, P, and Kendler, K. Genetic case-control studies in neuropsychiatry. *Arch Gen Psychiatry* 2001;58:1015–24.

131. Bierut, LJ, Heath, AC, Bucholz, KK, et al. Major depressive disorder in a community-based twin sample: Are there different genetic and environmental contributions for men and women? *Arch Gen Psychiatry* 1999;56:557–63.

132. Kendler, KS, Pedersen, N, Johnson, L, et al. A pilot Swedish twin study of affective illness, including hospital and population-ascertained subsamples. *Behav Genet* 1995;25:217–32.

133. Lyons, MJ, Eisen, SA, Goldberg, J, et al. A registry-based twin study of depression in men. *Arch Gen Psychiatry* 1998;55:468–72.

134. Kendler, KS, and Prescott, CA. A population-based twin study of lifetime major depression in men and women. *Arch Gen Psychiatry* 1999;56:39–44.

135. McGuffin, P, Katz, R, Watkins, S, and Rutherford, J: A hospital-based twin register of the heritability of DSM-IV unipolar depression. *Arch Gen Psychiatry* 1996;53:129–36.

136. Holmans, P, Zubenko, GS, Crowe, RR, et al. Genome-wide significant linkage to recurrent early-onset major depressive disorder on chromosome 15q. *Am J Hum Genet* 2004;74:1154–67.

137. Zubenko, GS, Maher, B, Hughes, HB, et al. Genome-wide linkage survey for genetic loci that influence the development of depressive disorders in families with recurrent early-onset major depression. *Am J Med Genet* 2003;123B:1–18.

138. Zubenko, GS, Hughes, HG, Stiffler, JS, et al. Sequence variations in CREB1 co-segregate with depressive disorders in women. *Mol Psychiatry* 2003;8:611–18.

139. McInnis, MG, Lan, TH, Willour, VL, et al. Genome-wide scan of bipolar disorder in 65 pedigrees: supportive evidence for linkage at 8q24, 18q22, 4q32, 2p12, and 13q12. *Mol Psychiatry* 2003;8:288–98.

140. Cichon, S, Schumacher, J, Muller, DJ, et al. A genome screen for genes predisposing to bipolar affective disorder detects a new susceptibility locus on 8q. *Hum Mol Genet* 2001;10:2933–44.

141. Liu J, Juo SH, Dewan A, et al. Evidence for a putative bipolar disorder locus on 2p13-16 and other potential loci on 4q31, 7q34, 8q13, 9q31, 10q21-24, 13q32, 14q21 and 17q11-12. *Mol Psychiatry* 2004;8:333–42.

142. Kelsoe, JR, Spence, MA, Loetscher, E, et al. A genome survey indicates a possible susceptibility locus for bipolar disorder on chromosome 22. *Proc Natl Acad Sci USA* 2001;98:585–90.

143. Dick, DM, Foroud, T, Flury, L, et al. Genome-wide linkage analyses of bipolar disorder: a new sample of 250 NIMH Genetics Initiative pedigrees. *Am J Hum Genet* 2003;73:107–14.

144. Potash, JB, Chiu, Y-F, MacKinnon, DF, et al. Familial aggregation of psychosis in a replication set of 69 bipolar pedigrees: *Am J Med Genet* 2003;116B:90–7.

145. Berrettini, W, Ferraro, T, Goldin, L, et al. Chromosome 18 DNA markers and manic depressive illness: evidence for a susceptibility gene. *Proc Natl Acad Sci USA* 1994;91:5918–21.

17 Pedigree Analyses and the Study of Chimpanzee (*Pan troglodytes*) Personality and Subjective Well-Being

Alexander Weiss and James E. King

CONTENTS

17.1 OVERVIEW

In this chapter we describe our research on the behavior genetics of personality and subjective well-being (SWB) in our chimpanzee cousins, a research program that we plan to extend to the other great ape species—orangutans, gorillas, and bonobos. Although we had been familiar with classic behavior genetics designs before initiating this research, we quickly realized that these approaches were not suitable for studying great apes in naturalistic and semi-naturalistic settings. Our subjects could not be selectively bred, they could not be conveniently grouped into sets of monozygotic and dizygotic twins, and their pedigrees did not resemble trees but, instead, kudzu. Fortunately, we stumbled upon two types of pedigree analysis that allowed us to hack through the familial wilderness and estimate the relative contributions of genetic and environmental effects on individual differences in chimpanzee personality and SWB. The first part of this chapter will describe reasons why the behavior–genetic study of chimpanzees is interesting and compelling. The second part will describe how multiple regression analysis can be used to estimate relevant variance components and will highlight the advantages and disadvantages of this method. The third part will describe the use of mixed-model equations and a restricted maximum likelihood analysis to estimate variance and covariance components. Finally, we will offer concluding thoughts and suggest other potential uses for these analyses.

17.1.1 CHIMPANZEE NATURAL HISTORY AND BEHAVIOR

An important reason for studying the behavior genetics of chimpanzee personality and SWB is that chimpanzees are our closest living nonhuman relatives. Indeed, we shared a common ancestor only 5 to 6 million years ago, and consequently we now share a little more than 99% of our genes.[1–3] Genetic analysis of chimpanzee personality thus enables us to do more than engage in armchair speculation of the hominid condition; we can investigate that condition in a chimpanzee model of early humans. That chimpanzees show evidence of some of the more complex behavioral and cognitive characteristics possessed by humans is an outgrowth of this genetic similarity. For example, several years of research by the Rumbaughs[4] demonstrated that, in contrast with the claims of some linguists,[5] chimpanzees and bonobos have language abilities previously thought to exist only in humans. Theory of mind[6,7] and moral behavior[8] have also been observed in chimpanzees, both in the field and laboratory. Culture, once touted as the highest human accomplishment, appears a feat that also is not above the abilities of the "lowly" chimpanzee.[9,10] These human–chimpanzee similarities in behavior and cognition appear to be evolutionary homologies, not independent convergence of similar traits in evolutionarily independent species.

17.1.2 HUMAN AND CHIMPANZEE PERSONALITY

There is now a consensus that in humans five broad dimensions—neuroticism, extraversion, openness to experience, agreeableness, and conscientiousness—underlie human personality variation.[11] This model of human personality has come to be known as the five-factor model (FFM). The domains of the FFM have biological origins since they are heritable,[12] mostly stable in adulthood,[11] and are found in multiple Western and non-Western cultures.[13]

Human SWB also appears to have biological origins because it is substantially heritable[14] and stable over time.[15] Additionally, the best predictors of human SWB are personality dimensions, especially neuroticism and extraversion.[16]

Research on chimpanzee personality indicates that chimpanzee personality and SWB are similar to their human counterparts. First, factor analysis of chimpanzee personality descriptors yields a six-factor solution; the first and largest of these factors was a broad, chimpanzee-specific factor that was primarily a blend of low neuroticism and agreeableness and high extraversion.[17] Since the item with the highest loading on this factor was *dominant*, the factor was designated as dominance. The remaining five factors—extraversion, dependability, agreeableness, emotional stability, and openness—are analogous to the five human domains. Also, similar to human personality findings,[11] chimpanzee personality factors were mostly stable over time.[41] In addition, recent human personality studies showing that the factor structure as well as age and sex differences generalize across cultures[13] led us to ask the same questions about chimpanzee personality. We found that, as in humans, the chimpanzee personality factor structure as well as age and sex differences generalized from zoo environments to an African chimpanzee sanctuary whose habitat closely resembled that of wild chimpanzees.[19]

17.1.3 HUMAN AND CHIMPANZEE SUBJECTIVE WELL-BEING

Similarities also exist between measures of human and chimpanzee SWB. One study that examined SWB in chimpanzees used a measure based on the major components of human SWB: the balance of positive and negative moods, pleasure derived from social interactions, success in achieving goals, and a global measure of happiness.[18] As with human measures of SWB, about half of the variance of chimpanzee SWB was predicted by personality, namely, dominance and dependability.[*,18] In addition to these findings, there was evidence that chimpanzee SWB is as stable over time as human SWB.[18]

17.1.4 STUDY GOALS

Because of the similarity of factor structures of humans and chimpanzees and, of course, the close genetic correspondence between the two species, we sought to answer a series of questions regarding chimpanzee personality and SWB. First, to what extent are individual differences in chimpanzee personality factors and SWB a reflection of genetic differences among individuals, i.e., heritable, and to what extent are these individual differences a manifestation of differences in the shared zoo environment and nonshared environmental effects? Additionally, with respect to SWB, we asked whether heritable or nonheritable characteristics of individuals' mothers contributed to individual differences in SWB. Finally, we wished to determine whether the correlation between personality and SWB in chimpanzees could be explained by common genetic or environmental factors.

17.1.4.1 Reasons for Using Pedigree Analysis Methods

Our primary obstacle was the low number of discrete family units in the sample. The relative absence of discrete family units dramatically reduced the number of genetically independent pairs, thereby precluding use of most conventional techniques for genetic analysis. Fortunately, approaches that consider the interrelationships between all possible pairs of individuals in a sample enabled us to surmount this obstacle.

17.1.4.2 Common Features of Pedigree Analysis Methods

These approaches incorporate pedigrees based on the identity of each "subject's" sire (father) and dam (mother), the environment, and phenotype scores for as many individuals and generations as possible. Pedigree analyses have been used for many years in the agricultural industry with samples that include hundreds or thousands of animals. Use of these pedigree analyses has been focused on enhancing agriculturally and economically important traits through selective breeding. Detailed pedigree records incorporated into official "studbooks" are maintained for several species that are well represented in zoo populations, including endangered species such as chimpanzees. For the studies described in this chapter, the chimpanzee

---

* The relationship between dependability and subjective well-being in chimpanzees may be accounted for by the fact that the chimpanzee dependability factor is comprised of several items defining human low-neuroticism.

studbook was the essential component to unraveling the behavior genetics of chimpanzee personality and SWB.[20] Since our studies, a new edition has been released.[21]

17.2 SYMMETRIC DIFFERENCES SQUARED

Symmetric differences squared (SDS) is an approach based on multiple regression analysis.[22] As with other pedigree analyses, all possible pairs of individuals in the sample are included in the analysis. Studies using simulated[23] and real[24] data sets indicate that the parameter estimates provided by SDS analysis tend to be more accurate than those provided by methods based on analysis of variance.

17.2.1 SQUARED PHENOTYPIC DIFFERENCES AND THE VARIANCE

The foundation of SDS is that the expected value of the squared phenotypic differences of all possible pairs of individuals in a sample of independent scores is equal to two times the variance of that phenotype in the sample. However, if phenotype scores of related individuals are correlated, the expected value of the squared phenotypic differences will be lower than that in the population. This is illustrated by the following equation:[22]

$$E(Y_i - Y_j)^2 = 2[\text{var}(Y) - \text{cov}(Y_i, Y_j)]$$

where $i < j$ and Y_i and Y_j are the phenotype scores for individuals i and j.

17.2.1.1 A Latent Variable Model for Estimating Variances and Covariances

Variance and covariance estimates can be estimated with a simple latent variable model (see Figure 17.1). In this model, phenotype scores are the result of additive genetic (A), zoo (Z), and nonshared environment effects plus error (E). Additive genetic effects will be correlated between individuals to the extent that two individuals are genetically related (R_{ij}, Wright's coefficient of relatedness). One thing to note is that, while computing R_{ij} is a relatively simple task in samples containing mostly nuclear family units, it is considerably more difficult to do with complex pedigrees, especially those where inbreeding has occurred. Fortunately, free software packages, e.g., Animal Breeder's Toolkit,[25] can compute the degree of relatedness between all pairs of individuals in a sample and provide inbreeding coefficients by which the degree of relatedness must be adjusted (see Appendix A in this chapter for more details).

The correlation of zoo effects is denoted rz_{ij} and equal to 1 if the pair lives in the same zoo enclosure and 0 if they do not. Finally, the nonshared environment effects plus error are assumed to be uncorrelated.[22] If we let a^2, z^2, and e^2 be the variances attributable to additive genetic, zoo, and error, respectively, it then follows that

$$\text{var}(Y) = a^2 + z^2 + e^2$$

$$\text{cov}(Y_i, Y_j) = R_{ij}a^2 + rz_{ij}\, z^2.$$

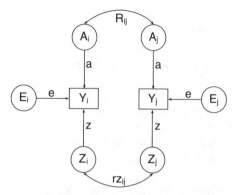

FIGURE 17.1 The correlation in trait Y for any given pair of animals, i and j, is described as a function of six latent variables with variances equal to 1. Latent variables include additive genetic (A), shared zoo (Z), and nonshared environment plus error (E) effects. Paths a, z and e are the effects of each variable on the trait of interest. R_{ij} is Wright's coefficient of relationship and rz_{ij} is a dummy-coded variable equal to 1 if a pair lives in the same zoo and 0 if it does not. (From *Behavior Genetics*, Weiss, A., King, J.E. and Figueredo, A.J. The heritability of personality factors in chimpanzees (*Pan troglodytes*), 2000, vol. 30, 213–221. With permission from Kluwer Academic Publishers.)

17.2.1.2 Using Multiple Regression to Solve for Variance Estimates

These values can then be substituted in the first equation[22] to yield a formula that can be solved as the linear regression equation

$$.5E(Y_i - Y_j)^2 = e^2 + a^2(1 - R_{ij}) + z^2(1 - rz_{ij}).$$

The squared differences among all possible pairs are regressed onto $(1 - R_{ij})$ and $(1 - rz_{ij})$. The unstandardized beta weight for the variable reflecting $(1 - R_{ij})$ is equal to the variance due to genetic effects (a^2) and the unstandardized beta weight for the variable reflecting $(1 - rz_{ij})$ is equal to the variance due to shared zoo effects (z^2). Finally, the intercept term is equal to the variance due to the nonshared environment plus error (e^2).[22]

17.2.1.3 Computing Proportions of Variance

With these estimates, it is a simple matter to compute heritability, the proportion of variance due to additive genetic effects.[22]

$$h^2 = a^2 / (a^2 + z^2 + e^2)$$

Substituting the a^2 in the numerator with z^2 or e^2 would yield the proportion of variance resulting from shared zoo and nonshared environment plus error effects, respectively.

17.2.2 RESULTS WHEN USING SYMMETRIC DIFFERENCES SQUARED

We used SDS to estimate the heritability of personality dimensions in 145 zoo-housed chimpanzees.[26] Approximately 63% of the dominance factor variance was heritable. In addition, 21% of the dependability variance was heritable, but was not statistically significant. Finally, virtually all of the remaining variance for all factors was accounted for by nonshared environment effects plus error; shared zoo effects were negligible for all six factors.[26] There are several possible reasons for the discrepancy between these findings and those of the behavior genetics of the FFM in humans, which indicate that all five domains are heritable.[12] First, the sample size was relatively small and it did not include twins. This problem may be especially important as a recent review of the heritability of personality literature indicates that some of the genetic variance in human personality is the result of nonadditive genetic effects, i.e., genetic effects reflecting interactions among genes, which are not easily assessed in non-twin designs.[27] A second potential reason for our inability to replicate the heritability of the five human-like personality traits may have been that dominance accounted for the lion's (or chimpanzee's) share of common factor variance among adjectival descriptors.

On the other hand, the dearth of shared or common environmental effects is relatively consistent with the results from behavior genetic analyses of human personality dimensions.[12] Specifically, shared zoo effects on chimpanzee personality are almost as nonexistent as shared family effects on human personality.[28]

17.2.3 ADDITIONAL CONSIDERATIONS

No one technique is perfect, so, when choosing any analysis technique, one should be aware of practical and methodological advantages and disadvantages. This is as true for quantitative genetic analysis as for conventional statistics.

17.2.3.1 Advantages of Symmetric Differences Squared

The major advantage of SDS is that, if the genetic relatedness is known for all possible pairs of individuals, there is no need for a specialized software package. We conducted our analysis using the PROC REG command in SAS,[29] but it could just as easily have been computed using the regression modules of any other statistics package.

17.2.3.2 Least Squares Estimation and Negative Variance Components

A possibly disquieting characteristic of SDS is that the least squares estimation procedure may produce negative variance estimates. The authors of this technique recommend that any negative variance estimates be treated as zeros.[22]

In unpublished analyses, we noticed that other statistical approaches that are based on restricted maximum likelihood analyses (described later) will estimate as zero the variance components that were negative in the SDS analysis.[26] Modern statistical packages allow one to solve regression equations via restricted maximum likelihood as opposed to least squares regression, thereby avoiding negative variance estimates.

17.2.3.3 Using Monte Carlo Procedures to Determine
Statistical Significance

The use of all possible pairs of subjects in the SDS procedure means that each subject will be present in $n-1$ pairs. This lack of independence means that standard procedures for assessing the statistical significance of these estimates will be incorrect. The statistical significance of these estimates can, however, be addressed by Monte Carlo simulation, in which the variable representing degree of relatedness is randomly assigned to subject pairs. Repeated iterations of the analysis can then be used to estimate the sampling distribution of these estimates. We used 1,000 iterations to estimate these sampling distributions.[30] The proportion of heritability estimates from the simulation that were greater than or equal to the estimates derived when the degrees of relatedness were properly assigned was used as an estimate of p.

17.2.4 CONCLUDING THOUGHTS

While SDS can be used to estimate genetic and environmental effects, it is probably best for single trait analyses, especially when sample sizes are small. The relative ease of implementation also makes it potentially useful for initial data exploration before progressing to other analyses.

17.3 MULTITRAIT DERIVATIVE FREE RESTRICTED
MAXIMUM LIKELIHOOD ANALYSIS

Multitrait derivative free restricted maximum likelihood (MTDFREML[31]) analysis is another means by which variance and covariance components can be estimated in samples with complex pedigrees. Like SDS, MTDFREML is based on all pairs of individuals. MTDFREML is based on a specific form of the mixed-model equation.[32]

17.3.1 FOUNDATIONS OF THE UNIVARIATE CASE

In the univariate case,** this approach is described in the following formula. It assumes that the phenotype of each individual, i, is composed of fixed (β) and random effects including, but not limited to, additive genetic (a); heritable maternal (m); nonheritable maternal (c); and nonshared environment effects plus error (e).[32]

$$y_i = \beta_i + a_i + m_i + c_i + e_i.$$

17.3.1.1 Henderson's Mixed Model Equation and the
Estimation of Variance and Covariance Components

A more general form of this equation can be used to describe phenotypic deviations of each individual from a sample or population mean. This general equation is based

---

** This approach can be used to estimate variances and covariance of and among multiple traits.[29,30]

on a series of vectors and matrices. The first of these vectors, β_i, is a vector of deviations from a grand mean attributable to fixed effects (zoos in our case). Other vectors include a, m, and c, which indicate the degree to which subjects' deviations from the zoo mean are the result of additive genetic, heritable maternal, and nonheritable maternal effects, respectively. The matrices include X and Z_a, which indicate each individual's zoo membership and identifying number, respectively. The final matrices are Z_m, and Z_c, which indicate the identifying number of each chimpanzee's mother—used to assess heritable maternal and nonheritable maternal effects, respectively.[***]

The products of these vectors and matrices and the residual term, e, which represents deviations resulting from nonshared environmental effects plus error, can be used to predict, y, a vector of phenotypes, in a sample or population.

$$y = X(\beta) + Z_a(a) + Z_m(m) + Z_c(c) + e$$

These summed and squared deviations are variances. If a given phenotype is heritable, the similarity of deviations of two individuals from the zoo mean will be dependent, in part, on how related they are. If some heritable characteristic of mothers contributes to a phenotype score of an offspring over and above the genetic effects based on the relatedness of mother and offspring, heritable maternal effects are said to be present. In this case, the similarity of deviations of two individuals from the zoo mean will be dependent, in part, on how related their mothers are. If some nonheritable characteristic of mothers has an effect on the phenotype of their offspring, i.e., there are nonheritable maternal effects, the similarity of deviations of two individuals will be more similar if they share the same mother than if they have different mothers, regardless of the relatedness of the two mothers.

With two $n \times n$ matrices, A, indicating how related all possible pairs are, and I, with 1s in the diagonal and 0s elsewhere, we can decompose the phenotypic variance by weighting additive genetic and heritable maternal variance by A and nonheritable maternal and nonshared environmental variance by I.

$$\sigma^2_y = \sigma^2_\beta + A\sigma^2_a + A\sigma^2_m + I\sigma^2_c + I\sigma^2_e$$

These variance components can then be used to compute the proportions of variance and covariance that arise from genetic or different environmental effects.

17.3.2 STUDY GOALS

We used MTDFREML[31] to solve for these variance and covariance components in a sample of 128 zoo chimpanzees that were rated on personality and SWB.[33] MTD-FREML uses an iterative procedure that tries different values for variance components until it achieves a set that minimizes the difference between the expected and predicted phenotype scores.[34,35] Our previous study indicated that dominance was the only

---

[***] If the source of these effects is the same, as when two types of maternal effects are investigated, the matrices will be identical.

personality trait that was heritable,[26] so we focused exclusively on the genetic and environmental sources of variances and covariance for dominance and SWB.[33]

17.3.2.1 Choosing a Model Using Akaike's Information Criterion

Along with the parameter estimates, MTDFREML provides a statistic $(-2log\lambda)$ indicating how well the model replicated the actual data. The lower this statistic, the better the model fit. We compared seven different models using two different approaches. The first approach was based on Akaike's information criterion (AIC).[36] AIC is based on the notion that model fit must be balanced with model parsimony and, hence, "penalizes" the degree of model fit by the number of parameters estimated (k).

$$AIC = -2log\lambda + 2k$$

The model with the lowest AIC, then, is the one that best balances model fit and parsimony. In our analysis this model was one in which 66% of the dominance and 40% of the SWB variance were heritable.[33] The heritability estimate for dominance was nearly identical to what we obtained in univariate analyses with SDS[26] and the heritability estimate for SWB was nearly identical to what we obtained using SDS and another estimation procedure, Method R.[37,38] The covariance between dominance and SWB was almost entirely due to shared genes. Furthermore, heritable maternal effects accounted for approximately 22% of the SWB, but none of the variance in dominance.[33] Finally, largely independent nonshared environmental effects plus error accounted for 27% of the remaining dominance and 35% of the remaining SWB variance.[33]

17.3.2.2 Comparing Nested Models Using Chi-Square Tests

The second approach, nested models comparisons, is based on the fact that the $-2log\lambda$ statistic is distributed as a chi-square. Hence, one can test whether parameters are significant by comparing the $-2log\lambda$ of models with and without those parameters.[39] If the $-2log\lambda$ difference between models is not significant, this indicates that the additional parameters were not contributing to the model beyond what would be expected by chance and, hence, the model with fewer parameters should be adopted.

In our analysis, comparing models using nested models comparison indicated that there was no significant difference between the model that we chose based on the comparison of the AIC statistics and a model that was identical except for not including heritable maternal effects. We therefore accepted this more parsimonious model that did not include heritable maternal effects as a plausible alternative.[33]

17.3.3 CONSIDERATIONS

As with SDS, using MTDFREML to solve for variance and covariance components has advantages and disadvantages. These should be considered carefully when planning a behavior genetic study in samples with complex pedigrees.

17.3.3.1 Advantages of Using Multitrait Derivative Free Restricted Maximum Likelihood Analysis

One of the main advantages is that the MTDFREML approach does not produce any negative variance components. Additionally, it estimates sources of covariance between two or more traits and even between different effects. Another advantage is that these analyses provide model fit parameters that can easily be used to compare the fit and parsimony of different models. Finally, there are several free programs that can be used to conduct these analyses. We used MTDFREML,[31] but there are similar programs such as DFREML.[40]

17.3.3.2 Drawbacks of Using Multitrait Derivative Free Restricted Maximum Likelihood Analysis

However, these programs are not as flexible as standard statistical packages. Thus, if one wanted to use the program to investigate gene × environment interactions, it would require considerable knowledge in FORTRAN programming. Another disadvantage is that even with small samples, a thorough analysis can take a long time. The author spent from 10:00 PM to 8:00 AM computing variances and covariances with the university's mainframe computer for the seven models we compared.[33]

17.4 CONCLUDING THOUGHTS AND OTHER USES FOR MULTITRAIT DERIVATIVE FREE RESTRICTED MAXIMUM LIKELIHOOD ANALYSIS

Pedigree analyses are a powerful means of addressing questions about genetic and environmental effects in samples with complex pedigrees. Certainly, these methods could also be applied to studies of human populations.

17.4.1 Other Applications of Multitrait Derivative Free Restricted Maximum Likelihood Analysis

Another potential use of these analyses is that, if one knows the heritability of a trait and the pedigree, one can obtain "breeding values" for individuals. Breeding values reflect the amount by which individuals' phenotypic values deviate from the sample mean because of additive genetic effects. Calculation of an individual's breeding value is dependent upon having phenotypic data from genetic relatives and knowing the heritability of the trait. Breeding values have been used extensively in the agricultural industry to increase the efficiency of selective breeding. Selectively breeding animals with the highest breeding values has been used, for example, to produce more milk with fewer cows in the United States.

17.4.1.1 Genetic Association Studies

Because breeding values reflect the genetic contribution to a given phenotype, they could be used to improve the power of genetic association studies. Genetic association studies typically compare individuals who are high in some phenotype with those who are low or average in the phenotype. Because the phenotype is a blend of additive and nonadditive genetic variance as well as environmental effects, the phenotype score is only an approximation of an individual's genetic predisposition for that trait. The use of breeding values in these studies would increase the likelihood of detecting associations.

17.4.1.2 Environmental Association Studies

Similarly, pedigree analyses can also output the degree to which individuals differ from their expected mean based on measured environmental effects. This potentially useful measure is the obverse of breeding values and would show individual differences in susceptibility to environmental modification of the trait being studied. An interesting question is whether this environmental susceptibility is itself heritable. Those studying the environmental determinants of behavior could attempt to associate different environmental events and conditions to these environmental susceptibility values and reduce the likelihood that genetic effects confound their findings.

17.4.1.3 Assessing Gene × Environment Interactions

Environmental susceptibility scores could be used to identify individuals who are particularly susceptible to environmental and social sources of stress that may lead to depression and other psychopathological disorders. The reasoning would be that those individuals who have high phenotypic psychopathology scores and high environmental susceptibility scores are particularly vulnerable to stressful social and environmental conditions. Two individuals with identical phenotypic psychopathology scores might differ markedly in their environmental susceptibility scores, thereby implying different underlying causes of psychopathology in the two individuals. Likewise, environmental susceptibility scores would be a basis for a new approach to individual differences in personality.

17.4.2 Concluding Thoughts

Although the classic twin method is powerful and invaluable, especially as it provides broad-sense heritability estimates, it is our hope that pedigree analyses will be more widely adopted. Pedigree analyses offer the ability to analyze data on a wider range of samples and can provide more accurate estimates of genetic variances and covariances because they use all possible relationships.

AUTHOR'S NOTE

We wish to thank Virginia Landau and the staff of the ChimpanZoo program and all the raters at the different zoos for making our research possible. We also wish to thank our past collaborators A. J. Figueredo and R. Mark Enns. Finally, we would like to thank R. R. McCrae and Antonio Terracciano for their comments on an earlier manuscript.

REFERENCES

1. Purvis, A. (1995). A composite estimate of primate phylogeny. *Philosophical Transaction of the Royal Society of London, Series B: Biological Sciences* 348, 405–421.
2. Sarich, V.M. and Wilson, A.C. (1967). Immunological time scale for hominid evolution. *Science* 158, 1200–1203.
3. Wilson, A.C. and Sarich, V.M. (1969). A molecular time scale for human evolution. *Proceedings of the National Academy of Sciences U.S.A.* 63.
4. Rumbaugh, D.M. and Savage-Rumbaugh, E.S. (1994). Language in comparative perspective. In *Animal Learning and Cognition*, N.J. Mackintosh, ed. (San Diego, CA: Academic Press), pp. 307–333.
5. Chomsky, A.N. (1972). *Language and the Mind* (New York: Harcourt Brace Jovanovich).
6. Hare, B., Call, J., and Tomasello, M. (2001). Do chimpanzees know what conspecifics know? *Animal Behaviour* 61, 139–151.
7. Whiten, A. (1997). The Machiavellian mindreader. In *Machiavellian Intelligence II: Extensions and Evaluations*, A. Whiten and R.W. Byrne, eds. (New York: Cambridge University Press), pp. 144–173.
8. de Waal, F.B.M. (1996). *Good Natured: The Origins of Right and Wrong in Humans and Other Animals.* (Cambridge, MA: Harvard University Press).
9. de Waal, F.B.M. (2001). *The Ape and the Sushi Master Cultural Reflections by a Primatologist* (New York: Basic Books).
10. Whiten, A., Goodall, J., McGrew, W.C., Nishida, T., Reynolds, V., Sugiyama, Y., Tutin, C.E.G., Wrangham, R.W., and Boesch, C. (1999). Cultures in chimpanzees. *Nature* 399, 682–685.
11. McCrae, R.R. and Costa, P.T., Jr. (2003). *Personality in Adulthood: A Five-Factor Theory Perspective* (New York: Guilford).
12. Bouchard, T.J., Jr. and Loehlin, J.C. (2001). Genes, evolution, and personality. *Behavior Genetics* 31, 243–271.
13. McCrae, R.R. (2001). Trait psychology and culture: Exploring intercultural comparisons. *Journal of Personality* 69, 819–846.
14. Lykken, D.T. and Tellegen, A. (1996). Happiness is a stochastic phenomenon. *Psychological Science* 7, 186–189.
15. Eid, M. and Diener, E. (2004). Global judgements of subjective well-being: Situational variability and long-term stability. *Social Indicators Research* 65, 245–277.
16. Diener, E. (1996). Traits can be powerful, but are not enough: Lessons from subjective well-being. *Journal of Research in Personality* 30, 389–399.
17. King, J.E. and Figueredo, A.J. (1997). The Five-Factor Model plus Dominance in chimpanzee personality. *Journal of Research in Personality* 31, 257–271.
18. King, J.E. and Landau, V.I. (2003). Can chimpanzee (*Pan troglodytes*) happiness be estimated by human raters? *Journal of Research in Personality* 37, 1–15.
19. King, J.E., Weiss, A., and Farmer, K.H. (2005). A chimpanzee (*Pan troglodytes*) analogue of cross-national generalization of personality structure: zoological Parks and an African Sanctuary. *Journal of Personality* 73, 389–410.
20. Fulk, R. (1999). *Chimpanzee Species Survival Plan: Master Plan 1998–1999*, North Carolina Zoological Park: Asheboro, NC.
21. Ross, S. (2003). *North American Regional Studbook for the Chimpanzee (Pan troglodytes)*, Lincoln Park Zoo: Chicago.

22. Grimes, L.W. and Harvey, W.R. (1980). Estimation of genetic variances and covariances using Symmetric Differences Squared. *Journal of Animal Science* 50, 632–644.

23. Bruckner, C.M. and Slanger, W.D. (1986). Symmetrical differences squared and analysis of variance procedures for estimating genetic and environmental variances and covariances for beef-cattle weaning weight: I. Comparison via simulation. *Journal of Animal Science* 63, 1779–1793.

24. Bruckner, C.M. and Slanger, W.D. (1986). Symmetrical differences squared and analysis of variance procedures for estimating genetic and environmental variances and covariances for beef-cattle weaning weight: II. Estimates from a data set. *Journal of Animal Science* 63, 1794–1803.

25. Golden, B.L., Snelling, W.M., and Mallinckrodt, C.H. (1992). *Animal Breeder's Toolkit: User's Guide and Reference Manual*, Colorado State University Agricultural Experiment Station: Fort Collins.

26. Weiss, A., King, J.E., and Figueredo, A.J. (2000). The heritability of personality factors in chimpanzees (*Pan troglodytes*). *Behavior Genetics* 30, 213–221.

27. Bouchard, T.J., Jr. (2004). Genetic influence on human personality traits: A survey. *Current Directions in Psychological Science* 13, 148–151.

28. Rowe, D.C. (1994). *The Limits of Family Influence* (New York: Guilford).

29. SAS Institute, I. (1999). *SAS/STAT User's Guide, Version 8* (Cary, NC: SAS Institute).

30. Cheverud, J.M., Falk, D., Vannier, M., Konigsberg, L., Helmkamp, R.C., and Hildebolt, C. (1990). Heritability of brain size and surface features in rhesus monkeys (*Macaca mulatta*). *Journal of Heredity* 81, 51–57.

31. Boldman, K.G., Kriese, L.A., Van Vleck, L.D., Van Tassel, C.P., and Kachman, S.D. (1995). *A Manual for Use of MTDFREML* [draft], Animal Research Station, United States Department of Agriculture: Clay Center, NE.

32. Henderson, C.R. (1975). Best linear unbiased estimation and prediction under a selection model. *Biometrics* 32, 423–447.

33. Weiss, A., King, J.E., and Enns, R.M. (2002). Subjective well-being is heritable and genetically correlated with Dominance in chimpanzees (*Pan troglodytes*). *Journal of Personality and Social Psychology* 83, 1141–1149.

34. Meyer, K. (1989). Restricted maximum likelihood to estimate variance components for animal models with several random effects using a derivative–free algorithm. *Genetics, Selection, Evolution* 21, 317–340.

35. Meyer, K. (1991). Estimating variances and covariances for multivariate animal models by restricted maximum likelihood. *Genetics, Selection, Evolution* 23.

36. Akaike, H. (1987). Factor analysis and AIC. *Psychometrika* 52, 317–332.

37. Weiss, A., King, J.E., and Enns, R.M. (1999). Can we breed a happier ape? The heritability of subjective well-being in zoo chimpanzees (*Pan troglodytes*). *Behavior Genetics* 29, 374.

38. Reverter, A., Golden, B.L., Bourdon, R.M., and Brinks, J.S. (1994). Method R variance-components procedure: Application on the simple breeding value model. *Journal of Animal Science* 72, 2247–2253.

39. Loehlin, J.C. (1998). *Latent Variable Models: An Introduction to Factor, Path, and Structural Analysis* (Mahwah, NJ: Lawrence Erlbaum Associates).

40. Meyer, K. (1998). *DFREML, 3.1 ed.*, Self Published: Armidale, NSW, Australia.Meyer, K. (1998). *DFREML, 3.1 ed.*, Self Published: Armidale, NSW, Australia.

41. King, J.E., Weiss, A., and Sisco, M.M. Age differences in chimpanzee personality, under review.

APPENDIX A: USING ANIMAL BREEDER'S TOOLKIT TO COMPUTE RELATEDNESS AMONG INDIVIDUALS WITHIN A PEDIGREE

The Animal Breeder's Toolkit (ABTK)[25] is one of several software packages that compute degrees of relatedness (R_{ij}) and inbreeding coefficients (f) in a sample. It is comprised of a package of commands that perform the necessary matrix algebra needed to compute these as well as other parameters. Recently, SAS[29] has incorporated a new procedure, INBREED, which performs the same function. There are other free and commercial products as well.

To compute the degree of relatedness with ABTK one must first have a space-delimited pedigree file in which the first column represents the stud or identity number of the individual, the second the identity number of his or her father, and the third the stud number of his or her mother. If either parent's identity is unknown, it should be represented by a ".". We have included a small section of the pedigree from a sample of chimpanzees at the Yerkes Regional Primate Center below:

```
   .
   .
   .

86    16    21
87    37    24
88     8    40
89     8    28
90    37    44
91     .     .
92    53    36
93     .    41
94    16    23
95    37    45
96     8    11

   .
   .
   .
```

The first step is to reorder and check the pedigree file:
```
chkord ped.old > ped.chk 2 > ped.bad
```
This command reorders the pedigree so that parents precede offspring and checks whether there are any problems, e.g., individuals that appear as both fathers and mothers. This procedure is required for computing the degrees of relatedness and it is generally a good idea to apply this to any pedigree file including those that will be used in other pedigree analyses such as MTDFREML. The `ped.old`[†] file is the

original space-delimited pedigree file. The `ped.chk` file is the properly ordered version of the pedigree file, i.e., where parents precede their offspring, which is required for the next steps. The `ped.bad` file will contain a listing of any errors, e.g., individuals that appear as mothers and fathers in the file.

The second step is to create a file containing the inverse numerator relationship matrix (A^{-1}) and a file containing inbreeding coefficients (f) for the subjects in the analysis from the reordered pedigree file that we created in the previous step.

`ainv -i ped.chk -o ped.inbreed -v ped.nrm`

where `ped.inbreed` and `ped.nrm` are the files containing the inbreeding coefficients for all subjects and the inverse numerator relationship matrix for all possible pairs of subjects, respectively.

The third step is to invert the inverse numerator relationship matrix:

`invert -i ped.nrm -o ped.output`

Here, `ped.nrm` is the file derived from the previous command and `ped.output` contains a matrix, **A**, indicating the degree of relatedness between all pairs of individuals.

Unfortunately, the format of the matrix might be difficult to use in a conventional statistics packages. Hence, our next two steps use ABTK to change the format of the `ped.output` file so that each row includes the stud numbers of the individuals within a pair and their degree of relatedness. The commands to do this are as follows:

`d2s ped.output ped.newoutput`

`s2t ped.newoutput`

Here, `ped.output` is the file we created in the prior step and `ped.newoutput` is the "tree" version of this file where each row represents a single pair of individuals.

If there is no inbreeding, i.e., the `ped.inbreed` file is empty; the previous step was the last. However, if there was some inbreeding, then the degree of relatedness for every pair of individuals in which an inbred animal was present needs to be adjusted by dividing the degree of relatedness by the following:

$$\sqrt{(f_i + 1)(f_j + 1)}$$

where f_i is equal to the inbreeding coefficient of one individual in the pair and f_j is equal to the inbreeding coefficient of the other individual of the pair. If an individual is not inbred, the inbreeding coefficient is equal to 0.

---

† These file names are only examples; any file name can be used.

18 The Elusive World of Personality Genes: Cherchez le Phenotype

Richard P. Ebstein, Rachel Bachner-Melman,
Jonathan Benjamin, and Robert H. Belmaker

CONTENTS

18.1 PERSONALITY

18.1.1 WHAT IS PERSONALITY AND HOW IS IT MEASURED?

The term "personality" typically is used in two different ways. One refers to the totality of a person's psychological nature. Personality in this sense is to be distinguished from physique (an individual is comprised of body and soul—if we separate out the explicitly spiritual we are left with body and personality), and, in a different context, from mental illness (we distinguish between someone's behavior during illness and her normal personality). For most authors, personality does not encompass all mental phenomena. Few would include simple perception, memory, and concentration as aspects of personality, and intelligence is also frequently not considered part of personality, although a recent book[1] on personality genetics did include a chapter on cognitive ability. All agree that the emotional aspects of the psyche are included in personality. This approach to personality considers the universal aspects of people, not how they differ from each other. Traditional theories of personality, of which Freud's remains the outstanding example, purport to explain what is man, "what makes him tick," to suggest a "physiology" of personality. If genomics aims to elucidate the entire genome of a species, and biochemistry and physiology aim to account for the functions of that genome and its protein products, and if genetics is limited to the study of the consequences of genetic *differences* between members of the species, then by analogy we can say that this first sense of "personality" is a "genomic" approach. Its subjects include the unconscious, motivation, instincts, defense mechanisms, the life cycle, interpersonal relations, conditioning and self-actualization. While this branch of psychology has had a major influence on twentieth-century culture, much of it is psychoanalytic, metaphoric, and untestable, it has hardly been studied by geneticists.

A second sense of the term "personality," and the one we will use throughout the remainder of this chapter, is *the individual psychological aspects of people that make them "recognizable," which is to say different from one another.* In this sense we say that Jones is miserly, Smith is gregarious, and, even, that Brown is "a character." The personalities of fictitious characters like Falstaff, Othello, and Scarlet O'Hara seem familiar to millions. However, this intuitive grasping of personality is based on one or two outstanding traits of each individual and there are no obvious or systematic relationships among the various examples of traits mentioned above. A general, empirical theory of personality differences must include a battery of dimensional variables, and the number of variables must be manageable. The attributes chosen should account for most of the variance in personality, and hopefully also make "clinical sense" to the psychologist, that is, somehow relate to what seem to be fundamental and universal aspects of personality in the first sense of the term as described above. It is moreover a part of the definition of a personality trait or a personality type that the description is relatively enduring and stable for the individual concerned, not a temporary condition.

Virtually all approaches to personality assessment used in genetic research have used either direct laboratory evaluation of dimensional behaviors or pencil-and-paper questionnaires that inventory dimensional personality traits. Laboratory assessments are objective but labor-intensive both for collecting data and for analyzing them. Furthermore, they sample a single moment in time. Self- or other report questionnaires may be more subjective, but have the advantage that they typically sample subjects' characteristics over extended periods.

The most widely used general pencil-and-paper personality inventories assume that rating an individual accurately on the important dimensions of personality gives a reasonable and economical description of that individual's personality. The number and makeup of dimensions chosen are typically derived by one of the following three approaches.

Approaches based on psychopathology derive the personality traits from psychiatric illnesses. Psychiatric illnesses are assumed to be extreme variants, and therefore the clearest expressions, of normal personality. Thus the Minnesota Multiphasic Personality Inventory[2] grades people on the degree to which they resemble patients diagnosed with hysteria, obsessive–compulsive disorder, schizophrenia, and so on. This approach is often reflected in popular discourse; we say that someone is compulsive, hysterical, and so on, but the approach has not been widely used in genetic research on human personality.

Theoretical approaches derive from a theory of personality in the first sense described above. Cloninger developed the Tridimensional Personality Questionnaire (TPQ)[3] based on a biological model employing the monoamines dopamine, noradrenaline, and serotonin, which function as neurotransmitters in subcortical brain systems. Animal and human findings from lesioning, imaging and pharmacological studies involving these pathways implicate serotonin (5-HT) pathways in behaviors devoted to avoiding harm or escaping punishment, and to the associated emotion of anxiety (called "harm avoidance" (HA) in the TPQ). Human subjects with high HA would be called "neurotic" on many other instruments; those with low HA are "stable" or "healthy" or "well adjusted." Similar studies implicate noradrenergic (NE) pathways in approach behaviors, which the TPQ calls "reward dependence" (RD). Human beings high on RD are sentimental and affectionate; those with low RD are tough and pragmatic. Finally dopaminergic (DA) pathways are implicated in drug use, sensation seeking and explorative behaviors, and emotions like curiosity and recklessness, which the TPQ calls "novelty seeking" (NS). Individuals with low NS are deliberate and frugal. The eight possible combinations of the two extremes of the distributions of these three dimensions yield personality constellations held to reflect clinically recognized personality disorders such as antisocial psychopathy and obsessive–compulsive personality disorder. A later version of the TPQ is the Temperament and Character Inventory (TCI).

The third approach derives from *everyday speech*. Natural language is considered to have proved itself during evolution to be a reasonably accurate and adaptive way of describing people. Personality inventories have been constructed by listing thousands of personality descriptors culled from dictionaries of everyday language and reducing them to sizable numbers by factor analysis. The most popular such inventory in clinical and academic use is the neopersonality inventory (NEO-PI)[4] which assesses five main personality factors. Neuroticism resembles TPQ HA. Extraversion is the degree of interest in other people and in social dominance. Openness to new ideas and experiences is the opposite of traditionalism. Agreeableness is a measure of how likeable an individual is. Conscientiousness assesses conscience, thoughtfulness, planning and order, and their opposites: spontaneity and lack of prudence. Extraversion and conscientiousness (with opposite signs) correlate with TPQ NS. While details of dimensions differ between instruments, a number of recent factor analyses concur that five main factors adequately describe human personality differences, and these have been termed "the big five."

Given the uncertainty inherent in measuring personality traits, it seems advisable, where possible, to study putative personality traits simultaneously with more than one instrument and/or laboratory assessment. Good correlations between apparently similar traits in two or more assessments, and converging findings of associations between a given polymorphism and those traits in two or more assessments, enhance our confidence in those findings. This is all the more necessary, given the difficulty in replicating molecular genetic findings in personality genetics, similar to the problems encountered in other complex phenotypes.

Additionally, personality geneticists study traits of particular interest, which may not be included in general personality inventories (see below).

18.2 HERITABILITY

Personality dimensions measured by self-report questionnaires such as the NEO and TPQ are moderately heritable (40–60%).[5–8]

18.3 MOLECULAR GENETICS

An important step in genetic studies of human personality consisted of several initial investigations in 1996 that suggested association between specific genes and personality traits assessed by self-report questionnaires. The two common polymorphisms first studied, the dopamine D4 receptor (DRD4)[9,10] and the serotonin transporter promoter region (5-HTTLPR),[11] continue to be the focus of many investigations relevant not only to narrow definitions of personality (e.g., extraversion/novelty-seeking, harm avoidance/neuroticism), but also to an ever widening circle of related phenotypes including substance abuse,[12–21] attention deficit,[22–26] shyness,[27] fear responses,[28] and depression.[29]

The associations between the DRD4 exon III repeat region and novelty seeking[30,31] and the 5-HTTLPR with neuroticism[32,33] have also been the subjects of recent meta-analyses. Although the DRD4 exon III repeat region is not associated overall with novelty seeking, a functional promoter region polymorphism upstream (-C521T) first studied in a Japanese population[34] shows more promising results following meta-analysis,[31] In an overall analysis of results with 5-HTTLPR (n = 5,629 subjects), Sen et al.[32] found suggestive evidence for an association between the 5-HTTLPR short allele (s) and increased anxiety-related personality trait scores (p = 0.087). However, they also found strong evidence for heterogeneity. This heterogeneity is largely explained by substantial variation between the studies in the inventory used. When the analysis was stratified by inventory type, there was a significant association between 5-HTTLPR and NEO neuroticism (p = 0.000016), a nonsignificant association between 5-HTTLPR and TCI/TPQ harm avoidance (p = 0.166), and no association between 5-HTTLPR and other anxiety-related personality traits (p = 0.944). Similar meta-analysis findings for 5-HTTLPR were reported by Schinka et al.[33] who suggest that "the success of future personality genetics research will be maximized by the use of personality measures from both the psychobiological and five-factor models." It is perplexing and frustrating that NEO neuroticism and TPQ harm avoidance, which at face value appear to measure very similar traits, produce such divergent results.

A parallel strategy is to look at these common polymorphisms across a range of related phenotypes, some of which might "capture" associations that are often hard to pin down when a single self-report questionnaire is used to inventory subjects. In the following section we discuss some results using the DRD4 and 5-HTTLPR polymorphisms as examples of how looking at extended, or broader, phenotypes may shed light on complex phenotype–genotype relationships in personality genetics.

18.4 RELATED PHENOTYPES

18.4.1 FIBROMYALGIA

Fibromyalgia syndrome (FM) is an under-diagnosed musculoskeletal condition of unknown etiology affecting more than 10% of patients attending general medical clinics and 15 to 20% attending rheumatology clinics.[35] It is seen predominantly in women. Patients complain that they ache all over. A number of other symptoms are often present, particularly fatigue, morning stiffness, sleep disturbance, paresthesias, and headaches. In 1999, a German group reported an association between this syndrome and the short 5-HTTLPR allele[36] consistent with the observed psychosocial profile of these women who score high on anxiety-related personality traits. Toward confirming this first report, we genotyped a group of 99 female fibromyalgia patients from two Israeli ethnic groups.[37] Additionally, we assessed each patient with the Tridimensional Personality Questionnaire (TPQ). The percentage of patients showing the short/short genotype is twice that observed in the control population (35% vs. 17%). The difference in 5-HTTLPR genotype frequencies between the fibromyalgia and control group was highly significant (chi-square = 25.31, $p = 0.00019$, df = 2). Additionally, between-subjects testing showed a highly significant effect of diagnosis on three of the TPQ personality traits. Novelty seeking is lower in FMS (F = 14.46, $p < 0.0001$), whereas harm avoidance (F = 31.89, $p < 0.0001$) and persistence (F = 41.38, $p < 0.0001$) are higher. No effect on reward-directed behavior was observed. Intriguingly, there was also a significant association between 5-HTTLPR genotype and the TPQ harm avoidance trait (F = 6.21, $p = 0.013$), an association we have consistently failed to observe[38] in our large nonclinical population using this personality inventory.

We were also prompted in these women to examine the DRD4 exon III repeat since as a group they score low on TPQ novelty seeking traits.[39] As we predicted, a significant *decrease* in the frequency of the 7 repeat genotype was observed in the fibromyalgia patients (chi-square = 29.91, $p < 0.008$, df = 14), consistent with low novelty seeking in these women. Although in the control population no association was detected between TPQ novelty seeking and the DRD4 7 repeat, in the phenotypically extreme, albeit smaller, FM group a significant association was observed between the 7 repeat and novelty-seeking (one-way ANOVA: F = 4.47, $p = 0.038$). Patients with at least one 7 repeat had higher novelty-seeking scores as is observed in some but not all nonclinical populations (subjects with at least one 7-repeat NS = 14.57 ± 3.26 SD, vs. "all others" NS = 12.72 ± 3.49 SD).

Consequently, by looking at a clinical group of women with an extreme personality profile, high harm avoidance and low novelty seeking we were able to verify the sometimes elusive associations between the short 5-HTTLPR allele and the DRD4 7 repeat, respectively. We consider these results "proof of principle" that reliance on a single personality measure in one population category may be inadequate to demonstrate small effect sizes of common polymorphisms modestly contributing a small fraction of the genetic variance in complex traits such as personality.

18.4.2 ALTRUISM

We have also examined an intriguing dimension of human personality, altruism, in a large nonclinical population for association with several common polymorphisms germane to a broader view of human personality than that examined with the usual inventories (TPQ or NEO).[40]

The paradox of human altruism, helping others and thereby reducing one's own fitness, has confounded evolutionary biologists since the days of Darwin. Not only is altruism a puzzle for evolutionary biologists but this trait has also perplexed psychologists who have questioned whether there is such a thing as an altruistic personality.[41,42] However, altruistic behavior is commonplace, and a unique feature of human altruism is that it extends beyond Hamilton's concept of "inclusive fitness,"[43] which explains altruistic acts by including helping genetically related individuals, and even beyond reciprocal altruism and reputation-based altruism.[44] Out of all the animals, only humans practice wholesale mutual aid among genetically unrelated individuals.

Although theoretical understanding of the evolutionary[45] and psychological mechanisms[46] that underlie altruism has greatly increased in the past several decades, almost nothing is known regarding specific genes contributing to this behavior. However, several twin studies[47–52] reported significant heritability of prosocial attitudes. As expected of a trait that is partially influenced by genes, prosocial or altruistic dispositions show individual differences with origins in early childhood and stability over developmental time.[53,54]

The mechanisms facilitating prosocial behavior and human altruism are of interest not only to evolutionary biology, but also to students of psychopathology. Psychiatric research has for obvious reasons focused on negative factors that adversely affect mental well-being. However, there is growing interest in positive or protective factors that play a role in promoting normal development. Prosocial or altruistic attitudes, such as cooperativeness, helpfulness, sharing, and being empathetic, are important elements that facilitate social networks thought to promote mental well-being. For example, prosocial behavior correlates with children's academic excellence, allows them to resist social pressures for antisocial activities, and to engage themselves with empathy in others' emotional experiences.[55] Additionally, positive social interactions undoubtedly eased by altruistic behaviors have been shown to exert powerful beneficial effects on health outcomes and longevity.[56] Importantly, self-report measures of altruism correlate with peers' and teachers' evaluation.[53,57]

Any account of prosocial behavior and altruism in humans should include the identification of particular genetic polymorphisms partially contributing to this trait. Toward the goal of identifying specific genes associated with altruism and prosocial attitudes, 354 nonclinical families with multiple siblings were inventoried for scores on the selflessness scale.[58] This questionnaire measures the propensity to ignore one's own needs and serve the needs of others, or in other words, altruism. Subjects were also inventoried on Cloninger's TPQ[59] since the reward subscale of this questionnaire taps into elements of human altruism, such as empathy. We examined two dopaminergic genes in these subjects that we hypothesized might contribute to prosocial or altruistic traits, based on the role a single variant of these genes plays in novelty seeking, which includes antisocial aspects, and in a childhood behavioral disorder, attention-deficit hyperactivity disorder (ADHD). This disorder is often comorbid with antisocial behavior such as conduct disorder and oppositional defiant disorder. The dopamine D4 receptor (DRD4) exon III 7 repeat (D4.7) and the DRD5 148 bp microsatellite variant both have been robustly associated with ADHD.[22,60] We reasoned that if one variant contributes to antisocial traits, then, conversely, the absence of this variant or the presence of other variants might contribute to altruistic behavior. Additionally, we examined three DRD4 promoter region SNPs.[34] We also genotyped 3 SNPs[61] in the insulin-like growth factor 2 gene (IGF2), an imprinted gene on chromosome 11p15.5 that is an attractive candidate because some studies connect growth factors of this class with survival of dopamine neurons specifically,[62–69] and with neural development overall.[70–75]

Significant association with scores on the selflessness scale was observed with the most common D4.4 allele (FBAT multivariate test, bi-allelic mode: chi-square = 14.06, p = 0.0008, df = 2). No association was observed with the second most common allele in this population, the D4.7 repeat, the risk allele in ADHD. Evidence was also observed showing a negative correlation between the DRD5 148 bp microsatellite repeat, the risk allele in ADHD (p = 0.027). Significant association was also observed with the IGF2 *ApaI* "G" allele (p = 0.006). Association was also observed between TPQ reward, a temperament that taps into some elements of altruism such as empathy, and the D4.4 repeat (p = 0.0016).

The demonstration that common polymorphisms pertinent to dopamine pathways are associated with altruistic or prosocial attitudes in humankind is consistent with a central role of the brain reward center in mediating this phenotype. This is one of the first reports to characterize specific genes acting as positive or protective factors hypothesized to promote normal behavioral development and mental well-being.

The provisional finding that the DRD4 exon III 4 repeat is associated with a positive personality trait, altruism, is intriguing though not altogether surprising. The DRD4 7 repeat has been the focus of extensive investigation both regarding ADHD and novelty seeking/impulsivity, behavioral phenotypes with considerable negative features. It makes sense therefore that the most common 4 repeat would be associated with a more positive phenotype such as prosocial or altruistic behavior.

Evidence from haplotype data indicates that the relatively new D4.7 repeat allele originated as a rare mutational event (or events) that nevertheless increased to high frequency in human populations by positive selection.[76,77] The estimated origin of this event is approximately ~40,000 to 50,000 years ago, a crucial period in the history of our species when, in one key wave of migration, there was a spread of modern humans out of Africa. Ding et al.[76] hypothesize that the currently observed frequencies of the D4.4 and D4.7 alleles of this two-allele system are an example of balanced selection. We conjecture that the balanced maintenance of both the D4.4 and D4.7 repeats is related to the need for diverse behavioral phenotypes in human populations partially determined by this gene, altruistic and prosocial (D4.4) versus a more aggressive, novelty seeking or perhaps even antisocial type (D4.7). Indeed, models suggest that populations composed of extremely altruistic individuals would be unstable.[44]

18.5 THE SOCIAL PERSONALITY

Although progress has been made in unraveling the molecular genetic architecture of individual personality traits,[30,31,78,79] little is known regarding the genetic basis of social behaviors (measures of the interaction between at least two individuals) in humans. Much more has been elucidated on the genetic control of social behavior in animals, including insects and lower vertebrates.[80–82] In particular, research over the past two decades has revealed the molecular mechanisms by which two peptide hormones, vasopressin and oxytocin, shape social behavior in species from fish to rodents.[83] However, despite speculation,[84] little evidence has been forthcoming linking these hormones to corresponding human behaviors.

Arginine-vasopressin (AVP) in rodents has been associated with the modulation of a broad range of behavioral phenotypes including social recognition and learning, affiliative behaviors, aggression, dominant–subordinate relationships, parental behavior, grooming, and categories of pair bonding such as monogamy.[85] Of particular interest for the present study, however, are affiliative behaviors that interact with but are distinct from reproductive pair bonding. For example, in squirrel monkeys vasopressin has been shown to modulate male–male interactions,[86] and in humans a facet of aggressive behavior.[87] Extrapolating from these studies, we suggest that vasopressin might modulate a range of human behaviors distinct from romantic pair bonding as observed in the prairie vole. Strengthening this notion is a recent study showing linkage between the *AVPR1A* receptor and autism, a behavior disorder of which dysfunctional social interaction is a core symptom.[88]

In our study,[89] we examined linkage between microsatellites in the vicinity of the *AVPR1A* receptor, which in animal studies regulates brain regional expression of this gene,[90] and self-report questionnaires designed to measure two aspects of complex social behavior, self-presentation[91] and sibling relationships.[92] We hypothesized that only these two social constructs, among those inventoried in this group of subjects, represent human equivalents of behaviors observed in animals and likely modulated by vasopressin. None of the other measures of behavior we inventoried were significantly associated with the vasopressin receptor except for abnormal eating behavior.[93]

We evaluated the perceived quality of the relationship between siblings using the Sibling Relationship Questionnaire (SRQ)[94] and assessed three dimensions including: (a) relative status/power (b) warmth/closeness, and (c) conflict. In a factor analysis of the original questionnaire,[94] the authors found the following non-orthogonal factor pattern: conflict is manifested as quarreling, competition and antagonism. Warmth/closeness is manifested as intimacy, prosocial behavior, companionship, admiration, perceived similarity and affection. Relative status/power refers to the degree and direction of asymmetry in the sibling relationship, one extreme indicating greater power by one sibling, and the other greater power by the other sibling.

Snyder[95] first proposed that people differ in the extent to which they monitor (observe and regulate) their social self and balance their inner directives as opposed to their perceptions of how others expect them to behave. Lennox and Wolfe[91] revised Snyder's original self-monitoring scale into the revised self-monitoring scale

(RSMS) and the concern for appropriateness scale (CFA), which together comprise the social psychological construct of self-presentation. The two have generally been found to be orthogonal[96,97] and more psychometrically sound than Snyder's scale.[98] The RSMS measures acquisitive self-presentation, or "getting ahead," an active and flexible social approach aimed at gaining power and status. The CFA measures a defensive and fearful social approach associated with conformity, aimed at gaining acceptance and approval and avoiding social threats. Whereas both self-presenta- tional styles reflect social orientations with a high degree of concern for social cues and social approval, the RSMS correlates positively with measures of social adap- tation such as self-esteem and extraversion,[97] whereas the CFA correlates negatively with such measures.

As discussed above, AVPR1A molds social behavior across the vertebrates, and it makes sense that self-presentation, a construct that characterizes the style (dom- inating, fitting in) an individual adopts in participating in group activities should also be linked to this receptor. We further hypothesized that the SRQ construct of conflict, and possibly that of power, would tap into the "aggressive" style of behav- iors that are also molded by vasopressin.

We also thought it of some interest to examine sibling relationships because such relationships are often the first dyadic social relationship that individuals experience, and the style people adopt with their siblings may have ramifications for their subsequent relationships with nonrelated persons. Indeed, we observe in this cohort a correlation between the CFA and two of the SRQ scales (conflict and warmth).

Suggestive linkage was observed between both microsatellites (RS1 and RS3) and the SRQ "Conflict" Scale (RS1: chi-square = 13.65, LOD = 2.96, $p = 0.0001$; RS3: chi-square = 14.54, LOD = 3.16, $p = 0.00007$) and the CFA "self-presentational style" (RS1: chi-square = 8.25, LOD = 1.79 $p = 0.002$; RS3: chi-square = 8.81, LOD = 1.91, $p = 0.002$).

The provisional linkage observed between microsatellites in the *AVPR1A* pro- moter region and scores on the CFA and the SRQ conflict scale constitutes the first substantial evidence that vasopressin, which mediates species-specific social behav- ior across the vertebrates, may play a similar role in humans. Although social behavior and the acquisition of social skills in humans constitute the cornerstone of society and culture, few if any genetic studies have attempted to relate individual differences in social skills to specific genes. Moreover, social skills are relevant to a variety of psychiatric disorders including autism, schizophrenia, and externalizing behavior problems in children. From an evolutionary perspective, it is not surprising that a hormone that almost universally affects a spectrum of social and affiliative behaviors in lower animals[85] has a parallel role in the human, a social animal highly dependent on group interactions for individual and species survival.

18.6 LINKAGE STUDIES

Four studies[99–102] have now identified a broad region on chromosome 8p that harbors a locus that contributes to individual differences in a personality trait that is a measure of emotional lability. Two studies found linkage to harm avoidance[99,101] and used identical self-report questionnaires, the TPQ, whereas the Fullerton et al.[100] report used Eysenck's psychoticism scale and found linkage to neuroticism.[103] Subjects who score high on harm avoidance can be described as worrying and pessimistic, fearful and doubtful, and shy and fatigable. High scorers on Eysenck's neuroticism dimension can be similarly described. The Cloninger et al.[99] genome scan used subjects recruited from families with an alcoholic proband as part of the COGA study.[104] The highest LOD score (3.2) using 291 microsatellite markers for a site to harm avoidance was near marker D8S1106 (8p 21-23) at 27 cM. The Fullerton British study genotyped 561 extremely discordant and concordant sibling pairs selected from a population of 34,580 families. The observed linkage to neuroticism near marker D8S277 (8 cM) gave a LOD score of 2.9.

In our own study[102] we have now genotyped altogether 24 microsatellite markers in 377 families. Using three methods (maximum likelihood binomial or MLB, MER-LIN and an associated one parameter model) we observed significant results (p-values from 0.002 to 0.0004) for linkage to harm avoidance in this region. A peak multipoint LOD score of 2.76 (p-value 0.0002) was obtained with the MLB method. The region-wide empirical p-value was 0.002 [0.001 – 0.0046]. Although the peak position varied somewhat according to the method (D8S1048 for MLB, D8S1463 for the two other methods), for three methods D8S1810 (~60 cM) is within 1 to 2 cM of the peak for harm avoidance. This marker is of particular interest since it is proximate (<0.5 cM) to the core haplotype that in several recent studies shows significant association with schizophrenia near neuroregulin 1.[105,106] Although association studies with microsatellite markers need to be interpreted cautiously, using the Haplotype Trend Regression test one marker, D8S499 (~60 cM), showed an empirical p-value of 2×10^{-5} for allele 3, which confers a decreased harm avoidance score.

Altogether, our current linkage and association results suggest the possibility that the same locus near the neuroregulin 1 gene on chromosome 8p confers risk for both an anxiety-related personality trait as well as schizophrenia. We hypothesize that this common genetic factor may contribute to emotional lability during early development which constitutes a predisposing factor for major psychosis.

Curtis et al.[108] recently reanalyzed the original dataset first reported by Cloninger et al.[99] using a new method of linkage analysis for quantitative traits that deals with extended pedigrees. As well as supporting linkage of HA to D8S549 as originally reported,[99] this method also produces an MALOD of 2.4 ($p = 0.002$) near DBH and several positive LOD scores for novelty seeking, the largest being MALODs of 3.1 ($p = 0.0004$) near D12S391 and 3.4 ($p = 0.0003$) near D17S1299. There is no support for linkage of novelty seeking or HA to the regions around DRD4 and 5-HTTLPR, respectively.

Neale et al.[107] analyzed a genome scan for neuroticism on a sample of 129 sib-pair families (113 with a single sibling pair, 18 with multiple sibling pairs) containing a total of 201 possible sibling pairs, ascertained for concordance on nicotine dependence. Using Merlin-Regress, and replicated peaks for neuroticism described by prior studies, on chromosomes 1 and 11 with LOD scores of 2.52 and 1.97 (p-value of 0.003 and 0.0108), respectively, and have some evidence for a novel finding on chromosome 12 with a LOD of 2.85 (p-value of 0.0014).

18.7 IMPORTANCE

Advances in the molecular genetics of human personality could have major implications for both psychology and psychiatry. In the field of psychology, alleles or allele–allele interactions affecting personality traits with large effect sizes would invite renewed explorations of the neurochemical or neuroanatomical effects involved. Discoveries of such effects could potentially revolutionize understanding of the physical bases of the complex and hitherto intractable phenotypes of personality traits. Such discoveries might also raise the specter of a pill for, or against, every emotion or trait, a sort of "plastic surgery of the psyche." History provides ample evidence of our readiness to use and abuse psychoactive substances. Thus far, however, findings have been limited to modest effect sizes on monoamine systems that were already predicted by knowledge outside the field of genetics.

In psychiatry, chromosomal regions or interactions between regions, or alleles or allele–allele interactions, affecting personality traits with large effect sizes would invite new approaches to genetics for mental illnesses. We have already alluded to the idea that mental illnesses may be extreme manifestations of normal personality traits, or of maladaptive combinations of these traits. In analogous fashion, idiopathic hypertension and adult-onset diabetes may represent harmful expressions of genes, or harmful interactions between genes, that originally evolved to code for proteins that enable normal blood pressure and carbohydrate metabolism. Modern environments may differ from those in which those genes were adaptive. Genes affecting personality traits may also have been selected when both the external environment and intrapersonal and interpersonal existence differed from modern experience. Discoveries of chromosomal regions or alleles affecting, for example, harm avoidance, invite investigations of those same regions or alleles in clinical anxiety disorders. In practice, findings so far have been so few that any promising leads probably will be followed up with initial investigations in all or almost all psychiatric disorders.

The region on chromosome 8 implicated in harm avoidance will undoubtedly be further scrutinized for linkage to psychiatric disorders. To the extent that future genetic findings in personality are unexpected, they may suggest novel approaches to psychiatric illnesses.

ACKNOWLEDGMENTS

This research was partially supported by the Israel Science Foundation founded by the Israel Academy of Sciences and Humanities (RPE).

REFERENCES

1. Benjamin, J., Ebstein, R.P. & Belmaker, R., H. *Molecular Genetics and the Human Personality*, 355 (American Psychiatric Publishing, Inc., Washington, D.C., 2002).
2. Butcher, J.N. *MMPI-2: Minnesota Multiphasic Personality Inventory-2: manual for administration, scoring, and interpretation*, ix, 212 p. (University of Minnesota Press, Minneapolis, 2001).
3. Cloninger, C.R. A unified biosocial theory of personality and its role in the development of anxiety states. *Psychiatr Dev* **4**, 167–226 (1986).
4. Costa, P.T., Jr. & McCrae, R.R. Stability and change in personality assessment: the revised NEO Personality Inventory in the year 2000. *J Pers Assess* **68**, 86–94 (1997).
5. Jang, K.L., Livesley, W.J. & Vernon, P.A. Heritability of the big five personality dimensions and their facets: a twin study. *J Pers* **64**, 577–91 (1996).
6. Heiman, N., Stallings, M.C., Hofer, S.M. & Hewitt, J.K. Investigating age differences in the genetic and environmental structure of the tridimensional personality questionnaire in later adulthood. *Behav Genet* **33**, 171–80 (2003).
7. Heath, A.C., Cloninger, C.R. & Martin, N.G. Testing a model for the genetic structure of personality: a comparison of the personality systems of Cloninger and Eysenck. *J Pers Soc Psychol* **66**, 762–75 (1994).
8. Bergeman, C.S. et al. Genetic and environmental effects on openness to experience, agreeableness, and conscientiousness: an adoption/twin study. *J Pers* **61**, 159–79 (1993).
9. Ebstein, R.P. et al. Dopamine D4 receptor (D4DR) exon III polymorphism associated with the human personality trait of Novelty Seeking. *Nat Genet* **12**, 78–80 (1996).
10. Benjamin, J. et al. Population and familial association between the D4 dopamine receptor gene and measures of Novelty Seeking. *Nat Genet* **12**, 81–4 (1996).
11. Lesch, K.P. et al. Association of anxiety-related traits with a polymorphism in the serotonin transporter gene regulatory region [see comments]. *Science* **274**, 1527–31 (1996).
12. Lusher, J.M., Chandler, C. & Ball, D. Dopamine D4 receptor gene (DRD4) is associated with Novelty Seeking (NS) and substance abuse: the saga continues. *Mol Psychiatry* **6**, 497–9 (2001).
13. Comings, D.E. et al. Studies of the 48 bp repeat polymorphism of the DRD4 gene in impulsive, compulsive, addictive behaviors: Tourette syndrome, ADHD, pathological gambling, and substance abuse. *Am J Med Genet* **88**, 358–68 (1999).
14. Mel, H. et al. Additional evidence for an association between the dopamine D4 receptor (D4DR) exon III seven-repeat allele and substance abuse in opioid dependent subjects: Relationship of treatment retention to genotype and personality. *Addiction Biology* **3**, 473–481 (1998).
15. Kotler, M. et al. Excess dopamine D4 receptor (D4DR) exon III seven repeat allele in opioid-dependent subjects. *Mol Psychiatry* **2**, 251–4 (1997).
16. Li, T. et al. Association analysis of the dopamine D4 gene exon III VNTR and heroin abuse in Chinese subjects. *Mol Psychiatry* **2**, 413–6 (1997).
17. Muramatsu, T., Higuchi, S., Murayama, M., Matsushita, S. & Hayashida, M. Association between alcoholism and the dopamine D4 receptor gene. *J Med Genet* **33**, 113–5 (1996).
18. Lerman, C. et al. Depression and self-medication with nicotine: the modifying influence of the dopamine D4 receptor gene [see comments]. *Health Psychol* **17**, 56–62 (1998).

19. Shields, P.G. et al. Dopamine D4 receptors and the risk of cigarette smoking in African-Americans and Caucasians. *Cancer Epidemiol Biomarkers Prev* **7**, 453–8 (1998).
20. Lerman, C. et al. Interacting effects of the serotonin transporter gene and neuroticism in smoking practices and nicotine dependence. *Mol Psychiatry* **5**, 189–92 (2000).
21. Hu, S., Brody, C.L., Fisher, C. et al. Interaction between the serotonin transporter gene and neuroticism in cigarette smoking behavior. *Mol Psychiatry* **5**, 181–8 (2000).
22. Faraone, S.V., Doyle, A.E., Mick, E. & Biederman, J. Meta-analysis of the association between the 7-repeat allele of the dopamine D(4) receptor gene and attention deficit hyperactivity disorder. *Am J Psychiatry* **158**, 1052–7 (2001).
23. Cadoret, R.J. et al. Associations of the serotonin transporter promoter polymorphism with aggressivity, attention deficit, and conduct disorder in an adoptee population. *Compr Psychiatry* **44**, 88–101 (2003).
24. Kent, L. et al. Evidence that variation at the serotonin transporter gene influences susceptibility to attention deficit hyperactivity disorder (ADHD): analysis and pooled analysis. *Mol Psychiatry* **7**, 908–12 (2002).
25. Retz, W., Thome, J., Blocher, D., Baader, M. and Rosler, M. Association of attention deficit hyperactivity disorder-related psychopathology and personality traits with the serotonin transporter promoter region polymorphism. *Neurosci Lett* **319**, 133–6 (2002).
26. Manor, I. et al. Family-based association study of the serotonin transporter promoter region polymorphism (5-HTTLPR) in attention deficit hyperactivity disorder. *Am J Med Genet* **105**, 91–5. (2001).
27. Arbelle, S. et al. Relation of shyness in grade school children to the genotype for the long form of the serotonin transporter promoter region polymorphism. *Am J Psychiatry* **160**, 671–6 (2003).
28. Hariri, A.R. et al. Serotonin transporter genetic variation and the response of the human amygdala. *Science* **297**, 400–3 (2002).
29. Caspi, A. et al. Influence of life stress on depression: moderation by a polymorphism in the 5-HTT gene. *Science* **301**, 386–9 (2003).
30. Kluger, A.N., Siegfried, Z. & Ebstein, R.P. A meta-analysis of the association between DRD4 polymorphism and novelty seeking. *Mol Psychiatry* **7**, 712–7 (2002).
31. Schinka, J.A., Letsch, E.A. & Crawford, F.C. DRD4 and novelty seeking: results of meta-analyses. *Am J Med Genet* **114**, 643–8 (2002).
32. Sen, S., Burmeister, M. & Ghosh, D. Meta-analysis of the association between a serotonin transporter promoter polymorphism (5-HTTLPR) and anxiety-related personality traits. *Am J Med Genet* **127B**, 85–9 (2004).
33. Schinka, J.A., Busch, R.M. & Robichaux-Keene, N. A meta-analysis of the association between the serotonin transporter gene polymorphism (5-HTTLPR) and trait anxiety. *Mol Psychiatry* **9**, 197–202 (2004).
34. Okuyama, Y. et al. Identification of a polymorphism in the promoter region of DRD4 associated with the human novelty seeking personality trait. *Mol Psychiatry* **5**, 64–9 (2000).
35. Buskila, D. Fibromyalgia, chronic fatigue syndrome, and myofascial pain syndrome. *Curr Opin Rheumatol* **13**, 117–27 (2001).
36. Offenbaecher, M. et al. Possible association of fibromyalgia with a polymorphism in the serotonin transporter gene regulatory region. *Arthritis Rheum* **42**, 2482–8 (1999).
37. Cohen, H., Buskila, D., Neumann, L. & Ebstein, R.P. Confirmation of an association between fibromyalgia and serotonin transporter promoter region (5- HTTLPR) polymorphism, and relationship to anxiety-related personality traits. *Arthritis Rheum* **46**, 845–7 (2002).

38. Ebstein, R.P. et al. No association between the serotonin transporter gene regulatory region polymorphism and the Tridimensional Personality Questionnaire (TPQ) temperament of harm avoidance. *Mol Psychiatry* **2**, 224–6 (1997).

39. Buskila, D., Cohen, h., Neuman, L. & Ebstein, R.P. An association between fibromyalgia and the dopamine D4 receptor exon III repeat polymorphism and relationship to novelty seeking personality traits. *Mol Psychiatry* **9**, 730–1 (2004).

40. Bachner-Melman, R. et al. Dopaminergic polymorphisms associated with self-report measures of human altruism: A fresh phenotype for the dopamine D4 receptor. *Mol Psychiatry* **10**, 333–5 (2005).

41. Gergen, K.-J., Gergen, M.-M. & Meter, K. Individual orientations to prosocial behavior. *Journal of Social Issues* **28**, 105–130 (1972).

42. Batson, C.D. *The Altruism Question: Toward a Social Psychological Answer*, ix, 257 p. (L. Erlbaum, Associates, Hillsdale, N.J., 1991).

43. Hamilton, W.D. The genetical evolution of social behaviour. I. *J Theor Biol* **7**, 1–16 (1964).

44. Sigmund, K. & Hauert, C. Primer: altruism. *Curr Biol* **12**, R270–2 (2002).

45. Fehr, E. & Fischbacher, U. The nature of human altruism. *Nature* **425**, 785–91 (2003).

46. Eisenberg, N. Prosocial behavior, empathy, and sympathy. In *Well Being: Positive Development across the Life Course. Crosscurrents in Contemporary Psychology* (eds. Bornstein, M.H. & Davidson, L.) 253–265 (Lawrence Erlbaum Associates, Publishers, Mahwah, NJ, 2003).

47. Emde, R.N. et al. Temperament, emotion, and cognition at fourteen months: the MacArthur Longitudinal Twin Study. *Child Dev* **63**, 1437–55 (1992).

48. Loat, C.S., Asbury, K., Galsworthy, M.J., Plomin, R. & Craig, I.W. X inactivation as a source of behavioural differences in monozygotic female twins. *Twin Res* **7**, 54–61 (2004).

49. Rushton, J.P., Fulker, D.W., Neale, M.C., Nias, D.K. & Eysenck, H.J. Altruism and aggression: the heritability of individual differences. *J Pers Soc Psychol* **50**, 1192–8 (1986).

50. Loehlin, J.C. & Nichols, R.C. *Heredity, Environment, & Personality: A Study of 850 Sets of Twins*, xii, 202 p. (University of Texas Press, Austin, 1976).

51. Matthews, K.-A., Batson, C.D., Horn, J. & Rosenman, R.-H. Principles in his nature which interest him in the fortune of others: The heritability of empathic concern for others. *Journal of Personality* **49**, 237–247 (1981).

52. Zahn-Waxler, C., Robinson, J.-L. & Emde, R.-N. The development of empathy in twins. *Developmental Psychology* **28**, 1038–1047 (1992).

53. Eisenberg, N. et al. Prosocial development in early adulthood: a longitudinal study. *J Pers Soc Psychol* **82**, 993–1006 (2002).

54. Eisenberg, N. et al. Consistency and development of prosocial dispositions: a longitudinal study. *Child Dev* **70**, 1360–72 (1999).

55. Bandura, A., Caprara, G.V., Barbaranelli, C., Gerbino, M. & Pastorelli, C. Role of affective self-regulatory efficacy in diverse spheres of psychosocial functioning. *Child Dev* **74**, 769–82 (2003).

56. Seeman, T.E. Health promoting effects of friends and family on health outcomes in older adults. *Am J Health Promot* **14**, 362–70 (2000).

57. Caprara, G.V., Barbaranelli, C., Pastorelli, C., Bandura, A. & Zimbardo, P.G. Prosocial foundations of children's academic achievement. *Psychol Sci* **11**, 302–6 (2000).

58. Bachar, E. et al. Rejection of life in anorexic and bulimic patients. *International Journal of Eating Disorders* **31**, 43–48 (2001).

59. Cloninger, C.R. A systematic method for clinical description and classification of personality variants. A proposal. *Arch Gen Psychiatry* **44**, 573–88 (1987).

60. Lowe, N. et al. Joint Analysis of the DRD5 Marker Concludes Association with Attention-Deficit/Hyperactivity Disorder Confined to the Predominantly Inattentive and Combined Subtypes. *Am J Hum Genet* **74**, 348–56 (2004).

61. Gaunt, T.R., Cooper, J.A., Miller, G.J., Day, I.N. & O'Dell, S.D. Positive associations between single nucleotide polymorphisms in the IGF2 gene region and body mass index in adult males. *Hum Mol Genet* **10**, 1491–501 (2001).

62. Quesada, A. & Micevych, P.E. Estrogen interacts with the IGF-1 system to protect nigrostriatal dopamine and maintain motoric behavior after 6-hydroxdopamine lesions. *J Neurosci Res* **75**, 107–16 (2004).

63. Shavali, S., Ren, J. & Ebadi, M. Insulin-like growth factor-1 protects human dopaminergic SH-SY5Y cells from salsolinol-induced toxicity. *Neurosci Lett* **340**, 79–82 (2003).

64. Stefansson, H. et al. Neuregulin 1 and susceptibility to schizophrenia. *Am J Hum Genet* **71**, 877–92 (2002).

65. Bach, L.A. & Leeding, K.S. Insulin-like growth factors decrease catecholamine content in PC12 rat pheochromocytoma cells. *Horm Metab Res* **34**, 487–91 (2002).

66. Offen, D. et al. Protective effect of insulin-like-growth-factor-1 against dopamine-induced neurotoxicity in human and rodent neuronal cultures: possible implications for Parkinson's disease. *Neurosci Lett* **316**, 129–32 (2001).

67. Clarkson, E.D. et al. IGF-I and bFGF improve dopamine neuron survival and behavioral outcome in parkinsonian rats receiving cultured human fetal tissue strands. *Exp Neurol* **168**, 183–91 (2001).

68. Guan, J. et al. N-terminal tripeptide of IGF-1 (GPE) prevents the loss of TH positive neurons after 6-OHDA induced nigral lesion in rats. *Brain Res* **859**, 286–92 (2000).

69. Zawada, W.M. et al. Growth factors improve immediate survival of embryonic dopamine neurons after transplantation into rats. *Brain Res* **786**, 96–103 (1998).

70. Hemberger, M. et al. H19 and Igf2 are expressed and differentially imprinted in neuroectoderm-derived cells in the mouse brain. *Dev Genes Evol* **208**, 393–402.

71. Pham, N.V., Nguyen, M.T., Hu, J.F., Vu, T.H. & Hoffman, A.R. Dissociation of IGF2 and H19 imprinting in human brain. *Brain Res* **810**, 1–8 (1998).

72. Mori, H. et al. Expression of mouse igf2 mRNA-binding protein 3 and its implications for the developing central nervous system. *J Neurosci Res* **64**, 132–43 (2001).

73. Albrecht, S. et al. Variable imprinting of H19 and IGF2 in fetal cerebellum and medulloblastoma. *J Neuropathol Exp Neurol* **55**, 1270–6 (1996).

74. Pera, E.M., Wessely, O., Li, S.Y. & De Robertis, E.M. Neural and head induction by insulin-like growth factor signals. *Dev Cell* **1**, 655–65 (2001).

75. Pera, E.M., Ikeda, A., Eivers, E. & De Robertis, E.M. Integration of IGF, FGF, and anti-BMP signals via Smad1 phosphorylation in neural induction. *Genes Dev* **17**, 3023–8 (2003).

76. Ding, Y.C. et al. Evidence of positive selection acting at the human dopamine receptor D4 gene locus. *Proc Natl Acad Sci U S A* **99**, 309–314 (2002).

77. Wang, E. et al. The Genetic Architecture of Selection at the Human Dopamine Receptor D4 (DRD4) Gene Locus. *Am J Hum Genet* **74**, 931–944 (2004).

78. Bouchard, T.J., Jr. & Loehlin, J.C. Genes, evolution, and personality. *Behav Genet* **31**, 243–73 (2001).

79. Ebstein, R.P., Benjamin, J. & Belmaker, R.H. Personality and polymorphisms of genes involved in aminergic neurotransmission. *Eur J Pharmacol* **410**, 205–214 (2000).

80. Giraud, T., Pedersen, J.S. & Keller, L. Evolution of supercolonies: the Argentine ants of southern Europe. *Proc Natl Acad Sci U S A* **99**, 6075–9 (2002).
81. Krieger, M.J. & Ross, K.G. Identification of a major gene regulating complex social behavior. *Science* **295**, 328–32 (2002).
82. Keller, L. & Parker, J.D. Behavioral genetics: a gene for supersociality. *Curr Biol* **12**, R180–1 (2002).
83. Young, L.J., Lim, M.M., Gingrich, B. & Insel, T.R. Cellular mechanisms of social attachment. *Horm Behav* **40**, 133–8 (2001).
84. Taylor, S.E. et al. Biobehavioral responses to stress in females: tend-and-befriend, not fight-or-flight. *Psychol Rev* **107**, 411–29 (2000).
85. Ferguson, J.N., Young, L.J. & Insel, T.R. The neuroendocrine basis of social recognition. *Front Neuroendocrinol* **23**, 200–24 (2002).
86. Winslow, J. & Insel, T.R. Vasopressin modulates male squirrel monkeys' behavior during social separation. *Eur J Pharmacol* **200**, 95–101 (1991).
87. Thompson, R., Gupta, S., Miller, K., Mills, S. & Orr, S. The effects of vasopressin on human facial responses related to social communication. *Psychoneuroendocrinology* **29**, 35–48 (2004).
88. Kim, S.J. et al. Transmission disequilibrium testing of arginine vasopressin receptor 1A (AVPR1A) polymorphisms in autism. *Mol Psychiatry* **7**, 503–7 (2002).
89. Bachner-Melman, R. et al. Linkage between vasopressin receptor AVPR1A promoter region microsatellites and measures of social behavior in humans. *Journal of Individual Differences* **36**, 451–60 (2005).
90. Hammock, E.A. & Young, L.J. Variation in the vasopressin V1a receptor promoter and expression: implications for inter- and intraspecific variation in social behaviour. *Eur J Neurosci* **16**, 399–402 (2002).
91. Lennox, R.-D. & Wolfe, R.-N. Revision of the Self-Monitoring Scale. *Journal of Personality and Social Psychology* **46**, 1349–1364 (1984).
92. Buhrmester, D. & Furman, W. Perceptions of sibling relationships during middle childhood and adolescence. *Child Dev* **61**, 1387–98 (1990).
93. Bachner-Melman, R. et al. Association between a vasopressin receptor AVPR1A promoter region microsatellite and eating behavior measured by a self-report questionnaire (Eating Attitudes Test) in a family-based study of a non-clinical population. *Int J Eat Disord* **36**, 451–60 (2004).
94. Furman, W. & Buhrmester, D. Children's perceptions of the qualities of sibling relationships. *Child Dev* **56**, 448–61 (1985).
95. Snyder, M. Self-monitoring of expressive behavior. *Journal of Personality and Social Psychology* **30**, 526–537 (1974).
96. Wolfe, R.N., Lennox, R.D. & Cutler, B.L. "Getting along" and "getting ahead": Empirical support for a theory of protective and acquisitive self-presentation. *Journal of Personality and Social Psychology* **50**, 356–361 (1986).
97. Miller, M.L., Omens, R.S. & Delvadia, R. Dimensions of social competence: Personality and coping style correlates. *Personality and Individual Differences* **12**, 955–964 (1991).
98. Day, D., Schleicher, D.J., Unckless, A.L. & Hiller, N.J. Self-Monitoring personality at work: A meta-analytic investigation of construct validity. *Journal of Applied Psychology* **87**, 390–401 (2002).
99. Cloninger, C.R. et al. Anxiety proneness linked to epistatic loci in genome scan of human personality traits. *Am J Med Genet* **81**, 313–7 (1998).

100. Fullerton, J. et al. Linkage analysis of extremely discordant and concordant sibling pairs identifies quantitative-trait Loci that influence variation in the human personality trait neuroticism. *Am J Hum Genet* **72**, 879–90 (2003).

101. Zohar, A.H. et al. Tridimensional personality questionnaire trait of harm avoidance (anxiety proneness) is linked to a locus on chromosome 8p21. *Am J Med Genet* **117B**, 66–9 (2003).

102. Dina, C. et al. Fine mapping of a region on chromosome 8p gives evidence for a QTL contributing to individual differences in an anxiety-related personality trait, TPQ harm avoidance. *Am J Med Genet B* 132, 104–8 (2004).

103. Eysenck, H.J. *The Scientific Study of Personality*, (Routledge & Kegan Paul, London, 1952).

104. Schuckit, M.A. et al. A genome-wide search for genes that relate to a low level of response to alcohol. *Alcohol Clin Exp Res* **25**, 323–9 (2001).

105. Stefansson, H. et al. Neuregulin 1 and susceptibility to schizophrenia. *Am J Hum Genet* **71**, 877–92 (2002).

106. Stefansson, H. et al. Association of neuregulin 1 with schizophrenia confirmed in a Scottish population. *Am J Hum Genet* **72**, 83–7 (2003).

107. Neale, B.M., Sullivan, P.F. & Kendler, K.S. A genome scan of neuroticism in nicotine dependent smokers. *Am J Med Genet B* 132, 65–9 (2005).

108. Curtis, D. Re-analysis of collaborative study on the genetics of alcoholism pedigrees suggests the presence of loci influencing novelty–seeking near D12S391 and D17S1299. *Psych Gen* **14**, 151–5 (2004).

19 Aggression: Concepts and Methods Relevant to Genetic Analyses in Mice and Humans

Stephen C. Maxson and Andrew Canastar

CONTENTS

SUMMARY

Aggression is a complex, social behavior involving at least two individuals. Some conceptual and methodological issues relevant to genetic analyses of aggression in mice and humans are considered.

In animals, there are several types of aggression. In male and female mice, these are offense, defense, infanticide, and predation. In females, there are also types of aggression associated with pregnancy and lactation. There are strain differences for each type, and genes have been identified for offense and defense. At present, 38 genes have been reported to affect some aspect of offense. There are also environmental effects that include prenatal maternal environment, postnatal maternal environment, postweaning environment, and test situation. All of these must be considered in critically evaluating experimental data and experimental design for the genetics of a type of aggression. The behaviors observed and measured are also a factor. Some of these issues are discussed in detail for offense in males.

Although there may be different types of aggression in humans, it is difficult to identify these. Consequently, aggression in humans is often treated as a single, unitary trait. This lumping together of different types of aggression may render difficult the interpretation and evaluation of genetic analyses for human aggression. Regardless, there is some evidence that individual differences in personality traits associated with aggressiveness are due to genetic variants. Also, three genes may have been shown to have effects on impulsive aggression, and there is a genotype interaction for one of these. However, there is no proof at present that criminal behaviors associated with physical assault are substantially heritable, other than as interactions of specific genes with specific environments.

19.1 INTRODUCTION

There are many definitions of aggression and of aggressive behavior. One of these is "overt behavior involving intent to inflict noxious stimulation to or behave destructively toward another organism."[28] Another of these is "any form of behavior directed toward the goal of harming or injuring another living being who is motivated to avoid such treatment."[3] These definitions specify broadly the behavioral domain considered in this chapter.

It has long been recognized that, at least in animals, there are many types of aggression or aggressive behavior. These differ in mechanism, eliciting stimuli, development, function, and phylogeny. Consequently, they may also differ in their genetics. That is to say, the same genes may not be causes of differences or act in development of each type of aggression.

Regardless, aggression has long been of interest as a behavioral trait for genetic analyses.[16,17] Although most of the research on genetics of aggression has been done with mice, much of the scientific and public interest in the genetics of aggression is sparked by concerns about the role of nature and nurture in human aggression. This includes the causes of not only adaptive types of aggression, but also aggression in psychopathologies, crime, and war.[5,28] Here, some conceptual and methodological issues involved in research on the genetics of aggression are considered, first for mice and then for humans.

19.2 MICE

In male and female mice, the four types of aggression are offense, defense, infanticide, and predation.[16] There are also female specific types of aggression which occur during pregnancy or lactation.[4] Between strain differences exist for each type of aggression,[10] and genes have been identified for offense and defense.[16,19,21,23]

Offense and defense by males and females (not pregnant or lactating) are agonistic behaviors. Infanticide and predation by males and females are not agonistic behaviors. The agonistic behaviors have the common function of adaptation to situations involving physical conflict between members of the same species. Each type of aggressive behavior can be characterized in mice by function and bite targets. These are listed in Table 19.1. For each type, there are also differences in mechanism and development. Thus, it may be expected that although some genes may affect all types of aggression, others will affect only one or a few kinds. Also, conceptual and methodological issues in identifying genes for each type of aggression may be different. Because most of the current research on the genetics of aggression in mice focuses on offense in males, this part of the chapter will consider this behavior in mice. Information on the other types of aggressive behavior in mice can be found in references 2, 6, 14, 16.

TABLE 19.1
Types of Mouse Aggression

Type	Bite Target	Function
Offense	Back, rump, tail	Obtaining and restraining resources
Defense	Face, shoulders	Self-protection from injury by others
Pregnancy	Anywhere	Prevent reproductive termination
Lactation	Flanks, neck	Prevent reproductive termination
Infanticide	Anywhere	Reproductive termination
Predation (insects)	Head, thorax	Food

Elsewhere, we have reviewed in detail the many aspects of the environment that can have an effect on offensive aggression in mice[17,18,21] and that must be considered in critically evaluating research results and in designing experimental studies on the genetics of offense. These include prenatal maternal environment, postnatal parental environment, postweaning environment, and test conditions. Also, the behavioral measures used are a factor. Effects of test conditions and behavioral measure will be considered in more detail.

Offense is obviously a social behavior involving at least two mice, and most genetic analyses of aggression in mice use dyadic tests. For genetic analyses, there are three types of dyadic tests. These are homogeneous set test, the panel of testers, and the standard opponent test. In the homogeneous set test, all encounters are between mice of the same genotype (strain or F1 hybrids). In the panel of testers, encounters are between mice of the same or different genotype (strains or F1 hybrids), and each experimental group of genotypes (strains or F1 hybrids) is tested against a panel of genotypes (strains or F1 hybrids). The panel of testees (experimental group) and panel of testers (opponent group) may be of the same or different array of genotypes (strains or F1 hybrids). Because in both the homogeneous set test and panel of testers, the genotypes of both individuals in the dyadic encounter must be known, only isogenic populations may be used in the two paradigms. Isogenic populations include inbred strains, recombinant inbred strains, consomic strains, recombinant congenic strains, congenic strains, coisogenic strains, and F1 hybrids of two such strains. With these two paradigms, only recombinant inbred strains, recombinant congenic, congenic (such as knockout mutants), and coisogenic strains (such as transgenes and insertional mutants) can be used to identify genes with effects on offense. In the standard opponent test, all encounters are between mice of one or another genotype and mice of a single, standard genotype (stock, strain, or F1 hybrid). Since only the genotype of the standard opponent must be known, the standard opponent test can be used with mice from isogenic or heterogenic populations. The latter include F2, F3, and other Fn hybrids, backcrosses, heterogeneous stocks, and selected lines. These can be used to map quantitative trait loci (QTLs) with effects on offense.

Strain or genetic differences in male offense can depend on the genotype not only of the testee (experimental group), but also of the tester (opponent group). For example, in a homogeneous set test, the mean number of fights/animal/week was higher for males of the BALB/cBy strain than for males of the C57BL/6By strain, whereas in a standard opponent test, the mean number of fights/animal/week was higher for males of the C57BL/6By than for BALB/cBy males. The standard opponent was of the CXBH recombinant inbred strain which fought rarely in homogeneous set tests. Also, effects of the male-specific part of the Y chromosome on percent of attacking males depended on the genotype of the opponent. An effect of the male specific part of the Y chromosome on offense was detected in homogeneous set tests but not with an inbred mouse strain (A/J) males as the standard opponent.

However, the same genetic effects are sometimes obtained with opponents of more than one type. For example, males mutant for Nos1 (nitric oxide synthase) are more aggressive than nonmutant males with opponents of two different genotypes. Similarly, males mutant for *Esr1* (α-estrogen receptor) are less aggressive than nonmutant males with either opponents of the same genotype or with olfactory bulbectomized Swiss-Webster males.

These opponent effects may be mediated by the chemosignals or behavior of the stimulus mouse. Testosterone-dependent chemosignals of one male are known to affect the aggressive behavior of another male. There are strain differences in the potency of these signals and in the receptivity to these signals. For example, different levels of aggression are elicited from a testee mouse in response to urine of three inbred strains daubed onto gonadectomized opponents. Strain differences in behavior of the tester or opponent mouse may also be involved in the differential elicitation of aggression by the tester or opponent. This has been shown for strains of opponents that are aggressive or passive.[13] Aggressive males elicit different qualities and quantities of fighting than passive males. There is also evidence that standard opponents of the same genotype but with different treatment to eliminate their attack and interactive behaviors elicit different patterns of offensive behaviors from testee mice. These treatments include olfactory bulbectomy, gonadectomy, and experience of repeated defeats.

Offense is a complex behavior consisting of four motor patterns. These are chase, sideways offensive posture, upright offensive posture, and attack. The latency, frequency, and duration of each behavior can be measured. The development of video recording and computer analysis of animal behavior, including offense, has made it possible to describe, record, and measure all aspects of the motor patterns of offense, to examine the temporal patterns of behavioral elements within and between episodes of offensive behaviors, and to assess total time allocated by a mouse to broad behavioral categories such as nonsocial activity, social investigation, defense, attack, threats, and displacement activity (digging and self-grooming). Such detailed analysis of offense is common in studies of drug but not of gene effects.

Many genetic studies of offense have either used a single composite score or a few measures, usually for the attack component, of this complex social behavior. This is perhaps necessary when mapping QTLs with segregating populations or for screening chemically induced mutants. Having many measures in QTL analysis can be a source of two problems. These are the possibility of more false positives and of selecting population extremes for DNA genotyping. The selection of single index measures should be based on phenotypic or genotypic correlations as estimated from inbred strains. There are several reports of phenotypic correlations among latency to attack, number of attacks, and accumulated attack time. Also, it has been reported that in tests with A/J males as standard opponents, the percentage of males that attack is correlated with many other measures of offense across several strains. However, once the QTL is identified and congenic strains for it are established, there should be complete analyses of this "gene's" effects on offense. This can and should always be done for congenic or coisogenic strains with knockout mutants, other mutants, or transgenes.

Just as drugs do not always affect all aspects of offense, genes may not always do so. For example, it has been reported that the proportion of mice attacking has a mode of inheritance across several strains and their hybrids different from that for number of attacks and accumulated attack time. Also, effects of genes on the Y chromosome on offense are dependent on the behavioral measure. In a standard opponent test, the CBA/H and NZB Y chromosomes differ in effect on proportion of males that attack at least once but not in effect on latency to attack. Similarly, the Nos1 mutant affects the frequency but not latency of attacks in a standard opponent test. Hence, in both types of research, complete ethological or behavioral analysis is essential.

These and other methodological issues must be considered in evaluating reports of gene effects on offense and in designing research to test gene effects on offense. To date, there is some evidence for effects of 38 genes on male offense.[17,21] These are listed in Table 19.2. Several Slc6a3[27] and Gnai2[25] genes have been studied intensively, leading to proposals for a role of genes expressed in the hippocampus,[11]

TABLE 19.2
Genes and Male Offense

Gene	Name	Chromosome
Adora1a	Adenosine1A receptor	1
Adora2a	Adenosine2A receptor	10
Adra2C	Adrenergic alpha 2C receptor	5
Ar	Androgen receptor	X
Avpr1b	Argenine vasopressin 1B receptor	1
B2m	Beta2-microglobulin	12
Bcr	Breakpoint cluster region	10
CamK2a	α-Calcium/calmodulin kinase II	18
Ckb	Brain creatinine kinase	12
Comt	Catecho-o-methyl transferase	16
Cyp19	Aromatase	9
Esr1	α-Estrogen receptor	6
Esr2	β-Estrogen receptor	12
Fyn	Fyn tyrosine kinase	10
Gad1	Glutamnic acid decarboxylase 65	9
Gdi-1	Guanosine diphosphate dissociation inhibitor	X
Girgeo-22	Gene trap ROSA-b-Geo 22	10

(Continued)

TABLE 19.2
Genes and Male Offense *(Continued)*

Gene	Name	Chromosome
Gnai2	Guanine nucleotide binding protein, Alpha inhibiting 2	9
Hrh1	Histamine 1 receptor	6
Htr1b	5-HT1B receptor	9
Il-6	Interleukin-6	5
Maoa	Monoamine oxidase A	X
Mm2	Membrane metalo enopeptidase	3
Ncam	Neural cell adhesion molecule	9
Nos1	Nitric oxide synthase 1	5
Nos3	Nitric oxide synthase 3	5
Nr3e1	Nuclear receptor family2 group E member	10
Oxt	Oxytocin	2
Penk	Enkephalin	4
Pet-1	Pet-1 ETS factor	1
Reg-2	Regulator of G-protein signaling	1
Slc6a3	Dopamine transporter	13
Slc6a4	Serotonin transporter	11
Sts	Steroid sulfatase	X/Y pseudoautosomal
Tacr1	Tachnykinine-1 receptor	6
Tgfa	Transforming growth factor α	6
Trpc2	TRP ion channel	7

steroid sulfatase and neurosteroids,[15] serotonergic system,[9] and effects of hormones on neuropeptide gene transcription[24] in mouse offense.

19.3 HUMANS

Most researchers believe that, as with animals, there are many types of human aggression. Although some have suggested that offense, defense, and predation are distinct types of aggression in humans, it is widely recognized that it is difficult to define the different types of aggression in humans. In part, this is because different motor patterns of offense, defense, and predation in humans may be masked by the use of extrinsic weapons. Other distinctions among types of human aggression include hostile vs. instrumental aggression[3] and, within the hostile category, between impulsive and nonimpulsive aggression.[19] Also, there are further distinctions within any of these categories. These are physical vs. verbal, active vs. passive, and direct vs. indirect. These different parsing of the aggressive phenotype could have an effect on the results of genetic analyses. But distinctions are rarely made in research on the genetics of human aggression. Rather, aggression often is treated as a single unitary trait. This lumping together of different types of aggression may render difficult genetic analyses. However, it has recently been suggested but not proven that all aggression in humans is of the defensive type.[1]

The methods for investigating human aggression are described by Baron and Richardson[3] and by Volavka.[28] The actual behavior may be directly observed or inferred from self- or archival reports. Observed behavior is frequently studied in controlled experiments in the laboratory. More recently, it has also been observed in normal social and cultural settings. Human aggression is not only investigated as behavior (for example, the Child Behavior Check List) but also as a personality trait. Some tests used are the Minnesota Multiphasic Personality Invertory (MMPI) and Buss-Durkee Hostility Inventory. The implication from results of these tests is that in the same interpersonal situation, some personality types are more prone than others to be aggressive.

Many genetic studies of human aggression have been for traits inferred from personality tests. These have often focused on impulsive or antisocial personality traits that are associated with aggression.[7,26] There is indication that some of the individual differences in such traits are due, at least in part, to genetic variation. Also, variants of the genes for monoamine oxidase A (MAOA), tryptophan hydroxylase, and the 5HT1B receptor may be associated with some aspects of impulsivity, including aggression. Recently, the effects of MAOA variants on measures of male aggression have been shown to be dependent on maltreatment in childhood[8] or heroin addiction.[12]

Other genetic studies have been concerned with antisocial or criminal behavior based on archival data. However, the data do not prove that variations in murder, assault, rape, or other offenses associated with physical attacks are substantially heritable. The failure to detect a nonzero heritability for criminal behavior involving physical attacks may be a function of the low incidence of criminal violence. Large sample sizes are required to detect nonzero heritability for behaviors of low incidence. These have not been available to date for criminal behaviors involving physical attacks. However, this may also be due to interactions of genotype and environment.[8,12]

All of this research should be critically evaluated in the context of the chapters in this volume on twin studies and family studies. Regardless, it is clear that in humans as in mice only part of the individual differences in any type of aggression is due to genetic variants.

19.4 MICE AND HUMANS

Candidate genes for effects on human aggression may be identified from research on animals, especially mice. We have written several times about the use of genetic variants with effects on mouse aggression as models for human aggression.[16,18,19,20,22] For these to be behavioral models, the aggressive trait in mice and humans must be the same. For example, genetic variants for offense in mice would be relevant to genetics of offense but not defense in humans. Thus, there is a great need to identify the categories of aggression that are the same in mice and humans as well as to investigate the genetics not only of escalated encounters but also conflict resolution.[22] Once this is done, it may be possible to use genetic variants, including single genes, with effects in mice to search for the genes with effects on each type of aggression in humans. Similar approaches in other invertebrate and vertebrate animals may also be of use in developing genetic models of human aggression.[22]

REFERENCES

1. ALBERT DJ, WALSH ML, JONIK RH: Aggression in humans: What is its biological foundation? *Neuroscience and Biobehavioral Reviews*, 1993, 17, 405–425.
2. ALLEVA E: Assessment of aggressive behavior in rodents. In: PM Conn (Ed): Paradigms for the study of behavior. *Methods in Neurosciences Vol. 14.* Academic Press, New York, 1993, pp 111–137.
3. BARON RA, RICHARDSON DR: *Human Aggression 2nd ed.* Plenum Press, New York, 1994.
4. BJORKQVIST K, NIEMELA P: *Of Mice and Women: Aspects of Female Aggression.* Academic Press, New York, 1992.
5. BOCK GR, GOODE JA: *Genetics of Criminal and Antisocial Behaviour.* John Wiley & Sons, New York, 1996.
6. BRAIN PF, HAUG M, KAMIS A: Hormones and different tests for aggression with particular reference to the effects of testosterone metabolites. In: J Balthazart, E Prove, R. Gilles (Eds): *Hormones and Behaviour in Higher Vertebrates.* Springer-Verlag, Berlin, 1983, pp 290–304.
7. CAREY G: Genetics and violence. In: AJ Reiss Jr., KA Miczek, JA Roth (Eds) *Understanding and Preventing Violence Vol. 2: Biobehavioral Influence.* National Academy Press, Washington, 1994, pp 21–58.
8. CASPI A, MCCLAY J, MOFFITT TE, MILL J, MARTIN J, CRAIG I W, TAYLOR A, POULTON R: Role of genotype in the cycle of violence in maltreated children. *Science*, 2002, 297, 851–854.
9. CHIAVEGATTO S, NELSON RJ: Interaction of nitric acid and serotonin in aggressive behavior. *Hormones and Behavior*, 2003, 44, 233–241
10. CRAWLEY JN, BELKNAP JK, COLLINS A, CRABBE JC, FRANKEL W, HENDERSON N, HITZEMANN RJ, MAXSON SC, MINER LL, SILVA AJ, WEHNER JM, WYNSHAW-BORIS A, PAYLOR R: Behavioral phenotypes of inbred mouse strains. *Psychopharmacology*, 1998, 132, 107–124.
11. Feldker DE, Datson N.A., Veenema AH, Proutski V, LATHOUWERS D , DE KLOET E R, VREUGDENHIL E: (2003b). Gene chip analysis of hippocampal gene expression profiles of short- and long-attack-latency mice: technical and biological implications. *Journal of Neuroscience Research*, 2003, 74, 701–716.
12. GERRA G, GAROFANO L, BOSARI S, PELLEGRINI C, ZAIMOVIC A, MOI G, BUSSANDRI M, MOI A, BRAMBILLA F, MAMELI A, PIZZAMIGLIO M, DONNINI C: Analysis of monoamine oxidase A (MAO-A) promoter polymorphism in male heroin-dependent subjects: behavioural and personality correlates. *J Neural Transm.* 2004, 111, 611–21.
13. HAHN ME, SCHANZ N: Issues in the genetics of social behavior: revisited. *Behavior Genetics*, 1996, 26, 463–470.
14. JONES SE, BRAIN PF: Performance of inbred and outbred laboratory mice in putative tests of aggression. *Behavior Genetics*, 1987, 17, 87–96.
15. LE ROY I, MORTAUD S, TORDJMAN S, DONSEZ-DARCEL E, CARLIER M, DEGRELLE H, ROUBERTOUX PL: Genetic correlation between steroid sulfatase concentration and initiation of attack behavior in mice. *Behavior Genetics*, 1999, 29, 131–136.
16. MAXSON SC: Potential genetic models of aggression and violence in males. In: P Driscoll (Ed) *Genetically Defined Animal Models of Neurobehavioral Dysfunctions.* Birkhauser, Boston, 1992, pp 174–188.

17. MAXSON SC: Methodological issues in genetic analyses of an agonistic behavior (offense) in male mice. In: D Goldowitz, D Wahlsten, RE Wimer (Eds): *Techniques for the Genetic Analysis of Brain and Behavior: Focus on the Mouse.* Elsevier, Amsterdam, 1992, pp 349–373.

18. MAXSON SC: Issues in the search for candidate genes in mice as potential animal models of human aggression. In: GR Bock and JA Goode (Eds): *Genetics of Criminal and Antisocial Behaviour.* John Wiley & Sons, New York, 1996, pp 21–30.

19. MAXSON SC: Homologous genes, aggression, and animal models. *Developmental Neuropsychology,* 1998, 14, 143–156

20. MAXSON SC: Mouse genes and animal models of aggression in humans. In: M Haug and R Whalen (Eds): *Brain, Behavior and Cognition.* American Psychological Association, Washington, 1998, pp 273–282.

21. MAXSON SC, CANASTAR A: Conceptual and methodological issues in the genetics of mouse agonistic behavior. *Hormones and Behavior,* 2003, 44, 258–262.

22. MAXSON SC, CANASTAR A: Genetic aspects of aggression in non-human animals. In, RJ Nelson (Ed.): *Biology of Aggression.* Oxford University Press, New York, pp 3–19.

23. MICZEK KA, MAXSON SC, FISH EW, FACCIDOMO, S: Aggressive behavioral phenotypes in mice. *Behavioral Brain Research,* 2001, 125, 167–181.

24. NOMURA M, MCKENNA E, KORACH KS, PFAFF, D, OGAWA, S: Estrogen receptor-β regulates transcript levels for oxytocin and arginine vasopressin in the hypothalamic paraventricular nucleus of male mice. *Molecular Brain Research,* 2002, 109, 84–94.

25. NORLIN EM, GUSSING F, BERGHARD A: Vomeronasal phenotype and behavioral alterations in G alpha i2 mutant mice. *Curr Biol.,* 2003, 13, 1214–1219.

26. RHEE SH, WALDMAN ID: Genetic and environmental influences on antisocial behavior: a meta-analysis of twin and adoption studies. *Psychol Bull.,* 2002, 128, 490–529.

27. RODRIGUIZ RM, CHU R, CARON MG, WETSEL WC: Aberrant responses in social interaction of dopamine transporter knockout mice. *Behav Brain Res.,* 2004, 148, 185–198.

28. VOLAVKA, J: *Neurobiology of Violence.* American Psychiatric Press, Washington, D.C., 1995.

20 Genetic Analysis of Emotional Behaviors Using Animal Models

André Ramos and Pierre Mormède

CONTENTS

SUMMARY

The aim of this chapter is to introduce the reader to the universe of experimental approaches available for the genetic study of emotionality, the focus being on the use of animal models. First, some preliminary conceptual issues will be discussed and then an overview of the main methods of measuring and interpreting the so-called emotional behaviors will be presented. Finally we present the classical and modern strategies to determine the genetic basis of variance in emotionality-related traits.

20.1 INTRODUCTION

Studying the genetic aspects of emotions, in either humans or animal models, is a particularly challenging task for three main reasons: emotions are difficult to *define*; emotions are difficult to *measure*; and emotions are difficult to *genetically analyze*.

291

These three types of difficulties, as well as some of the experimental methods designed to overcome them, will be discussed here. Without intending to be exhaustive, we will introduce throughout the chapter some findings related to the search for genes influencing emotional responses of laboratory rodents.

20.2 DEFINITIONS

The interpretation of terms derived from the word *emotion* may vary among authors, but most researchers agree that emotions are subjective experiences, that they appear under nonordinary circumstances and that they involve behavioral and physiological changes.[46] In the field of experimental psychobiology, the concept of *emotionality* has been used widely for laboratory animals, as defined by Hall.[28]

> The term *emotionality* is defined as the state of being emotional. This state consists of a group of organic, experiential and expressive reactions and denotes a general upset or excited condition of the animal. Emotionality can be thought of as a trait ... The reader is warned against interpreting emotionality as a thing or faculty. It is merely a convenient concept for describing a complex of factors. ... the term *emotion* is conventionally used to designate the experience which results from emotional stimulation.

For being subjective, emotion and emotionality cannot be directly measured. Hence, scientists often attempt to infer such subjective states from their measurable behavioral manifestations. In operational terms, the defecation rate in the open field test (see below) has long been used as *the* measure of emotionality in the rat. Because locomotor activity in this same test often covaries negatively with the defecation rate,[17] an animal with high defecation rate and low locomotion in the open field is usually considered as highly emotional. Nevertheless, a number of experimental data show that the correlation between these two variables is not always confirmed and that individual animals or strains may present, for example, a high defecation rate together with a high locomotor activity.[45] In addition, pharmacological studies do not provide unequivocal support for the use of open field measures as suitable indices of anxiety,[36] a psychological trait or state which is related to emotionality.

Therefore, although the term *emotionality*, like *stress*, may be convenient to describe a broad category of psychological and biological processes, interindividual or interstrain differences must be defined in a more analytical way. Those differences are not only quantitative (more or less emotional), but their behavioral and neuroendocrine expressions also may vary in nature (qualitative differences). For instance, highly emotional animals may display either freezing or fleeing behavior, depending on the wiring between the central emotional state and behavioral outputs. Therefore, there is a growing perception among researchers that emotionality has in fact several different dimensions, instead of being a unitary and linear construct.[46]

20.3 MEASUREMENTS

A number of paradigms have been used to measure "emotional" behaviors of experimental animals. Basically, these experimental paradigms consist of exposing animals to aversive stimuli while observing and analyzing their behavior. The aversive stimuli may vary in nature from being essentially physical (electrical footshocks, food deprivation, submersion in water, etc.) to those considered to be mainly psychological (novel environments, strongly illuminated areas, open spaces, exposure to the smell of predators, heights, social instability, etc.). Another important variable is the ability of the animal to avoid or escape the aversive stimulus. Most experimental situations use unconditioned responses to a number of novel environments, or learning paradigms.

20.3.1 UNCONDITIONED STRESS RESPONSES, FORCED EXPOSURE

The **open field test** is most widely used in behavioral research. The apparatus consists of a large arena, round or square, typically 1 m wide for rats, surrounded by a wall to prevent escape; the floor is usually marked with lines to allow the quantification of locomotion. The illumination is variable, and frequently intense to increase its aversiveness. Behavioral measures include ambulation (peripheral and in the center), rearings (on the walls and in the open space), grooming, freezing, as well as defecation and urination (measures of autonomic nervous system activation). The association of low ambulation with a high rate of defecation, as mentioned above, has been the operational definition of emotionality in the rat, but this unitary measure of emotionality has been criticized. Indeed, many studies show that the emotionality construct, as measured in the open field, is made of several independent factors or components, the two main factors being "motor discharge" or "activity" (including locomotion and rearing), and "autonomic balance" or "emotional reactivity" (measured by the number of fecal boli). Urination primarily would be an index of territorial marking.[54,46] Locomotion can be further divided in central (away from the walls) and peripheral (when the animals can touch the walls), the former being more related with fear and anxiety and the latter with general activity.[46] Strain differences in defecation and locomotor activity were described in mice[53,54,62] and in rats[45] (Figure 20.1).

Many other **novel environment tests** are used to characterize the emotional reactivity of animals. Some of them are smaller with a low level of illumination to reduce the aversiveness of the environment, which favors the locomotor activity of the animals. These are known as **activity cages**, and are usually equipped to record automatically the movements of the animals. General motor activity can also be recorded by various devices without moving the animals out of their home cage (for instance with infrared beams, induction coils, or implanted devices).

Alternatively, the environment can be made more complex to favor other aspects of behavior and allow the analysis of the different components of activity in novel environments, such as **exploratory activity** (hole-board, introduction of objects in the open field, perforated walls, etc.), or **motivational conflicts** (presence of food, water, or conspecifics). One such test is the **social interaction test**[21] based on the observation that the time spent by pairs of male rats in performing social behaviors varies with the aversiveness of the environmental conditions. **Hyponeophagia** is a conflict test where hungry rats are exposed to food placed in the center of a brightly lit novel environment. Highly emotional rats are expected to show longer latency to

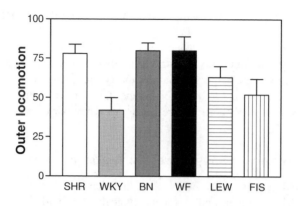

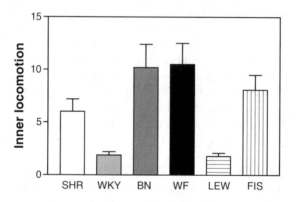

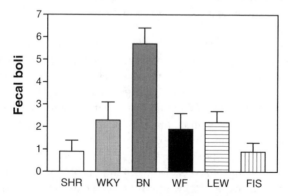

FIGURE 20.1 Open field. Locomotion in the peripheral (outer) and central (inner) part and defecation score measured in 12 males each from six rat strains (SHR = Spontaneously Hypertensive Rat; WKY = Wistar Kyoto; BN = Brown Norway; WF = Wistar Furth; LEW = Lewis; FIS = Fischer 344). (Reprinted from *Behav. Brain Res.,* 85, Ramos, A. et al., A multiple test study of anxiety-related behaviours in six inbred rat strains, 57–69, Copyright 1997, with permission from Elsevier.)

approach food and have a lower food consumption.[42,62] These tests have been developed mainly for the screening of anxiolytic drugs.

Among other tests of reactivity, we can cite the **acoustic startle reflex**[26,62] and **ultrasonic pup vocalizations**, elicited during the first 2 weeks of life by isolating the pups from their mother,[52] but this list is far from complete.

20.3.2 UNCONDITIONED STRESS RESPONSES, CHOICE TESTS

These tests consist of two (or more) compartments differing in their level of aversiveness. The animal can freely move from one compartment to the other(s). For instance, the **black and white box** is made of a small, dark compartment and a large, brightly illuminated area connected by a small passage or door through which the animal can move freely. Transitions between compartments and general locomotion are recorded. Anxiolytic drugs usually increase both measures, antidepressants being inactive.[12] Many variants of this test have been described as **emergence tests**, the animal being introduced in a "safe" compartment (usually small and dark, where the animal can be left for some time to habituate) and the time to emerge into a more aversive compartment (usually larger, brightly illuminated and unexplored previously) is measured.

The **elevated plus maze** is based on studies showing that rats seem to experience more fear when exposed to open, elevated alleys than when exposed to enclosed alleys, as the result of fear naturally induced in rodents by open spaces and/or elevated areas. This is the most popular test used nowadays for the screening of anxiolytic drugs in rats.[29,61] The apparatus has four elevated arms 50 cm long and 10 cm wide arranged in a crosslike disposition, with two opposite arms enclosed in high walls and two arms open. At their intersection, a central platform gives access to any of the four arms. Rats are placed on the central platform, and for 5 min, total locomotion or exploration is measured as the total number of arm entries, whereas the percentage of entries in the open arms is inversely related to fear, since it is increased by some anxiolytic treatments (benzodiazepine, but not serotonin-related drugs) and reduced by anxiogenic drugs such as caffeine, yohimbine, and amphetamine. Indeed, rats spend more time in the enclosed arms and show more behavioral and physiological signs of fear when confined to the open arms than when confined to the closed arms: decreased locomotion and higher defecation rate and plasma corticosterone response. Strain differences have been described in mice[62] and in rats[45] (Figure 20.2).

20.3.3 CONDITIONING EXPERIMENTS

Most of these experiments are based on the administration of aversive stimuli, most frequently foot shocks. In **active avoidance paradigms**, experimental animals can escape or avoid the unconditioned stimulus by performing the adequate response programmed by the experimenter: for instance bar-pressing or shuttling from one side of the cage to the other. The learning of the response can be impaired by previous administration of unavoidable shocks. This paradigm, known as **learned helplessness**, is used as a model of depression for pathophysiological and pharmacological

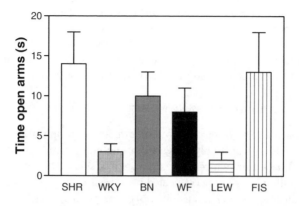

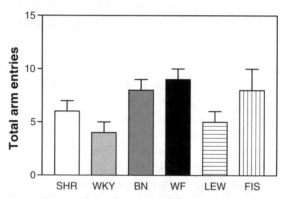

FIGURE 20.2 Elevated plus maze. Time spent in the open arms and number of total arm entries measured in 12 males each from six rat strains. (Reprinted from *Behav. Brain Res.*, 85, Ramos, A. et al., A multiple test study of anxiety-related behaviours in six inbred rat strains, 57–69, Copyright 1997, with permission from Elsevier.)

studies. Another popular paradigm for psychopharmacological studies is the **behavioral despair**, or **forced swimming, test**, where a rat learns in a first exposure that there is no escape possible from a small water tank and eventually displays "depressed behavior" upon subsequent testing.[43]

20.3.4 INTERPRETING EMOTIONALITY THROUGH A MULTIVARIATE APPROACH

Understanding the psychological significance of a given "emotional behavior" is not simple. Very often, different individuals vary for a series of behaviors in a seemingly inconsistent way, that is, an animal (or group of animals) may appear highly fearful in relation to one specific behavior (e.g., defecation) but not to another (e.g., ambulation). For an example, see the behavior of BN rats in Figures 20.1 and 20.2. Therefore, to increase the reliability of the study of emotional reactivity in laboratory animals, different behaviors can be measured in one single test, or a set of different tests can be applied to the same group of animals. In both cases, results normally suggest that different behavioral reactions and different experimental contexts may be associated to different dimensions of emotionality.

Single test situations. The multivariate or factorial analysis has been successfully applied to distinguish several dimensions of emotional reactivity in experimental animals in single experimental designs. The elevated plus maze paradigm is taken as an example because it is probably the best documented. Cruz and collaborators[15] measured 13 behavioral variables in the elevated plus maze from 30 male Wistar rats. The first and most important two factors are seen as representing variables of anxiety and locomotion. Percentage of open arm entries, time in open arms, and time in the closed arms (negatively related) are the variables with the highest loadings on the first factor. These measures were changed to one direction by administration of anxiolytic drugs and to the opposite direction by anxiogenic drugs and were therefore thought to measure anxiety. The second factor reflects mainly the number of entries in the closed arms (locomotion). The total number of arm entries, which is normally used as the main index of locomotion in the plus maze, is an ambiguous measure since this variable loads simultaneously on both anxiety and locomotion factors, appearing to be more contaminated by emotional responses than the variable "number of closed arm entries." Finally, time spent in the center of the plus maze loaded on a third factor.

Multiple test situations. This kind of analysis is most powerful to investigate the data obtained from the same animals in different test situations. For instance, Trullas and Skolnick[62] studied the behavior of 16 inbred mouse strains in the open field and the plus maze. The variables that saturated factor 1 (44.5% of total variance) were all related to time spent or number of crosses into the open arms of the elevated plus maze. Since the time spent in the open arms correlated with both the acoustic startle reflex and hyponeophagia in a novel environment, factor 1 was interpreted as reflecting the level of anxiety of the animal. Variables related to ambulation, both in the open field and plus maze, loaded on factor 2 (27.5% of total variance), which appears to measure general activity. Here again, a third factor (10.3% of total variance) was defined almost exclusively as time spent in the central platform of the plus maze. A similar analysis was conducted in our laboratory on the data obtained for 19 variables from four behavioral tests: open field, plus maze, black and white box, and social interaction.[45] The subjects were male and female rats from six inbred strains. The first factor (36.6% of total variance) was also defined by variables classically associated with anxiety (locomotion in the central part of the open field, entries and time spent in the open arms of the plus maze, behavior in the black and white box). The second factor (30.3% of total variance) was defined by measures of locomotion (total and outer locomotion in the open field, locomotion in the closed arms of the plus maze). Interestingly enough, time of social interaction in a novel environment loaded on a third factor (18.3% of total variance), together with defecation rates in the open field and black and white box. Such an independence between behaviors measured in different "anxiety tests," such as the plus maze and social interaction, has already been described by File.[21] These data show that the three measures of open field behavior—outer locomotion, inner locomotion and defecation—load on different factors and are therefore independent for the most part.

This multifactorial approach brings a number of interesting conclusions. For example, in many of these experiments, the first two principal axes of variability are related, respectively, to the level of emotional stimulation and to the amount of locomotor activation in the apparatus. This reminds us of the two main factors extracted by Eysenck[19] from numerous descriptive studies of human personality: emotional stability vs. instability (neuroticism) and extraversion vs. introversion. Of course, this approach is still an oversimplification of the reality, as shown, for instance, by the independence of the results obtained in different "tests of anxiety." Complexity pops up as soon as it is looked for.

20.4 GENETIC ANALYSIS

A great variability can be seen in "emotional" behaviors among individual animals. Different experimental strategies have been used throughout the past 50 years to analyze this variability and identify its causes. For the remainder of this chapter, these strategies will be presented while some illustrative experimental data will be used as examples. Differences between inbred strains raised in the same environment, selection experiments and quantitative genetic studies have shown that genetic factors have a major influence on interindividual differences, although they interact with environmental influences, especially during the prenatal and neonatal period. Molecular genetic studies, on the other hand, are now blossoming and should soon unravel the molecular bases for these differences in the various aspects of emotional reactivity.

20.4.1 SELECTION EXPERIMENTS

We have shown examples of strain differences for emotional behaviors in Figures 20.1 and 20.2. If these traits are heritable, selective breeding should be effective to modify the value of the character in the population. Many examples show that indeed this is the case for numerous reactivity traits as measured:

In the open field: Maudsley reactive (MR) and Maudsley nonreactive (MNRA) rat strains, selected on their defecation rate;[3] Floripa H and L rats selectively bred for high and low locomotion in the central aversive area;[50] strains of mice selected for their divergent locomotor activity from an F3 generation of a cross between BALB/c and C57BL/6.[17]

In the elevated plus-maze: High anxiety-related behavior (HAB) and low anxiety-related behavior (LAB) rat lines.[33]

In the ultrasonic vocalization (USV) test: High- and low-USV rat lines selected for contrasting rates of USV response to isolation in 10-day old pups.[4]

In activity tests: Naples high-excitability (NHE) and Naples low-excitability (NLE) rat lines,[57] Wistar-Kyoto hyperactive (WKHA) rat strain,[32] Tsukuba high-emotional (THE) and Tsukuba low-emotional (TLE) rat strains.[25]

In active avoidance tests: Roman high and low avoidance (RHA/RLA),[2] Syracuse high and low avoidance (SHA/SLA)[5] lines of rats, Koltushi high and low avoidance (KHA/KLA),[56] high and low avoidance animals (HAA/LAA)[40] strains of rats.

Those strains/lines have been used extensively for quantitative genetic analysis and are now being used in molecular genetic studies.

20.4.2 QUANTITATIVE GENETIC ANALYSIS

Most phenotypes related to emotional and stress responses are polygenically regulated. The aim of quantitative genetic analysis is to give information about the structure of genetic effects on the trait under study. One frequently calculated parameter is *heritability*, the fraction of the observed variance which is caused by differences in heredity. Much more information can be obtained, as it can be seen in the chapters on quantitative genetics. These analyses are based on the study of populations from various genetic crossings. The study of families takes advantage of the similarities between related individuals that share part of their genes.

The *classical analysis* method involves the parental, F1, and F2 generations (and eventually backcross) used by Mendel. Parental populations are frequently inbred strains, produced by brother–sister mating for more than 20 generations to obtain a high degree of homozygosity. This was done for instance for open field and plus maze behavior in rats,[47] open field behavior in pigs,[18,38] exploratory behavior in mice,[13] ultrasonic vocalizations in mice[27,51] and avoidance behavior in rats.[8]

The *diallel design* compares several inbred strains and all possible F1 hybrid crosses. Examples can be found for open-field behavior in mice.[13,14,54]

20.4.3 GENETIC LINKS BETWEEN DIFFERENT TRAITS

In highly inbred strains and their F1 hybrids, individuals are genetically identical to one another. Variability within these populations must be caused by nongenetic factors. In the F2+ and backcross populations, genes assort and the individuals differ from each other genetically as well as environmentally. We have seen before that this property is used to study the genetic architecture of the traits under study. Another use of recombinant populations is to study the genetic links between different traits. If two traits are correlated in both nonsegregating and segregating populations, they have a high probability of being controlled by the same segregating unit (and vice versa).

A classic example is the study by Hendley and co-workers[30–32] of the relationship between hypertension and hyperactivity in the spontaneously hypertensive rat (SHR). This strain was selected from the Wistar-Kyoto strain (WKY) for high blood pressure, and is also behaviorally more active in novel as well as familiar environments. Although many reasons could be put forward to justify a genetic link between these phenotypes,[63] Hendley showed that these traits segregated independently[30] and she was able to raise two new strains of rats—one hypertensive but normoactive (WKHT), and another normotensive but hyperactive (WKHA)—by inbreeding selected brother–sister pairs from the segregating F2 population.[31,3] As compared with the WKY, the WKHA strain also shows a lower emotional reactivity, as measured by neuroendocrine responses to novel environment exposure.[7] However, when animals from the segregating F2 population (WKY×WKHA) were studied, no correlation between these different parameters was found.[10] These data show (1) that the selection for a high locomotor activity in a novel environment independently coselected animals with a low emotional reactivity, and (2) that the various measures of emotional reactivity are not necessarily linked genetically.

Therefore, one should be careful when interpreting simple associations between traits even in selectively bred lines (especially if no replicate lines are available), because independent traits may appear to be correlated (false positives) due to undesirable factors such as inbreeding and genetic drift.[11] Nevertheless, selective breeding is certainly a powerful method to either suggest or corroborate genetic correlations between emotional traits. For example, behaviors measured in the open field and elevated plus maze seem to be genetically linked in some studies but not in others. Accordingly, the inbred rat strains Lewis and SHR (which were not intentionally selected for any emotional behavior) differ for their scores of exploration in both the central area of the open field and the open arms of the plus maze. In both cases, Lewis rats seem more fearful than their SHR counterparts.[45,47,49] These consistent findings suggest that the anxiety-related measures of the two tests are genetically linked. In one factor analysis involving various inbred strains, such a link was indeed confirmed.[45] However, another study using F2 rats derived from the Lewis–SHR intercross showed open field and plus maze measures to be unrelated.[47] Following these seemingly contradictory results, two independent selection experiments came to corroborate the existence of a genetic correlation between open field and plus maze variables. HAB and LAB rats selected for high and low avoidance of the plus maze open arms also showed high and low avoidance of the central area of the open field, respectively.[35] The same pattern was observed in the Floripa H and L rat lines, selected for central locomotion in the open field and differing also in their plus-maze behavior.[50]

In conclusion, genetic links between emotional behaviors do not follow an "all or nothing" type of situation. Instead, of all genes having some effect on emotionality, some are likely to influence specific responses in specific tests, whereas others have pleiotropic influences on a variety of behaviors. Consequently, the genetic overlap between these behaviors will depend on the populations involved and are expected to be most frequently partial only.

20.4.4 THE QTL CHASE

The principles and methods of QTL (quantitative trait loci) analysis are presented in Chapter 5 of this book. Briefly, the aim of the strategy is to map chromosome regions that contain genes affecting a quantitative trait (e.g., behavior) through the use of polymorphic genetic markers, the most popular ones being the microsatellites, which are simple-sequence nucleotide repeats regularly spaced on the genome and highly polymorphic. The populations normally used are segregating generations (F2 or backcross) derived from two inbred strains, or a panel of recombinant inbred (RI) strains. The detection of QTLs associated with differences in emotional behaviors is a first step to the identification of the genes responsible for genetic variation. This knowledge will allow the analysis of the functional pathways involved in behavioral variability.

Flint et al.[24] were the first to report, in 1995, QTLs associated with emotional behaviors in mice. In rats, the first QTLs for emotionality were mapped in 1996 by Moisan et al.[37] (locomotor activity in a novel environment, WKY × WKHA F2 intercross) and in 1999 by Ramos et al., using an F2 intercross from Lewis and SHR inbred strains.[48] These initial studies very often identified loci that were rather specific for one or a few behavioral responses and that were not replicated across different strains. Both these aspects brought some doubts about the psychological significance of these traits and the biological interest of these loci. More recent studies showed that some QTLs are likely to harbor genes with more general emotional effects, whereas others affect specific behaviors that may correspond to particular forms of fear and anxiety or even reflect other unrelated traits. For example, an elegant work by Turri et al.[64] showed that a QTL on mouse chromosome 15 has a broad range of behavioral effects that are consistent with a presumed influence on fear or anxiety, but not on general locomotion. On the other hand, a QTL on chromosome 1, which was thought to be anxiety-related after having been replicated in different studies,[23] was shown in fact to have an overall influence on exploratory behaviors in both aversive and nonaversive situations. Finally, one locus on chromosome 6 showed a very specific effect on a single variable from one particular test, the elevated plus maze. In rats, a multivariate approach applied to QTL mapping pointed to similar conclusions by identifying two loci (on chromosomes 1 and 19) for very specific behaviors and one locus (on chromosome 5) affecting several anxiety-related measures, such as avoidance responses, fear conditioning, and behavior in the plus maze and open-field tests.[20] QTLs with pleiotropic effects on numerous fear-related measures are seen as good candidates for harboring genes expected to be relevant for human anxiety.[23]

Some QTLs for emotionality may have pleiotropic effects on other types of behaviors, potentially related to pathologies that can be comorbid with anxiety disorders, such as alcoholism. In 1998, the group of Lucinda Carr, using the selected rat lines P (alcohol-preferring) and NP (alcohol-nonpreferring), identified a powerful QTL for alcohol consumption on the mid-portion of chromosome 4.[1,6] A few months later, a QTL located near this chromosome region was shown to have a strong influence on the locomotion in the inner part of the open field, an index of emotionality, in Lewis/SHR F2 rats.[48] In spite of having their peaks more than 15 cM apart, the confidence intervals of these two QTLs overlapped. The existence of a QTL on chromosome 4 for inner locomotion in the open field (named *Ofil1*) was corroborated by a second study involving two selected rat lines derived from the Lewis and SHR strains.[39] More recently, another whole-genome QTL analysis based on the F2 intercross of two other rat strains that contrast for alcohol intake, namely HEP (high-ethanol preferring) and WKY (Wistar-Kyoto) led to the identification of a significant locus on chromosome 4, approximately midway between the two above-mentioned QTLs (Figure 20.3), that strongly affects ethanol preference.[59]

Considering that anxiety is one of the predisposing factors involved in alcohol abuse,[41] the combination of findings described above led to the hypothesis that these different QTLs might correspond in fact to one single locus with pleiotropic effects on alcohol drinking and emotionality. This hypothesis has been corroborated so far. The latest data from two rat lines derived from HEP and WKY, which were selected

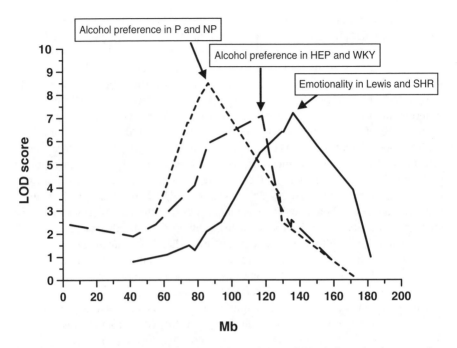

FIGURE 20.3 QTL map. Approximate positions of three QTLs independently mapped on rat chromosome 4 for: inner locomotion in the open field (emotionality) in Lewis and SHR rats;[48] preference for 10% ethanol in P and NP rats;[1,6] preference for 5% ethanol in HEP and WKY rats.[59] Physical map positions were estimated by comparing linkage data from the original publications and marker coordinates (in base pairs) from the Rat Genome Database (http://rgd.mcw.edu/).

according to their genotypes (HEP/HEP or WKY/WKY) at this QTL, showed that this locus influenced not only free or forced alcohol intake, but also locomotion in the center of the open field.[60] Interestingly, when the same experimental approach was applied to a Lewis/SHR intercross (unpublished data), this chromosome region was also found to affect both alcohol- and open-field–related behaviors. Moreover, another QTL study has associated this same genomic region with corticosterone levels in rats, a well-known stress-related hormone.[44] Considering that Lewis and SHR rats also differ in alcohol drinking,[16] further studies shall demonstrate whether the QTLs revealed by these three distinct genetic models (P/NP, Lewis/SHR, HEP/WKY) represent different loci that happen to pertain to the same linkage group or if they correspond to real pleiotropic genes. If the latter situation were true, then the close investigation of this chromosome region may shed some light on the molecular relationship between emotional reactivity and alcoholism.

Besides the classical methods for identifying QTLs for emotional behaviors, which involve phenotyping and genotyping of very large segregating populations, a new strategy recently has been applied with success. It is based on a panel of chromosome substitution strains (CSSs), each of which carries a single chromosome from a donor strain on the genome background of a host strain. The effectiveness of this strategy was shown, for example, by the replication of a QTL for open-field behavior located on mouse chromosome 1.[58] Two other alternative strategies, high-resolution mapping using commercially available outbred animals and quantitative complementation testing, were recently used to dissect this specific mouse QTL into smaller, more precise components as well as to identify one of its underlying genes.[65]

20.4.5 Gene Targeting and Transgenesis

A vast literature on the behavioral characterization (including emotionality) of knockout, knockin and transgenic animals was produced in the past 10 years. In fact, the manipulation of individual genes that are potentially involved in the control of fear-related responses became one of the main tools used in the genetic study of anxiety. Since a large number of genes coding for neuronal messengers, receptors, enzymes, and signaling molecules started to be either inactivated, overexpressed, or modified, new integrative hypotheses have been proposed to explain the molecular pathways involved in the pathogenesis of anxiety. Discussing that approach is beyond the scope of this chapter and the interested readers are encouraged to read the recent reviews on this topic.[9,22,34,55,66]

REFERENCES

1. BICE P, FOROUD T, BO R, CASTELLUCCIO P, LUMENG L, LI T–K, CARR LG: Genomic screen for QTLs underlying alcohol consumption in the P and NP rat lines. *Mamm Genome*, 1998, 9:949–955.
2. BIGNAMI G: Selection for high rates and low rates of avoidance conditioning in the rat. *Anim Behav*, 1965, 13:220–221.
3. BROADHURST PL: The Maudsley reactive and nonreactive strains of rats : A survey. *Behav Genet*, 1975, 5:299–319.
4. BRUNELLI SA, HOFER MA, WELLER A: Selective breeding for infant vocal response: a role for postnatal maternal effects? *Dev Psychobiol*, 2001, 38:221–228.
5. BRUSH FR: Genetic determinants of individual differences in avoidance learning: behavioral and endocrine characteristics. *Experientia*, 1991, 47:1039–1050.
6. CARR LG, FOROUD T, BICE P, GOBBETT T, IVASHINA J, EDENBERG H, LUMENG L, LI T-K. A quantitative trait locus for alcohol consumption in selectively bred rat lines. *Alcohol Clin Exp Res*, 1998, 22:884–887.
7. CASTANON N, HENDLEY ED, FAN X-M, MORMÈDE P: Psychoneuroendocrine profile associated with hypertension or hyperactivity in spontaneously hypertensive rats. *Am J Physiol*, 1993, 265:R1304–R1310.
8. CASTANON N, PEREZ-DIAZ F, MORMEDE P: Genetic analysis of the relationships between behavioral and neuroendocrine traits in roman high and low avoidance rat lines. *Behav Genet*, 1995, 25:371–384.
9. CLÉMENT Y, CALATAYUD F, BELZUNG C: Genetic basis of anxiety-like behaviour: A critical review. *Brain Res Bull*, 2002, 57:57–71.
10. COURVOISIER H, MOISAN MP, SARRIEAU A, HENDLEY ED, MORMEDE P: Behavioral and neuroendocrine profile of stress reactivity in the WKHA/WKY inbred rat strains : a multifactorial and genetic analysis. *Brain Res*, 1996, 743:77–85.

11. CRABBE JC: Animal Models in Neurobehavioral Genetics: Methods for Estimating Genetic Correlation. In BC JONES, P MORMÈDE (Eds): *Neurobehavioral Genetics: Methods and Applications*. CRC Press, Boca Raton, 1999, pp 121–138.

12. CRAWLEY JN: Neuropharmacologic specificity of a simple animal model for the behavioral actions of benzodiazepines. *Pharmacol Biochem Behav*, 1981, 15:695–699.

13. CRUSIO WE, SCHWEGLER H, VAN ABELEEN JHF: Behavioral responses to novelty and structural variation of the hippocampus in mice. I. Quantitative–genetic analysis of behavior in the open-field. *Behav Brain Res,* 1989, 32:75–80.

14. CRUSIO WE, VAN ABELEEN JHF: The genetic architecture of behavioural responses to novelty in mice: a quantitative-genetic analysis. *Heredity*, 1986, 56:55–63.

15. CRUZ APM, FREI F, GRAEFF FG: Ethopharmacological analysis of rat behavior on the elevated plus-maze. *Pharmacol Biochem Behav*, 1994, 49:171–176.

16. DA SILVA GE, RAMOS A, TAKAHASHI RN: Comparison of voluntary ethanol intake by two pairs of rat lines used as genetic models of anxiety. *Braz J Med Biol Res*, 2004, 37:1511–1517.

17. DEFRIES JC, GERVAIS MC, THOMAS EA: Response to 30 generations of selection for open-field activity in laboratory mice. *Behav Genet*, 1978, 8:3–13.

18. DÉSAUTÉS C, BIDANEL JP, MORMEDE P: Genetic study of behavioral and pituitary-adrenocortical reactivity in response to an environmental challenge in pigs. *Physiol Behav*, 1997, 62:337–345.

19. EYSENCK HJ: Psychopathology and Personality: Extraversion, Neuroticism and Psychoticism. In A GALE, JA EDWARDS (Eds): *Individual Differences and Psychopathology*. Academic Press, London, 1997, pp 13–30.

20. FERNÁNDEZ-TERUEL A, ESCORIHUELA RM, GRAY JA, AGUILAR R, GIL L, GIMÉNEZ-LLORT L, TOBEÑA A, BHOMRA A, NICOD A, MOTT R, DRISCOLL P, DAWSON GR, FLINT J: A quantitative trait locus influencing anxiety in the laboratory rat. *Genome Res*, 2002, 12:618–626.

21. FILE SE: The biological basis of anxiety. In HY MELTZER, D NEROZZI (Eds): *Current Practices and Future Developments in the Pharmacotherapy of Mental Disorders*. Elsevier Science Publishers B.V. Amsterdam, 1997, pp 159–165.

22. FINN DA, RUTLEDGE-GORMAN MT, CRABBE JC: Genetic animal models of anxiety. *Neurogenetics*, 2003, 4:109–135.

23. FLINT J: Analysis of quantitative trait loci that influence animal behavior. *J Neurobiol*, 2003, 54:46–77.

24. FLINT J, CORLEY R, DEFRIES JC, FULKER DW, GRAY JA, MILLER S, COLLINS AC: A simple genetic basis for a complex psychological trait in laboratory mice. *Science*, 1995, 269:1432–1435.

25. FUJITA O, ANNEN Y, KITAOKA A: Tsukuba High- and Low-emotional strains of rats (*rattus norvegicus*) : an overview. *Behav Genet*, 1994, 24:389–415.

26. GLOWA JR, HANSEN CT: Differences in response to an acoustic startle stimulus among forty-six rat strains. *Behav Genet*, 1994, 24:79–84.

27. HAHN ME, HEWITT JK, ADAMS M, TULLY T: Genetic influences on ultrasonic vocalizations in young mice. *Behav Genet*, 1987, 17:155–166.

28. HALL CS: Emotional behavior in the rat I. Defecation and urination as measures of individual differences in emotionalitlty. *J Comp Psychol*, 1934, 18:385–403.

29. HANDLEY SL, MCBLANE JW: An assessment of the elevated X-Maze for studying anxiety and anxiety-modulating drugs. *J Pharmacol Toxicol Method*, 1993, 29:129–138.

30. HENDLEY ED, ATWATER DG, MYERS MM, WHITEHORN D: Dissociation of genetic hyperactivity and hypertension in SHR. *Hypertension*, 1983, 5:211–217.

31. HENDLEY ED, OHLSSON WG: Two new inbred rat strains derived from SHR : WKHA, hyperactive, and WKHT, hypertensive, rats. *Amer J Physiol*, 1991, 261:H583–H589.

32. HENDLEY ED, WESSEL DJ, VAN HOUTEN J: Inbreeding of Wistar-Kioto rat strain with hyperactivity but without hypertension. *Behav Neural Biol*, 1986, 45:1–16.

33. LANDGRAF R, WIGGER A: High vs. low anxiety-related behavior rats: an animal model of extremes in trait anxiety. *Behav Genet*, 2002, 32:301–314.

34. LESCH K-P: Molecular foundation of anxiety disorders. *J Neural Transm*, 2001, 108:717–746.

35. LIEBSCH G, MONTKOWSKI A, HOLSBOER F, LANDGRAF R: Behavioural profiles of two Wistar rat lines selectively bred for high or low anxiety-related behaviour. *Behav Brain Res*, 1998, 94:301–310.

36. LISTER RG: Ethologically-based animal models of anxiety disorders. *Pharmacol Ther*, 1990, 46:321–340.

37. MOISAN MP, COURVOISIER H, BIHOREAU MT, GAUGUIER D, HENDLEY ED, LATHROP M, JAMES MR, MORMEDE P: A major quantitative trait locus influences hyperactivity in the WKHA rat. *Nat Genet*, 1996, 14:471–473.

38. MORMEDE P, DANTZER R, BLUTHE R, CARITEZ J: Differences in adaptive abilities of three breeds of chinese pigs. *Genet Sel Evol*, 1984, 16:85–102.

39. MORMÈDE P, MONEVA E, BRUNEVAL C, CHAOULOFF F, MOISAN M-P: Marker-assisted selection of a neuro-behavioural trait related to behavioural inhibition in the SHR strain, an animal model of ADHD. *Genes Brain Behav*, 2002, 1:111–116.

40. OHTA R, MATSUMOTO A, HASHIMOTO Y, NAGAO T, MIZUTANI M: Behavioral characteristics of rats selectively bred for high and low avoidance shuttlebox responses. *Cong Anom*, 1995, 35:223–229.

41. PANDEY SC: Anxiety and alcohol abuse disorders: a common role for CREB and its target, the neuropeptide Y gene. *Trends Pharmacol Sci*, 2003, 24:456–460.

42. PARE WP: Hyponeophagia in Wistar Kyoto (WKY) rats. *Physiol Behav*, 1994, 55:975–978.

43. PORSOLT RD, BERTIN A, JALFRE M: "Behavioural despair" in rats and mice: strain differences and the effects of imipramine. *Eur J Pharmacol*, 1978, 51:291–294.

44. POTENZA MN, BRODKIN ES, JOE B, LUO X, REMMERS EF, WILDER RL, NESTLER EJ, GELERNTER J: Genomic regions controlling corticosterone levels in rats. *Biol Psychiatry*, 2004, 55:634–641.

45. RAMOS A, BERTON O, MORMEDE P, CHAOULOFF F: A multiple test study of anxiety-related behaviours in six inbred rat strains. *Behav Brain Res*, 1997, 85:57–69.

46. RAMOS A, MORMEDE P: Stress and emotionality : a multidimensional and genetic approach. *Neurosci Biobehav Rev*, 22:33–57, 1998.

47. RAMOS A, MELLERIN Y, MORMÈDE P, CHAOULOFF F: A genetic and multi-factorial analysis of anxiety-related behaviours in Lewis and SHR intercrosses. *Behav Brain Res*, 1998, 96:195–205.

48. RAMOS A, MOISAN M-P, CHAOULOFF F, MORMÈDE C, MORMÈDE P: Identification of female-specific QTLs affecting an emotionality-related behavior in rats. *Mol Psychiatry*, 1999, 4:453–462.

49. RAMOS A, KANGERSKI AL, BASSO PF, SANTOS JES, ASSREUY J, VENDRUSCOLO LF, TAKAHASHI, RN: Evaluation of Lewis and SHR rat strains as a genetic model for the study of anxiety and pain. *Behav Brain Res*, 2002, 129:113–123.

50. RAMOS A, CORREIA EC, IZÍDIO GS, BRÜSKE GR: Genetic selection of two new rats lines displaying different levels of anxiety-related behaviors. *Behav Genet*, 2003, 33:657–668.

51. ROUBERTOUX P, CARLIER M, DEGRELLE H, HAAS-DUPERTUIS MC, PHIL-LIPS J, MOUTIER R: Co-segregation of intermale aggression with the pseudoauto-somal region of the *Y* chromosome in mice. *Genetics*, 1994, 135:225–230.

52. ROUBERTOUX P, MARTIN B, LEROY I, BEAU J, MARCHALAND C, PEREZ-DIAZ F, COHEN-SALMON C, CARLIER M: Vocalizations in newborn mice: genetic analysis. *Behav Genet*, 1996, 26:427–437.

53. ROYCE JR, CARRAN A, HOWARTH E: Factor analysis of emotionality in ten inbred strains of mice. *Multivariate Behav Res*, 1970, 5:19–48.

54. ROYCE JR, HOLMES TM, POLEY W: Behavior genetic analysis of mouse emo-tionality. III.The diallel analysis. *Behav Genet*, 1975, 5:351–372.

55. RUDOLPH U, MÖHLER H: Analysis of $GABA_A$ receptor function and dissection of the pharmacology of benzodiazepines and general anesthetics through mouse genetics. *Annu Rev Pharmacol Toxicol*, 2004, 44:475–498.

56. RYZHOVA LY, KULAGIN DA, LOPATINA NG: Correlated variability of the loco-motor activity and emotionality in rats under selection program for high and low active avoidance value. *Genetika*, 1983, 19:121–125.

57. SADILE AG, CERBONE A, LAMBERTI C, CIOFFI LA: The Naples high (NHE) and low excitable (NLE) rat strains: progressive report. *Behav Brain Res*, 1984, 12:228–229.

58. SINGER JB, HILL AE, BURRAGE LC, OLSZENS KR, SONG J, JUSTICE M, O'BRIEN WE, CONTI DV, WITTE JS, LANDER ES, NADEAU JH: Genetic dis-section of complex traits with chromosome substitution strains of mice. *Science*, 2004, 304: 445–448.

59. TERENINA-RIGALDIE E, MOISAN M-P, COLAS A, BEAUGÉ F, SHAH KV, JONES BC, MORMÈDE P: Genetics of behaviour: phenotypic and molecular study of rats derived from high- and low-alcohol consuming lines. *Pharmacogenetics*, 2003, 13:543–554.

60. TERENINA-RIGALDIE E, JONES BC, MORMEDE P: Pleiotropic effect of a locus on chromosome 4 influencing alcohol drinking and emotional reactivity in rats. *Genes Brain Behav*, 2003, 2:125–131.

61. TREIT D: Animal models for the study of anti-anxiety agents : A review. *Neurosci Biobehav Rev*, 1985, 9:203–222.

62. TRULLAS R, SKOLNICK P: Differences in fear motivated behaviors among inbred mouse strains. *Psychopharmacology*, 1993, 111:323–331.

63. TUCKER DC, JOHNSON AK: Behavioral correlates of spontaneous hypertension. *Neurosci Biobehav Rev*, 1981, 5:463–471.

64. TURRI MG, DATTA SR, DEFRIES J, HENDERSON ND, FLINT J: QTL analysis identifies multiple behavioral dimensions in ethological tests of anxiety in laboratory mice. *Curr Biol*, 2001, 11:725–734.

65. YALCIN B, WILLIS-OWEN SAG, FULLERTON J, MEESAQ A, DEACON RM, RAWLINS JNP, COPLEY RR, MORRIS AP, FLINT J, MOTT R: Genetic dissection of a behavioral quantitative trait locus shows that Rgs2 modulates anxiety in mice. *Nat Genet*, 2004, 36:1197–1202.

66. WOOD SJ, TOTH M: Molecular pathways of anxiety revealed by knockout mice. *Mol Neurobiol*, 2001, 23:101–119.

21 Genetic Analysis of Food Search Behavior in the Fruit Fly (*Drosophila melanogaster*)

Amsale T. Belay and Marla B. Sokolowski

CONTENTS

INTRODUCTION

Food search behavior is a complex trait influenced by many internal and external factors. It can involve sensory modalities such as smell, taste, and vision; motor functions such as locomotion; internal environmental cues such as the degree of starvation or satiation; and external environmental cues such as the quality, quantity, and distribution of the food supply. All of this information is likely integrated in the brain and an output (e.g., feed or search more) is then generated. In the Sokolowski laboratory, we use the fruit fly *Drosophila melanogaster* as our model organism to genetically dissect the components of food search behavior. In particular, we are interested in understanding the biological basis for naturally occurring variation in larval and adult food search behaviors as well as the biochemical pathways that contribute to these variations. In this chapter, we discuss the fruit fly *D. melanogaster* as a model organism for behavior–genetic analysis. We compare and contrast the polygenic analysis of naturally occurring behavioral variation and the single-gene mutant analysis approaches and discuss food search behavior in the larval and adult fruit fly *D. melanogaster*. We conclude by discussing how this research can be extended to additional organisms.

21.1 THE FRUIT FLY *D. MELANOGASTER* AS A MODEL ORGANISM

D. melanogaster offers many advantages for behavior and neurogenetic analysis. It expresses complex patterns of behavior including learning, courtship, circadian rhythms, food search, and locomotion.[23,45] In addition, a superb genetic database of information has been collected over more than 80 years (see reference 21   for an excellent concise tour of *Drosophila* genetic technology and reference 2 for all you may want to know about the fly). In addition to its fully sequenced genome, *Drosophila* has a wonderful array of molecular, neurobiological, and genomic tools (see http://flybase.bio.indiana.edu/ for database of the *Drosophila* genome and related tools). Moreover, it has relatively few neurons and neural networks, making it feasible to identify neural substrates underlying behavioral variation. Finally, findings from *Drosophila* are readily extended to additional organisms because mechanisms underlying behavior often show functional homologies between species.[24]

The life cycle of the fly is rapid (10–14 days at 25°C), and thousands of genetically identical flies are bred in less than a month, making it amenable for behavioral analysis of genetically identical individuals reared under controlled environmental conditions. The mated female lays hundreds of eggs on the food (fly medium in the laboratory, and fruit in nature) that hatch after 24 h (at 25°C). The larva (all three instars) forages by feeding on yeast with its mouth hooks and moves by extending its anterior and retracting its posterior end. Midway through the third instar, the larva leaves the food in search of a pupation site. After 5 days of larval life, the larva pupates for 4 days during which it metamorphoses into an adult fly. The fly then emerges from the pupal case and searches for food and mates.

21.2 POLYGENIC AND SINGLE-GENE ANALYSIS

Polygenic and single-gene analyses are useful approaches for the genetic dissection of behavior in *Drosophila*. Historically, single-gene mutant analysis addresses questions concerning the mechanisms involved in behavior, whereas polygenic analysis concerns itself with questions about the evolution of behavior. However, the two approaches are valid and complementary, as we illustrate later in the chapter. Using *Drosophila* food search behavior as an example, we describe how the genetic contribution to a naturally occurring behavior polymorphism was first determined using a polygenic strategy and later refined using the single-gene mutant approach. Below we discuss some widely used techniques for polygenic analysis of behavioral traits and follow this with a discussion of single-gene mutant analysis. We then compare and contrast these approaches.

21.2.1 POLYGENIC ANALYSIS

The type of polygenic inheritance involved in naturally occurring variation ranges from a single major gene with minor modifiers to that of many genes, all with small, equal, and additive effects on the phenotype. When few genes are involved, it is possible in *D. melanogaster* to determine their chromosomal location. When many genes are involved quantitative trait loci (QTL) mapping and whole genome microarrays are used to identify genes underlying behavioral variations.

Until recently, the identification of these genes was considered a daunting task. However, recent advances make the fly an attractive organism for identifying the genes underlying QTL. These include a collection of highly polymorphic molecular markers with known physical map locations, deficiency chromosome stocks made on an isogenic background, as well as numerous strains with mutations in a single locus. Detailed methods for the analysis of quantitative traits are found in Falconer and Mackay,[13] Mackay,[26] and Mather and Jinks.[27] QTL analysis has been applied to many *Drosophila* traits such as olfactory behavior,[14] courtship,[19] longevity,[28] bristle number,[29] environment-dependent survival,[50] and wing shape.[51] The QTL intervals are first narrowed down using quantitative deficiency complementation tests and finally through complementation tests with strains carrying mutations in candidate genes.[1] Association analysis between the single nucleotide polymorphisms (SNPs) and the behavioral variation enables determination of the within-gene molecular alterations responsible for the natural variation in behavior. Together these approaches identified *dopa decarboxylase* (*ddc*)[12] and *shuttle craft* (*stc*)[33] as genes underlying QTLs for *Drosophila* life span.

Whole genome transcriptional profiling using micoarrays is another approach used to identify genes involved in natural variation in behavioral traits.[46] Toma et al.[49] investigated the transcription profiles of two artificially selected strains of flies that differ in their response to gravity. Microarray analyses of these strains revealed that 250 genes differed in the RNA expression between the positive and negative geotaxis strains. Preexisting single-gene mutants were then used to test and validate the functional relevance of these genes to geotaxis behavior. Studies such as these link polygenic and single-gene mutant approaches in the analysis of complex behaviors.

21.2.2 SINGLE-GENE ANALYSIS

The genetic dissection of behavior in *Drosophila* using single-gene mutant analysis provides insight into the mechanistic bases of behavior. In single-gene analysis, mutations are induced in normal (wild-type) behaving strains using mutagens such as ionizing radiation, ethyl methane sulfonate (EMS), or P-element mutagenesis, each of which has its own advantages and disadvantages (discussed in Reference 21). After mutagenesis is performed, the behavior of the larva or fly is screened with the aim of later identifying X-linked or autosomal mutations affecting the phenotypes of interest. The intent is to alter normal function by mutation (mutate away from the wild-type phenotype) and determine, through comparisons of wild-type and mutant function, the genetic factors essential for the expression of the

normal wild-type phenotype. Mutant screens can also generate conditional mutations that are dependent on factors such as temperature, light, or diet. Conditional
mutations are useful for studies of the developmental and/or acute functions of a
gene. In subsequent analyses, mutants can also be mutated (mutation can be generated in a mutant genetic background) to find new mutations that suppress or
enhance the original phenotype.[35]

Numerous single-gene mutations that affect behavior (e.g., learning and memory,
courtship, circadian rhythms, olfaction, phototaxis) have been identified and characterized in *D. melanogaster.*[22,32,45,48] In the past, it was unusual for single-gene
behavioral mutants to be generated on defined genetic backgrounds. Once a mutant
is generated, it is important to know (1) how the genetic background affects the
expression of the behavior, and (2) whether the mutant phenotype is selected against
in the laboratory.[11] If so, careful control of genetic background, whether through
making strains homozygous or continuous backcrossing to a well-defined control
strain, should be carried out.

One area of research that has been virtually ignored is the study of whether a
gene originally identified by mutant analyses is variable in natural populations (an
exception is the work of Costa and colleagues[6]). This can be accomplished by
analyzing natural populations to determine if identified SNPs show associations with
natural variation in behavior.

21.2.3 Polygenic vs. Single-Gene Analysis of Behavior

Most naturally occurring individual differences in behavior that have a genetic
component are polygenically based; that is, the behavioral differences can be attributed to more than one genetic factor. This contrasts with mutant forms that are
generated and screened for in the laboratory. Mutant screens enable us to identify
more loci that affect a given phenotype than studying natural variants.

Most single-gene mutants would not survive in nature because their phenotypes
are usually more severe than natural behavioral variants. Many of these induced
mutants have deleterious effects on fitness due to the pleiotropic effects of their
mutation. One can get around this problem by selectively studying hypomorphic
mutations that cause a small reduction in the amount of gene product. These kinds
of mutations tend to exhibit the behavioral alterations but not the other pleiotropic
phenotypes.[20] Furthermore, such subtle mutations are likely representative of natural
variants that have been selected under natural conditions. It is important to note that
in the laboratory, an induced mutant strain would accumulate heritable differences
in minor genes as a result of genetic drift and/or natural selection in the laboratory.
These minor genes may not be directly related to the expression of the mutant
phenotype. In contrast, naturally occurring variants likely have more subtle effects
on the phenotype and have been "incorporated" into the genome over many generations through selection. Thus, by definition, naturally occurring variants have a
polygenic rather than a single-gene basis. Little is known about the molecular basis
of genes implicated in natural behavioral variation. Questions for the future include:
Do differences in normal behavior result from alterations in the structural or regulatory part of the proteins encoded by these genes? Which "behavioral" genes are

likely to vary in nature? How, at the molecular level, does genetic background influence the expression of the behavioral phenotype?

21.3 FOOD SEARCH BEHAVIOR

Most of our studies on food search behavior have focused on the larval stage of *Drosophila*. The larva's sensory system[47] and behaviors are simple relative to those of the adult fly. The larva can sense light,[36] smell,[5] and taste[47] as well as modify its locomotion depending on abiotic factors such as humidity, food gradients, and light. As mentioned previously, the larva spends most of its life feeding and moving in search of food. Sokolowski[41] developed a larval assay to measure locomotion in the presence of food (a yeast and water paste) that she called a foraging behavior assay. Generally speaking, when animals determine that the food source is a good one, they slow down their rate of locomotion and exhibit high turning rates.[3] In contrast, when food is absent or when the food quantity is not adequate, the rate of locomotion is high and larvae move in more of a straight-line pattern. Foraging behavior, of the larval[18] and the adult fly,[3] can be modified by how satiated the animal is prior to the behavioral test. The behavior can also be modified by external environmental factors such as food quality, food patch size, and interpatch distance. The manner in which these factors modify the food search behavior is predictable. If a fly or larva is starved, it will accept lower-quality food (or smaller patch sizes).

In the early 1980s Sokolowski[41] discovered a naturally occurring polymorphism in larval food search behavior. Third instar *D. melanogaster* larvae collected from single fruits in a pear orchard show a bimodal distribution in their movement patterns while foraging on yeast: 70% of the larvae forage as rovers, exhibiting foraging trails that are longer and straighter than those of the sitter larvae (30% of the population).[41,44] Importantly, this difference in behavior is only exhibited in the presence of food (yeast paste). In the absence of food, the foraging paths of both morphs are long, relatively straight, and do not significantly differ in length.[34] Thus, the rover/sitter difference is conditional on the presence of a nutritive environment and is not due to a general sluggishness in the sitter larvae.[43] When food is unlimited, their development times and growth curves do not differ.[18] The length of the foraging trails are modifiable by the environment (e.g., by modifying patch quality or the degree of starvation prior to the test). However, the relative differences in behavior between the rover and sitter morphs are maintained within a given environment.[18]

Food search behavior in the adult fly is assayed by measuring the distance the female walks postfeeding in an arena with a sucrose drop.[34] After feeding, rover adults walk significantly farther than do sitters, and like larvae they do not differ in their locomotion in the absence of food. When more than one food patch is present, both rover larvae and adults move between patches in search of food, whereas sitters move to the nearest patch and remain feeding nearby. We have shown that there is a selective advantage for the rovers in crowded conditions, whereas sitters do better in uncrowded conditions. Thus, density-dependent selection may have played a role in the evolution of the rover/sitter polymorphism. The olfactory behavior of rover and sitter larvae is wild-type and does not differ in response to yeast.[40]

As behavioral geneticists, we are, of course, interested in determining whether there is a genetic component to the difference in behavior between the rover and sitter morphs. Polygenic analysis (including chromosome substitutions, 16 reciprocal cross-analysis, and compound autosome analysis[42]) shows that this trait is affected by a single major gene named *foraging (for)* located on the second pair of chromosomes in *D. melanogaster*.[7–9] Further localization using standard recombination mapping techniques was extremely difficult due to the overlapping nature of this quantitative behavioral phenotype, and the pleiotropic effects that marker genes and genetic background have on these behavioral phenotypes.[42] To map *for*, we devised a new version of the single-gene mutant approach which we term *lethal tagging*. Briefly, we tagged the behavioral trait (difficult to map) with a recessive lethal mutation (easy to map), proved that the tag was very close to *for*, and then mapped the lethal tag, which enabled subsequent mapping of *for*. In this way, *for* was localized to region 24A3–5 on the polytene chromosome map.[8–10] Osborne et al.[30] then cloned *for* and proved that *for* is synonymous with the *dg2* gene in *Drosophila* that encodes one of two cGMP-dependent protein kinases (PKG).[25] They showed that *for* mutants map within *dg2*, rover fly heads have significantly higher PKG enzyme activity than do naturally occurring sitters or sitter mutants, and that rover larvae have significantly higher PKG activity in their central nervous systems (CNS) than do sitters. The abundance of *dg2* RNA is also higher in the naturally occurring rovers than in sitters, indicating that the difference between the morphs likely results from differences in *dg2* expression. The strongest proof that *for* is *dg2*, however, came when using transgenic sitter larvae that had four copies of rover CDNA from *dg2*. The behavior of these transgenic sitter larvae did not differ from rover, nor did the PKG enzyme activities in their larval CNS. This was the final proof that *for* is *dg2*. Our future work involves studying where and when in the fly higher levels of *for*-PKG need to be expressed to produce rover behavior. We are currently mapping the tissue distribution of *for*-PKG in order to identify cell types where higher level of *for*-PKG is sufficient for rover-like behavior.

Osborne et al.[30] provided the first rigorous demonstration of the molecular basis for a naturally occurring difference in behavior. The significance of this study is the demonstration that small changes in a kinase (rovers have only a 12% increase in PKG enzyme activity in their heads compared with sitters) can account for large differences in naturally occurring behavior. These subtle differences at the molecular level may be representative of a way that behavior is modulated in natural populations.

21.4 ROLE OF THE *FORAGING* GENE IN OTHER ORGANISMS

The fact that many genes identified in the fly have structural and functional homologies to vertebrate genes suggests that genetic discoveries in the fruit fly can contribute to our general understanding of evolutionarily conserved developmental, behavioral, and physiological processes. As such, the mammalian counterparts of PKG might play a role in the regulation of eating in mammals. Indeed, *for* has a human counterpart called *PRKG1* that has high sequence homology to *for* and conservation of most of the *for* splice sites.[15,31] We are currently investigating whether a genetic polymorphism found in the 3'UTR of the *PRKG1* mRNA is associated with obesity in a sample of morbidly obese and normal-weight humans. In the future, it will be of interest to investigate the association of other recently identified *PRKG1* SNPs in humans with alterations in food-related behaviors.

Carrying on with the hypothesis that PKG's role in foraging is conserved across the animal kingdom, Ben-Shahar et al.[4] investigated the role of *for*-PKG in the honeybee, *Apis mellifera*. Following the cloning of *Amfor*, an ortholog of the *Drosophila foraging* gene, they showed that the age-related transition by honeybees from in-hive work (nursing) to out-of-hive work (foraging) is associated with an increase in the expression of the honeybee *foraging* gene, *Amfor*. Furthermore, they demonstrated that this effect is not a byproduct of the nurses being 2 to 3 weeks younger than foragers. They generated same-age nurses and foragers by colony manipulation and found that precocious foragers still had higher expression of *Amfor* RNA and PKG activity. Additionally, cGMP treatment of in-hive workers elevated PKG activity and caused precocious foraging behavior. Thus, the same gene is involved in food-related behaviors in the honeybee and the fly; however, the regulation of the gene differs. *Drosophila for* is involved in allelic variation in behavior in the fly, whereas the honeybee *Amfor* gene is involved in behavioral plasticity during the lifetime of the individual. While other genes are involved in the switch from nurse to forager in the honeybee,[52] it was sufficient to manipulate the levels of *for*-PKG to cause a change in the percentage of forager honeybees.

Scheiner et al.[37,38] used lessons learned from the honey bee to ask function-related questions in the fruit fly. Interestingly, sensory responsiveness in honey bees has been correlated with different aspects of foraging behavior and different types of learning. Scheiner et al.[39] hypothesized that the fly *for* gene might also influence sensory responsiveness and nonassociative learning in *Drosophila*. They used techniques originally developed for honeybee studies to investigate the role of *for*-PKG in adult fruit fly sensory responsiveness. A fruit fly extends its proboscis when its foreleg is stimulated with sucrose. This is called a proboscis extension response (PER). Scheiner and colleagues measured sucrose responsiveness of food, but not water-deprived flies. They elicited PER following stimulation of sucrose receptors found on the terminal segment of the foreleg known as the tarsus. Habituation of the PER occurred when, after repeated stimulations, the flies no longer gave a PER in response to sucrose stimulation. The results showed that rovers are more responsive to sucrose than sitters and sitter mutants, and that this was true for flies exposed to short or long periods of food deprivation. rovers also showed less habituation than sitters and sitter mutants; rovers and sitters with similar sucrose responsiveness were used for these habituation studies. Thus *for*-PKG affects sensory responsiveness to sucrose and habituation in *D. melanogaster*. Like rover flies, forager bees show higher levels of sensory responsiveness than do nurse bees and sitter flies.

Genetic manipulation of PKG in the nematode *C. elegans*, reveals two modes of locomotion behavior in the presence of food: roaming and dwelling.[17] Roamers have high-speed locomotion with infrequent turns, while dwellers show low-speed locomotion with frequent turns. The worm ortholog of the *foraging* gene, *egl-4*, affects this feeding behavior. Interestingly, contrary to the case in *Drosophila* and the honey bee, lower levels of PKG in *C. elegans* is associated with increased roaming behavior. It is of interest to know whether PKG plays a role in feeding behavior of other organisms. Recently, Fitzpatrick and Sokolowski[15] constructed protein phylogenies using 32 PKG sequences that include 19 species and proposed five different evolutionary histories that can explain the link between PKG and feeding behaviors in fruit flies, honeybees and worms. Studies such as these start to bridge the gap between genetic model organisms and organisms not traditionally used for genetic analyses.[16]

Sokolowski and collaborators have implicated a cGMP signal transduction pathway in the regulation of food search behavior in flies and honeybees. The next step is to identify additional components of this food search pathway. This can be accomplished by using: (1) the single-gene mutant approach to generate mutations in genes that interact with *for*, and (2) the QTL analysis approach to localize naturally occurring variants that act as minor modifiers of foraging behavior in *Drosophila*. We are currently mapping genes identified from these types of studies.

The rover/sitter model system is being developed to understand both the mechanistic and evolutionary significance of behavioral variation. With this in mind, we are on the road to developing an understanding of how food search behavior is affected at the level of the gene, molecule, nervous system, organism, population, and species—in both the laboratory and in nature.

ACKNOWLEDGMENTS

This chapter is dedicated to the memory of Dr. Lionel Peypelut, a dear friend and colleague. The research described here from the Sokolowski laboratory was funded by NSERC, CIHR and the Canada Research Chairs Program to M.B.S.

REFERENCES

1. Anholt, R.R. and Mackay, T.F.C., Quantitative genetic analysis of complex behaviors in *Drosophila, Nat. Rev. Genet.*, 5:838–849, 2004.
2. Asburner, M., Hawley, S., and Golic, K., *Drosophila: A Laboratory Handbook*, Cold Spring Harbor Laboratory, Cold Spring Harbour, New York, 2004.
3. Bell, W.J., *Searching Behavior: The Behavioral Ecology of Finding Resources*, Chapman & Hall, New York, 1991.
4. Ben-Shahar, Y., Ribichon, A., Sokolowski, M.B., and Robinson, G.E., Influence of gene action across different time scales on behavior, *Science*, 296:741–744, 2002.
5. Carlson, J., Olfaction in *Drosophila*: from odor to behavior, *Trends. Genet.*, 12:175–180, 1996.
6. Costa, R., Peixoto, A.A., Barbujani, G., and Kyriacou, C.P., A latitudinal cline in *Drosophila* clock gene, *Proc. R. Soc. Biol. Sci.* 250:43–49, 1992.
7. de Belle, J.S. and Sokolowski, M.B., Heredity of rover/sitter: alternative foraging strategies of *Drosophila melanogaster, Heredity*, 59:73–83, 1987.
8. de Belle, J.S. and Sokolowski, M.B., Heredity of rover/sitter foraging behavior in *Drosophila melanogaster*, genetic localization to chromosome-2L using compound autosomes, *J. Insect Behav.*, 2:291–299, 1989.

9. de Belle, J.S., Hilliker, A.J., and Sokolowski, M.B., Genetic localization of *foraging (for)*: a major gene for larval behavior in *Drosophila melanogaster, Genetics*, 123:157–164, 1989.

10. de Belle, J.S., Sokolowski, M.B., and Hilliker, A.J., Genetic analysis of the *foraging* microregion of *Drosophila melanogaster, Genome*, 36:94–101, 1993.

11. de Belle, J.S. and Heisenberg, M., Expression of *Drosophila* mushroom body mutations in alternative genetic backgrounds: a case study of the mushroom body miniature gene (*mbm*), *Proc. Natl. Acad. Sci. U.S.A.*, 93(18):9875–9880, 1996.

12. De Luca, M., Roshina, N.V., Geiger-Thornsberry, G.L., Lyman, R.F., Pasyukova, E.G., and Mackay, T.F.C., *Dopa-decarboxylase* affects variation in *Drosophila* longevity, *Nat. Genet.* 34:429–433, 2003.

13. Falconer, D.S. and MacKay, T.F.C., *Introduction to Quantitative Genetics*, 4th ed., Longman, New York, 1996.

14. Fanara, J.J., Robinson, K.O., Rollmann, S.M., Anholt, R.R., and Mackay, T.F.C, *Vanaso* is a candidate quantitative trait gene for *Drosophila* olfactory behavior, *Genetics*, 162(3):1321–1328, 2002.

15. Fitzpatrick, M.J. and Sokolowski, M.B., In search of food: exploring the evolutionary link between cGMP-dependent protein kinase (PKG) and behavior, *Integr. Comp. Biol.*, 44:28–36, 2004.

16. Fitzpatrick, M.J., Ben-Shahar, Y., Smid, H.M., Vet, L.E.M., Robinson, G.E., and Sokolowski, M.B., Candidate genes for behavioral ecology, *Trends Ecol Evol*, 20:96–104, 2005.

17. Fujiwara, M., Sengupta, P., and McIntire, S.L., Regulation of body size and behavioral state of *C. elegans* by sensory perception and the EGL-4 cGMP-dependent protein kinase, *Neuron*, 36:1091–1102, 2002.

18. Gleason, J.M., Nuzhdin, S.V., and Ritchie, M.G., Quantitative trait loci affecting a courtship signal in *Drosophila melanogaster, Heredity*, 89:1–6, 2002.

19. Graf, S.A. and Sokolowski, M.B., The rover/sitter *Drosophila* foraging polymorphism as a function of larval development, food patch quality and starvation, *J. Insect Behav.*, 2:301–313, 1989.

20. Greenspan, R.J., A kinder, gentler genetic analysis of behavior: dissection gives way to modulation, *Curr. Opin. Neurobiol*, 7:805–811, 1997.

21. Greenspan, R.J., *Fly Pushing, The Theory and Practice of Drosophila Genetics*, Cold Spring Harbor Laboratory Press, Cold Spring Harbour, New York, 2004.

22. Greenspan, R.J., E Pluribus Unum, Ex Uno Plura: Quantitative- and single-gene perspectives on the study of behavior, *Annu. Rev. Neurosci.*, 27:79–105, 2004.

23. Hall, J.C., The mating of a fly, *Science*, 264: 1702–1714, 1994.

24. Harmer, S.L., Panda, S., and Kay, S.A., Molecular bases of circadian rhythms, *Annu. Rev. Cell Dev. Biol.*, 17:215–253, 2001.

25. Kalderon, D and Rubin, G.M., cGMP-dependent protein kinase genes in *Drosophila*, *J. Biol. Chem.*, 264: 10738–10748, 1989.

26. Mackay, T.F.C., Quantitative trait loci in *Drosophila, Nat. Rev. Genet.* 2:11–20, 2001.

27. Mather, K. and Jinks, J.L., *Biometrical Genetics*, 3rd ed., Chapman & Hall, New York, 1982.

28. Nuzhdin, S.V., Pasyukova, E.G., Dilda, C.L., Zeng, Z.-B., and Mackay, T.F.C., Sex-specific quantitative trait loci affecting longevity in *Drosophila melanogaster, Proc. Natl. Acad. Sci. U.S.A.*, 94:9734–9739, 1997.

29. Nuzhdin, S.V., Dilda, C.L. and Mackay, T.F.C., The genetic architecture of selection response: inference from fine-scale mapping of bristle number quantitative trait loci in *Drosophila melanogaster, Genetics*, 153:1317–1331, 1999.

30. Osborne, K., Robichon, A., Burgess, E., Butland, S., Shaw, R.A., Coulthard, A., Pereira, H.S., Greenspan, R.J., and Sokolowski, M.B., Natural behavior polymorphism due to a cGMP-dependent protein kinase of *Drosophila, Science,* 277:834–836, 1997.

31. Ostravik, S., Natarajan, V., Tasken, K., Jahnsen, T., and Sandberg, M., Characterization of the human gene encoding the type I alpha and type I beta cGMP-dependent protein kinase (PRKGI) 1997, *Genomics,* 42:311–318, 1997.

32. Pak, W.L. and Leung, H.T., Genetic approaches to visual transduction in *Drosophila melanogaster, Receptors Channels,* 9(3):149–167, 2003.

33. Pasyukova, E.G., Roshina, N.V., and Mackay, T.F.C., Shuttle craft: a candidate quantitative trait gene for *Drosophila melanogaster* lifespan, *Aging Cell,* 3:297–307, 2004.

34. Pereira, H.S. and Sokolowski, M.B., Mutations in the larval *foraging* gene affect adult locomotory behavior after feeding in *Drosophila melanogaster, Proc, Natl. Acad. Sci. U.S.A.,* 90:5044–5046, 1993.

35. Pereira, H.S., MacDonald, D.E., Hilliker, A.J., and Sokolowski, M.B., *Chaser (Csr),* a new gene affecting larval foraging behavior in *Drosophila melanogaster, Genetics,* 140:263–270, 1995.

36. Sawin, E.P., Harris, L.R., Campos, A.R., and Sokolowski, M.B., Sensorimotor transformation from light reception to phototactic behavior in *Drosophila* larvae, *J. Insect Behav.,* 7:553–567, 1994.

37. Scheiner, R. Erber, J., and Page, R.E., Tactile learning and the individual evaluation of the reward in honey bees (*Apis mellifera* L.), *J. Comp. Physiol.,* 185:1–10, 1999.

38. Scheiner, R. Erber, J., and Page, R.E., The effects of genotype, foraging role and sucrose responsiveness on the tactile learning performance of honey bees (*Apis mellifera* L.), *Neurobiol. Learn. Mem.,* 76:138–150, 2001.

39. Scheiner, R., Sokolowski, M.B. and Erber, J., Activity of cGMP-dependent protein kinase (PKG) affects sucrose responsiveness and habituation in *Drosophila melanogaster, Learn. Mem.,* 11:303–311, 2004.

40. Shaver, S.A., Varnam, C., Hilliker, A.J., and Sokolowski, M.B., The *foraging* gene affects adult but not larval olfactory related behavior in *Drosophila, Behav. Brain Res.,* 95:23–29, 1998.

41. Sokolowski, M.B., Foraging strategies of *Drosophila melanogaster:* a chromosomal analysis, *Behav. Genet.,* 10:291–302, 1980.

42. Sokolowski, M.B., Genetic analysis of behavior in the fruit fly, *Drosophila melanogaster.* In *Techniques for the Genetic Analysis of Brain and Behavior,* D. Goldwitz, D. Wahlsten, and R.E. Winer, Eds., Elsevier Science Publishers, 1992.

43. Sokolowski, M.B. and Hansell, K.P., The *foraging* locus: behavioral tests for normal muscle movement in rover and sitter *Drosophila melanogaster* larvae, *Genetica,* 85:205–209, 1992.

44. Sokolowski, M.B., Pereira, H.S., and Hughes, K., Evolution of foraging behavior in *Drosophila* by density dependent selection, *Proc. Natl. Adac. Sci. U.S.A.,* 94:7373–7377, 1997.

45. Sokolowski, M.B., *Drosophila*: genetics meets behavior, *Nat. Rev. Genet.,* 2:879–890, 2001.

46. Sokolowski, M.B. and Wahlsten, D., Gene-Environment interaction and complex behavior. In *Methods in Genomic Neuroscience,* H. Chin, and S. Moldin, Eds., CRC Press, New York, 2002.

47. Stocker, R.F., The organization of the chemosensory system in *Drosophila melanogaster*: a review, *Cell Tissue Res.,* 273:3–26, 1994.

48. Stocker, R.F., *Drosophila* as a focus in olfactory research: mapping of olfactory sensilla by fine structure, odor specificity, odorant receptor expression, and central connectivity, *Microsc. Res. Tech.*, 55(5):284–296, 2001.
49. Toma, D.P., White K.P., Hirsch, J., and Greenspan, R.J., Identification of genes involved in *Drosophila melanogaster* geotaxis, a complex behavioral trait, *Nat. Genet.*, 31:349–353, 2002.
50. Wang, M.H., Lazebny, O., Harshman, L.G., and Nuzhdin, S. Environment-dependent survival of *Drosophila melanogaster*: a quantitative genetic analysis, *Aging Cell*, 3:133–140, 2004.
51. Weber, K., Eisman, R., Morey, L., Patty, A., Sparks, J., Tausek, M., and Zeng, Z.B., An analysis of polygenes affecting wing shape on chromosome 3 in *Drosophila melanogaster*, *Genetics*, 153(2):773–786, 1999.
52. Whitefield, C.W., Cziko, A.M., and Robinson, G.E., Gene expression profiles in the brain predict behavior in individual honey bees, *Science*, 302(5643):296–299, 2003.

22 Genetic and Molecular Analyses of *Drosophila* Courtship Behavior

Jean-Marc Jallon

CONTENTS

22.1 INTRODUCTION

Drosophila melanogaster is a marvelous material for behavioural genetic analysis. First, it has been shown to be highly amenable for genetic analysis. It has a reduced developmental time (about 9 days at 25°C) and a compact genome (four pairs of chromosomes including three pairs of autosomes and one pair of sex chromosomes). The classic genetic tools such as visible markers and balancer chromosomes can be efficiently used together with new technology to elucidate molecular mechanisms. The complete genome sequence has been known for 5 years[1] and can be compared with those of other model organisms that appear in the literature. Large parts of those of other *Drosophila* species are also available[2] at http://flybase.bio.indiana.edu.

As far as behavior is concerned, *Drosophila* behaves in two states, which follow each other in postembryonic development, as a crawling larva first and then as a flying insect. The imaginal behaviors are quite sophisticated, involving instinctive and plastic facets. Mating behavior is the most complex.

22.2 MATING BEHAVIOR OF *DROSOPHILA MELANOGASTER*, VISUAL AND ACOUSTIC STIMULI

Contrary to butterflies, which attract each other from a distance, *Drosophilae* usually meet on the feeding spot attracted by decomposing fruit volatiles. There the mating behavior of *D. melanogaster* is a sequence of fixed action patterns including orientation by a male toward a female, its tapping of the female cuticle, male wing display generating the courtship song, following of the female by the male, its licking the female's genitalia, rejection or acceptance of the male by the female, male mounting and copulation upon female's acceptance.[3,4]

The orientation by males appears to be guided primarily by visual cues as shown in studies with several types of visual mutants.[5] Among them, a most spectacular deviation from wild-type behavior was observed with the blind no retention potential A males: once they have touched a female, they get excited but are unable to follow the partner and go on singing in the place where "she" had been.[6] In species of the auraria complex, the orientation behavior can be quantified more objectively—male neck movements—than in *D. melanogaster* but in all cases it is triggered across a window by moving flies whichever their sex.[7,8] Moreover, experiments described in the next section strongly suggest that chemical cues stimulate courtship behavior in synergy with visual cues.[9]

Acoustic signals are mainly produced by the vibration of one or two male wings in which indirect flight muscles play a major role thoroughly studied by Ewing.[10] They are perceived by the Johnston's organ, on the antennal second segment, which is stimulated by movements of the terminal arista, as shown by surgical and mutational experiments. The love song of *D. melanogaster* consists of two different elements known as sine song and pulse song. In the pulse song, the time between successive pulses, called the interpulse interval (IPI), is approximately 35 msec for *D. melanogaster* and 50 msec for the sibling species *D. simulans*. Playback experiments have shown the importance of the IPI duration to stimulate the mating speed of pairs involving a wingless male and a winged female.[11] But other acoustic parameters that differ between species, such as the pulse song burst duration or sine song bout length, may play a role. Actually, at early ages, the two species share similar IPI values, close to those of *D. simulans*, but maturation establishes the species specificity of this character.

Single-gene mutations affecting song have led to interesting results. In the old mutant cacophony induced by the chemical mutagen EMS, pulse song showed a longer average IPI (45 msec—very different from the wild-type value of 35 msec and close to that of *D. simulans*). Males, then, are less successful at mating when they bear this mutation in an X gene.[12] Dissonance, another mutation induced by the chemical mutagen EMS was found also to affect pulse song—polycyclic and irregular —and was shown to be an allele of the mutant no-on transient, identified for visual defects and which encodes an RNA binding protein.[13] Another song mutation, croaker, was induced by P insertion mutagenesis. Such mutant males initiate courtship after a long latency and also generate polycyclic pulse songs with long IPI values. Moreover their flying ability is much reduced, suggesting a general defect in the motor system.[14] Other P insertions created mutations affecting the sine song frequency including Beethoven in a chromosome II gene encoding a protein with ATPase activity. Several sex determination genes, when mutated, also showed song abnormalities. Altogether, 17 single genes were involved.[15] Another approach with quantitative genetics took advantage of the natural variation within and between species.[16,17] Several studies of *D. melanogaster* interstrain variations showed additive effects of chromosomes II and III to control IPI, with a larger effect of the third chromosome. But only one candidate gene, *tipE*, another channel gene on 3L, was clearly included in a quantitative trait locus (QTL)—actually, in the major QTL of chromosome III. A recent QTL study of IPI differences in backcross hybrids between *D. simulans* and *D. sechellia*, another species of the melanogaster subgroup, has shown similar contributions of autosomal chromosomes to IPI and localized four QTL on 2R and two on 3R. In this study, three candidate genes, maleless, croaker, and fruitless are related to QTL.[15]

The courtship behavior may be subject to associative modification. Immature virgin females and recently fertilized females stimulate males as much as mature virgin females, but both reject courters. While virgins usually escape from the courting male by decamping, the copulated females extrude their ovipositor which blocks genital contact. The rejection behavior in fertilized females is triggered by a peptide which is produced in the male accessory gland and transferred to females during copulation.[18] Males that have previously courted fertilized females show a reduced level of courtship toward a virgin female.[19]

22.3 PHEROMONES AND BIOSYNTHETIC GENES

Pheromones have been involved in *Drosophila* courtship for a long time,[20] but analytical experiments have started more recently—mainly in my group. Olfactometric tests have shown that *Drosophila melanogaster* CantonS males were not attracted from a distance by conspecific females, as for moths, but that females could be attracted by males.[9] This led us to look first for female pheromones similar to the aphrodisiac contact pheromones discovered in *Musca domestica*,[21] and Venard and Jallon[22] showed that one male did not react to a female cuticular extract but that two males courted each other in the presence of a female extract, more or less intensively depending on the dose of extract. This suggested that female cuticular pheromones could indeed stimulate male courtship, but that such chemical cues acted in synergy with visual cues.[6]

In the CantonS strain of *D. melanogaster*, female contact pheromones were identified in a complex mixture of long-chain hydrocarbons present in the cuticle of both sexes: (Z)-7, (Z)-11, heptacosadiene (7,11 HD), and homologous dienes are characteristic of mature females.[23] However, while such abundant dienes can stimulate CantonS male courtship to copulation,[24,25] a larger spectrum of long-chain hydrocarbons, present in mature or immature flies, is able to stimulate early courtship behaviors such as male wing vibrations.[26] This is especially the case of unsaturated hydrocarbon (HC) with 27+/–2 carbons bearing at least one double bond in position (Z)-7, including (Z)-7 pentacosene present in mature flies of both sexes but not of (Z)-7 tricosene.[27] Thus chemical sex appeal involves more than one type of molecule and is rather a complex mixture of synergistic compounds, HC signatures, whose proportions might vary between populations like the pheromone blend of moths.[28,29] But the more volatile pheromones hypothetized by Shorey and Bartell[20] and Savarit et al.[30] have not yet been identified (see also next section).

As far as the male attractivity evidenced in olfactometric tests is concerned, it should be due to male rich compounds, which are more volatile than female cuticular compounds. Two of them have to be considered: 7T, the most abundant HC in mature CantonS males,[23] and cis-vaccenyl acetate (cVA), a non-HC compound produced in the ejaculatory bulb.[31] There are two opposite behavioral roles played by cVA: at low dose it is an aggregation pheromone for both sexes,[32] and at high dose it is a male courtship inhibitor.[33] Further behavioral experiments in CantonS have shown that 7T was also able to stimulate females and inhibit other male courtships.[34–38] These cuticular HC also play a critical role in the sexual isolation of the sympatric D. melanogaster and D. simulans and related species.[39,40] Indeed 7,11 HD is absent from the cuticle of D. simulans females, whereas 7T is abundant in both sexes of most of its populations. 7T stimulates D. simulans male courtship in a dose-dependent way,[34] which is inhibited by 7,11 HD.[40] 7T inhibits CantonS male courtship, which is stimulated by 7,11 HD.

Most of the HC information about D. melanogaster presented above was deduced from studies performed with flies from the well-known laboratory strain CantonS. Later, a rich polymorphism of these cuticular HCs was discovered during a large-scale study involving 85 populations collected around the world. This study showed that sub-Saharan African females markedly differed from those of most other populations, including CantonS. Such females, like those collected in the primary forest of Tai (Ivory Coast) also have 7,11 HD—but 10 times less, plus a large amount of its position isomer (Z)-5, (Z)-9 heptacosadiene (5,9 HD), which is rare in Canton S females. This chemical difference is recognized by CantonS males which prefer homotypic females while Tai males make little difference between the two types of females.[34] Moreover males of the various populations showed marked differences in their content of 7T which increased with the latitude of their collection site, while the content of the homologous (7P) decreased.[41] Females, both equatorial and temperate showed a clear preference for males of their population.[42] It is thus suggested that behavioural responses of males or females of different populations might have dose-response curves to specific chemical component differing with genome differences between populations.[34,35,43,44]

Chromosome exchanges between both diene morphs showed that the 7,11 HD/5,9 HD ratio was mainly controlled by third chromosome loci.[45] Later recombinations experiments with marker genes and detailed deficiency studies linked this variation to a single segregating factor in cytological position 87 C-D of chromosome III,[46] located close to gene desat1.[47] On the other hand, metabolic studies using radiolabelled potential precursors had shown that cuticular HC could be synthetized de novo by flies which could synthetize, elongate, desaturate and decarboxylate fatty acids to make their pheromones.[48] Taking advantage of the high degree of conservation observed among known desaturase genes, Wicker-Thomas et al.[47] cloned desat1, the first gene of desaturase enzyme in Drosophila. This gene with four introns shows only one open reading frame, although it has a complex pattern of transcription with some sex differences.[49,50] Its heterologous expression in yeast could complement the mutant *ole1*, deficient in desaturase activity, and established the D9 specificity of the *Drosophila* enzyme as well as its preference for palmitic acid as substrate.[51] Insertions of various P transposons in the promoter region of *desat1* showed more or less decrease in the level of all 7 unsaturated HC in both sexes including 7T and 7,11HD. As their excision often reverted the wild-type HC phenotype, the *desat1* enzyme is clearly involved in a common step of biosynthesis, the desaturation which transforms palmitic acid into palmitoleic acid (C16:1,D9/w7).[25]

Although the 7,11 HD/5,9 HD polymorphism was first hypothetized to be linked to an allelic variation in the *desat1* gene with each allele producing a different desaturase isoform, the reality was more complex. Although the Tai *desat1* gene coded a polypeptide chain with 3 amino acid substitutions compared to that of CantonS, the yeast transformation results with either coding sequence were similar, infirming the simple hypothesis. Later a mini-dissection of the locus around the *desat1* gene revealed the presence of two more desaturase genes, *desat2* and *desat3*. *Desat2* was characterized in the Tai strain. Its uncorrupted open reading frames (ORF) encodes a polypeptide with 65% identity with that of *desat1* but a smaller amino domain. The heterologous expression of *desat2* in yeast also complemented the deficient mutant *ole1*, suggesting again a D9 specificity, but a different substrate; myristoleic acid (C14:1,D9/w5) was mainly produced instead of palmitoleic acid.[51] This position w5 for HC is abundant mainly in the African strains and there only in females. Actually *desat2* can be expressed only in African females that make 5,9HD. These facts support a critical role for both *desat1* and *desat2* in the first desaturation of linear saturated fatty acids. Another study has shown that the knock-down of the *desat2* gene in ancestral populations during their migrations in Africa was linked to the deletion of 16 bp in the promoter region.[52]

To finally make dienes, monoenic fatty acids have to go through a second desaturation, taking place in females of the *D. melanogaster* species and probably not in *D. simulans*. As the enzyme specificity is usually restricted to a certain type and position of bond along the chain and a substrate either saturated or monoenic, other desaturase genes were searched for. Indeed the published sequence of *D. melanogaster* genome suggested the existence of more desaturase genes, one on chromosome II supposed to be more specific for males and four more on chromosome III.[1,49] Among these latter genes, one in the cytological region 68 had been shown to be possibly important to make female dienes.[53] Chertemps et al.[54] then completely characterized a new desaturase gene, called *desatF*, which is expressed in *D. melanogaster* females but not in conspecific males. Using the interference RNA technique, they were able to knock down this gene almost completely. A dramatic result was observed in the cuticle of transformed females: the level of 7,11 HD was decreased by more than 80% while that of all 7-monoenes was much increased. This clearly shows that *desatF* adds a second double bond to monoenic fatty acids to make dienic ones. The description of another group of biosynthetic genes has been initiated, those coding for fatty acid elongases.[55]

Another approach to try to localize genes controlling HC production was quantitative genetics QTL analyses of HC were performed in flies of recombinant lines produced between *D. simulans* and *D. sechellia*, another member of the *melanogaster* subgroup restricted to the Seychelles archipelago. As already mentioned, the former species has 7T as its major cuticular compound in both sexes of most populations while the latter species shows a marked sexual dimorphism with much 7,11HD in mature females as in temperate *D. melanogaster* females. Separate QTL analyses of 7T and 7,11HD evidenced two regions for each trait. Both are associated with one region on the right arm of chromosome III, while the others are either on the left arm of that chromosome (7,11HD) or on chromosome X (7T). Interestingly, it was noted that *desat1/2* on one hand and *desatF* on the other hand were found to be located near the QTL shared by 7T and 7,11HD and the QTL specific for 7,11HD, respectively.[15]

22.4 GUSTATORY AND OLFACTORY RECEPTORS

The sex pheromones of female moths, which are very volatile, are perceived by male antennal olfactory sensillae. In *D. melanogaster* male antennae had been considered less important for the female excitatory messages, as antennaless homozygous males were known to court intensively wild type females whichever the number of their antennae.[56] But it has been suggested that they might receive male inhibitory inputs as a large percentage of the mutant males without antennae courted each other.[8] Moreover it has been shown with mutants that the main class of antennal olfactory sensillae —basiconicae—were not necessary for the detection of female phero- mones.[57] On the other hand, there were evidences that gustatory organs could perceive excitatory female compounds, when the male raises his foreleg to tap the epicuticle and then vibrates one wing or when he licks the female genitalia with his proboscis before attempting copulation. For example Venard et al.[58] have coated male legs with a thin layer of dental wax; when all legs had been coated, no wing vibration toward females could be observed, but if only five legs had been treated, the response varied in intensity depending on the untreated leg; it was quasi-null if a hindleg was untreated and strong if it was a foreleg. This suggested that some of the gustatory hairs present on the foreleg and not in the other pairs of legs might perceive cuticular compounds. Scanning electron microscopy and crystal violet diffusion experiments did detect a set of typical gustatory sensillae displaying a marked sexual dimorphism.[58] Other gustatory sensillae, which have been described on the proboscis, are used when the male licks the female genitalia in a later courtship step.[59]

Many new and exciting information resulted from the molecular studies which followed the discovery of olfactory receptors by Buck and Axel[60] and the completion of the *Drosophila* genome sequencing.[1] Two large families of about 60 genes coding for chemoreceptors with seven transmembrane domains and coupled with G proteins were discovered by several groups,[61–63] about half involved in olfaction *Or*, and half involved in gustation *Gr*. Then an important functional work has been initiated using the GAL4-UAS method designed by Brand and Perimon[64] and which makes use of a yeast transcription factor GAL4 which directs the expression of a gene of inter- est*—for example a given *Gr-or* a reporter gene only in the target cells defined by the gene* promoter. In this way Bray and Amrein[65] have characterized *Gr68a* which is expressed only in males and only in about 20 of the male specific chemosensillae borne by the forelegs. Then the knockdown of this *Gr68a* gene by RNA interference in the very neurons where their expression is restricted and the inactivation of the same neurons by tetanus toxin have both led to a marked decrease of male courting activity; moreover, both manipulations reduced occurrences of specific courtship steps such as wing vibrations and attempted copulations. This strongly suggested that *Gr68a* was indeed a good candidate receptor to attach a female contact phero-mone such as 7,11 HD; direct binding experiments still have to be done. The fact that these transformed males are still able to perform part of the courtship suggests the involvement of other types of *Gr*, possibly involved in binding of other contact pheromones.[66] Other elements of the pheromone transduction cascade have also been described, two chemosensory proteins *CheA29a* and *CheB42a* which share some characteristics with oderant pheromone binding protein (OBP/PBP) and are expressed only in the front legs of males as *Gr68a*[67] and one ion channel subunit, *ppk25*. This gene is expressed not only in legs, but also in antennae and several of its mutations strongly affect the courtship behaviour of males.[68]

In the olfactory system, it has been clearly shown that most of the 62 *Or* genes were expressed alone in olfactory sensory neurons, while the situation looked more complex for *Gr* in gustatory neurons.[69–71] Another difference is the projection pattern of the sensory neurons. While most olfactory sensory neurons expressing a given *Or* converge to a single glomerulus, one functional subunit of the brain antennal lobes, a similar structure to the vertebrate olfactory bulb, gustatory sensory neuron primary projections are not in the brain but in different parts of the nervous system and thus much less well characterized. For example all gustatory neurons of both forelegs converge to the anterior part of the thoracic ganglion. In a pioneer study, Possidente and Murphey[72] have described the sexual dimorphism of these projections.

Only 2 out of about 50 glomeruli[73–74] have been found to display a dimorphism between males and females VA1v and mainly DA1, which is 62% larger in males than in females.[75] These glomeruli are anatomically homologous to the more dra-matically dimorphic glomeruli of moths. In males of these insects there is a macro-glomerulus whose function is to integrate the complexity of the female pheromone bouquet. Interestingly these two large structures are targets for the fruitless tran-scription factor which also target only one or two other glomeruli.[69] The olfactory sensory neurons which converge toward the *Drosophila* dimorphic glomeruli start in trichoid sensillae like those sensitive to pheromones in moths;[2] they express olfactory receptors which have been identified as respectively Or47b and Or67d.

Could cVA not be a possible ligand for one of these olfactory receptors? cVA has been shown to be detected by a subset of trichoid sensillae of the antennal third segment[58,61] and one piece of its transduction machinery has been identified. Flies lacking the gene *lush*, which codes for its OBP, display much less aggregation ability.[67] This gene has been originally identified because some of its mutations lead to specific attraction toward high concentrations of ethanol and other short-chain alcohols which are avoided by wild-type flies. Actually, lush is expressed in about 150 trichoid sensillae on the ventral-lateral surface, in both males and females, whose sensory neurons project into the VA1 and DA1 glomeruli.[76] It is possible that another ligand might be the volatile female pheromone hypothetized by Shorey and Bartell.[20]

22.5 SEX DETERMINATION CASCADE AND THE COURTSHIP BEHAVIOR MASTER GENE FRUITLESS

The chromosomal sex of *D. melanogaster* is determined by the relative number of X and A chromosomes (A being any autosome, either II, III, or IV). The presence of a Y chromosome is not as important as it is in higher vertebrates. But there is a set of genes which control each other in a hierarchical way as shown in the Baker group and the Nothiger group. The X-linked gene, *sex-lethal*, controls a series of autosomal genes including two *transformer* genes (*tra* on chromosome III and *tra-2* on the second) and the *doublesex* gene (III). Female somatic development requires first the activation of *sex-lethal* whose products regulate the splicing of *tra* main transcript. This *tra* product is necessary for femaleness during one critical period during development. *Tra* encodes another RNA binding protein which regulates the expression of a very small number of genes, among them *doublesex* which is then expressed in a female specific transcript. In the absence of the femizing *tra* product, a male specific *doublesex* transcript is produced. The misexpression of *doublesex* might result in the transformation of both chromosomal males and females into intersexes.[77–79]

The use of gynandromorphs has suggested that neural circuits for male and female behaviors might coexist in the brain. Such sex mosaics are produced with unstable ring-X chromosome (Xr), the early loss of which produces clones of cells, either female (X*/Xr) or male (X*/O). Using sex-linked markers (such as the asterisk,*) to infer the sex genotypes of internal tissues and a detailed analysis of the sexual behaviors either typically male or female, it was possible to identify the tissue the sex of which correlates with the sex of behavioral patterns. They were called behavioral foci by Hotta and Benzer.[80] The focus for male behavior includes a limited part of the head. A more refined mapping using internal histochemical markers localized this focus in the dorsal posterior brain, an era which includes the mushroom bodies, structures already involved in the sensory information integration. But the focus controlling female behaviors, such as receptivity and oviposition, was mapped in the dorsal anterior part of the brain which contains the neuroendocrine cells.[81] The two foci commanding sex specific behaviors thus seemed distinct.[81] Another female important focus, the sex-appeal focus, was very far from the brain; this internal structure was in the ventroposterior region of the blastoderm fate map, pointing to fat bodies and oenocytes.[82]

More recently it became possible to produce sex mosaics where in a fly of one sex the other sex was expressed only in a brain tissue, thanks to the GAL4-UAS method introduced earlier.[64] Actually, a group of brain cells could be feminized in males through a tissue restricted transcription of the major feminizing gene, tranformer.[83] Some of the transformed males courted wild-type males as continuously as wild-type females. The parallel activation of the reporter gene B-galactosidase allowed the identification of the target tissues, a few functional glomeruli of the antennal lobes and part of their projections in the mushroom bodies.[84] The observed bisexual behavior might be linked either to the dysfunction of the male inhibition center or to the switching on of a female specific activation center, as a consequence of the *tra* induced feminization.

Flies with a female caryotype but which are mutant for *tra* develop and behave as males. In such flies, a female phenotype could be restored by a transgene that carried the female-specific cDNA of *tra* under the control of a heat-shock promoter. This transgene could also transform caryotypic male animals into sterile females. Raised at 25°C, both such transformed XX and XY displayed typical male courtship while at the same time, rich in dienes, they were attractive to males. When the expression of *tra* was forced by heat shock, applied during a limited period around puparium formation, male behavior was abolished and replaced by female behavior. Thus sexual behavior is irreversibly programmed during a critical period as a result of the activity switch of *tra*.[85]

Another strategy used a collection of mutations which affected all male courtship behavior and mapped in a small region of the third chromosome, at the *fruitless(fru)* locus. The first *fru* mutant (*fru 1*) was induced by x-rays and described behaviorally first as a male-sterile mutant, then as a male courter of both sexes.[86] Actually, it corresponded to a small deletion together with a chromosomal inversion in the interval 90C–91B.[87] Later four additional alleles (*fru 2, fru 3, fru 4* and *fru-satori*) have been obtained by P element insertion mutagenesis and showed different behavioral characteristics. For example *fru-satori* males do not court females but can court males; *fru2-4* males court both males and females; *fru 4* males do not perform love song whichever the courted sex.[4,8,88] Although a recognizable insertion like P made this gene cloning possible, molecular studies were laborious as *fru* had several transcripts and several promoters were identified. Homologies with members of the BTB family of transcriptional regulators that also contains zinc-finger motifs were found. In the regulatory region, the presence of three putative binding sites for the *transformer* gene product, similar to those found in *doublesex* clearly showed that *fru* was a neural sex-determination switch downstream of *transformer*, occupying in the sex determination cascade a position parallel to *doublesex*.[89,90]

While transcripts from three promoters are common for both sexes, transcripts from promoter P1 are controlled by the sex determination cascade and produce in males a specific form *fru*M that is longer. A series of elegant experiments have allowed Demir and Dickson[91] to demonstrate that male splicing specifies the main part of male courtship behavior. They also showed that this mode of splicing is sufficient to have an otherwise normal female court as a male.[91] But how is the gene organizing the male neurocircuit which performs the male courtship program?

A transgene, in which the GAL4 coding sequence had been introduced into the *fru*M coding sequence, was constructed by two groups and used to direct a detailed search of *fru* target cells.[92,93] Most of the peripheral sensory systems were concerned, that is about 20 to 30 gustatory neurons of both the prothoracic legs and the proboscis but also more than 100 to 200 neurons of the olfactory and acoustic parts of the antennae. The olfactory neurons projected primarily to 3–4 glomeruli of the antennal lobes as reported earlier.

In the central part of the nervous system, *fru*M is expressed in a few scattered groups of cells. One group is a bilateral cluster of about 60 neurons in the suboesophageal ganglion where *fru*M expression could be blocked with the RNA interference technique.[94] Surprisingly, such transformed males started to court wild-type females ten times more quickly than wild types, apparently bypassing the sensory steps of early courtship such as orientation and tapping usually occurring in wild type males; but they quickly displayed all later steps, often all at the same time,[94] suggesting that these suboesophageal neurons might inhibit the courtship start until sufficient and adequate sensory stimuli allow it.

The question of sexual dimorphism in the number of and positions of brain cell clusters involved in courtship behavior has also been investigated. Until recently, clear differences had been found only in the abdominal ganglion.[72,95] In 2005, Stockinger et al.,[93] using a fluorescent reporter gene directed by a specific P [GAL4] element inserted in the *fru* gene, have only observed larger clusters in the male superior protocerebrum. But, with a slightly different technique, Kimura et al.[96] could identify two more groups of sexually dimorphic neuron clusters: in the medulla of the optic lobes there were a few more interneurons only in males; marked differences in a brain region dorsal to the antennal lobes were also characterized. There, on each side, a cluster of about 30 neurons was characterized in males while there were not more than 5 in females. Moreover, these neurons also displayed clear sex differences in their projection patterns. These interneurons seem to start in the suboesophageal ganglion where gustatory informations—possibly contact phero-mone inputs—collected by tarsal and labellar gustatory neurons might be integrated and finish in the superior lateral protocerebrum.

Transformer and Fruitless are thus remarkable examples of switch genes able to transform a complex behavioral program such as courtship. The behavioral trans-formation is linked to structural modifications in the nervous system. In both cases, they are not fully understood but seem limited to a few clusters of neurons. The action of such master genes can be studied in more depth thanks to all the sophis-ticated tools that *Drosophila* provides for cellular and molecular analyses.

REFERENCES

1. Myers, E.W., Sutton, G.G., Delcher, A.L. et al., A whole genome assembly of Drosophila, *Science*, 287, 2196, 2000.
2. Ashburner, M., Golic, K. and Hawley, S., *Drosophila, A Laboratory Handbook*, 2nd ed., Cold Spring Harbor Laboratory Press, Cold Spring Harbor, 2005.
3. Bastock, M. and Manning, A., The courtship of *Drosophila melanogaster*, *Behaviour*, 8, 85,1955.
4. Hall, J.C., The mating of a fly, *Science*, 264, 1702, 1994.
5. Markow, T.A., Behavioral and sensory basis of courtship success in *Drosophila melanogaster*, *Proc. Natl. Acad. Sci. USA*, 84, 6200, 1987.
6. Jallon, J.M. and Hotta, Y., Nonchemical messages of the female *Drosophila melan-ogaster*, in *Taniguchi Symposia in Biophysics No7*, Hotta, Y., Tokyo University Press, Tokyo, 1983, 136.
7. Oguma, Y. et al., Courtship behavior and sexual isolation between *Drosophila auraria* and *D. triauraria* in darkness and in light, *J. Evol. Biol.*, 9, 803, 1996.
8. Yamamoto, D., Jallon, J.M. and Komatsu, A., Genetic dissection of sexual behavior in *Drosophila melanogaster*, *Annu. Rev. Entomol.*, 42, 551, 1997.
9. Venard, R. and Jallon, J.M., Evidence for an aphrodisiac pheromone of female *Drosophila*, *Experientia*, 36, 211, 1980.

10. Ewing, A., Functional aspects of *Drosophila* courtship, *Biol. Rev.*, 58, 275, 1983.
11. Rybak, F., et al., Acoustic communication in *Drosophila melanogaster* courtship: are pulse- and sine-frequencies important for courtship success? *Can. J. Zool.*, 80, 987, 2002.
12. Schilcher, F.V., A mutation which changes courtship song in *Drosophila melanogaster, Behav. Genet.*, 7, 251, 1977.
13. Kulkarni, S.J., Steinlanf, A.F. and Hall, J.C., The dissonance mutant of courtship song in *Drosophila melanogaster, Genetics*, 118, 267, 1988.
14. Yokokura, T., Ueda, R. and Yamamoto, D., Phenotypic and molecular characterization of croaker, a new mating behavior mutant of *Drosophila melanogaster, Japan J. Genet.*, 70, 103, 1995.
15. Gleason, J. and Ritchie, M., Do quantitative trait loci for a courtship song difference between *Drosophila simulans* and *D. sechellia* overlap with candidate genes and intraspecific QTLs?, *Genetics*, 166, 1303, 2004.
16. Mackay, T.F. et al., Genetics and genomics of *Drosophila* mating behavior, *Proc. Natl. Acad. Sci. USA*, 102, 6622, 2005.
17. Gleason, J. et al., A quantitative trait loci for cuticular hydrocarbons associated with sexual isolation between *Drosophila simulans* and *D. sechellia*, *Genetics*, 171, 1, 2005.
18. Chen, P.S., Stumm-Zollinger, E., Aigaki, T., Balmer, J., Bienz, M. and Bohlen, P., A male accessory gland peptide that regulates reproductive behavior of female *Drosophila melanogaster, Cell*, 54, 291, 1988.
19. Gailey, D.A., Jackson, F.R. and Siegel, R.W., Conditioning mutations in *Drosophila melanogaster* affect an experience-dependent behavioral modification in courting males, *Behav. Genet.*, 106, 613, 1984.
20. Shorey, H.H. and Bartell, R., Role of volatile sex pheromone in stimulating male courtship behavior in *Drosophila melanogaster, Anim. Behav.*, 18, 159, 1970.
21. Carlson, D.A. et al., Sex attractant pheromones of the house fly: isolation, identification and synthesis, *Science*, 174, 76, 1971.
22. Venard, R., Attractants in the courtship behavior of *Drosophila melanogaster*, in *Development and Neurobiology of Drosophila*, Siddiqi, O., Babu, P., Hall, L.M. and Hall, J.C., Plenum Press, New York, 1980, 457.
23. Antony, C. and Jallon, J.M., The chemical basis for sex recognition in *Drosophila melanogaster, J. Insect. Physiol.*, 28, 873, 1982.
24. Arienti, M., Analyse de la variabilité de quelques mécanismes impliqués dans le comportement sexuel de populations differentes de *Drosophila melanogaster,* Ph.D. thesis, Université de Paris XI, Orsay, 1993.
25. Ueyama, M. et al., Mutations in the *desat1* gene reduces the production of courtship-stimulatory pheromones through a marked effect on fatty acids in *Drosophila melanogaster, Insect Biochem. Mol. Biol.,* 35, 911, 2005.
26. Pechine, J.M., Antony, C. and Jallon, J.M., Precise characterization of cuticular compounds in young *Drosophila, J. Chem. Ecol.*, 14, 1071, 1988.
27. Antony, C. et al., Compared behavioural responses of male *Drosophila melanogaster* (Canton S) to natural and synthetic aphrodisiacs, *J. Chem. Ecol.*, 11, 1617, 1985.
28. Klun, J.A. and Maini, S., Genetics basis of an insect communication system in the European corn borer, *Environ. Entomol.*, 11, 1084, 1982.
29. Jallon, J.M. and Pechine, J.M., Une autre race chimique de *Drosophila melanogaster* en Afrique, *CR. Acad. Sci.*, Paris, 309, 1551, 1989.
30. Savarit, F. et al., Genetic elimination of known pheromones reveals the fundamental chemical basis of mating and isolation in *Drosophila, Proc. Natl. Acad. Sci. USA*, 96, 9015, 1999.

31. Butterworth, F.M., Lipids in *Drosophila*, a newly detected lipid in the male, *Science*, 163, 1256, 1969.

32. Bartelt, R.J. et al., *Cis*-vaccenyl acetate acts as an aggregation pheromone in *Drosophila melanogaster*, *J. Chem. Ecol.*, 11, 1747,1986.

33. Jallon, J.M., Antony, C. and Benamar, O., Un anti-aphrodisiaque produit par les males de *Drosophila melanogaster*, *C. R. Acad. Sci.*, Paris, 292, 1147, 1981.

34. Jallon, J.M., A few chemical words exchanged by *Drosophila* during courtship, *Behav. Genet.*, 14, 441, 1984.

35. Scott, D. and Jackson, L.L., Interstrain comparison of male predominant aphrodisiacs in *Drosophila melanogaster*, *J. Insect Physiol.*, 34, 863, 1988.

36. Scott, D., Genetic variation for female mate discrimination in *Drosophila melanogaster*, *Evolution*, 48, 112, 1994.

37. Ferveur, J.F. and Sureau, G., Simultaneous influence on male courtship of stimulatory and inhibitory pheromones produced by live sex-mosaics *Drosophila melanogaster*, *Proc. R. Soc. London B*, 263, 967, 1996.

38. Grillet, M., Dartevelle, L. and Ferveur, J.F., A Drosophila male pheromone affects female sexual receptivity, *Proc. R. Soc. London B*, 273, 315–323, 2006.

39. Cobb, M. and Jallon, J.M., Pheromones, mate recognition and courtship stimulation in the *Drosophila melanogaster* species sub-group, *Anim. Behav.*, 39, 1058, 1990.

40. Coyne, J.A., Crittenden, A.P. and Mah, K., Genetics of a pheromonal difference contributing to reproductive isolation in *Drosophila*, *Science*, 265, 1461, 1994.

41. Rouault, J., Capy, P. and Jallon, J.M., Variations of male cuticular hydrocarbons with geoclimate variables: an adaptative mechanism in *Drosophila melanogaster*?, *Genetica*, 110, 117, 2001.

42. Haerty, W. et al., Reproductive isolation in natural populations of *Drosophila melanogaster*, *Genetica*, 116, 215, 2002.

43. Sureau, G. and Ferveur, J.F., Co-adaptation of pheromone production and behavioural responses in *Drosophila melanogaster* males, *Genet. Res.*, 74, 129, 1999.

44. Marcillac, F., Grosjean, Y. and Ferveur, J.F., A single mutation alters production and discrimination of *Drosophila* pheromones, *Proc. R. Soc. London B*, 272, 303, 2004.

45. Ferveur, J.F. et al., World-wide variation in *Drosophila melanogaster* sex pheromones: behavioural effects, genetic bases and potentially evolutionary consequences, *Genetica*, 97, 73, 1996.

46. Coyne, J.A.,Wicker-Thomas,C. and Jallon, J.M., A gene responsible for a cuticular hydrocarbon polymorphism in *Drosophila melanogaster*, *Genet. Res.*, 73, 189, 1999.

47. Wicker-Thomas, C., Henriet, C. and Dallerac, R., Partial characterization of a fatty acid desaturase gene in *Drosophila melanogaster*, *Insect Biochem. Mol. Biol.*, 27, 963, 1997.

48. Pennanec'h, M. et al., Incorporation of fatty acids into cuticular hydrocarbons of male and female *Drosophila melanogaster*, *J. Insect Physiol.*, 43, 1111, 1997.

49. Jallon, J.M. and Wicker-Thomas, C., Genetic studies on pheromone production in Drosophila, in *Insect Pheromone Biochemistry And Molecular Biology*, Blomquist, G.J. and Vogt, R.G, Elsevier, Amsterdam, 2003, 253.

50. Marcillac, F. et al., A mutation with major effects on *Drosophila melanogaster* sex pheromones, *Genetics*, 171, 1617, 2005.

51. Dallerac, R.C. et al., Delta 9 desaturase gene with a different substrate specificity is responsible for the cuticular diene hydrocarbon polymorphism in *Drosophila melanogaster*, *Proc. Natl. Acad. Sci.* USA, 97, 9449, 2000.

52. Takahashi, A.S. et al., The nucleotide changes governing cuticular hydrocarbon variation and their evolution in *Drosophila melanogaster*, *Proc. Natl. Acad. Sci. USA*, 98, 3920, 2001.

53. Wicker-Thomas, C. and Jallon, J.M., Role of Enhancer of zeste on the production of *Drosophila melanogaster* pheromonal hydrocarbons, *Naturwissenschaften*, 87, 76, 2000.

54. Chertemps, T. et al., Female-specific desaturase gene responsible for diene hydrocarbon biosynthesis and courtship behaviour in D. melanogaster, *Insect Molec. Biol.*, in press, 2006.

55. Chertemps, T. et al., A new elongase selectively expressed in *Drosophila* male reproductive system, *Biochem. Biophys. Res. Commun.*, 333, 1066, 2005.

56. Begg, M. and Hogben, L., Chemoreceptivity of *Drosophila melanogaster*, *Proc. R. Soc. London B*, 133, 1, 1946.

57. Stocker, R.F. and Gendre, N., Courtship behavior of *Drosophila*, genetically and surgically deprived of basichonic sensilla, *Behav. Genet.*, 19, 371, 1988.

58. Venard, R., Antony, C. and Jallon, J.M., *Drosophila* Chemoreceptors, in *Neurobiology Of Sensory Systems*, Singh, N. and Strausfeld, N., Eds., Plenum Press, New York, 1989, 377.

59. Nayak, S.V. and Singh, R.N., Sensilla on the tarsal segments and the mouthparts of adult *Drosophila melanogaster*, *Int. J. Insect Morph. Embryol.*, 12, 273, 1983.

60. Buck, L. and Axel, R., A novel multigene family may encode odorant receptors: a molecular basis for odor recognition, *Cell*, 65, 175, 1991.

61. Clyne, P.J., Warr, C.G. and Carlson, J.R., Candidate taste receptors in *Drosophila*, *Science*, 287, 1830, 2000.

62. Scott, K. et al., A chemosensory gene family encoding candidate gustatory and olfactory receptors in *Drosophila, Cell*, 104, 661, 2001.

63. Dunipace, L., Meister, S., McNealey, C. and Amrein, H., Spatially restricted taste receptors in the *Drosophila* gustatory system, *Curr. Biol.*, 11, 822, 2001.

64. Brand, A.H. and Perimon, N., Targeted gene expression as a means of altering cell fates and generating dominant phenotypes, *Development*, 117, 401, 1993.

65. Bray, S. and Amrein, H., A putative *Drosophila* pheromone receptor expressed in male-specific yaste neurons is required for efficient courtship, *Neuron*, 39, 1019, 2003.

66. Thorne, N. et al., Taste perception and coding in *Drosophila*, *Cur. Biol.*, 14, 1065, 2004.

67. Xu, P.X. et al., *Drosophila* OBP LUSH is required for activity of pheromone-sensitive neurons, *Neuron*, 45, 193, 2005.

68. Lin, H. et al., A *Drosophila* DEG/EnaC channel subunit is required for male response to female pheromones, *Proc. Natl. Acad. USA*, 102, 12831, 2005.

69. Fishilevitch, E. and Vosshall, L.B., Genetic and functional subdivision of the *Drosophila* antennal lobe, *Cur. Biol.*, 15, 1548, 2005.

70. Couto, A., Alenius, M., and Dickson, B.J., Molecular, anatomical and functional organization of the *Drosophila* olfactory system, *Cur. Biol.*, 15, 1535, 2005.

71. Amrein, H. and Thorne, N., Gustatory perception and behavior in *Drosphila melanogaster*, *Cur. Biol.*, 15, 673, 2005.

72. Possidente, D.R. and Murphey, R.K., Genetic control of sexually dimorphic axon morphology on *Drosophila* sensory neurons, *Dev. Biol.*, 132, 448, 1989.

73. Stocker, R., The organization of the chemosensory system in *Drosophila melanogaster*, *Cell Tissue Res.*, 275, 3, 1994.

74. Laissue, P.P. et al., Three-dimensional reconstruction of the antennal lobe in *Drosophila melanogaster*, *J. Comp. Neurology*, 405, 543, 1999.

75. Kondoh, Y. et al., Evolution of sexual dimorphism in the olfactory brain of Hawaiian *Drosophila, Proc. R. Soc. Lond.*, 270, 1005, 2003.
76. Kim, M.S., Repp, A. and Smith, D.P., LUSH Odorant-Binding Protein mediates chemosensory responses to alcohols in *Drosophila melanogaster, Genetics*, 150, 711, 1998.
77. Nothiger, R. and Steinman-Zwicky, M., Sex determination in *Drosophila, Trends in Genetics*, 1, 209, 1985.
78. Baker, B., Sex in flies: the splice of life, *Nature*, 340, 521, 1989.
79. Christiansen, A.E. et al., Sex comes in from the cold: the integration of sex and pattern, *Trends in Genetics*, 18, 510, 2002.
80. Hotta, Y. and Benzer, S., Courtship in *Drosophila* mosaics: sex-specific foci for sequential action patterns, *Proc. Natl. Acad. Sci. USA*, 73, 4154, 1976.
81. Hall, J.C., Control of male reproductive behavior by the central nervous system of *Drosophila*: dissection of a courtship pathway by genetic mosaics, *Genetics*, 92, 437, 1979.
82. Jallon, J.M. and Hotta, Y., Genetic and behavioural studies of female sex appeal in *Drosophila, Behav. Genet.*, 9, 257, 1979.
83. McKeown, M., Belote, J.M. and Boggs, R.T., Ectopic expression of the female transformer gene leads to female transformation of chromosomally male *Drosophila, Cell*, 53, 887, 1988.
84. Ferveur, J.F. et al., Genetic feminization of brain structures and changed sexual orientation in male *Drosophila, Science*, 267, 902, 1995.
85. Arthur, B.I. et al., Sexual behavior in *Drosophila* is irreversibly programmed during a critical period, *Cur. Biol.*, 8, 1187, 1998.
86. Hall, J.C., Courtship among males due to a male-sterile mutation in *Drosophila melanogaster, Behav. Genet.*, 8, 125, 1978.
87. Gailey, D. and Hall, J.C., Behavior and cytogenetics of fruitless in *Drosophila melanogaster:* different courtship defects caused by separate closely linked lesions, *Genetics*, 121, 773, 1989.
88. Villella, A., Extended reproductive roles of the fruitless gene in *Drosophila melanogaster* by behavioural analysis of new fruitless mutants, *Genetics*, 147, 1011, 1997.
89. Ito, H. et al., Sexual orientation in *Drosophila* is altered by the satori mutation in the sex-determination cascade gene fruitless that encodes a zinc finger protein with a BTB domain, *Proc. Natl. Acad. Sci. USA*, 93, 9687, 1996.
90. Ryner, L.C. et al., Control of male sexual behavior and sexual orientation in *Drosophila* by the fruitless gene, *Cell*, 87, 1079, 1996.
91. Demir, S. and Dickson, B.J., fruitless splicing specifies male courtship behaviour in *Drosophila, Cell*, 121, 785,2005.
92. Manoli, D.S. et al., Male-specific fruitless specifies the neural substrates of *Drosophila* courtship behaviour, *Nature*, 436, 395, 2005.
93. Stockinger, P. et al., Neural circuitry that governs *Drosophila* male courtship behavior, *Cell*, 121, 795, 2005.
94. Manoli, D.S. and Baker, B., Median bundle neurons coordinate behaviours during *Drosophila* male courtship, *Nature*, 430, 564, 2004.
95. Lee, G. and Hall, J.C., Abnormalities of male-specific FRU protein and serotonin expression in the CNS of fruitless mutants in *Drosophila, J. Neurosc.*, 21, 513, 2001.
96. Kimura, K.I. et al., Fruitless specifies sexually dimorphic neural circuitry in the *Drosophila* brain, *Nature*, 438, 229, 2005.

23 A New Era for *Drosophila* Learning and Memory Studies

Daniel Comas, Guillaume Isabel and Thomas Préat

CONTENTS

INTRODUCTION

Drosophila melanogaster is one of the most intensively studied organisms in biology, and it serves as a model system for the investigation of many cellular, developmental and behavioral processes common to other species, including humans. This holds true in particular for brain studies, as *Drosophila* central nervous system is made of neurons and glia that operate on the same fundamental principles as their mammalian counterparts. Thus, most neurotransmitters are identical in flies and humans, and despite the fact that the *Drosophila* brain has only 100,000 cells,[1] it produces complex behaviors and sustains various forms of learning and memory. Besides studies of fundamental brain properties, fly models are being developed for a variety of neurodegenerative disorders, and the field is beginning to harness the power of *Drosophila* genetics to dissect pathways of disease pathogenesis and identify potential targets for therapeutic intervention. This approach is possible because about 50% of human genes have a *Drosophila* ortholog.[2] In addition, transgenic strategies that allow to introduce human genes into *Drosophila* continue to expand the list of modeled diseases, which now includes Parkinson's disease, Alzheimer's disease, Huntington's disease and several spinocerebellar ataxias.[3,4]

During the past few decades, the pool of molecular genetics techniques that apply to the fruit fly has increased enormously. It has been a long time since the selection of randomly and chemically produced mutants was the most common technique to analyze genes function. Nowadays, the extensive use of transposable element-induced mutations allows a much faster identification and study of genes involved in a given developmental or physiological pathway. Once combined with the systematic generation of high-resolution deletions, this global approach should eventually provide researchers with a complete mutant collection (see Table 23.1).[5,6]

Moreover, new techniques have arisen that permit one to directly disturb the expression of a defined gene, without the need of performing a mutagenesis. Rong[7-9] and colleagues have developed the technique of homologous recombination in *Drosophila*. This technique, however, remains heavy. The development of the interference RNA (RNAi) combined with the P[GAL4] enhancer-trap system allows one to inhibit the expression of any gene of interest in almost any group of cells—for example, in a particular brain circuit.[10]

Drosophila genome is one of the first multicellular eukaryotic genomes to be completely sequenced, together with *Caenorhabditis elegans*.[11,12] Genes prediction has been carried out by two different consortiums (Berkeley *Drosophila* Genome Project, BDGP, and Heidelberg FlyArray) with some interesting small differences in the total gene number and limits.[13,14] The results of these international collaborations have been made public in different Web sites (Table 23.1). Flybase offers most of the information available for each of the 14,000 to 16,000 predicted *Drosophila* genes (Humans are predicted to have about 20,000 to 25,000 protein-coding genes).[15,16] One can find in Flybase not only the sequence of the different transcripts and proteins, but also the different mutant alleles and phenotypes, predicted protein domains, molecular function, similarities within other species, publications referring to this gene, and more.

The whole-genome sequencing strategy has opened a new era of molecular strategies such as the studies of the transcriptome and protein interactome.[17,18] Several papers report the use of microarrays to study physiological pathways involved in circadian rhythms, immunity, sex differences, or memory.[19-24] Recently, White and collaborators used probe sequences tiled throughout the genome.[14] More than 170,000 oligonucleotides were designed against each exon, intron, and intergenic region. Their purpose was not only to detect the differential expression of all genes and their alternative mRNAs throughout development, but also to identify new, transcribed regions.

Giot and colleagues[25] published a two-hybrid–based, protein–protein interaction map of the fruit fly proteome. They produced a draft map of 7,048 proteins and 20,405 interactions, each with an associated rate of confidence. The complete description of the proteome (expression pattern of all proteins and their interactions) is thought to provide the mechanistic basis for much of the physiology.

The exponential growth in the volume of accessible biological information has generated a plurality of voices surrounding the annotation of genes and their products. The Gene Ontology project (see Table 23.1) seeks to provide a unified set of structured vocabularies to describe gene products in any organism. This work includes building three extensive ontologies to describe molecular function, biological process, and cellular component, and providing a community database resource that supports the use of these ontologies. The annotated genome sequence

TABLE 23.1
Drosophila's Web Resources

Web Site	Brief Description	What to Search?
FLYBASE http://www.flybase.org	A database of the *Drosophila* genome. The core Internet resource for *Drosophila* researchers.	Genes, mutants, stocks, researchers, publications, etc.
BDGP (BERKELEY *DROSOPHILA* GENOME PROJECT) http://www.fruitfly.org	Consortium of the *Drosophila* Genome Center to generate and maintain biological annotations of its genome sequence. 1. Producing gene disruptions using P element-mediated mutagenesis. 2. Characterizing the sequence and expression of cDNAs. 3. Developing informatics tools that support the experimental process, identify features of DNA sequence, and allow them to present up-to-date information about the annotated sequence to the research community.	*Drosophila* sequence, new releases, expression patterns, cDNA and EST, natural transposable elements, gene disruption, comparative genomics, methods, etc.
The Heidelberg FlyArray http://www.hdflyarray.zmbh. uni-heidelberg.de	A database of the *Drosophila* genome. A comparison between BDGP, Heidelberg gene prediction, and the *in situ* hybridisation patterns for Heidelberg unique genes.	Genes, expression patterns, phylogenetics, P-element insertions, Fly Arrays, etc.
Drosophila Genomics Resource Center http://www.dgrc.cgb.indiana. edu	1. Collecting and distributing DNA clones, 2. collecting and distributing cell lines, 3. manufacturing and distributing microarrays, 4. assisting scientists to use these materials, 5. developing and testing emerging genomics technologies.	Cell lines, clones, vectors, microarrays, etc.
DroSpeGe http://www.species.flybase. net/	*Drosophila* species genomes. A service of FlyBase and Genome Sequencing Centers. This service provides a preview of *Drosophila* genome data, with genome maps and BLAST sequence search, for several *Drosophila* species.	Comparison between *Drosophila* species genomes.
Gene Ontology http://www.geneontology.org	Collaborative effort to address the need for consistent descriptions of gene products in different databases. Consortium to produce a controlled vocabulary that can be applied to all organisms even as knowledge of gene and protein roles in cells is accumulating and changing. There are three organizing principles of GO: molecular function, biological process and cellular component.	Biological process and cellular compartment of your protein of interest.

(Continued)

TABLE 23.1
Drosophila's Web Resources (*Continued*)

Bloomington http://www.flystocks.bio.indi ana.edu	The main *Drosophila* stock center: with a collection of P-element strains, EP strains, EMS, natural produced mutants, etc.	Mutant *Drosophila* strains.
Szeged *Drosophila* Stock Center http://www.expbio.bio. u-szeged.hu/fly	EPs *Drosophila* stock collection. *Drosophila* strains in order to overexpress genes.	EPs strains.
Tucson Stock Center http://www.stockcenter.arl.ar izona.edu/	The Tucson *Drosophila* stock center provides subcultures of approximately 270 different *Drosophila* species.	*Drosophila* species.
DrosDel http://www.drosdel.org.uk/	*Drosophila* deletion stock center. Construction of a new *Drosophila* deletion collection using RS (re-arrangement screen) P-elements and a European *Drosophila* network.	*Drosophila* deletion collection
Flytrap http://www.fly-trap.org/	HTML-based gene expression database.	A database of P(Gal4) strains and its expression within the adult brain.
GETDB http://www.flymap.lab.nig.ac .jp/%7Edclust/getdb.html	*Drosophila* stock center. GETDB is an integrated database compiling insertion site, expression pattern and mutant phenotype of a large collection (6966) of Gal4 enhancer-trap lines generated and analyzed by NP Consortium.	Gal4 enhancer-trap strains and its expression pattern.
Exelixis *Drosophila* Stock Collection http://www.*Drosophila*.med. harvard.edu	*Drosophila* stock center. New collection stocks generated by Exelixis and distributed to Bloomington and Harvard.	*Drosophila* P-element strains.
GENEXEL http://www.genexel.com/eng /htm/ genisys.htm	*Drosophila* stock center. GeniSys database of *Drosophila* EP lines.	EP strains (charge).
P-SCREEN Database http://www.flypush.imgen.bc m.tmc.edu/pscreen	Another P-element *Drosophila* database from the BDGP gene disruption project.	*Drosophila* P-element strains.
Flytrap http://www.flytrap.med.yale. edu	GFP Protein Trap database. Flytrap acts as a central data repository for *Drosophila* transgenic lines being created using an intron protein-trap strategy.	Micrographs, DNA sequence and mapping information from the inserted P-element, and expression patterns of the tagged proteins.

FlyMine http://www.flymine.org	An integrated database for *Drosophila* and *Anopheles* genomics. The FlyMine project is building an integrated database of genomic, expression and protein data for *Drosophila* and *Anopheles* and making this available to the worldwide research community.	Gene comparisons between *D. melanogaster, A. gambiae* and *C. elegans.*
FlyRNAi http://www.flyrnai.org	*Drosophila* RNAi screening center. A library of double-stranded RNAs directed against all predicted open reading frames. This resource can now be used to conduct high-throughput cell-based RNAi screens to identify genes involved in various assays.	A facility center to do genome-wide RNAi screens and the creation of a database of information.
Flybrain http://www.flybrain. uni-freiburg.de	*Drosophila*'s brain. An online atlas and database of the *Drosophila* nervous system.	Atlas of the *Drosophila* Brain, Genetic Dissection of the Brain, Developmental Studies, The Dissectable Brain, 3D Models of the Brain.
FlyView http://www.flyview. uni-muenster.de	*Drosophila* Image database. Image database on *Drosophila* development and genetics, especially on expression patterns of genes (enhancer-trap lines, cloned genes).	Compare images on the computer screen and search for special patterns at different developmental stages.
FlyMove http://www.flymove. uni-muenster.de	Development image database. Internet resource to study *Drosophila* development.	Images and videos of *Drosophila* development.
Interactive fly http://www.sdbonline.org/fly /aimain/1aahome.htm	*Drosophila* development. A cyberspace guide to *Drosophila* genes and their roles in development.	Information about gene role during development. Biochemical pathways, organs, images and brain genes.
CURAGEN Corporation http://www.portal.curagen.com/ cgibin/interaction/ flyHome.pl	The world's first comprehensive protein interaction map for a multicellular organism. Description of the *Drosophila* two-hybrid interactome pathways.	Interactions between *Drosophila* proteins.
Homophila http://www.superfly.ucsd. edu/homophila	Human disease to *Drosophila* gene database. Homophila utilizes the sequence information of human disease genes in order to determine if sequence homologs of these genes exist in the current *Drosophila* sequence database.	*Drosophila* homologs of human diseases genes.

of *Drosophila melanogaster* and of *Drosophila pseudoobscura*, together with the associated biology, provides the foundation for a new era of sophisticated *in silico* and *in vivo* functional studies.

23.1 VISUALIZING NEURONAL CIRCUITS THANKS TO GENETICS

A powerful mean of revealing neuronal architecture in the *Drosophila* brain (Figure 23.1) is the GAL4/UAS enhancer-trap technique[26–28] that enables a selective activation of any cloned gene in a wide variety of tissues and cells (Figure 23.2). An enhancer-trap element is a transposon containing an exogenous gene, such as the yeast transcriptional activator GAL4. Insertions that occur close to a transcriptional enhancer cause the GAL4 gene to be expressed in a pattern reflecting the enhancer's spatio–temporal regulatory properties. Thousands of lines carrying an insertion of the P[GAL4] element are available. With a single cross, flies are created that carry a second P-element with a reporter gene inserted downstream of GAL4 binding sites: the upstream activating sequences (UASs). The reporter gene is expressed in the same cells as GAL4. This flexible system allows, for example, one to label a particular group of cells if the reporter gene encodes the green fluorescent protein (GFP) (Figure 23.2). Using this approach, Yang and colleagues[29] characterized intrinsic cells of the *Drosophila* memory center as the mushroom bodies (MBs) (Figure 23.1). Rather than

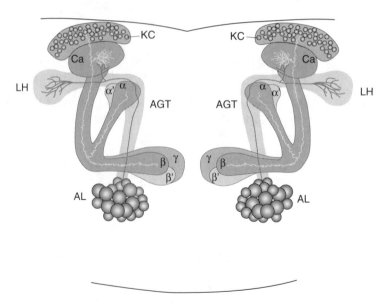

FIGURE 23.1 (SEE COLOR INSERT FOLLOWING PAGE 236) *Drosophila* olfactory memory system. Olfactory sensory neurons project to the antennal lobes (ALs). From there, projection neurons project through the antenno-glomerular tract (AGT) and connect mushroom body (MB) dendrites localized in the calyx (Ca), as well as the lateral horn (LH). Each MB is composed of about 2,500 neurons, the Kenyon cells (KC). Three types of KC project in five lobes: α/β, α'/β' and γ.

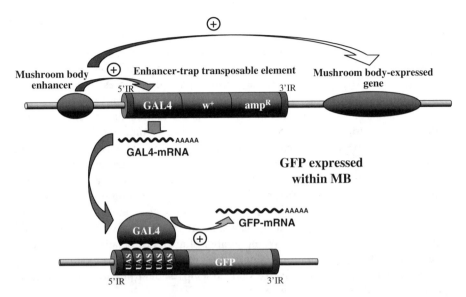

FIGURE 23.2 (SEE COLOR INSERT) The GAL4/UAS system. When an enhancer-trap transposable element is inserted near transcription enhancers that control expression in a given structure, the GAL4 gene that is contained in the P-element is expressed in the same structure. If this fly contains also a reporter gene downstream of the UAS sequences, GAL4 will transcribe the reporter.

seeing homogenous neurons, they found the 5,000 MB neurons to be compound neuropils in which parallel subcomponents exhibit discrete patterns of gene expression.[30,31] Parallel channels of information flow, perhaps with different computational properties, subserve different roles (see below).

Using the enhancer-trap system, Ito and colleagues[32] searched for MBs extrinsic neurons. They used a reporter construct encoding the presynaptic protein *neuronal synaptobrevin-green fluorescent protein*. They showed that output MB neurons are scarce, and that, surprisingly, few MB extrinsic neurons project to the deutocerebrum, the premotor pathway immediate modifier of behavior.

23.2 IMAGING BRAIN ACTIVITY

One major caveat of *Drosophila* central brain studies is that direct electrophysiological analysis is scarce,[33] due mainly to the small size of neuron cell bodies (less than 5 μm in diameter). To circumvent this difficulty, two partially alternative approaches have been followed: analysis of learning memory mutants at the neuro–muscular junction,[34,35] which leaves out the complexity of brain physiology, as well as the analysis of isolated MB neurons in culture.[36] Those experimental systems can provide interesting molecular and cellular information, but they are inadequate for assessing neuronal function at the level necessary for a global understanding of memory systems.

Odor processing occurs in a complex-tissue environment, and identification of the repertoire of brain cell assemblies involved in olfactory memory requires visualization of the network activity at high spatial and temporal resolution, in preparations as intact as possible. Optical neural activity recordings allow study of active brains with micrometer-spatial resolution, and activity-sensitive fluorescent probes have been recently used in *Drosophila*. Interestingly, those sensors are proteins and therefore their expression can be driven to specific subsets of neurons with the GAL4/UAS system. Several sensors have been successfully brought to *Drosophila* that monitor the local change of pH that accompanies neurotransmitter release,[37] or changes in the intracellular calcium concentration that provide a valuable indicator of electrical activity (Aequorin;[38] Cameleon;[39] Camgaroo;[40] G-CaMP[41,42]). Using the G-CaMP reporter and two-photon microscopy, stereotyped odor-evoked patterns have been observed in the antennal lobe glomeruli[41] and in the MBs.[42]

The ultimate goal of imaging studies is to build a functional map of cell assemblies encoding memory in different regions of *Drosophila* brain, by comparing the activity of trained and naïve animals, in normal flies or memory mutants. A first step achieved recently as a transient change in the spatial code was observed in the antennal lobe of wild-type flies 3 minutes after olfactory associative conditioning.[37] The field of *Drosophila* brain imaging is only emerging, and the extensive use of various activity-sensors and diverse optical setups should provide us with major information within the next few years.

23.3 LEARNING AND MEMORY MUTANTS

We do not intend here to review in detail all *Drosophila* learning and memory genes or the various conditioning protocols.[43–46] Rather, we would like to show how recent genetic technologies greatly facilitate the identification and study of learning and memory genes. The first *Drosophila* memory mutants were issued from random chemical mutagenesis using ethyl methane sulfonate (EMS), a potent mutagen. But because EMS induces mostly subtle DNA alterations, the identification of the mutated gene is often difficult. The first step consists of genetically mapping the mutation by complementation test, using *Drosophila*'s deletion collection. However, this only defines a broad area covering many other genes. On the contrary, the use of a marked transposable P-element allows one to easily identify the gene of interest.

In the first behavioral screen, about 4,000 fly stocks carrying EMS-induced mutations were tested for their ability to learn in an olfactory conditioning assay.[47,48] The first *Drosophila* learning and/or memory mutants were *dunce* (*dnc*) and *rutabaga* (*rut*). Biochemical defects were observed in *rut* (reduced cAMP level) and *dnc* (increased cAMP level).[49] Nevertheless, the cloning of the responsible gene took years and was facilitated by the use of new P-induced alleles. For example, description of P-element insertions within 200 nucleotides of where the *rut* transcription started, identified *rut* as the structural gene for the Ca^{2+}/Calmodulin-responsive adenylate cyclase.[50,51] New EMS-induced *dnc* mutants came from a screen for female sterility mutants.[52] The *dnc* gene was identified by recombinational mapping of *dnc* mutations with restriction site polymorphisms as genetic markers.[53]

The first *amnesiac* (*amn*) mutant was identified in a screen for flies with affected memory.[48] However, the *amn* gene was cloned as a second site suppressor of the *dnc* female sterility phenotype from a P-element-induced allele, and has been repeatedly isolated since.[54–56] The *amn* encodes a putative adenylate-cyclase activating peptide.

Another learning and/or memory mutant is *radish* (*rsh*).[57] The *rsh* gene was localized within a 180-kb interval in the 11D-E region of the X chromosome, and several candidate genes were identified.[57] Recently, Chiang and colleagues[58] reported that the responsible gene for *rsh* phenotype was a phospholipase A2. However a second team has reported that *rsh* encodes a novel protein with possible nuclear localization motifs.[46,59] This discrepancy illustrates the difficulty linked to working with EMS-induced behavioral mutants.

P-element–based behavioral screens for learning and memory mutants have also been performed. Various mutants were issued from these mutagenesis, including *nalyot* (a myb-related Adf1 transcription factor),[60,61] *leonardo* (a zeta isoform of the 14-3-3 protein),[62] and *volado* (two splice variants of an α-integrin).[63]

Dubnau and colleagues[24] recently performed a behavioral screen for long-term memory mutants, in parallel to microarray experiments aimed to select genes with altered expression after long-term memory training. This work led to the identification of proteins involved in mRNA processing and translation.[24]

We recently described *crammer* (*cer*), a gene involved specifically in the setting up of long-term memory.[64] The P[GAL4] *cer* strain has a reduced long-term memory but a normal short-term and middle-term memory. Interestingly, in the wild-type strain, *cer* is transiently underexpressed three hours after long-term memory training. As the Cer peptide is an inhibitor of cysteine proteinases, the decrease in its expression shortly after intensive training must lead to a transient activation of its cysteine proteinase(s) target(s).[64] Altogether these works demonstrate the power of the forward-genetic approach: learning or memory mutants are first characterized without knowing in advance the molecular function of the gene involved.

Last, some genes have been shown to be involved in *Drosophila* learning or memory due to function similarity to well-described genes in other species. For example, NF1, a GTPase-activating protein for Ras, is linked to the human disease neurofibromatosis that sometimes leads to learning disabilities.[65] *Drosophila NF1* mutants show a defect in olfactory learning.[66] Notch, a critical component of an evolutionarily conserved signaling mechanism that regulates neural development, is involved in hippocampal synaptic plasticity and in learning and memory in mice.[67,68] It was recently shown that Notch signaling is required for long-term memory formation in adult *Drosophila*.[69,70]

23.4 TEMPORAL CONTROL OF GENE EXPRESSION

When a gene is mutated, the biochemical function of the protein may be affected, or the transcription level of the normal mRNA may be decreased. The resulting biochemical defect may occur during development and/or adult life. Thus, a learning or memory defect or both may be due to structural brain defects rather than to a specific physiological alteration. This issue is crucial as even a faint developmental defect may have major consequence on brain function.

Two recent techniques, developed by the laboratory of Ronald Davis, enable one to discriminate a role in brain development from a role in memory formation per se. TARGET and Gene-Switch[71] are variants of the invaluable GAL4/UAS expression system. TARGET, which stands for Temporal and Regional Gene Expression

Targeting, uses a modified temperature-sensitive yeast, GAL80 transcription factor (GAL80^{ts}), to repress GAL4 activity (Figure 23.3). At low temperature (typically 18°C), the functional GAL80^{ts} protein inhibits GAL4 transcriptional activity. At 30°C, GAL80^{ts} becomes inactive and allows the GAL4-mediated transcription (Figure 23.3). This tool can be used to rescue the defect of a memory mutant by expressing the normal protein in specific regions of the adult brain, while preventing its expression during development. Conversely, if the UAS downstream gene expresses a double-stranded RNA (RNAi), one will be able to lower the expression of a given gene in specific adult circuits (Figure 23.3). Thus, the TARGET technique allows one to increase or decrease the expression of a precise gene in a temporal and regional manner.

Gene-Switch uses a different mechanism to temporally control GAL4 activity.[72] The GAL4 DNA-binding domain was fused to a mutated progesterone receptor ligand-binding domain and part of NF-B p65 activation domain. This synthetic transcription factor is active only when the antiprogestin mifeprestone (RU486) is present. When flies are fed RU486, Gene-Switch is on and the UAS-transgene is expressed. As with the GAL4/UAS system, region-specific Gene-Switch activity is accomplished by the use of a specific enhancer.[73,74]

Finally, we recently developed a new GAL4 tool using an inducible chimerical protein that carries the suppressor-of-hairy-wing repressor domain fused to the

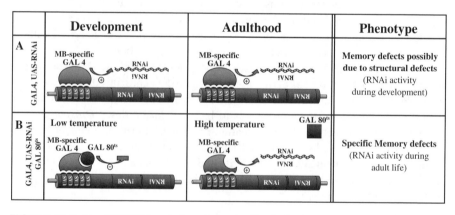

FIGURE 23.3 The TARGET technique helps in discriminating between a developmental gene action from an adult physiological effect. A sequence encoding an inverted repeat RNA (RNAi) is placed under upstream activating sequence (UAS) control. (A) GAL4 activates RNAi expression during development and adulthood. A memory defect could be caused indirectly by structural defects occurred during development. (B) The TARGET system solves this problem. Flies are maintained at low temperature during their developmental stages. At this temperature, GAL80^{ts} inhibits the transcription activity of GAL4, so no RNAi is expressed. When adults emerge, they are transferred at permissive temperature (30°C); the GAL80^{ts} protein becomes ineffective and the GAL4 activates the transcription of the RNAi in adult MBs. The putative memories defects are due to the decreased concentration of the mRNA targeted by the RNAi.

GAL4 DNA-binding domain. This tool allows also a temporally controlled repression of genes located near a P[UAS] insertion.[75]

23.5 A TOOL TO BUILD ANATOMO-FUNCTIONAL MAPS

A sophisticated approach to disturb neuronal circuits, based on a rapid and reversible blockage of synaptic transmission was developed by Kitamoto.[76] The *shibire (shi)* gene encodes a microtubule-associated GTPase, Dynamin, which is involved in endocytosis and is essential for synaptic vesicle recycling and maintenance of the readily releasable pool of synaptic vesicles.[77] The temperature-sensitive allele, shi^{ts1}, is defective in vesicle recycling at restrictive temperature (>29°C), resulting in a rapid blockage of synaptic transmission. The shi^{ts1} mutation has a dominant effect because it blocks chemical synapses even in the presence of a normal shi^+ allele. Expression of shi^{ts1} tool can therefore be used within the GAL4/UAS system. The GAL4/UAS-shi^{ts1} approach is very powerful as it allows inhibiting particular brain circuits at a precise time, for example during conditioning or memory retrieval. Thus, it has been established that abolishing vesicle-mediated secretion from the dorsal-paired medial (DPM) cells, which normally express *amn,* phenocopies *amn* mutant memory defect.[78]

The tools described so far allow a temporal control of gene expression and neuronal activity. Their recent development has opened the way for a dynamic analysis of memory systems.

23.6 LOCALIZATION OF OLFACTORY MEMORY IN *DROSOPHILA* MBs

As mentioned, the insect brain contains a pair of prominent and characteristically shaped neural centers known as MBs. In *Drosophila*, MBs are composed of three main classes of neurons whose axons divide to form two vertical lobes (α and α') and three median ones (β, β'and γ)[30] (Figure 23.1). MBs are a specialized neuropile involved in processing and storing multimodal sensory information.[79] In the 1970s and 1980s, the function of the MBs was assessed in different insect species with the classical interventionism approaches—cooling and ablation.[80,81] Over the years, MBs have been implicated in olfactory learning and memory, and in a variety of complex functions including courtship, motor control, and spatial recognition.

In *Drosophila*, a single associative-learning trial (named below *the short protocol*), consisting of an odor (the conditional stimulus, or CS) accompanied by electric shocks (the unconditional stimulus, or US), induces olfactory learning and memory. Brain mutants with MBs structural defects were isolated in the 1980s. The use of an unlimited number of mutant animals with the same anatomical defect without interventionism helped in highlighting the role of MBs in olfactory learning and memory.[82] However, one caveat of this approach is that the anatomical defects are not specific to the MBs.[83] An alternative approach was used to generate flies without MBs.[83] Hydroxyurea, an antimitotic agent, was fed to newly hatched wild-type larvae. At that early developmental stage, only five neuroblasts—the neurons' precursor cells—are mitotically active within each brain hemisphere: the four that generate the MBs, and one in the antennal lobe. Thus, hydroxyurea treatment leads to viable adult flies with almost no MBs. These flies showed no olfactory memory.[83] The essential role of MBs in olfactory memory is further outlined by the the high expression in the MBs of many of the proteins involved in learning and memory such as Dnc,[84] Rut_{51} or DCO.[85] Even though Amn_{48} is not expressed in the MBs, it is expressed by the dorsal-paired median neurons that project onto the MBs.[78]

Thus, *Drosophila* MBs are required for olfactory learning and memory, and insights from the first mutants have indicated that the cAMP pathway plays a key role for memory establishment. Nevertheless, it remained to be proven that cAMP metabolism was strictly required in MBs for correct learning and memory. Thanks to the GAL4/UAS system,[26] Connolly and colleagues[86] disrupted MBs cAMP signaling by expressing a constitutively activated G-protein. Permanent adenylyl cyclase activation led to an impaired associative memory.

That memory is impaired after ablation or functional disruption of a brain structure does not necessarily imply that the memory trace itself is localized within this structure. To localize short-term memory (STM), it was necessary to rescue the memory abilities of a mutant by expressing the corresponding protein in a specific brain structure. In a first step, Zars and colleagues[87] expressed Rut Ca^{2+}/Calmodulin adenylyl cyclase in MBs to restore a normal learning capacity. However, in this experiment, Rut was expressed in the MBs at the adult stage and also during development. Thus, the behavioral rescue might have been due to the correction of an MB developmental defect, indirect cause of the learning defect. Thanks to the TARGET and SWITCH methods (see above), McGuire and colleagues[88] showed that the presence of Rut in adult MBs alone was sufficient to rescue *rut* memory defect. The current view is that Rut adenylyl cyclase could act during learning as a coincidence detector for the US and CS.

Are the MBs required during the acquisition, consolidation, or retrieval phase? It was possible to address this question by expressing the temperature sensitive Shi^{ts1} protein specifically in MBs neurons to transiently disrupt synaptic neurotransmission.[89,90] It was shown that the synaptic outputs of MB neurons are required during retrieval of the STM but not during acquisition or consolidation. All together, these data indicate that a STM trace can be localized in MBs.

In addition to STM, *Drosophila* can display long-term memory (LTM) after a spaced and repeated conditioning (long protocol).[91] LTM can last several days and depends on *de novo* protein synthesis. Are MBs implicated in *Drosophila* LTM? To answer this question we analyzed the *alpha-lobes absent* (*ala*) mutant that shows a peculiar MB phenotype. Ten percent of *ala* individuals possess all five MBs lobes, 36% lack the horizontal β and β' lobes, and 4.5% lack vertical α and α' lobes (the remaining sub-populations presented different MB phenotypes in the left and the right hemispheres).[92] *ala* mutant flies were trained with the short or the long protocol, and we analyzed separately the brains of flies that had made the correct and the wrong choices during the memory test, to calculate the memory score of each class of *ala* mutants. Flies lacking α/α' lobes displayed no LTM although their

STMs were wild-type. Flies with all lobes present or without β/β' lobes had a normal STM and LTM. Thus, MBs are necessary to perform LTM, and more particularly the α/α' vertical lobes.[92] By expressing Shi[ts1] in α/β lobes, it was further shown that α lobes outputs are required during LTM retrieval.[93]

If MBs play a pivotal role in *Drosophila* olfactory learning and memory, a recent study suggests that brain structures located outside the MBs also participate in olfactory memory.[94] We have recently shown that the *Drosophila* brain is asymmetric, as a small structure expressing Fasciclin II—the asymmetric body—is present only in the right hemisphere. Interestingly, about 7% of Canton-Special wild-type flies present a bilateral structure. Those symmetric flies showed no LTM at 4 days, suggesting that asymmetry may be required for generating, maintaining, or retrieving LTM.[94] The asymmetric body is located near the central complex, a structure that connects brain hemispheres. However, the functional links between the asymmetric body and other brain circuits remain to be sorted out.

23.7 DYNAMIC OF OLFACTORY MEMORY PHASES IN *DROSOPHILA*

The short protocol induces two labile phases: STM, which is disrupted in mutants affected for cAMP metabolism and lasts about 30 min; and middle-term memory (MTM), which is disrupted in *amn* and lasts for a few hours. STM and MTM are anesthesia sensitive as they are erased if flies are cooled down to 4°C after conditioning. This property suggests that STM and MTM are sustained by electrical brain activities.

Drosophila also displays two long-lasting memory phases, anesthesia-resistant memory (ARM)[95] and LTM.[96] ARM is generated by the short protocol or several massed trials. ARM can still be detected after a few days.[91] This memory is disrupted in *rsh* mutants.[57] The molecular pathways involved in ARM formation remain largely unknown, in part because *rsh* gene identity is debated (see above). The atypical protein kinase M (aPKM) is a persistently active truncated isoform of atypical protein kinase C (aPKC). Over-expression of aPKM enhances memory after massed conditioning, but not after spaced training.[97] Moreover, inhibition of aPKM disrupts consolidated memory after massed conditioning.[97] It is therefore conceivable that aPKC is a molecular support of ARM.

LTM is induced by the long protocol and measured for at least one week.[91] Like ARM, LTM is anesthesia resistant.[96] It is disrupted by a cAMP response binding protein (dCREB2-b) repressor,[98,99] likely due to the inhibition of gene expression required to establish LTM. Yin and colleagues[100,101] reported that flies over-expressing a particular CREB isoform (dCREB2-a) generated LTM after a single training cycle. However, this work could not be replicated, and it was shown that the original dCREB2-a transgenic flies carried an accidental mutation that produced a truncated protein with no DNA binding domain.[99] Moreover, adult induction of the correct CREB2-a isoform led to lethality.[99]

What is the dynamic of memory phases in *Drosophila*? We proposed recently a model that involves two parallel memory pathways, one with cAMP dependent

STM/MTM, and the other with ARM (Figure 23.4A). Indeed, *dnc* and *rut* retain a significant level of early memory,[102] suggesting that an adenyl cyclase-Rut independent learning might exist. Moreover, ARM levels in *rut* and *amn* are near normal,[57,93] while their labile memories are strongly affected. Thus ARM does not seem to depend on STM/MTM. Instead a second learning process could give rise to STM2 phases and later to ARM (Figure 23.4).

What are the relationships between ARM and LTM? To answer that question, the *ala* mutant was trained with the long protocol, and the memory of flies lacking vertical α/α' lobes was measured at 30 minutes and 5 hours after the training. Thirty-minute memory was normal, but, surprisingly, 5-hour memory was close to zero. Memory performance was normal at 5 hours when flies without vertical lobes were trained with the short protocol[93] (Figure 23.5). Why does a longer training give rise to a weaker memory? *ala* flies display no LTM because they lack the vertical lobes, the center for LTM. These flies show a normal ARM at 5 hours after the short protocol, but they show no ARM after the long protocol. This result suggests that ARM is erased after LTM conditioning. Thus the consolidated memory phases generated by olfactory conditioning are exclusive (Figure 23.6).[93] Why is ARM erased after LTM conditioning? We propose that ARM could act as a gating mechanism for LTM formation, avoiding a heavy cascade of gene expression in absence of intensive, spaced conditioning. Despite the relative simplicity of the *Drosophila*

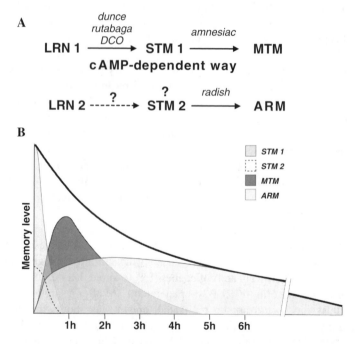

FIGURE 23.4 (A) Model of associative memory phases and (B) temporal dynamics of memory phases generated by a single cycle of conditioning (short protocol). LRN: learning; STM: short-term memory; MTM: middle-term memory; ARM: anesthesia-resistant memory.

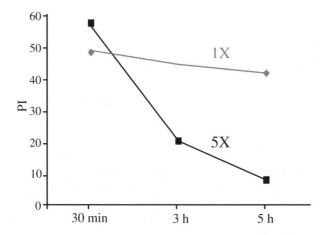

FIGURE 23.5 In flies without MB alpha lobes that normally sustain long-term memory, the long protocol decreases memory performance at 5 hours in comparison with the short protocol. Grey line: short protocol; black line: long protocol.

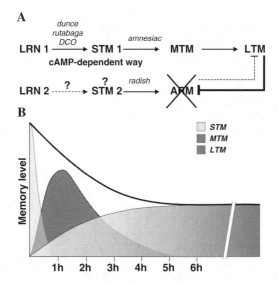

FIGURE 23.6 (A) Model of associative memory phases and (B) temporal dynamics of memory phases generated by five spaced cycles of conditioning (long protocol). LRN: learning; STM: short-term memory; MTM: middle-term memory; ARM: anesthesia-resistant memory; LTM: long-term memory.

brain, this model suggests a cognitive complexity more frequently associated with mammalian models. It supports the idea that *Drosophila* is a valid model in which to study some of the molecular and cellular mechanisms involved in normal or pathological human memory.[3]

REFERENCES

1. Shimada, T. et al. Analysis of the distribution of the brain cells of the fruit fly by an automatic cell counting algorithm, *Physica A* 350, 144, 2005.
2. Rubin, G.M. et al. Comparative genomics of the eukaryotes, *Science* 287 (5461), 2204, 2000.
3. Shulman, J.M. et al. From fruit fly to bedside: translating lessons from *Drosophila* models of neurodegenerative disease, *Curr Opin Neurol* 16 (4), 443, 2003.
4. Bier, E. *Drosophila*, the golden bug, emerges as a tool for human genetics, *Nat Rev Genet* 6 (1), 9, 2005.
5. Parks, A.L. et al. Systematic generation of high-resolution deletion coverage of the *Drosophila melanogaster* genome, *Nat Genet* 36 (3), 288, 2004.
6. Thibault, S.T. et al. A complementary transposon tool kit for *Drosophila melanogaster* using P and piggyBac, *Nat Genet* 36 (3), 283, 2004.
7. Rong, Y.S. and Golic, K.G. Gene targeting by homologous recombination in *Drosophila*, *Science* 288 (5473), 2013, 2000.
8. Rong, Y.S. et al. Targeted mutagenesis by homologous recombination in D. *melanogaster*, *Genes Dev* 16 (12), 1568, 2002.
9. Rong, Y.S. and Golic, K.G. A targeted gene knockout in *Drosophila*, *Genetics* 157 (3), 1307, 2001.
10. Piccin, A. et al. Efficient and heritable functional knock-out of an adult phenotype in *Drosophila* using a GAL4-driven hairpin RNA incorporating a heterologous spacer, *Nucleic Acids Res* 29 (12), E55, 2001.
11. Adams, M.D. et al. The genome sequence of *Drosophila melanogaster*, *Science* 287 (5461), 2185, 2000.
12. Misra, S. et al. Annotation of the *Drosophila melanogaster* euchromatic genome: a systematic review, *Genome Biol* 3 (12), R1, 2002.
13. Hild, M. et al. An integrated gene annotation and transcriptional profiling approach towards the full gene content of the *Drosophila* genome, *Genome Biol* 5 (1), R3, 2003.
14. Stolc, V. et al. A gene expression map for the euchromatic genome of *Drosophila melanogaster*, *Science* 306 (5696), 655, 2004.
15. Venter, J.C. et al. The sequence of the human genome, *Science* 291 (5507), 1304, 2001.
16. Consortium, I.H.G.S. Finishing the euchromatic sequence of the human genome, *Nature* 431 (7011), 931, 2004.
17. Callahan, C.A. et al. Control of neuronal pathway selection by a *Drosophila* receptor, *Nature* 376 (6536), 171, 1995.
18. Hoskins, R.A. et al. Heterochromatic sequences in a *Drosophila* whole-genome shotgun assembly, *Genome Biol* 3 (12), R1, 2002.
19. Claridge-Chang, A. et al. Circadian regulation of gene expression systems in the *Drosophila* head, *Neuron* 32 (4), 657, 2001.
20. McDonald, M.J. and Rosbash, M. Microarray analysis and organization of circadian gene expression in *Drosophila*, *Cell* 107 (5), 567, 2001.
21. De Gregorio, E. et al. Genome-wide analysis of the *Drosophila* immune response by using oligonucleotide microarrays, *Proc Natl Acad Sci* USA 98 (22), 12590, 2001.
22. Girardot, F., Monnier, V. and Tricoire, H. Genome wide analysis of common and specific stress responses in adult *Drosophila melanogaster*, *BMC Genomics* 5 (1), 74, 2004.
23. Jin, W. et al. The contributions of sex, genotype and age to transcriptional variance in *Drosophila melanogaster*, *Nat Genet* 29 (4), 389, 2001.

24. Dubnau, J. et al. The staufen/pumilio pathway is involved in *Drosophila* long-term memory, *Curr Biol* 13 (4), 286, 2003.

25. Giot, L. et al. A protein interaction map of *Drosophila melanogaster*, *Science* 302 (5651), 1727, 2003.

26. Brand, A.H. and Perrimon, N. Targeted gene expression as a means of altering cell fates and generating dominant phenotypes, *Development* 118 (2), 401, 1993.

27. Wilson, C. et al. P-element-mediated enhancer detection: an efficient method for isolating and characterizing developmentally regulated genes in *Drosophila*, *Genes Dev* 3 (9), 1301, 1989.

28. Bellen, H.J. et al. P-element-mediated enhancer detection: a versatile method to study development in *Drosophila*, *Genes Dev* 3 (9), 1288, 1989.

29. Yang, M.Y. et al. Subdivision of the *Drosophila* mushroom bodies by enhancer-trap expression patterns, *Neuron* 15 (1), 45, 1995.

30. Crittenden, J.R. et al. Tripartite mushroom body architecture revealed by antigenic markers, *Learn Mem* 5 (1–2), 38, 1998.

31. Strausfeld, N.J., Sinakevitch, I. and Vilinsky, I. The mushroom bodies of *Drosophila melanogaster*: an immunocytological and golgi study of Kenyon cell organization in the calyces and lobes, *Microsc Res Tech* 62 (2), 151, 2003.

32. Ito, K. et al. The organization of extrinsic neurons and their implications in the functional roles of the mushroom bodies in *Drosophila melanogaster* Meigen, *Learn Mem* 5 (1–2), 52, 1998.

33. Wilson, R.I., Turner, G.C. and Laurent, G. Transformation of olfactory representations in the *Drosophila* antennal lobe, *Science* 303 (5656), 366, 2004.

34. Zhong, Y. and Wu, C.F. Altered synaptic plasticity in *Drosophila* memory mutants with a defective cyclic AMP cascade, *Science* 251 (4990), 198, 1991.

35. Renger, J.J. et al. Role of cAMP cascade in synaptic stability and plasticity: ultra-structural and physiological analyses of individual synaptic boutons in *Drosophila* memory mutants, *J Neurosci* 20 (11), 3980, 2000.

36. Wright, N.J. and Zhong, Y. Characterization of K+ currents and the cAMP-dependent modulation in cultured *Drosophila* mushroom body neurons identified by lacZ expression, *J Neurosci* 15 (2), 1025, 1995.

37. Yu, D., Ponomarev, A. and Davis, R.L. Altered representation of the spatial code for odors after olfactory classical conditioning; memory trace formation by synaptic recruitment, *Neuron* 42 (3), 437, 2004.

38. Rosay, P. et al. Synchronized neural activity in the *Drosophila* memory centers and its modulation by amnesiac, *Neuron* 30 (3), 759, 2001.

39. Fiala, A. et al. Genetically expressed cameleon in *Drosophila melanogaster* is used to visualize olfactory information in projection neurons, *Curr Biol* 12 (21), 1877, 2002.

40. Yu, D. et al. Detection of calcium transients in *Drosophila* mushroom body neurons with camgaroo reporters, *J Neurosci* 23 (1), 64, 2003.

41. Wang, J.W. et al. Two-photon calcium imaging reveals an odor-evoked map of activity in the fly brain, *Cell* 112 (2), 271, 2003.

42. Wang, Y. et al. Stereotyped odor-evoked activity in the mushroom body of *Drosophila* revealed by green fluorescent protein-based Ca2+ imaging, *J Neurosci* 24 (29), 6507, 2004.

43. Waddell, S. and Quinn, W.G. Flies, genes, and learning, *Annu Rev Neurosci* 24, 1283, 2001.

44. Dubnau, J., Chiang, A.S. and Tully, T. Neural substrates of memory: from synapse to system, *J Neurobiol* 54 (1), 238, 2003.

45. Heisenberg, M. Mushroom body memoir: from maps to models, *Nat Rev Neurosci* 4 (4), 266, 2003.

46. Davis, R.L. Olfactory learning, *Neuron* 44 (1), 31, 2004.

47. Quinn, W.G., Harris, W.A. and Benzer, S. Conditioned behavior in *Drosophila melanogaster*, *Proc Natl Acad Sci USA* 71 (3), 708, 1974.

48. Quinn, W.G., Sziber, P.P. and Booker, R. The *Drosophila* memory mutant amnesiac, *Nature* 277 (5693), 212, 1979.

49. Livingstone, M.S., Sziber, P.P. and Quinn, W.G. Loss of calcium/calmodulin responsiveness in adenylate cyclase of rutabaga, a *Drosophila* learning mutant, *Cell* 37 (1), 205, 1984.

50. Levin, L.R. et al. The *Drosophila* learning and memory gene rutabaga encodes a Ca2+/Calmodulin-responsive adenylyl cyclase, *Cell* 68 (3), 479, 1992.

51. Han, P.L. et al. Preferential expression of the *Drosophila* rutabaga gene in mushroom bodies, neural centers for learning in insects, *Neuron* 9 (4), 619, 1992.

52. Mohler, J.D. Developmental genetics of the *Drosophila* egg. I. Identification of 59 sex-linked cistrons with maternal effects on embryonic development, *Genetics* 85 (2), 259, 1977.

53. Davis, R.L. and Davidson, N. Isolation of the *Drosophila melanogaster* dunce chromosomal region and recombinational mapping of dunce sequences with restriction site polymorphisms as genetic markers, *Mol Cell Biol* 4 (2), 358, 1984.

54. Feany, M.B. and Quinn, W.G. A neuropeptide gene defined by the *Drosophila* memory mutant amnesiac, *Science* 268 (5212), 869, 1995.

55. Moore, M.S. et al. Ethanol intoxication in *Drosophila*: Genetic and pharmacological evidence for regulation by the cAMP signaling pathway, *Cell* 93 (6), 997, 1998.

56. Toba, G. et al. The gene search system. A method for efficient detection and rapid molecular identification of genes in *Drosophila melanogaster*, *Genetics* 151 (2), 725, 1999.

57. Folkers, E., Drain, P. and Quinn, W.G. Radish, a *Drosophila* mutant deficient in consolidated memory, *Proc Natl Acad Sci USA* 90 (17), 8123, 1993.

58. Chiang, A.S. et al. radish encodes a phospholipase-A2 and defines a neural circuit involved in anesthesia-resistant memory, *Curr Biol* 14 (4), 263, 2004.

59. Folkers, E., Waddell, S. and Quinn, W.G. Personal communication, 2004.

60. Boynton, S. and Tully, T. latheo, a new gene involved in associative learning and memory in *Drosophila melanogaster*, identified from P element mutagenesis, *Genetics* 131 (3), 655, 1992.

61. DeZazzo, J. et al. Nalyot, a mutation of the *Drosophila* myb-related Adf1 transcription factor, disrupts synapse formation and olfactory memory, *Neuron* 27 (1), 145, 2000.

62. Skoulakis, E.M. and Davis, R.L. Olfactory learning deficits in mutants for leonardo, a *Drosophila* gene encoding a 14-3-3 protein, *Neuron* 17 (5), 931, 1996.

63. Grotewiel, M.S. et al. Integrin-mediated short-term memory in *Drosophila*, *Nature* 391 (6666), 455, 1998.

64. Comas, D., Petit, F. and Preat, T. *Drosophila* long-term memory formation involves regulation of cathepsin activity, *Nature* 430 (6998), 460, 2004.

65. Shen, M.H., Harper, P.S. and Upadhyaya, M. Molecular genetics of neurofibromatosis type 1 (NF1), *J Med Genet* 33 (1), 2, 1996.

66. Guo, H.F. et al. A neurofibromatosis-1-regulated pathway is required for learning in *Drosophila*, *Nature* 403 (6772), 895, 2000.

67. Costa, R.M., Honjo, T. and Silva, A.J. Learning and memory deficits in Notch mutant mice, *Curr Biol* 13 (15), 1348, 2003.

68. Wang, Y. et al. Involvement of Notch signaling in hippocampal synaptic plasticity, *Proc Natl Acad Sci USA* 101 (25), 9458, 2004.
69. Presente, A. et al. Notch is required for long-term memory in *Drosophila*, *Proc Natl Acad Sci USA* 101 (6), 1764, 2004.
70. Ge, X. et al. Notch signaling in *Drosophila* long-term memory formation, *Proc Natl Acad Sci USA* 101 (27), 10172, 2004.
71. Leung, B. and Waddell, S. Four-dimensional gene expression control: memories on the fly, *Trends Neurosci* 27 (9), 511, 2004.
72. Burcin, M.M. et al. Adenovirus-mediated regulable target gene expression in vivo, *Proc Natl Acad Sci USA* 96 (2), 355, 1999.
73. Osterwalder, T. et al. A conditional tissue-specific transgene expression system using inducible GAL4, *Proc Natl Acad Sci USA* 98 (22), 12596, 2001.
74. Roman, G. et al. P[Switch], a system for spatial and temporal control of gene expression in *Drosophila melanogaster*, *Proc Natl Acad Sci USA* 98 (22), 12602, 2001.
75. Pascual, A., Huang, K.L. and Preat, T. Conditional UAS-targeted repression in *Drosophila*, *Nucleic Acids Res* 33 (1), e7, 2005.
76. Kitamoto, T. Conditional modification of behavior in *Drosophila* by targeted expression of a temperature-sensitive shibire allele in defined neurons, *J Neurobiol* 47 (2), 8–92, 2001.
77. Chen, M.S. et al. Multiple forms of dynamin are encoded by shibire, a *Drosophila* gene involved in endocytosis, *Nature* 351 (6327), 583, 1991.
78. Waddell, S. et al. The amnesiac gene product is expressed in two neurons in the *Drosophila* brain that are critical for memory, *Cell* 103 (5), 805, 2000.
79. Li, Y. and Strausfeld, N.J. Morphology and sensory modality of mushroom body extrinsic neurons in the brain of the cockroach, *Periplaneta americana*, *J Comp Neurol* 387 (4), 631, 1997.
80. Menzel, R., Erber, J. and Masuhr, T. Learning and memory in the honeybee, in *Experimental Analysis of Insect Behaviour*, Barton-Brown, L. Springer, Berlin, Germany, 1974, pp. 195.
81. Erber, J., Masuhr, T. and Menzel, R. Localization of short-term memory in the brain of the bee *Apis mellifera*, *Physiol Entomol* 5, 343, 1980.
82. Heisenberg, M. et al. *Drosophila* mushroom body mutants are deficient in olfactory learning, *J Neurogenet* 2 (1), 1, 1985.
83. de Belle, J.S. and Heisenberg, M. Associative odor learning in *Drosophila* abolished by chemical ablation of mushroom bodies, *Science* 263 (5147), 692, 1994.
84. Nighorn, A., Healy, M.J. and Davis, R.L. The cyclic AMP phosphodiesterase encoded by the *Drosophila* dunce gene is concentrated in the mushroom body neuropil, *Neuron* 6 (3), 455, 1991.
85. Skoulakis, E.M., Kalderon, D. and Davis, R.L. Preferential expression in mushroom bodies of the catalytic subunit of protein kinase A and its role in learning and memory, *Neuron* 11 (2), 197, 1993.
86. Connolly, J.B. et al. Associative learning disrupted by impaired Gs signaling in *Drosophila* mushroom bodies, *Science* 274 (5295), 2104, 1996.
87. Zars, T. et al. Localization of a short-term memory in *Drosophila*, *Science* 288 (5466), 672, 2000.
88. McGuire, S.E. et al. Spatiotemporal rescue of memory dysfunction in *Drosophila*, *Science* 302 (5651), 1765, 2003.
89. Dubnau, J. et al. Disruption of neurotransmission in *Drosophila* mushroom body blocks retrieval but not acquisition of memory, *Nature* 411 (6836), 476, 2001.

90. McGuire, S.E., Le, P.T. and Davis, R.L. The role of *Drosophila* mushroom body signaling in olfactory memory, *Science* 293 (5533), 1330, 2001.

91. Tully, T. et al. Genetic dissection of consolidated memory in *Drosophila*, *Cell* 79 (1), 35, 1994.

92. Pascual, A. and Preat, T. Localization of long-term memory within the *Drosophila* mushroom body, *Science* 294 (5544), 1115, 2001.

93. Isabel, G., Pascual, A. and Preat, T. Exclusive consolidated memory phases in *Drosophila*, *Science* 304 (5673), 1024, 2004.

94. Pascual, A. et al. Neuroanatomy: brain asymmetry and long-term memory, *Nature* 427 (6975), 605, 2004.

95. Quinn, W.G. and Dudai, Y. Memory phases in *Drosophila*, *Nature* 262 (5569), 576, 1976.

96. Isabel, G. and Preat, T. Unpublished results, 2005.

97. Drier, E.A. et al. Memory enhancement and formation by atypical PKM activity in *Drosophila melanogaster*, *Nat Neurosci* 5 (4), 316, 2002.

98. Yin, J.C. et al. Induction of a dominant negative CREB transgene specifically blocks long-term memory in *Drosophila*, *Cell* 79 (1), 49, 1994.

99. Perazzona, B. et al. The role of cAMP response element-binding protein in *Drosophila* long-term memory, *J Neurosci* 24 (40), 8823, 2004.

100. Yin, J.C. et al. A *Drosophila* CREB/CREM homolog encodes multiple isoforms, including a cyclic AMP-dependent protein kinase-responsive transcriptional activator and antagonist, *Mol Cell Biol* 15 (9), 5123, 1995.

101. Yin, J.C. et al. CREB as a memory modulator: induced expression of a dCREB2 activator isoform enhances long-term memory in *Drosophila*, *Cell* 81 (1), 107, 1995.

102. Tully, T. and Quinn, W.G. Classical conditioning and retention in normal and mutant *Drosophila melanogaster*, *J Comp Physiol [A]* 157 (2), 263, 1985.

24 Behavioral Genetics in the Nematode *Caenorhabditis elegans*

Gert Jansen and Laurent Ségalat

CONTENTS

24.1 SUMMARY

This chapter provides an overview of the behavioral studies that can be undertaken on an extremely simple organism, the nematode *Caenorhabditis elegans*. First, we will briefly describe this animal model, which has become increasingly popular for molecular and cellular biology studies, and then we will present a few examples of behavioral studies that are conducted on this organism. We aim to show that this model provides paradigms for questions central to many behaviors, and that they can be addressed at a single cell/single gene resolution.

24.2 *C. ELEGANS* AS A MODEL ORGANISM

24.2.1 *C. ELEGANS* HAS A VERY SIMPLE NERVOUS SYSTEM

The nematode *Caenorhabditis elegans* is a free-living round worm present in the soils of temperate climates (Wood, 1988; Riddle et al., 1997). For laboratory use, it can be grown easily on petri dishes seeded with *Escherichia coli* (Figure 24.1). Its length is about 1 mm as an adult and it has 959 somatic cells. Among those, 302 of them are neurons, which can be divided into 116 classes on the basis of their morphology (Figure 24.2). With a laser beam, it is possible to destroy a given neuron in an anaesthetized animal to assess the involvement of this neuron in a behavior or a function. Moreover, the—almost invariant—connectivity of all neurons has been established by 3D reconstruction of serially sectioned animals.

At the molecular level, the nervous system of *C. elegans* shares a common organization with the far-distant vertebrates: its main excitatory transmitter is

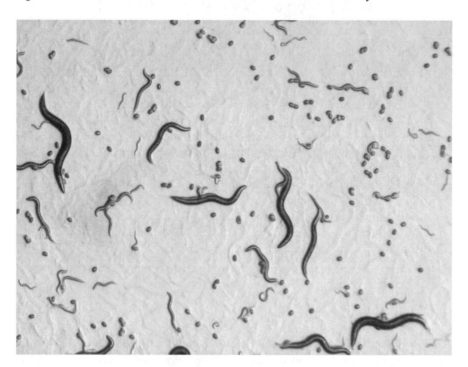

FIGURE 24.1 *C. elegans* on a culture plate. Worms of various stages can be seen, including eggs, larvae and adults. Adults are approximately 1 mm long.

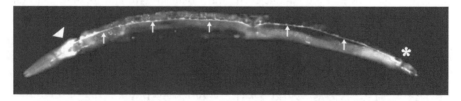

FIGURE 24.2 The nervous system of *C. elegans*. Most or all neurons of *C. elegans* are visualized by expressing a *gpc-2::GFP* construct. Weakly, staining of muscle cells can be seen. Arrows indicate the ventral nerve cord, a triangle indicates staining in the anterior ganglia, an asterisk indicates the posterior ganglia.

glutamate, its main inhibitor is GABA, the neuromuscular synapse is mainly cholinergic, and neuronal activity is regulated in part by two major modulators that display a wide spectrum of effects in the CNS of vertebrates: dopamine and serotonin. A variety of peptides—albeit less numerous than in mammals—also act as neuromodulators.

24.2.2 *C. ELEGANS* IS WELL SUITED FOR GENETIC STUDIES

Two intrinsic properties make *C. elegans* an excellent system for genetics: its rapidity to grow (3 days/generation) and the small size of its genome (100 Mbases). Animals are generally hermaphrodites with the capability to self-fertilize, an invaluable advantage when one has to perform large genetic screens. Males can also be generated at will, for the purpose of crosses. Over the years, hundreds of genetic markers useful for genetic mapping have been described (Riddle et al., 1997). Besides the classical genetics approach (from mutant to gene), one can also use reverse genetics (from gene to mutant), including techniques to generate knockout worms, RNAi and transgenesis. The cloning of mutations is facilitated by the fact that the genome of *C. elegans* is entirely covered by yeast artificial chromosomes (YACs) and cosmids. Furthermore, in 1998, *C. elegans* became the first animal of which the genome was entirely sequenced (The *C. elegans* Sequencing Consortium, 1998). The analysis of the sequence revealed an extensive conservation of protein sequences between this organism and vertebrates. In 1998, the RNAi technique—which is a fast and efficient way to inactivate a gene of interest—has been introduced in *C. elegans* (Fire et al., 1998). This technique has now been extended to most animal models.

In 2002, Sydney Brenner, John Sulston and Bob Horvitz were awarded the Nobel Prize for Medicine and Physiology for the introduction of *C. elegans* as a model organism in biology and for their studies that have elucidated key molecules that regulate programmed cell death.

24.3 BEHAVIORS OF *C. ELEGANS* AND MUTATIONS AFFECTING THEM

Despite its simplicity, more than 20 behaviors can be studied on the worm *C. elegans*. For the sake of simplicity, we will focus in this chapter on only six behaviors, representative of the work that can be done with *C. elegans*, and we will try to point out their relevance to other models. The behaviors described below are:

Chemosensory behavior and its plasticity
Egg laying
Regulation of locomotion in response to food
Ethanol resistance
Social feeding behavior
Memory

24.3.1 CHEMOSENSORY BEHAVIOR AND ITS PLASTICITY

The detection of various compounds in our environment is essential for life, it protects us from toxic compounds and enables us to find food and judge its quality. These principles also hold true for *C. elegans*. The nematode responds to many chemical cues from its environment, including water soluble and volatile attractants and repellents (reviewed in Mori and Oshima, 1997). Detection of these compounds is primarily mediated by 11 pairs of amphid chemosensory neurons in the head. Ablation of these neurons with a laser microbeam has identified the specific sensory functions for each of these neurons (Figure 24.3). Olfactory cues are detected by the AWA, AWB, AWC, ADL, and ASH neurons; the nose of the worm, of which the AWA and AWC cells mediate chemo-attraction. Chemotaxis to water soluble attractants, such as salts, is mediated by five pairs of amphid neurons, the ASE, ASG, ASI, ASK, and ADF neurons, of which the ASE neurons are most important. The ADL and ASH neurons mediate avoidance of water-soluble repellents.

The development of a robust odorant chemotaxis assay has made it possible to identify many genes involved in olfactory signaling using forward genetic screens. More recently, this approach has been complemented by reverse genetic studies. Together, these studies have given us a very good idea about the basic signaling mechanisms used in odorant detection in the AWA and AWC neurons. Since *C. elegans* has only few olfactory neurons, expression of only one olfactory receptor per cell (as is the case in mammals) would not be sufficient. Instead, the nematode expresses many G protein coupled receptors (GPCRs) in each olfactory neuron. With these receptors *C. elegans* can discriminate between different odorants detected by the same cells and adapt to an odorant while remaining responsive to another.

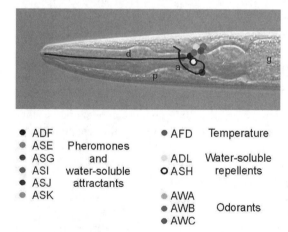

FIGURE 24.3 (SEE COLOR INSERT FOLLOWING PAGE 236) DIC photograph of the head of a worm. The outline of the pharynx (p) can be seen and the beginning of the gut (g). The approximate positions of the amphid sensory neuron cell bodies are indicated with colored circles. The dendrite (d) and an axon (a) of one of the cells are drawn. Below, the names of the corresponding cells are given, and the compounds perceived by these cells.

The exact molecular mechanisms that make this possible are not completely clear, but part of the specificity arises from a functional asymmetry between the left and the right AWC cell (Wes and Bargmann, 2001). In addition, odorant detection seems modulated by a complex G protein signaling network, involving two to three redundant stimulatory Gα subunits and an inhibitory Gα in each neuron (Lans et al., 2006).

As in mammals, binding of an odorant to a GPCR and activation of heterotrimeric G proteins is thought to activate downstream effector molecules. In the AWA cells, G proteins probably activate TRPV channels resulting in a Ca^{2+} influx. In the AWC cells, G proteins induce an increase in intracellular cGMP, mediated by guanylyl cyclases, leading to opening of a cyclic nucleotide-gated channel. These studies have shown that *C. elegans* uses quite similar sensory signaling pathways as mammals.

Prolonged exposure of nematodes to an odorant reduces chemotaxis to that odorant (Mori and Oshima, 1997). This behavior is odorant specific, modulated by food and stress and depends on the odorant concentrations used during pre-exposure and chemotaxis. This behavior provides a good basis to study olfactory adaptation, desensitization or perhaps even olfactory learning. However, although several genes have been identified that play a role in olfactory plasticity, it has been quite difficult to identify additional genes. The main difficulty is that this behavior is dependent on many factors and hence relatively variable. However, the careful dissection of the behavioral variables of olfactory plasticity over the past decade, the current knowledge of olfactory signaling, and the development of novel techniques to study gene function and identify cells involved should bring novel insights into the molecular mechanisms of olfactory plasticity.

C. elegans shows strong chemo-attraction to salts. In nature, the detection of salts is essential for salt homeostasis and salts are probably cues for food. In the laboratory, salt detection is mainly tested using two assays. In one assay, animals are placed on a salt gradient and tested for chemotaxis to the highest concentration (Ward, 1973). This single worm assay has enabled the identification of the sensory neurons involved in salt detection using laser ablation techniques. The detection of NaCl is mainly mediated by the ASE chemosensory neurons, but a residual response is mediated by the ADF, ASG, and ASI neurons. A more recent study confirmed the importance of the ASE neurons for chemotaxis to NaCl, and revealed a functional asymmetry between the left and the right cell: ASE left is primarily sensitive to sodium, whereas ASE right is primarily sensitive to chloride and potassium (Pierce-Shimomura et al., 2001). In an alternative assay, the quadrant assay, animals are given a choice between two salt concentrations, by putting them on a plate divided into four quadrants that can each be filled independently with salt containing agar (Wicks et al., 2000). This assay provides a robust method to screen for additional chemotaxis mutants.

In rodents and flies, the main mechanism of salt detection involves ion influx through degenerin/epithelial Na+ channels (DEG/ENaC), leading to membrane depolarization and neurotransmitter release. These ENaC channels can be blocked specifically with the drug amiloride. In rodents, 75% of the response to salt can be blocked with amiloride; in humans the amiloride sensitivity is less pronounced, indicating that also other salt detection mechanisms exist. In *C. elegans*, chemotaxis to salts is not blocked by amiloride. Perhaps the worm can help to elucidate the molecular mechanisms of salt detection.

C. elegans is attracted to salt concentrations ranging from 0.1 to 100 m*M*. High salt concentrations result in avoidance behavior. This latter response is probably due to a general avoidance of high osmotic strength, mediated by the nociceptive neurons ASH and ADL. Forward and reverse genetic approaches have identified several genes involved in salt detection in *C. elegans*. These include a cGMP-gated channel and several Ca^{2+} activated proteins, indicating that cGMP and Ca^{2+} are important second messengers in salt detection. Surprisingly, our recent analysis of these mutants in the quadrant assay showed that these animals still respond to 25–100 m*M* salt, indicating that another independent "high salt" detection pathway exists (Hukema et al., 2006). Thus far, no molecules involved in this pathway have been identified.

Recently, we have adapted the quadrant assay to study the plasticity of the response to salt. In this assay, animals are pre-exposed to relatively high, but attractive concentrations of NaCl (100 m*M*), and subsequently tested for chemotaxis to NaCl. Animals pre-exposed to NaCl are no longer attracted to NaCl but even avoid it. This behavior, which we call gustatory plasticity, is time and concentration dependent, reversible and partially salt specific (Jansen et al., 2002). The finding that prolonged exposure to NaCl results in avoidance behavior suggests that gustatory plasticity is more than adaptation or desensitization. We used two methods to identify sensory neurons involved in this process. We tried to rescue the plasticity defect of a G protein subunit mutant, *gpc-1*, by expressing the *gpc-1* gene in specific cells. In addition, we interfered with cell function by expressing a dominant mutant version of an ion channel in specific sensory neurons. Using these techniques, we have shown that gustatory plasticity requires a balance of antagonistic sensory inputs, involving attractive signals via the ASE salt detection neurons, and aversive signals via the ASI chemosensory neurons and the nociceptive neurons ASH and ADL (Hukema et al., 2006).

In a candidate-gene approach, we analyzed gustatory plasticity of 66 mutant strains (Hukema et al., submitted). In this study, we identified several signaling cascades and neurotransmitters involved in gustatory behavior. Taking into account the cellular circuitry and the expression patterns of the various signaling molecules, we propose a model for the molecular and cellular mechanisms that regulate gustatory plasticity. We are currently performing additional cell-specific rescue experiments to test whether the different signaling molecules indeed function in the cells as proposed in our model. First, the gustatory response requires salt detection transmitted by a cGMP and Ca^{2+} pathway and another unknown "high salt detection" pathway. These salt-attraction signals are probably mediated by the ASE neurons. This signal is balanced by an unknown signal detected by the ASI neurons, transmitted by G proteins, probably activating a thus far unknown guanylyl cyclase. cGMP subsequently activates a cGMP-gated channel resulting in a Ca^{2+} influx and activation of several proteins, ultimately leading to a salt-aversion signal. Another aversion signal is provided by the nociceptive neurons. At least two G alpha subunits and a G gamma subunit function in these neurons, transmitting an unknown environmental signal to activate heteromultimeric TRP channels.

At present it is unclear in which cells these signals are integrated, but we show that intercellular signaling requires glutamate, serotonin and dopamine. Integration of these signals in interneurons, leading to the activation of motoneurons, requires the Go/Gq signaling network. Also the environmental signals that modulate the response to NaCl have not been identified, but our results show that gustatory plasticity depends on salt concentration and exposure time. In addition, our preliminary data suggest that also food signals modulate gustatory plasticity. We feel that the analysis of gustatory plasticity in *C. elegans* provides a very good model to unravel the molecular mechanisms of behavioral plasticity and the integration of several different sensory cues.

24.3.2 EGG LAYING

How can a variety of neuronal inputs of different weights be converted in an all-or-none decision? The egg-laying behavior of *C. elegans* is an interesting paradigm to study this form of integration.

The anatomy of the egg-laying system is extremely simple. The opening of the vulva muscles is controlled by a pair of bilateral motoneurons (called HSN) running from the head to the vulva, where they innervate the vulva muscles. These neurons are serotonergic, and deliver pulses of serotonin directly on the vulva muscles. However, even under favorable conditions, eggs are not laid continuously, but in bursts. This is caused by fluctuations between an inactive state, during which eggs are retained in the uterus, and an active state, during which eggs are laid at short intervals. This constitutes a higher order of organization, which is controlled by a neuropeptide. This neuropeptide is related to FMRF-amide; it is not known if its physiological concentration varies over time.

Like most animals, worms have developed throughout evolution a behavior to secure their progeny at their best; they will lay their eggs only if they perceive the environment as favorable (Figure 24.4). This means that the inputs of the sensory neurons critical for the decision-making process have to be integrated to deliver an all-or-none output on the neurons driving the liberation of eggs. The sensory neurons sensing the parameters critical for this behavior are almost all known. They are olfactory, chemosensory, thermosensory, and touch sensory neurons, mainly located in the head. Thanks to the description of all neuronal connections, it is possible to infer in which head interneurons integration probably takes place. In other systems, one would successfully dissect this behavior through electrophysiology. Unfortunately, despite recent progress, this technique is still rarely used because of technical difficulties. But genetic tools can circumvent the lack of electrophysiological data. To study the regulation of egg laying by the environmental clues, a genetic approach is to perturb the regulation by appropriate mutations. One can imagine mutations in which animals would not lay eggs when they normally should, or mutations in which animals would lay eggs when they should not. In fact, both classes of mutations exist (called *egl-d* and *egl-c*, respectively).

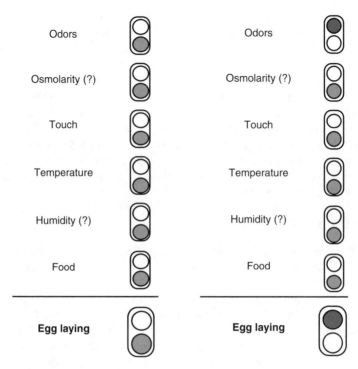

FIGURE 24.4 Egg laying as a model of integration. The egg-laying behavior is the result of the integration of various sensory input. Negative stimuli (such as bad odors, vibrations, cold, etc.) block the egg-laying process (right side of the figure). When all the sensory inputs are appropriate (left side) egg-laying occurs. The question marks indicate stimuli that are thought to influence egg laying although they have not been experimentally demonstrated. The advantage of the *C. elegans* model is that it is amenable to the dissection of the stimuli at a one cell/ one gene resolution.

Many causes can lead to an *egl-d* phenotype, including the inability of the animal to lay eggs (for instance, morphological defects of the vulva). To distinguish which *egl-d* mutants are truly unable to lay eggs, from those who could lay eggs but suffer from misregulation, these mutants are tested for their response to serotonin, a strong stimulator of egg laying, which acts directly on the vulva muscles. The *egl-d* mutants who are responsive to serotonin must possess a functional egg-laying apparatus. Therefore, their *egl-d* phenotype is supposed to be a misregulation mechanism. Almost a dozen genes produce such mutants; unfortunately, few have been identified yet. Interestingly, several *egl-c* mutants have abnormally short HSN processes, which prevents the formation of synapses between HSNs and head neurons. This suggests that the HSN neurons are indeed in a repressed (hyperpolarized) state under unfavorable conditions and are derepressed in favorable conditions. The molecular mechanisms involved are still unknown. The use of cameleon proteins, which are GFP-based indicators of intracellular calcium, will certainly be helpful tools to further understand this process.

24.3.3 REGULATION OF LOCOMOTION IN RESPONSE TO FOOD

Dopamine and serotonin are important modulators of various behavioral responses of many different animals. Also in humans these biogenic amines have been implicated in many behaviors and diseases. In *C. elegans* serotonin and dopamine modulate several behaviors, including egg laying and locomotion. Careful genetic dissection of these behaviors has led to the identification of several genes that mediate the effects of serotonin and dopamine.

In 2000, Sawin et al. described a new paradigm to analyze behavioral plasticity. They found that *C. elegans* modulates its locomotory rate when it encounters food. Well-fed animals slow down when they encounter food. This behavior was called the basal slowing response. Surprisingly, starved animals slow down even more, called the enhanced slowing response. From previous studies it was known that many responses to food are mediated by serotonin and dopamine. Therefore, the authors tested if animals with defects in serotonin and/or dopamine synthesis were affected in either the basal or the enhanced slowing responses. Testing several of such mutants revealed that the basal slowing response required dopamine but not serotonin. Applying exogenous dopamine to the mutant *C. elegans* cultures could restore these defects. Next, laser ablation experiments were used to identify the neurons involved in this response. *C. elegans* has 8 dopaminergic neurons, 4 CEP neurons, 2 ADEs, and 2 PDEs. Laser ablation of subsets of these neurons showed that these 3 types of neurons function redundantly in the regulation of the basal slowing response. Interestingly, only the neurons that are in contact with bacteria are required for this behavior. Sawin et al. found that mechanosensory stimuli probably activate the dopaminergic neurons.

Sawin et al. (2000) showed that the enhanced slowing response required serotonin, synthesized by the NSM neurons. The authors found no indication that additional serotonergic neurons were involved, but a residual enhanced slowing response could be observed in animals in which the NSMs were ablated, indicating that also other cells are involved in this response. It is unclear what signals induce the enhanced slowing response. Interestingly, the enhanced slowing response could be blocked by using serotonin antagonists and potentiated by the drug fluoxetine (Prozac), a selective serotonin reuptake inhibitor.

Using the enhanced slowing response, the group of Horvitz has performed genetic screens to identify genes that regulate this behavior. Thus far, two genes have been reported, *mod-1* and *mod-5* (mod stands for modulation of locomotion defective). MOD-1 is a serotonin-gated chloride channel and MOD-5 is a serotonin reuptake transporter (Ranganathan et al., 2000, 2001). Clearly, these proteins play important roles in the modulation of behavior by serotonin in *C. elegans*. The importance of serotonin reuptake transporters (SERT) was already known from previous studies in other organisms, including humans. The importance of the serotonin gated chloride channel in humans remains to be determined. Characterization of such channels from mammals might provide novel targets for the developmental of human pharmaceutical drugs.

24.3.4 Ethanol Resistance

Invertebrates have recently emerged as additional models of drug abuse. Although the range of testable behaviors is more limited (for instance, invertebrates cannot be tested in addiction tests in which animals self-administer cocaine or other stimulatory agents), small animals such as *C. elegans* are clearly sensitive to the effects of molecules such as ethanol, cocaine, and nicotine. Since the molecular architecture of vertebrates and invertebrates is quite similar, it is therefore tempting (and potentially rewarding) to use the powerful approach of genetics to understand the molecular effects of psychoactive agents at the gene level. Since growing these animals is cheap and easy, robust statistics can be obtained by testing large numbers.

When exposed to ethanol, *C. elegans* reacts in two phases. A first rapid phase of excitation consists of increased head movements and locomotion. Then, there is a progressive lack of coordination, followed by immobility and unresponsiveness to prodding.

Using the classical genetics approach of phenotypic screens, Morgan and Sedensky (1995) identified nine genes in which mutations modified the resistance to ethanol. Interestingly, these mutations also changed the sensibility to the volatile anesthetics enflurane and isoflurane. One of the genes (*slo-1*) has been cloned and turned out to encode a potassium channel of the BK family. When the *slo-1* gene is inactivated, animals are apparently normal, but they become highly resistant to ethanol (Davies et al., 2003). Conversely, mutations in other genes lead to increased ethanol sensitivity (such as *gas-1*, which encodes a component of the mitochondrial electron transport chain).

Natural isolates of *C. elegans* display various tolerance to ethanol. Interestingly, in these strains ethanol sensitivity correlates with sensitivity to anesthetics, and surprisingly, to the social behavior called clumping (see Section 24.3.5). Since the clumping behavior is mediated in part by the neuropeptide Y receptor (NPY), it is tempting to speculate that NPY may also be involved in ethanol resistance. Indeed, three laboratory strains carrying mutations in the NPY gene have been shown to be more resistant than the reference strain to ethanol in an acute tolerance assay (Davies et al., 2004). This is of particular interest since NPY is known to be a modulator of alcohol consumption in rodents and humans. In rats, abnormal or low NPY activity can promote high alcohol drinking. In humans, a population study recently showed that polymorphisms on the peptide sequence of NPY could be associated with high average alcohol consumption.

Recently, microarray analysis was applied to ethanol response in *C. elegans*. This technique allows a global survey of genes whose expression levels are modified upon experimental conditions. 230 out of the 19,000 genes of *C. elegans* are either up- or down-expressed following exposure to ethanol. Although such global experiments are not devoid of background noise, they provide a list of candidate genes, which can later be investigated in more details.

In conclusion, the combination of these various approaches should improve the knowledge of the mechanisms underlying ethanol intoxication in *C. elegans*, and provide a molecular model for better understanding the effects of ethanol in mammals.

24.3.5 Social Feeding Behavior

Many animal species live in groups, or aggregate. This simple form of social behavior depends on environmental conditions, such as the presence of food or predators or the season, and varies between individuals of a species and between species. Different natural isolates of *C. elegans* feed on bacteria either in groups or alone, named social or solitary feeders, respectively. This finding not only enables the dissection of the molecular mechanisms that govern social behavior, but also how these mechanisms have evolved.

Solitary feeders slow down when they encounter food and disperse on a bacterial lawn, while social feeders move rapidly on a bacterial lawn and aggregate where food is most abundant. These natural occurring behavioral differences depend on a one amino-acid difference in a neuropeptide receptor, NPR-1 (de Bono and Bargmann, 1998). The receptor contains a phenylalanine at position 215 (215F) in the social strains and a valine at this position (215V) in solitary strains. Loss-of-function of the *npr-1* gene resulted in strong aggregation, indicating that NPR-1 represses social feeding. Sequencing the *npr-1* gene from three different *C. elegans* species showed that the social NPR-1 215F must have been the ancestral form, indicating that the standard laboratory *C. elegans* strain N2, commonly referred to as *wild type* actually shows mutant solitary behavior due to mutation of the *npr-1* gene (Rogers et al., 2003).

The most likely candidate ligands of the NPR-1 receptor are encoded by 22 "FMR Famide and related peptide" (FaRP) genes, *flp-1* to *flp-22*. Two groups independently tested the 59 neuropeptides encoded by these FaRP genes for activation of the NPR-1 215F or NPR-1 215V receptors, either in Xenopus oocytes or the *C. elegans* pharynx (Rogers et al., 2003) or in mammalian CHO cells (Kubiak et al., 2003). These assays showed that *flp-18* and *flp-21* encoded FaRPs could activate the NPR-1 receptor. The NPR-1 215V receptor variant, found in solitary nematodes, displayed a stronger response to the peptides than the 215F variant. These findings are in agreement with the behavioral data, which show that the NPR-1 215V receptor is more active in repressing social feeding, resulting in solitary feeding behavior. These findings were confirmed by testing *flp-21* loss-of-function mutants and over-expression animals (Rogers et al., 2003).

Using various genetic methods de Bono and colleagues (2002) identified several other proteins that regulate this behavior and they identified cells involved. First, they tested mutants with defective responses to various chemical compounds. Mutations in the TRPV cation channel subunit genes *osm-9* and *ocr-2* and mutations in the *odr-4* and *odr-8* genes, required for the correct localization of particular receptors to cilia, abolished the social feeding behavior of *npr-1* animals. Using cell-specific expression of the *ocr-2* and *odr-4* genes in subsets of the chemosensory neurons and ablation of these neurons using a laser microbeam, de Bono et al. implicated the ADL and ASH nociceptive neurons in social feeding. Finally, their genetic analysis showed that the signal transduced by the nociceptive neurons is antagonized by other sensory neurons. Together, these experiments suggest a model in which aversive signals from bacteria stimulate the ADL and ASH nociceptive neurons, which promote social feeding, while signals from unidentified neurons repress social feeding.

Second, Coates and de Bono (2002) follow up on the finding that NPR-1 regulates social feeding. By carefully analyzing the expression pattern of the *npr-1* gene, using a *npr-1::GFP* fusion construct, and subsequent cell specific expression of the NPR-1 215V gene in candidate cells in *npr-1* animals they implicated the AQR, PQR and URX neurons in the regulation of social feeding. This finding was confirmed by genetically blocking the electrical activity of these neurons. In addition to inhibition of social feeding via NPR-1 in these cells, social feeding is stimulated by signaling through a cGMP-gated channel. The AQR, PQR and URX neurons are directly exposed to the body fluid, which seems to function analogously to blood. Together, these findings suggest a model in which the body cavity neurons, AQR, PQR and URX, regulate the choice between solitary and social feeding by integrating NPR-1 mediated signals, which inhibit aggregation, and cGMP mediated signals, which stimulate aggregation (Coates and de Bono, 2002). At present, it is unclear where these signals originate.

24.3.6 MEMORY

Although it may sound incredible to many mammalian neurobiologists, worms can display some forms of associative memory. The most popular assay, described in this section, is based on the storage of temperature information. Thirty years ago, when *C. elegans* genetics was still in its infancy, Hedgecock and Russel (1975) showed that when worms are placed in a thermal gradient in absence of food, they move to the temperature where they last were in presence of food (isothermal tracking). It is thought that, when in the wild, recollection of environmental parameters such as temperature and hygrometry can help them to find their food. On the laboratory bench, it is just a paradigm showing that the animals are able to associate the presence of food with the sensation of temperature and to store this information. In absence of food, the isothermal tracking behavior is kept for several hours (Figure 24.5); then the worms will cross isotherms randomly to seek food at other temperatures. If food is encountered, a new acquisition period begins.

By laser-ablation experiments, Mori and collaborators (1995) have identified two (possibly three) neurons as being critical for the memory process. These neurons (AIY and AIZ) are integrating interneurons, which receive synapses from most head sensory neurons (including AFD, a sensory neuron specialized in sensing temperature). Animals in which AIY or AIZ are deleted have peculiar phenotypes: AIY-deleted worms are cryophilic (they are attracted by colder temperatures) and AIZ-deleted worms are thermophilic, indicating that these neurons probably have antagonistic effects in wild-type worms. Since both AIY and AIZ synapse on RIA (another integrating interneuron downstream of AIY and AIZ), this suggests a mechanism by which the temperature information could be encoded by the relative strength of AIY and AIZ connections on RIA. When the sensory neuron AFD is ablated, the worms are either cryophilic, or atactic (do not feel the temperature). Thanks to the numerous mutants available in *C. elegans*, some of these ablation experiments can be reproduced genetically, which then provide the immense advantage of having unlimited numbers of modified animals

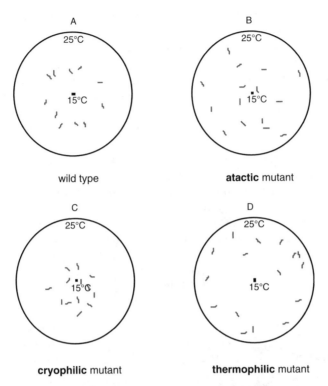

FIGURE 24.5 Associative learning assay and mutant classes. When placed in a temperature gradient without food (here 15°C at the center and 25°C at the edge of the dish), worms will migrate to the temperature they last encountered food (A). Several mutants have been isolated which are atactic (B), cryophilic (C), or thermophilic (D). Some of these mutations mimic phenotypes obtained after key neurons have been ablated, disturbing the balance between cold-driving and heat-driving neurons.

(Figure 24.5). For example, mutations in the *ttx-3* gene specifically disable the AIY neuron due to axonal defects.

How does the worm compare the ambiant temperature (Tamb) to the cultivation temperature (Tcult)? To test whether AFD has a role in this process, an easy way would be to record the activity of AFD (and other neurons) in various conditions. Unfortunately, it is not possible to stick an electrode in a single worm neuron (cell diameter = a few micrometers) without severely perturbing it. Samuel and collaborators (2003) have elegantly got around this difficulty by using a pH-sensitive green fluorescent protein localized in the synaptic vesicles of AFD. Since the internal side of the synaptic vesicles is more acidic than the extracellular environment, the fluorescence levels can be correlated with synaptic release. By doing so on immobilized live animals, the authors could come to the conclusion that AFD synaptic release is high if Tamb and Tcult are different, and lower if they are close. Therefore, the single AFD neuron encodes a direct comparison between actual and memorized temperature. It is likely that this process involves regulation of gene expression in AFD, but the molecular targets are still unknown. A schematic view of the neuronal circuitry associated with this form of learning is shown in Figure 24.6.

The picture is clearer in the AIY interneuron, which naturally expresses the neuron-specific calcium sensor (NCS-1) protein. In mammals, NCS-1 is involved in long-term potentiation. When worms are deprived of NCS-1 by a mutation, their behavior becomes cryophilic and resembles that of AIY-deleted animals, indicating that NCS-1 is a key component of AIY function. More interestingly, overexpression of NCS-1 in AIY increases performance levels, accelerates learning, and produces a memory with slower extinction (Gomez et al., 2001).

Serotonin is likely to be the main inductor of this form of memory in *C. elegans* for several reasons: (1) serotonin levels are high when animals are fed and low when animals are starved; (2) serotonin is released in the head by two neurosecretory cells, the NSM, which have sensory endings in the pharynx, or the muscle that pumps food in *C. elegans*. Thus, NSM release in the body cavity is directly related to the presence of food; (3) exogenous serotonin can mimic the presence of food in the

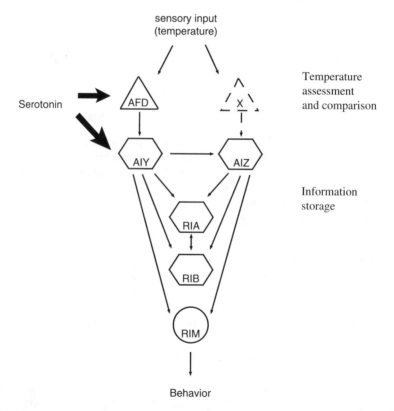

FIGURE 24.6 Neuronal circuitry underlying associative behavior. Neurons are represented according to their class (triangle: sensory, hexagon: interneuron, circle: motorneuron). Neuron X (dashed) has been postulated but is not identified. Only the major connections are drawn. During the conditioning phase, the signal is memorized by the action of serotonin, which may act on sensory or interneurons. See main text for details.

learning process: animals grown without food but in the presence of serotonin will go to this temperature when tested. It is particularly interesting to note that serotonin, one of the oldest neurohormones in evolution, has conserved its role of memory inductor from roundworms (*C. elegans*) to mollusks (Aplysia) and mammals (reviewed in Mayford and Kandel, 1999).

Over the years, several mutants have been identified in phenotypic screens after random mutagenesis. These mutants have been picked because they are defective for the memory assay. Some of them are cryophilic (e.g., *ttx-1*) and some of them are thermophilic (e.g., *tax-6*), suggesting again that the memory process results from a balance between thermo- and cryophilic inputs. The genes corresponding to these mutants have been identified in some cases, and this has revealed that proteins involved in information storage in *C. elegans* are common to neuronal activity. For instance, *tax-2* and *tax-4* encode both subunits of a cyclic nucleotide-gated ion channel and are expressed in the AFD neuron. Such channels have been implicated in olfaction and photoreception in mammals. Electrophysiological properties of the TAX-2/TAX-4 channel suggest that calcium influx is part of the signal transduction cascade that encodes the temperature information.

24.4 CONCLUSION

Despite its simplicity, the nematode *C. elegans* possesses in its CNS many of the neuronal features that have been selected and conserved throughout evolution because they confer a selective advantage for the bearer. These are the internal clocks, the possibility to habituate or to sensitize to a stimulus the need for coordination and integration, and, of course, the faculty to learn from experience, *that is* memory. The molecular description of the mechanisms underlying these neuronal properties is still far from complete in any organism.

As a model, *C. elegans* offers the advantage of being integrated. This means that it is possible to study a behavior at the level of the whole animal, and yet to describe it with a cellular and molecular resolution. The possibility to destroy a single cell to assay its role in a function is a powerful tool. But the strength of the *C. elegans* model resides in the power of genetics, which unambiguously relates molecules and biological functions, and which is an invaluable way to identify the genes relevant to a behavior.

In the few examples described in this chapter, it is striking to discover how much neuronal molecular mechanisms have been conserved through evolution.

REFERENCES

Coates, JC and de Bono, M. (2002) Antagonistic pathways in neurons exposed to body fluid regulate social feeding in *Caenorhabditis elegans*. *Nature,* 419, 925–929.

Davies, AG, Pierce-Shimomura, JT, Kim, H, VanHoven, MK, Thiele, TR, Bonci, A, Bargmann, CI, and McIntire, SL. (2003) A central role of the BK potassium channel in behavioral responses to ethanol in *C. elegans. Cell.,* 115, 655–66.

Davies, AG, Bettinger, JC, Thiele, TR, Judy, ME, and McIntire, SL. (2004) Natural variation in the npr-1 gene modifies ethanol responses of wild strains of *C. elegans. Neuron.,* 42, 731–43.

de Bono, M and Bargmann, CI. (1998) Natural variation in a neuropeptide Y receptor homolog modifies social behavior and food response in *C. elegans. Cell.,* 94, 679–689.

de Bono, M, Tobin, DM, Davis, MW, Avery, L and Bargmann, CI. (2002) Social feeding in *Caenorhabditis elegans* is induced by neurons that detect aversive stimuli. *Nature,* 419, 899–903.

Fire, A, Xu, S, Montgomery, MK, Kostas, SA, Driver, SE and Mello, CC. (1998) Potent and specific genetic interference by double-stranded RNA in *Caenorhabditis elegans. Nature,* 391, 806–811.

Gomez, M, De Castro, E, Guarin, E, Sasakura, H, Kuhara, A, Mori, I, Bartfai, T, Bargmann, CI, Nef, P. (2001) Ca^{2+} signaling via the neuronal calcium sensor-1 regulates associative learning and memory in *C. elegans. Neuron,* 30, 241–248.

Hedgecock, EM and Russel, RL (1975) Normal and mutant thermotaxis in the nematode *Caenorhabditis elegans. Proc. Nat. Acad. Sci.,* 72, 4061–4065.

Hukema, RK, Rademakers, S, Burghoorn, J and Jansen, G. Antagonistic sensory cues generate gustatory plasticity in *Caenorhabditis elegans.* Submitted.

Jansen, G, Weinkove, D and Plasterk, RHA. (2002) The G-protein gamma subunit gpc-1 of the nematode *C. elegans* is involved in taste adaptation. *EMBO J.,* 21, 986–994.

Kubiak, TM, Larsen, MJ, Nulf, SC, Zantello, MR, Burton, KJ, Bowman, JW, Modric, T and Lowery, DE. (2003) Differential activation of "social" and "solitary" variants of the *Caenorhabditis elegans* G protein-coupled receptor NPR-1 by its cognate ligand AF9. *J. Biol. Chem.,* 278, 33724–33729.

Lans, H, Rademakers, S and Jansen, G. (2006) A network of stimulatory and inhibitory Gα subunits regulates olfaction in *Caenorhabditis elegans. Genetics,* 167, 1677–1687.

Mayford, M, and Kandel, ER. (1999) Genetic approaches to memory storage. *Trends Genet.,* 15, 463–70.

Morgan, PG, and Sedensky, MM. (1995) Mutations affecting sensitivity to ethanol in the nematode, *Caenorhabditis elegans. Alcohol Clin Exp Res.,* 19:1423–9.

Mori, I, and Oshima, Y. (1995) Neural regulation of thermotaxis in *Caenorhabditis elegans. Nature,* 376, 344–348.

Mori, I, and Oshima, Y. (1997) Molecular neurogenetics of chemotaxis and thermotaxis in the nematode *Caenorhabditis elegans. Bioessays,* 19, 1055–1064.

Pierce-Shimomura, J.T., Faumont, S., Gaston, M.R., Pearson, B.J. and Lockery, S.R. (2001). The homeobox gene lim-6 is required for distinct chemosensory representations in *C. elegans. Nature,* 410, 694–698.

Ranganathan, R, Cannon, SC and Horvitz, HR. (2000). MOD-1 is a serotonin-gated chloride channel that modulates locomotory behaviour in *C. elegans. Nature,* 408, 470–475.

Ranganathan, R, Sawin, ER, Trent, C and Horvitz, HR. (2001). Mutations in the *Caenorhabditis elegans* serotonin reuptake transporter MOD-5 reveal serotonin-dependent and –independent activities of fluoxetine. *J. Neurosci.,* 21, 5871–5884.

Riddle, TBDL, Meyer, BJ and Priess, JR eds. (1997) *C. elegans II.* Cold Spring Harbor Laboratory Press, Plainview, New York.

Rogers, C, Reale, V, Kim, K, Chatwin, H, Li, C, Evans, P and de Bono, M. (2003) Inhibition of *Caenorhabditis elegans* social feeding by FMRFamide-related peptide activation of NPR-1. *Nat. Neurosci.,* 6, 1178–1185.

Samuel, AD, Silva, RA, and Murthy, VN. (2003) Synaptic activity of the AFD neuron in *Caenorhabditis elegans* correlates with thermotactic memory. *J Neurosci.,* 23, 373–6.

Sawin, ER, Ranganathan, R and Horvitz, HR. (2000) *C. elegans* locomotory rate is modulated by the environment through a dopaminergic pathway and by experience through a serotonergic pathway. *Neuron.,* 26, 619–631.

The C. elegans Sequencing Consortium (1998) Genome sequence of the nematode *C. elegans:* A platform for investigating biology. *Science,* 282, 2012–2018.

Ward, S. (1973) Chemotaxis by the nematode *Caenorhabditis elegans:* identification of attractants and analysis of the response by use of mutants. *Proc. Natl. Acad. Sci. USA* 70, 817–821.

Wes, PD and Bargmann, CI. (2001) *C. elegans* odour discrimination requires asymmetric diversity in olfactory neurons. *Nature,* 410, 698–701.

Wicks, SR, de Vries, CJ, van Luenen, HGAM and Plasterk, RHA (2000) CHE-3, a cytosolic dynein heavy chain, is required for sensory cilia structure and function in *Caenorhabditis elegans. Dev. Biol,.* 221, 295–307.

Wood, W.B, ed. (1988) *The Nematode Caenorhabditis elegans.* Cold Spring Harbor Laboratory Press, Plainview, New York.

AUTHORS' NOTE

We apologize to the authors whose work could not be cited due to lack of space.

25 Genetics, Behavior, and Brain Dopamine Systems

Robert Hitzemann, Shannon McWeeney, and John Belknap

CONTENTS

25.1 INTRODUCTION

Even for the reader unfamiliar with the topic of this chapter but familiar with dopamine and behavior, it is obvious that the topic of genes, behavior and brain dopamine systems cannot be covered in detail. Rather, the goal of the chapter is to provide the reader with an introduction and some examples, largely extracted from work in the authors' laboratories. Previous reviews (Hitzemann et al., 1995; Hitzemann, 1998) can be referenced for important background material. The focus of this chapter will be largely on D$_2$ dopamine receptors. The reasons for this emphasis include the longstanding emphasis on D$_2$ receptors in the etiology and/or expression of a variety of behaviors including schizophrenia, substance abuse and alcoholism (Hitzemann, 1998). The other reason for this emphasis is the amazing variability among individuals in D$_2$ receptor density and in the response to drugs which either stimulate or block these receptors. The reasons for this variability remain unclear, but there is ample evidence to suggest that genetic factors have an important role. Here we provide a clinical example to illustrate the variation in receptor density and drug response. Twenty-three normal controls were administered 0.5 mg/kg of methylphenidate iv. after being told they would receive either placebo or methylphenidate; behavioral effects were measured before and after the injection and included self-ratings of pleasant and unpleasant drug effects. On another day, the subjects were scanned for D$_2$/D$_3$ receptor density using positron emission tomography (PET) and ^{11}C-raclopride as the ligand. Twelve of the 23 subjects (52%) reported the drug effects as very pleasant, 9 subjects (39%) reported the drug effects as unpleasant and 2 subjects (9%) reported no marked behavioral effects (Volkow et al., 1999). Importantly, the individuals who reported the drug effects as pleasant had lower levels of D_2 receptor availability when compared with the "unpleasant" group (Figure 25.1). Although the average difference was small (~20%), it was similar to the persistent differences reported between normal controls and withdrawn chronic alcoholics, cocaine, methamphetamine and heroin addicts (see, e.g., Volkow et al. 1993). These differences were nominally thought to be associated with receptor down-regulation as a result of drug-induced dopamine release. However, the data described in Figure 25.1 suggest that low D_2 receptor availability could be a risk factor for drug abuse since presumably, the "pleasant" group would be more likely to try the drug again. The data in Figure 25.1 also illustrate that in this sample the range of D_2 receptor availability was more than 100%. For other clinical samples, including postmortem samples, the range of D_2 receptor availability has been similar and in some cases even greater (Hitzemann et al., 1998).

The question of whether these large differences are genetic or environmental cannot be easily answered in human populations; however, among mice it is clear that the genetic effect is substantial (Jones et al., 1999; Hitzemann et al., 2003). The

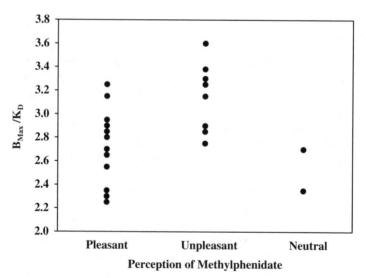

FIGURE 25.1 The relationship between D_2 dopamine receptor binding potential and the response to a methylphenidate challenge (data adapted from Volkow et al., 1999). The binding potential (B_{Max}/K_D) for the receptor was estimated in 23 normal controls using positron emission tomography (PET) and ^{11}C-raclopride as the PET ligand. Note that with the binding potential calculation a value of 1.0 indicates that there is no specific binding; thus, the range of binding potentials in this figure varies by >100%. Subsequently, the subjects were challenged with methylphenidate (0.5 mg/kg, iv.) and were asked to rate their response to the challenge on a visual analog scale; responses varied from very pleasant to neutral to unpleasant.

remainder of this chapter focuses on these and related data which have contributed to our "genetic" understanding of brain dopamine systems and behavior.

25.2 A BEHAVIORAL APPROACH—THE CATALEPSY PHENOTYPE

The typical antipsychotic drugs such as haloperidol exert both their therapeutic and untoward side effects by blocking the D_2 class of receptors, which includes the D_2, D_3, and D_4 receptors. The major untoward side-effects (largely, if not entirely generated by blockade of the D_2 receptors) are known as extrapyramidal symptoms (EPS) which have many of the same features as Parkinson's disease. In mice catalepsy is the murine equivalent of EPS; although different approaches have been used to measure catalepsy (e.g., Kanes et al., 1993; Fowler et al., 2001), all approaches assess some aspect of drug-induced immobility in the absence of sedation. Fink et al. (1982) appear to have been the first to note that two inbred mouse strains (BALB/cJ and CBA/J) differed markedly in terms of haloperidol-induced catalepsy (BALB/cJ was more sensitive). These authors also noted that the BALB/cJ mice had a higher D_2 receptor density which was counter to their expectation that a lower receptor density would make the animals more sensitive to the effects of haloperidol. Kanes et al. (1993) expanded the number of strains examined to 8 and was able to show a >30-fold difference in sensitivity between the most sensitive strain (BALB/cJ–ED_{50} = 0.3 mg/kg) and least sensitive strain (LP/J – ED_{50} = 10 mg/kg); it was also observed that there was very little variation among the inbred strains in terms of the catalepsy response induced by the D_1 receptor antagonist, SCH 23390. These data suggested that the variation in haloperidol response was not due to a general deficit in extrapyramidal function. Pharmacokinetic differences were also eliminated as a potential cause. The number of standard inbred strains examined was subsequently expanded to 15 strains; in addition, 25 strains of the BXD recombinant inbred panel were also phenotyped (Figure 25.2) (Hitzemann et al., 1995; Kanes et al., 1996). The expansion of the standard inbred panel only placed additional strains between the extreme genotypes. The development of the BXD RI panel is described elsewhere in this volume. Figure 25.2 illustrates that in the BXD RI panel no RI strain was more sensitive than the DBA/2J strain, but several strains were significantly less sensitive than the C57BL/6J strain. This phenomenon, sometimes called transgressive segregation, must in the RI panel be associated with genetic effects. For example, the DBA/2J strain may contribute some nonresponsive alleles. Thus, in a mapping study using a C57BL/6J × DBA/2J intercross, it would not be unexpected to detect a QTL in which the DBA/2J allele is associated with non-response, and, indeed, such a QTL has been detected (Kanes et al., 1996; Patel and Hitzemann, 1999).

The RI panel is useful to look for genetically correlated traits and these correlations may in turn be informative about the underlying mechanisms. However, before looking at these correlations it is necessary to introduce WebQTL. WebQTL is an assembly of databases and tools that exploits sophisticated gene mapping and related statistical methods to rapidly perform analyses of gene expression and

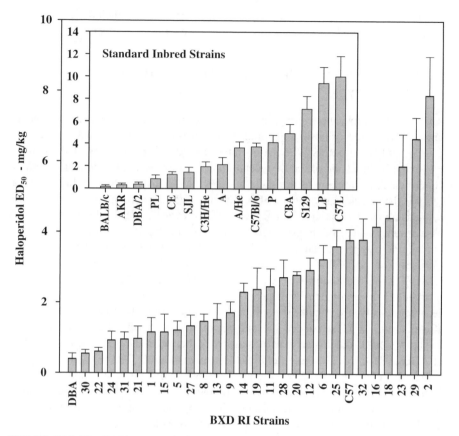

FIGURE 25.2 The ED50 values for haloperidol-induced catalepsy among a panel of standard-inbred mouse strains and among a panel of the BXD recombinant inbred strains (data adapted from Hitzemann et al. 1995). A full description of the catalepsy response and the methods used to determine the ED50 value is found in Kanes et al. (1993).

behavioral data (Chesler et al., 2004). The databases focus on the original BXD RI panel (Taylor 1978), although data for some of the newer BXD RI strains are available. Important for the immediate discussion is that the database contains more than 650 published phenotypes, with the largest single segment focusing on drug-related behavioral phenotypes. The catalepsy data illustrated above have been entered into the database (ID# 10336). Querying the database for phenotypes that are correlated with catalepsy revealed significant correlations with cocaine-induced (15 mg/kg) stereotypy (ID# 10297) and ventral midbrain D_2 dopamine receptor density (ID# 10270) (Jones et al., 1999). These data are illustrated in Figures 25.3 and 25.4. The data in Figure 25.3 show that the RI strains most sensitive to the haloperidol effect are also the strains most sensitive to the effects of cocaine ($r = -0.68$, $p < 0.0002$); we interpret these data to show that if we could determine the traits associated with increased haloperidol sensitivity, some of these would also be associated with increased cocaine sensitivity (or vice versa). Given that both drugs

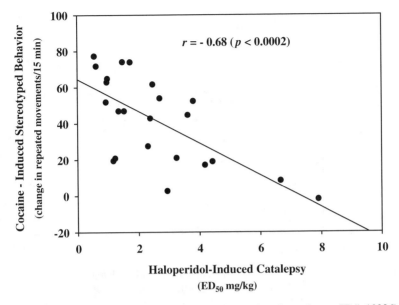

FIGURE 25.3 The relationship between haloperidol-induced catalepsy (ID# 10336) and cocaine (15 mg/kg)-induced stereotypy (ID# 10297) in a panel of the BXD recombinant inbred strains. The ID values are those that reference the phenotypes at WebQTL (www.WebQTL.org); the data entered into the analysis may be downloaded from this site. The catalepsy data are taken from Kanes et al. (1996) and the cocaine data are taken from Jones et al. (1999).

exert their pharmacological effects largely via effects on brain dopamine systems, such an association cannot be unexpected. Figure 25.4 illustrates a potential mechanism. Here we see that the animals most sensitive to haloperidol are those that have the highest density of D_2 receptors—a result consistent with that reported by Fink et al. (1982). Data not shown is the correlation between receptor density and cocaine response ($r = 0.68$, $p < 0.001$). Thus, we conclude that the density of ventral midbrain D_2 receptors (presumably autoreceptors) is associated both with haloperidol and cocaine response. These relationships are not entirely intuitives, but the data illustrate the power of the approach and the cumulative value of the RI data. Furthermore, we again see the wide variation in D_2 receptor density and the apparent association with a drug response. It could be argued that the variation in receptor density is simply an epiphenomenon reflecting a difference in the number of dopamine neurons in the ventral midbrain. Data on the number of neurons in the ventral tegmental area are also found in the database (ID# 10223). For the RI panel there is no significant correlation between receptor density and neuron number ($p > 0.3$). Finally, it should be noted that for the *striatum* there are 12 different measures of D_2 receptor density in the database; none of these are significantly correlated to catalepsy, confirming earlier results from selective breeding studies (Hitzemann et al., 1991; Qian et al., 1992; 1993).

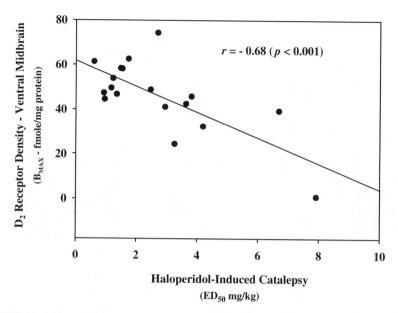

FIGURE 25.4 The relationship between haloperidol-induced catalepsy (ID# 10336) and ventral midbrain D_2 dopamine receptor density (ID# 10270) in a panel of the BXD recombinant inbred strains. Details as in the legend to Figure 3. Receptor density data taken from Jones et al. (1999).

25.3 FROM TRANSPORTER AND RECEPTOR DENSITY TO BEHAVIORAL PHENOTYPES

In the previous section we began with a behavioral phenotype (catalepsy) that genetically led us to another behavioral phenotype (cocaine response) and one measure of D_2 receptor density. In this section, we begin with dopamine transporter and receptor density and ask what behaviors show strong genetic correlations. Perhaps, not surprisingly we are again led to cocaine response. Janowsky et al. (2001) examined the DA transporter (DAT) binding of RTI-55 to 21 of the BXD RI strains. Using a data-mining approach similar to that described here, the authors concluded that DAT density was significantly correlated with cocaine and methamphetamine-induced locomotor activation and thermic responses (both hypo- and hyperthermia); however, the binding was not correlated with stimulant behaviors related to sensitization, reward, voluntary consumption, *stereotypy* or seizures. One example of a significant association is shown in Figure 25.5; RTI-55 binding (ID# 10234) is significantly ($r = 0.69$, $p < 0.0005$) correlated with the cocaine-induced (30 mg/kg) difference score (cocaine treatment – saline treatment) in vertical rearing movements (ID #10310) (Jones et al., 1999). Under the conditions used by these authors cocaine decreased rearing, with the most significant decrease found in those RI strains with the lowest DAT density. In contrast, the strains with the lowest DAT density are the strains that exhibit the most marked cocaine-induced locomotor activation (ID# 10489) (Phillips et al., 1998). For the data shown in Figure 25.6, the

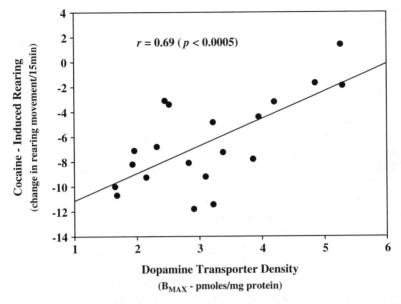

FIGURE 25.5 The relationship between striatal dopamine transporter density (ID# 10234) and cocaine (30 mg/kg)-induced changes of vertical rearing movements (ID# 10310) in a panel of the BXD recombinant inbred strains. Transporter density data taken from Janowsky et al. (2001) and the cocaine data are taken from Jones et al. (1999).

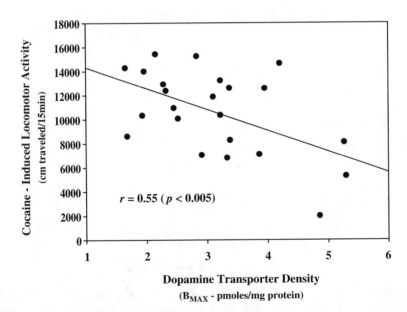

FIGURE 25.6 The relationship between striatal dopamine transporter density (ID# 10234) and cocaine (40 mg/kg)-induced changes in locomotor activity (ID# 10224) in a panel of the BXD recombinant inbred strains. Cocaine data are taken from Phillips et al. (1998).

cocaine (40 mg/kg) response was also indexed as a difference score. Figures 25.5 and 25.6 illustrate a familiar pattern in behavioral research—as some behaviors increase, others decrease. Figures 25.5 and 25.6 also illustrate another theme of this chapter, the marked genetic variation within the brain dopamine system; DAT density, here calculated as the B_{Max} differs by more than 300% among the RI strains. This variation is not significantly associated with the number of DA neurons in the ventral tegmental area ($r = 0.13$) (ID# 10223) or the substantia nigra zona compacta ($r = -0.07$) (ID# 10224) (data not shown). Here it should be noted that unlike DAT density, among the RI strains variation in the number of dopamine neurons is relatively modest (~40%).

Measures of RI strain *striatal* D_2 receptor density in the WebQTL database were obtained independently in two studies (Jones et al., 1999; Hitzemann et al., 2003) that used different techniques and ligands to assess receptor binding. Given that there were only 14 RI strains in common to both studies, the agreement between the studies is relatively good. Figure 25.7 illustrates the agreement for the dorsal striatum ($r = 0.67$, $p < 0.008$) (ID# 10254 vs. ID# 10220); the strains with low and high density were readily identified in both studies—the strains with intermediate receptor density were not as reliable. Also note the marked variation in receptor density among the RI strains: ~100% in Hitzemann et al. (2003) and >300% in Jones et al. (1999). Using the binding data for the dorsomedial striatum (Hitzemann et al., 2003), the database was queried as was done for DAT. The most significant behavioral correlation ($r = 0.74$, $p < 0.0004$) was for a cocaine (45 mg/kg) response (here

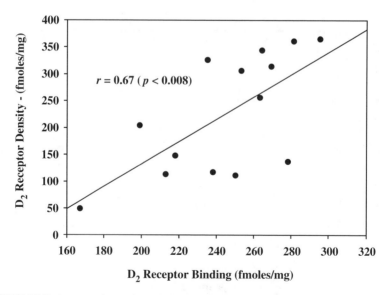

FIGURE 25.7 A comparison of two independent measures of striatal D_2 dopamine receptor density in a panel of the BXD recombinant inbred strains. Data were taken from Hitzemann et al. (2003) [ID#10220] and Jones et al. (1999) [ID# 10254]. The two studies were completely independent; both studies used 125I-epipride as the ligand but differed in strategy—autoradiography (Hitzemann et al., 2003) vs. membrane binding (Jones et al., 1999).

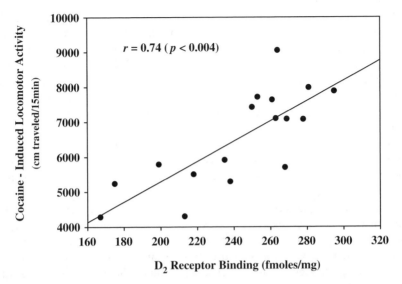

FIGURE 25.8 The relationship between striatal D_2 dopamine receptor density (ID# 10220) and cocaine (45 mg/kg)-induced locomotor activation (ID# 10318) in a panel of the BXD recombinant inbred strains. Receptor binding data are taken from Hitzemann et al. (2003); the cocaine response data are taken from Jones et al. (1999).

a difference score measure of locomotor activity- ID# 10318) (Jones et al., 1999) (Figure 25.8); the analogous data for Phillips et al. (1998)—see above—was also significantly correlated to receptor density ($r = 0.55$, $p < 0.03$) (data not shown).

Overall, the examples cited above illustrate how it is possible to use the RI strains to dissect the genetic relationships among dopamine systems and behavioral phenotypes. Importantly, it was found that individual cocaine responses are associated with different aspects of the brain dopamine system.

25.4 D_2 DOPAMINE RECEPTOR DENSITY AND *DRD2* EXPRESSION

In the previous sections we illustrated the marked differences among the RI strains in D_2 receptor density. It would be natural to assume that the variance (>300% in one study) is due to marked differences in *Drd2* expression. Some years ago, we and others concluded that there was a significant mismatch between receptor density and gene expression (e.g., Qian et al., 1992; 1993). Individual differences in D_2 receptor density were not associated with differences in *Drd2* expression; however, it was possible to show that the receptor gradients within the striatum (rostral- caudal, ventral-dorsal and lateral-medial) did parallel to a large extent the pattern of gene expression.

With the availability of brain gene expression data for the RI strains, it has been possible to return to the paradox of D_2 receptor density and gene expression with substantially greater statistical power. However, regardless of which measures of receptor density and gene expression from the WebQTL database are compared, the results are the same: there appears to be no relationship (see, e.g., Figure 25.9). Thus we again conclude that most of the variance in receptor density is associated with some factor or factors different from *Drd2* expression. Mapping all 12 measures of D_2 receptor density revealed moderate associations ($p < 0.005$) on Chr 15 and Chr 16 but none on Chr 9 near the *Drd2* locus (data not shown); in contrast, mapping *Drd2* expression reveals a moderate signal on Chr 9 near the gene locus (LOD ~2) and weaker signals on other chromosomes but none on Chr 15 and 16 (data not shown).

Given the mismatch between receptor density and gene expression, the question arises as to whether *Drd2* gene expression data per se is of some value to our understanding of the relationships among genetics, brain dopamine systems and behavior. We answer this question positively for several reasons. First, the possibility exists that receptor density and gene expression do tract together in some brain region that has not been measured. Areas that come to mind with potentially very salient behavioral effects include the prefrontal cortex, the central nucleus of the amygdala and the bed nucleus of the stria terminalis. Our attempts to measure receptor binding in these areas with quantitative autoradiography have (to date) not been successful. Second, as noted above, there is evidence that *Drd2* expression is important for explaining the receptor gradients within the brain and this perhaps is part of a larger pattern of brain organization that is important to the issue at hand—genetics, dopamine and behavior. Third, there is compelling evidence to show that

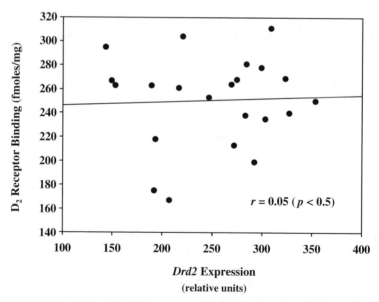

FIGURE 25.9 The relationship between *Drd2* expression and D_2 dopamine receptor density (ID# 10220) in a panel of the BXD recombinant inbred strains. Receptor binding data are taken from Hitzemann et al. (2003); *Drd2* expression data are from the Affymetrix U74A array and are posted at the WebQTL site (www.WebQTL.org).

Drd2 expression is strongly associated with behaviors for which there is strong evidence for dopamine modulation (Hitzemann et al., 2003). A query to the WebQTL database for phenotypes correlated to *Drd2* expression reveals that the strongest association ($r = 0.74$, $p < 0.00004$) is for ethanol-induced conditioned place preference (CPP)—time spent on the ethanol paired floor (ID# 10090) (Cunningham, 1995). CPP is considered to be a measure of a drug's hedonic or rewarding properties. These data are illustrated in Figure 25.10. Other behavioral phenotypes within the top 10 correlations include morphine two-bottle choice (ID# 5968) ($r = 0.60$, $p < 0.003$) and ethanol two-bottle choice (ID# 10479). Although the database is over-populated with drug-related phenotypes, the chance that 3 drug reward/consumption phenotypes would appear in the top 2% of all possible correlations is very small. The possibility must be considered that the associations with *Drd2* expression result from the close linkage of *Drd2* to other genes. In this regard, we have previously reported that ethanol two-bottle choice is highly correlated to *Ncam* expression (Hitzemann et al., 2003); *Ncam* is 96 kb from *Drd2*.

An alternative hypothesis is that *Drd2* is coexpressed with a family of genes and that it is the collective action of the gene family which modulates behavior. To test the first component of this hypothesis we turned to a C57BL/6J × DBA/2J F_2 brain gene-expression dataset. The data set contains expression results (Affymetrix 430 A and B arrays) for 56 F_2 animals formed from the reciprocal F_1 hybrids and is balanced for males and females. Genotypic data for this F_2 dataset have been obtained for >300 microsatellite markers (or on average 1 marker/5 cM). Both the genotypic

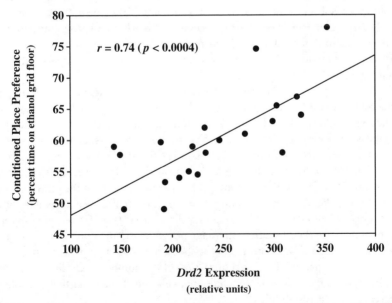

FIGURE 25.10 The relationship between *Drd2* expression and ethanol conditioned place preference (ID# 10090) in a panel of the BXD recombinant inbred strains. See legend to Figure 25.9 for details on expression data. Conditioned place preference data are taken from Cunningham (1995).

and phenotypic data have been posted at the WebQTL site. *Drd2* expression was significantly ($p < 10^{-6}$) associated with 38 unique transcripts (Table 25.1); assuming a total of 30,000 unique transcripts on both the A and B arrays, one would predict only 1 chance association. The SymAtlas database at http://www.gnf.org was queried to determine if the transcripts listed in Table 25.1 were specifically expressed in the striatum. For some transcripts data were available only for human brain expression. Expression in the striatum was rated on a four-point scale ranging from under-expressed to markedly and uniquely over-expressed in the striatum (compared to other brain regions). Fourteen of the 38 transcripts (37%) were moderately to significantly over-expressed in the striatum and these were by and large the transcripts most significantly associated with *Drd2* expression. Figure 25.11 illustrates the correlation between *Drd2* and preproenkephalin (*Penk1*) expression and *Drd2* and regulator of G-protein signaling 9 (*Rgs9*) expression. *Drd2* expression varied 49% across the F_2 sample while *Penk1* and *Rgs9* expression varied 94 and 73%, respectively. It should be noted (Table 25.1) that none of the genes significantly associated with *Drd2* expression were significantly differentially expressed between the B6 and D_2 strains. There are numerous implications to this paradox but importantly it suggests that screening gene lists for significant differences between the parental strains will significantly underestimate the number of genes showing differential expression in a segregating population.

The data in Figure 25.11 were collected from a whole brain sample; the question arises as to whether similar results would be obtained for striatal gene expression. Striatal data for an F_2 sample are not currently available; however, striatal data are available for a subset of the BXD RI strains. Figure 25.12 illustrates the correlation between *Drd2* and *Penk1* striatal expression in the RI strains; the correlation ($r = 0.6$) is similar to that seen in the F_2 sample, likely reflecting the balance between the increased sample power in the F_2 sample vs. the increased specificity of the striatal sample.

One could argue that strongest associations seen in Table 25.1 are simply an artifact related to differences in the size and/or number of cells in the striatum. Since only half the brain was used for the expression analysis, it was possible to determine if striatal size covaried with expression—no significant association was detected (data not shown). To determine if there is a relationship between gene expression and cell density requires a detailed stereological analysis. However, data are available in the BXD RI series for the number of cholinergic neurons in the striatum (Dains et al., 1996). A significant correlation ($r = 0.61$, $p < 0.001$) was detected between *Drd2* expression and the number of cholinergic neurons (ID# 10106) but only in the most rostral aspect of the striatum (data not shown); in more caudal areas, no significant correlation was detected but it should be noted that gene expression was not measured in a parallel fashion to measures of the number of cholinergic neurons. Thus, in lieu of other data, we conclude that cell density may partially account for the associations seen in Table 25.1.

Walker et al. (2004) have analyzed a rat brain gene expression dataset using a strategy somewhat similar to that described here. These authors asked the question what genes are coexpressed with *Drd1*. The results of their analysis (summarized in figure 2 of Walker et al., 2004) indicated that there was a large group of genes

TABLE 25.1
Transcripts Significantly ($p < 10^{-6}$) Associated with *Drd2* Expression[a]

Affymetrix ID#	Array	Gene Symbol	r	Striatal Expression
1455725	A	H3f3b	−0.62	−
1449373	A	Dnajc3	−0.61	−
1449314	A	Zfpm2	−0.60	−
1448113	A	4930403J22Rik	−0.60	−
1435250	A	2810013E07Rik	−0.60	−
1425023	A	Usp3	−0.60	−
1433582	A	1190002N15Rik	−0.60	−
1451864	A	Cacng8	0.60	−
1455085	B	1700086L19Rik	0.60	−
1424490	A	2410005H09Rik	0.61	+
1419066	A	Ier5l	0.61	+
1418782	A	Rxrg	0.62	+
1448790	A	Sema6b	0.62	+
1450944	A	Cspg4	0.62	+
1424132	A	Hras1	0.62	−
1434153	B	Shb	0.62	+
1455629	B	Drd1a	0.63	+++
1455609	B	C030025P15Rik	0.64	+
1417804	A	Rasgrp2	0.64	+
1456640	B	Sh3rf2	0.64	−
1455564	B	Bcr	0.64	++
1416050	A	Scarb1	0.64	−
1455701	B	Snx26	0.65	+
1422705	A	Tmepai	0.65	+++
1418881	A	Efcbp2	0.66	+++
1448327	A	Actn2	0.67	−
1442166	B	Cpne5	0.68	++
1432184	B	2610204M08Rik	0.69	+++
1460710	A	Adora2a	0.73	+++
1427343	A	Rasd2	0.74	+++
1455296	B	Adcy5	0.74	−
1427523	A	Six3	0.74	++
1455190	A	Gng7	0.75	+++
1451331	A	Ppp1r1b	0.76	+++
1418691	A	Rgs9	0.77	+++

(Continued)

TABLE 25.1

Transcripts Significantly ($p < 10^{-6}$) Associated with _Drd2_ Expression[a]
(Continued)

Affymetrix ID#	Array	Gene Symbol	r	Striatal Expression
1423544	A	_Ptpn5_	0.79	+++
1427038	A	_Penk1_	0.80	+++
1449420	A	_Pde1b_	0.88	++
1418950	A	_Drd2_	1.00	+++

[a] Brain transcript expression was measured in a sample of 56 C57BL/6J × DBA/2J F_2 animals using the Affymetrix 430A and 430B arrays. Data were analyzed using position dependent nearest neighbor algorithm (Zhang et al. 2003). _Drd2_ expression was then correlated with the expression of all other transcripts on the arrays; the threshold for significance was set at $p < 10^{-6}$. At this threshold, the expectation of a chance correlation is <1, assuming that there are 30,000 unique transcripts on the A and B arrays. Striatal expression data was taken from the GNF database and was semi-quantitatively judged from under-expressed (−) to highly over-expressed (+++).

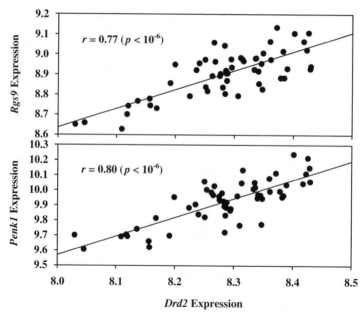

FIGURE 25.11 The relationship between _Drd2_ expression and the expression of _Penk1_ or _Rgs9_ in a sample of C57BL/6J × DBA/2J F2 animals. Details are found in the legend to Table 25.1.

coexpressed with _Drd1_ in the striatum (including the nucleus accumbens) and perhaps not surprisingly, their list overlaps with the genes listed in Table 25.1. Common genes in both lists were the following: _Drd1, Drd2, Penk 1, Ppp1r1b, Pde1b, Adcy5, Adora2a_ and _Rgs9_. Thus, data from two independent sources and two different species, mutually confirm a set of genes that includes _Drd2_ which are coexpressed in the striatum.

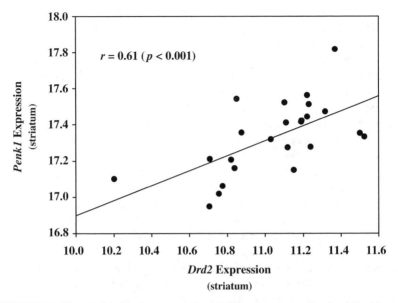

FIGURE 25.12 The relationship between striatal *Drd2* expression and striatal *Penk1* expression in panel of the BXD recombinant inbred strains. Expression data were obtained using the Affymetrix 430 2.0 array; data provided by Rosen G. and Williams R. (unpublished observations).

25.5 REGULATION OF *DRD2* EXPRESSION

As noted by Lee et al. (2003) "detailed analysis of the transcription control mechanisms that regulate dopamine receptor genes has revealed complex patterns involving both activators and repressors" (Minowa et al. 1992a, 1992b, 1993, 1994, Lee et al., 1996, 1997). Hwang et al. (2001) have identified a dopamine receptor regulating factor gene (*Drrf*) (alias: Kruppel-like factor 16 [*Klf16*]; basic transcription element binding protein 4 [*Bteb4l*]) which binds to GC and GT boxes in the promoters of both *Drd1* and *Drd2* and displaces Sp1 and Sp3 from these binding sites. Querying the databases noted above to determine if there is an association between *Drrf* and *Drd2* expression reveals no significant association ($r = 0.08$). Since *Drd1* and *Drd2* are co-expressed, factors known to regulate *Drd1* expression such as Mrf1(Meis2), TGIF, Zic2 and Sp3 (Yang et al., 2000a, 2000b) also become candidates for the regulation of *Drd2* expression. Of special note is the role Zic2 may have in the tissue specific distribution of *Drd1* expression (Yang et al., 2000b). On the basis of existing data, we have been unable to detect an association between the expression of these transcription factors and *Drd1* or *Drd2*.

An alternative approach to understanding gene co-expression is to ask if the key genes listed in Table 25.1 have common transcription factor binding motifs. There are several algorithms available to address this issue. Here we summarize the results using the oPOSSUM algorithm which is publicly available at http://www.cisreg.ca/cgi-bin/oPOSSUM/opossum. oPOSSUM is a system for determining the over-representation of transcription factor binding sites within a set of co-expressed genes as compared with a pre-compiled background set. The background data set was compiled by identifying all strict one-to-one human/mouse orthologs from the Ensembl database. The orthologs were aligned and the conserved noncoding regions were identified. The conserved regions 5 kb upstream and downstream from the transcription start site are scanned for common transcription factor sites among a group of coexpressed genes. Some genes from a coexpressed group are excluded from the analysis, e.g., if the conserved promoter regions could not be identified in a satisfactory manner. Of the 14 striatal coexpressed genes noted above, 7 were included in the analysis: *Drd2, Rgs9, Pde1b, Nptp, Penk1, CopineV* and *Adora2a*. The transcription factor binding site most enriched in this group (as compared with the reference group) was NF-kappaB [NF-κB] ($p < 0.0001$); four of the top ten enriched motifs were from the Rel/NF-kappaB family. These data cannot be unexpected. Previous studies have shown that both *Drd2* and *Penk1* expression are regulated by NF-κB (O'Neill and Kaltschmidt, 1997; Fiorentini et al., 2002). Other genes expressed in brain and known to be regulated by NF-κB include neural cell adhesion molecule (Simpson et al., 2000), inducible nitric oxide synthase (Madrigal et al., 2001), μ-opioid receptors (Kraus et al., 2003), brain derived neurotrophic factor (Lipsky et al., 2001), calcium/calmodulin dependent protein kinase II (Kassed et al., 2004) and amyloid presursor protein (Grilli et al., 1995). Thus, we must conclude that although the striatal coexpressed genes all have the NF-κB motif, the cause of the *striatal* specific expression must be associated with other factors. However, as noted by Meffert and Baltimore (2005), there is increasing evidence that the NF-κB family of transcription factors are involved in the regulation of neuronal activity-dependent transcription and *behavior.*

25.6 CONCLUSIONS

The intent of this chapter was to provide the interested reader with an introduction to the topic of genetics, brain dopamine systems and behavior. The focus was almost entirely on data collected in mice; however, some data suggest that these results can be extrapolated to other species, including man. Key findings which have been replicated numerous times are (a) the marked genetic differences in behavioral responses to dopamine agonists and antagonists, (b) the marked genetic variation in at least some elements of the brain dopamine systems—notably the D_2 dopamine receptor and the dopamine transporter and (c) the relationships between "a" and "b." The reasons for the receptor and transporter variation(s) are unclear but some strategies for dissecting these remarkable genetic effects are presented.

ACKNOWLEDGMENTS

The author would like to thank the many students, fellows and collaborators that have contributed in some way to the research described in this chapter. Special thanks to Glen Rosen, Rob Williams and colleagues for the BXD recombinant inbred striatal gene expression data. This work was supported in part by grants from the US Public Health Service—AA 11034, AA 13484 and MH 51372—and support from the Department of Veterans Affairs Medical Research Program. The Gene Network WebQTL site featured prominently in the review was funded in part by MH P20-62009 to Rob Williams.

REFERENCES

Chesler, E.J. et al., WebQTL: rapid exploratory analysis of gene expression and genetic networks for brain and behavior, *Nat Neurosci*, 7, 485, 2004.

Cunningham, C.L., Localization of genes influencing ethanol-induced conditioned place preference and locomotor activity on BXD recombinant inbred mice, *Psychopharmacol*, 120, 28, 1995.

Dains, K., Hitzemann, B., and Hitzemann, R., Genetics, neuroleptic response and the organization of cholinergic neurons in the mouse striatum, *J Pharmacol Exp Ther*, 279, 1430, 1996.

Fink, J.S., Swerdloff, A., and Reis, D.J., Genetic control of dopamine receptors in the mouse caudate nucleus: relationship on cataleptic response to neuroleptic drugs, *Neurosci Lett*, 32, 301, 1982.

Fiorentini, C. et al., Nerve growth factor regulates dopamine D2 receptor expression of prolactinoma cell lines via p75NGFR-mediated activation of nuclear factor -B, *Molecular Endocrinology*, 16, 353, 2002.

Fowler, S.C., Zarcone, T.J., and Vorontsova, E., Haloperiodol-induced microcatalepsy differs in CD-1, BALB/c, and C57BL/6 mice, *Exp Clin Psychopharmacol*, 9, 277, 2001.

Grilli, M. et al., Identification and characterization of a kappa B/Rel binding site in the regulatory region of the amyloid precursor protein gene, *J Biol Chem*, 270, 26774, 1995.

Hitzemann, R.J. et al., On the selection of mice for haloperidol response and non-response, *Psychopharmacol*, 103, 244, 1991.

Hitzemann, R. et al., Genetics and the organization of the basal ganglia, *Int Rev Neurobiol*, 7, 485, 1995.

Hitzemann, R. The regulation of D2 dopamine receptor expression, *Mol Psych*, 3, 198, 1998.

Hitzemann, R. et al., Dopamine D2 receptor binding, Drd2 expression and the number of dopamine neurons in the BXD recombinant inbred series: genetic relationships to alcohol and other drug associated phenotypes, *Alcohol Clin Exp Res*, 27, 1, 2003.

Hwang, C.K. et al., Dopamine receptor regulating factor, DDRF: a zinc finger transcription factor, *Proc Natl Acad Sci USA*, 98, 7558, 2001.

Janowsky, A. et al., Mapping genes that regulate density of dopamine transporters and correlated behaviors in recombinant inbred mice, *J Pharmacol Exp Ther*, 298, 634, 2001.

Jones, B.C. et al., Quantitative-trait loci analysis of cocaine-related behaviors and neurochemistry, *Pharmacogenetics*, 9, 607, 1999.

Kanes, S., Hitzemann, B.A., and Hitzemann R.J., On the relationship between D2 receptor density and neuroleptic-induced catalepsy among eight inbred mouse strains, *J Pharmacol Exp Ther*, 267, 538, 1993.

Kanes, S. et al., Mapping the genes for haloperidol-induced catalepsy, *J Pharmacol Exp Ther*, 277, 1016, 1996.

Kassed, C.A. et al., Injury-induced NF-B activation in the hippocampus: implications for neuronal survival, *FASEB J*, 18, 723, 2004.

Kraus, J. et al., The role of nuclear factor kappaB in tumor necrosis factor-regulated transcription of the human μ-opioid receptor gene, *Mol Pharmacol*, 64, 876, 2003.

Lee, S-H., Minowa, M.T., and Mouradian, M.M., Two distinct promoters derive transcription of the human D1A dopamine receptor gene, *J Biol Chem*, 271, 25292, 1996.

Lee, S-H. et al., Tissue-specific promoter usage in the D1A dopamine receptor gene in brain and kidney, *DNA Cell Biol*, 16, 1267, 1997.

Lee, S-H. et al., Genomic organization and promoter characterization of the murine dopamine receptor regulator factor (DRRF) gene, *Gene*, 304, 193, 2003.

Lipsky, R.H. et al., Nuclear factor kappaB is a critical determinant in N-methyl-D-aspartate receptor-mediated neuroprotection, *J Neurochem*, 78, 254, 2001.

Madrigal, J.L. et al., Inducible nitric oxide synthase expression in brain cortex after acute restraint stress in regulated by nuclear factor B-mediated mechanisms, *J Neurochem*, 76, 532, 2001.

Meffert, M.K. and Baltimore, D., Physiological function for brain NF-kappaB, *Trends in Neurosci*, 28, 37, 2005.

Minowa, M.T., Characterization of the 5' flanking region of the human D1A dopamine receptor gene, *Proc Natl Acad Sci USA*, 89, 3045, 1992a.

Minowa, M.T., Characterization of the 5' flanking region of the human D2 dopamine receptor gene, *Biochemistry*, 31, 8389, 1992b.

Minowa, M.T., Minowa, T., and Mouradian, M.M., Activator region analysis of the human D1A dopamine receptor gene, *J Biol Chem*, 268, 23544, 1993.

Minowa, M.T., Minowa, M.T., and Mouradian, M.M., Negative modulator of the rat D2 dopamine receptor gene, *J Biol Chem*, 269, 11656, 1994.

O'Neill, L.A.J. and Kaltschmidt, C., NF-kappaB: a crucial transcription factor for glial and neuronal cell function, *Trends Neurosci*, 20, 252, 1997.

Patel, N.V. and Hitzemann, R.J., Detection and mapping of quantitative trait loci for halo-peridol-induced catalepsy in a C57BL/6J x DBA/2J F2 intercross, *Behav Genet*, 29, 303, 1999.

Phillips, T.J., Huson, M.G., and McKinnon, C.S., Localization of genes mediating acute and sensitized locomotor responses to cocaine in BXD-By recombinant inbred mice, *J Neurosci*, 18, 3023, 1998.

Qian, Y., Hitzemann, B., and Hitzemann, R., D1 and D2 dopamine receptor distribution in the neuroleptic non-responsive (NNR) and neuroleptic responsive (NR) lines of mice—a quantitative receptor autoradiographic study, *J Pharmacol Exp Therap*, 261, 341, 1992.

Qian, Y. et al., D1 and D2 dopamine receptor turnover and D2 mRNA levels in the neuroleptic responsive (NR) and neuroleptic non-responsive (NNR) lines of mice, *J Pharmacol Exp Ther*, 267, 1582, 1993.

Simpson, C.S. and Morris, B.J., Regulation of neuronal cell adhesion molecule expression by NF-kappaB, *J Biol Chem*, 275, 16879, 2000.

Taylor, B.A., Recombinant inbred strains: use in gene mapping, in *Origins of Inbred Mice*, Morse, H.C., Ed., Academic Press, New York, 1978, pp 428–438.

Volkow, N.D. et al., Decreased dopamine-D(2) receptor availability is associated with reduced frontal metabolism in cocaine abusers, *Synapse*, 14, 169, 1993.

Volkow, N.D. et al., Prediction of reinforcing responses to psychostimulants in humans by brain dopamine D2 receptor levels, *Am J Psych*, 156, 1440, 1999.

Walker, J.R. et al., Applications of a rat multiple tissue gene expression data set, *Genome Research*, 14, 742, 2004.

Yang, Y. et al., Three-amino acid extension loop homeodomain proteins Meis2 and TGI differentially regulate transcription, *J Biol Chem*, 275, 20734, 2000a.

Yang, Y. et al., ZIC2 and Sp3 repress Sp1-induced activation of the human D1A dopamine receptor gene, *J Biol Chem*, 275, 38863, 2000b.

Zhang, L., Miles, M.F., and Aldape, K.D., A model of molecular interactions in short oligo-nucleotide array, *Nat Biotechnol*, 21, 818–821, 2003.

26 Natural Genetic Variation of Hippocampal Structures and Behavior—an Update

*Hans-Peter Lipp, Irmgard Amrein,
Lutz Slomianka, and David P. Wolfer*

CONTENTS

ABSTRACT

This chapter analyzes the natural genetic variability of the so-called intra/infrapy-ramidal mossy fiber (IIP-MF) projection in mice and rats, which terminates upon the basal dendrites of hippocampal pyramidal cells, and its relations to behavior. The analysis included thus far the following steps: (i) identification of structural traits sensitive to selective breeding for extremes in two-way avoidance, (ii) testing

389

the robustness of the associations found by studying individual and genetic correlations between hippocampal traits and behavior, (iii) confirming causal relationships by manipulating the structural variable in inbred (isogenic) strains, thereby eliminating the possibility of genetic linkage, (iv) further ruling out the possibility of spurious associations by studying the correlations between the hippocampal trait and other behaviors known to depend on hippocampal functioning, (v) searching for behavioral correlates of the IIP-MF variation in behaviors thought to be unrelated to hippocampal function, (vi) testing the effects of natural selection on differential IIP-MF distributions in naturalistic environments, and (vii), studying the variability of the IIP-MF distribution in small mammals with differential lifestyles. Taken together, the data imply that the hippocampus mediates a considerable number of behaviorally distinct processes, a view in accordance with the connectivity of this structure. The common denominator may be that variations of the IIP-MF entail differences in the stability of parallel hippocampal processing.

The advantages of this approach are: (i) hippocampal traits can be manipulated simply by controlled breeding (no manipulative approach can match this experimental finesse thus far); (ii) it identifies natural regulation sites for behavior; and (iii) it offers an easy way to discover possible behavioral functions of the hippocampus. The disadvantages are: (i) the study of the underlying causality depends on the detection of genes responsible for the trait; (ii) the behavioral effects are typically smaller than after invasive manipulations, and this entails a greater sensitivity to environmental influences; (iii) correlative studies often face traditionally more skepticism than invasive manipulations; and (iv) the rising costs of maintaining breeding colonies of mice for classical behavior genetic studies is slowly suffocating this approach.

26.1 INTRODUCTION

In any mammalian species, no brain is like another, and every individual behaves differently. Some of this variability obviously is dependent on the environment. On the other hand, a considerable portion must be caused by biochemical and structural variations of the brain that originate from both exogenous influences and actions of genes during brain development. This chapter will deal with the biology of traits, meaning the study of individual structural or behavioral differences that lie within the quantified normal range of a population, in contrast to abnormal, i.e., pathological, differences that are most obvious in so-called neurological mutants.

Since behavioral traits appear to be complex, they have been attributed to many genes, more precisely, to mutated alleles, which were thought to contribute small effects adding to a phenotype. More recently, gene-targeting techniques allow for selective inactivation of genes thought to be important for memory and learning and other aspects of behavior. Such studies have revealed a remarkable capacity of the genome and the brain to compensate for targeted mutations. However, such compensation is variable, and this variable penetrance of targeted deletions often results in changes resembling a quantitative trait. Hence, it appears reasonable to assume that at least part of the observed behavioral variability of genetic origin is not based on the additive action of hundreds of genes but on the effects of a few major genes, and, perhaps, on single locus effects.

A theoretical problem is whether such single locus effects could be specific for behavioral talents and motivational traits. Since activation of genes is thought to precede brain development, this would seem unlikely. Most genes regulating brain development are probably affecting many different neuronal populations, and their effects must thus be pleiotropic. Likewise, mutation of genes controlling the basic physiology of neurons and glial cells must show pleiotropic effects. There is an exception however. Genes that are activated late during development can modulate only the final steps of brain differentiation,[1] a period during which only a few forebrain systems undergo their final formation. The known late-forming systems include portions of the olfactory bulb, the hippocampal formation, in particular the dentate gyrus, and parts of the cerebellum. Thus, late-acting alleles that influence the connective organization, receptor regulation, and neurochemistry of these structures are likely to be fairly specific for learning, cognition, and motivation. Such mutated alleles will cause behavioral changes not because of a particular nucleic acid sequence but by virtue of timing. Another mechanism for behavioral selectivity is possible if brain structures known to modulate behavior show morphological or biochemical features not found elsewhere in the brain.

26.2 THE HIPPOCAMPAL MOSSY FIBER PROJECTION

A structure that shows both late differentiation and morphological/biochemical specificity is the dentate gyrus of the hippocampal formation. It contains a secondary proliferation zone for neurons, remote from the normal periventricular matrix. The neurons (the granule cells) send axons called mossy fibers (MFs) to the pyramidal cell layer of hippocampal region (CA3), to the hilus of the dentate gyrus (CA4), and recurrent collaterals to the granule cell layer (Figure 26.1). The axons to CA3 terminate in a layer above the pyramidal cells (the suprapyramidal mossy fiber layer, SP-MF), and in a much narrower band and more patchily on the basal dendrites, the intra/ infrapyramidal mossy fiber projection (IIP-MF). Barber et al. first reported genetic variation of the IIP-MF projection for inbred mouse strains.[2] Since the terminals of the mossy fibers can be visualized with clarity by means of Timm's stain, the area of their distribution on brain sections can be measured easily by means of stereology or digital image analysis.

Structural differences between inbred mouse strains imply that variations of the IIP-MF projection must have reduced adult plasticity—a desirable feature for studying structural variations related to behavior. One should notice that this reduced plasticity appears to be more pronounced in the CA3 region, where the mossy fiber axons terminate in highly specialized synaptic complexes. Recurrent collaterals of the same mossy fibers making more simple synapses on granule and basket cells of the dentate gyrus are known for reactive sprouting after injuries in rats and humans and for lifelong growth in guinea pigs. Clearly, systems showing pronounced adult growth and plasticity are less convenient if one wishes to discover hereditary covariations between behavioral and brain traits. Thus, the IIP-MF projection to CA3

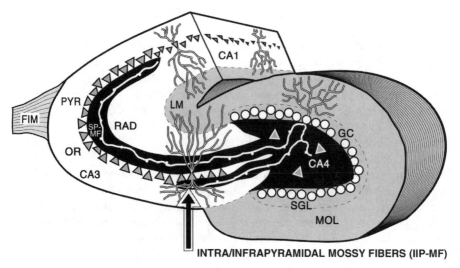

INTRA/INFRAPYRAMIDAL MOSSY FIBERS (IIP-MF)

FIGURE 26.1 Schematic view of a cross-section of the rodent hippocampal formation. The terminal fields of the hippocampal mossy fiber projection appear in black. Abbreviations: CA1, CA3, and CA4, hippocampal subfields; FIM, fimbria fornicis; GC, granule cells; LM, stratum lacunosum-moleculare; MOL, molecular layer of dentate gyrus; OR, stratum oriens; PYR, stratum pyramidale; RAD, stratum radiatum; SGL, supragranular layer of dentate gyrus.

appears to be a promising candidate structure for testing whether its variability might be related to hippocampus-dependent behaviors.

26.3 SELECTIVELY BRED STRAINS AS TOOLS FOR TESTING BRAIN–BEHAVIOR RELATIONS

A methodological problem in finding behavioral correlates of a brain trait is selecting appropriate behavioral tests. Unfortunately, they are numerous and can be varied infinitely. Thus, finding correlations remains a matter of chance. A more economical approach is to use mouse or rat strains that have been selectively bred for extremes in a behavior thought to be governed by a particular brain region. There are many examples of behavioral divergence after so-called artificial selective breeding (for a short review see Lipp and Wolfer[3]). For example, rats have been bred for high vs. low two-way avoidance conditioning, a behavior often improved by hippocampal lesions. These strains, in this case the roman high and low avoidance rats,[4] are useful to decide quickly whether a particular trait (which must be known to show genetic variation) differentiates after selective breeding. If so, it is a candidate for a brain system underpinning the behavior; if not, the system investigated is probably not relevant for causing individual differences in behavior. Investigations of a sample of RHA and RLA rats showed, indeed, a significant difference in two of the three MF projection fields, namely the size of the CA4 projection and the IIP-MF projection, which appeared enlarged in the poor avoiders (Figure 26.2).

These results confirmed a potential role of the IIP-MF variations, but posed the question whether it might not be the size difference of CA4, which was somehow

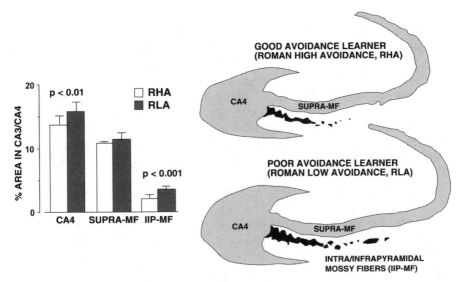

FIGURE 26.2 Selective breeding for extremes in two-way avoidance learning of rats and differentiation of the mossy fiber projections (modified after Lipp and Wolfer, 1995).[3] Note that the percentage of the IIP-MF in Figs. 26.3, 26.4, and 26.8 have been calculated in relation to the entire area of CA3/CA4 as measured by manual planimetry used for the older studies. In more recent publications, the use of digital morphometry permitted to measure only the three mossy fiber subfields. Hence, Figs. 26.5, 26.6, 26.7, and 26.9 give only the percentage of IIP-MF in relation to the size of the suprapyramidal mossy fiber layer, correcting in a simple way for differences in hippocampal size. Since the size of the suprapyramidal MF projection is a good covariate of the total area of CA3/CA4, the two indices are strongly correlated, however.

related to behavior, or another hidden covariate in the brain—the eternal question in correlative brain research. Because selective breeding for a trait acts presumably on more than one brain mechanism, the observed association might simply be dependent on another unknown system in the brain that shows genetic covariation with the MFs, the latter being irrelevant for the behavior. In the worst case, differences between selectively bred strains might be based simply on genetic drift. In such a case, small breeding populations will become increasingly homozygous for alleles. With bad luck, one line will then be homozygous for alleles facilitating avoidance learning, and the other line for alleles impairing it, without any functional relation to the selective breeding process. Hence, brain–behavior associations found in selectively bred strains must be analyzed further in order to test whether the association was spurious.

26.4 VERIFYING ASSOCIATION BETWEEN TRAITS: GENETIC AND PHENOTYPIC CORRELATIONS

The most easy way to delineate potential correlations between behavioral and brain traits is to study a sample of inbred mouse strains known to show differences in these traits. Genetic correlations can then be computed between the strain means of either trait. These values can be obtained from publications or can be assessed by studying a group of individuals. In the case of the IIP-MF and two-way avoidance learning, it was found that the rank orders of the two variables showed a very strong negative correlation [$r = -0.96$, $p < 0.01$, (Figure 26.3a)]. Again, poorly avoiding strains showed the largest IIP-MF projections, but there was no correlation with the size of the other mossy fiber fields. Genetic correlations between strain means are useful when the behavioral variable shows large epigenetic (environment-dependent) variability, blurring the influence of the brain variable. Their weakness is that they depend on a small number of strains that appear as data points in the correlations: one outlier strain can annihilate an existing real correlation, or an unlucky choice of strains can fake it.

This problem can be solved by studying phenotypic correlations, that is, correlations using individuals in which both traits are measured. Two extreme strains in terms of IIP-MF and two-way avoidance learning were intercrossed, namely DBA/2 and C3H and the resulting F2 generation of 51 animals was then tested and their hippocampi morphometrized.[4] In such a sample, other alleles potentially affecting two-way avoidance or IIP-MF projections are scrambled during meiosis and therefore might mask a weak phenotypic correlation. Nonetheless, the analysis revealed again a strong negative correlation [$r = -0.82$, $p < 0.001$, (Figure 26.3b)]. Thus, the association between avoidance learning and IIP-MF as observed in rats and inbred strains was confirmed again, and turned out to be insensitive against potentially confounding alleles. In other words, the correlation was functionally robust. This step of the analysis clarified also causal relations: the distribution of the IIP-MF is established in the postnatal period between days 10 to 20 and remains rather stable afterward. Because the variability of the IIP-MF was obtained "experimentally" (by means of meiotic scrambling of alleles), and because the effects of this manipulation predict adult behavior, one is no longer dealing with a simple correlation without causal implications (for example length of arm vs. length of leg), but with a regression of behavior on a structural variable. Nonetheless, phenotypic correlations have two weak points. One is genetic linkage (more precisely, linkage disequilibrium), in which case a chromosomal segment may contain both an allele relevant for avoidance learning and another one for the size of the IIP-MF projection. One may note that this is also a problem in gene targeting.[5] A more subtle point is that producing hybrids between strains with behavioral extremes often results in hybrid vigor: the filial generations outperform both parental strains. The reasons for hybrid vigor are unknown, but it may improve complex learning behavior to a point where it can mask strain-specific factors. See also Wolfer et al. for an example of this problem in knockout mice.[6]

26.5 SOLUTIONS FOR THE LINKAGE PROBLEM

This issue cannot always be solved. The prerequisite is to have an inbred strain in which the brain trait can be manipulated experimentally, or nearly identical substrains that show differences in the brain trait. For the IIP-MF behavior relation, both approaches were feasible. The IIP-MF distribution can be enlarged experimentally by means of postnatal thyroxin injections during the critical growth period of the

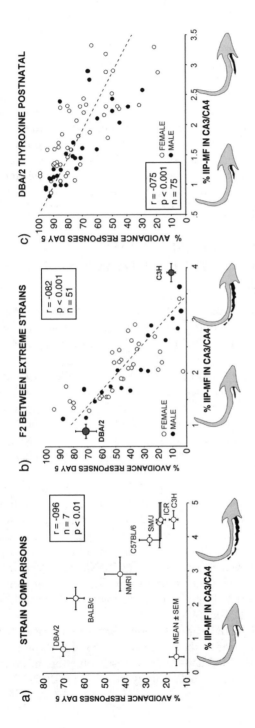

FIGURE 26.3 Variations of the infrapyramidal mossy fiber projection and avoidance learning in mice (modified after Lipp et al., 1989).[4] (a) genetic correlation between the strain means of the IIP-MF projection and avoidance learning at the last day of training (day 5); (b) phenotypic correlation between IIP-MF projection and avoidance learning in individual mice from an F2 cross between the inbred strains DBA/2 and C3H; (c) developmentally induced variation of the IIP-MF by means of thyroxin injections and adult two-way avoidance learning.

MFs. This treatment was applied to pups of the mouse strain DBA/2, a good avoidance learner with scanty IIP-MF projections. It resulted in considerable, partially dose-dependent variability of the IIP-MF projection that remained strongly and negatively correlated with the avoidance learning of the adult individuals [$r = -0.75$, $p < 0.0001$[4] (Figure 26.3c)]. Because all animals of this strain are isogenic, chromosomal linkage cannot account for the correlation between the two traits. Rather, it appears that size variations of the IIP-MF predict two-way avoidance learning, regardless of whether the structural variation has been caused by genetic or epigenetic factors. A further confirmation was then found by the group of Crusio reporting subline differentiation in radial maze learning (that is, other hippocampus-dependent behavior, see below) in the mouse strain C57BL/6.[7] This difference was found to be associated with a difference in the extent of the IIP-MF projection. Since the sublines can differ at only a few loci at best, these findings are a strong argument against chromosomal linkage.

26.6 EXTENDING THE BEHAVIORAL ANALYSIS: CORRELATIONS WITH OTHER HIPPOCAMPUS-DEPENDENT TASKS

The studies in avoidance learning suggested that this talent was somehow dependent on intrahippocampal size variations of the mossy fibers, but it was not clear whether there might be a hidden variable in the brain outside the hippocampal formation that was responsible for the learning differences. This variable might be concomitantly activated by genetic factors or thyroxin fluctuations, both of them activating the growth of the hippocampal IIP-MF. Such pleiotropy appeared not very likely, but nonetheless made control studies necessary. Thus, samples of mice were tested for other so-called hippocampal behavior, that is, lesion-induced impairment of learning or exploration in spatial test situations. If they would show correlations with the IIP-MF projection again, a spurious correlation with a nonhippocampal variable should become an untenable hypothesis.

Correlations between the extent of the IIP-MF projection and performance in several of these paradigms indeed were found. In most, superior performance was now associated with extended IIP-MF projections. For example, mice with extended IIP-MF projections made fewer working memory errors in a relatively small radial maze[8] (also see below). This was found in form of a genetic correlation (Figure 26.4a) and after developmental stimulation of IIP-MF growth by means of thyroxin (Figure 26.4 b, c). Likewise, an inbred rat strain with superior radial maze performance showed enlarged IIP-MF projection, while another strain with smaller IIP-MF was superior in multiple T-maze learning.[9]

Selective breeding for differential rearing (a form of mouse exploratory behavior in the open field assumedly mediated by the hippocampus also) showed an associated increase in the extent of the IIP-MF,[8] and similar observations were made in mice selectively bred for high and low open-field activity[10] (see Figure 26.5).

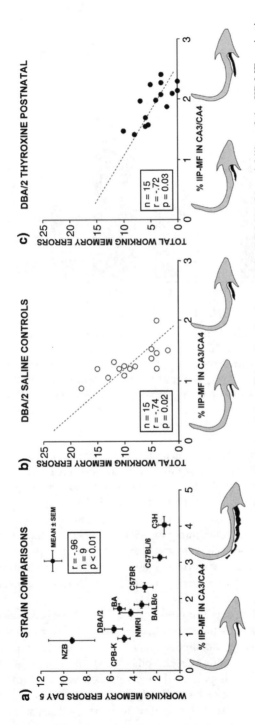

FIGURE 26.4 (a) Genetic correlation between IIP-MF and radial maze learning (strain means ± S.E.M.); (b) variability of the IIP-MF projection as observed after postnatal saline (sham) injections. Note that the total error scores are higher than those of the thyroxin-injected mice, and that a correlation can be found; (c) reduced error scores of mice having received postnatal injections of thyroxin showing a similar correlation (modified after Lipp and Wolfe, 1995).[3]

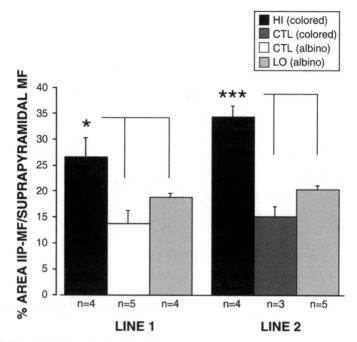

FIGURE 26.5 Activity and IIP-MF in mice selectively bred for differential open-field activity. Highly active mice show larger IIP-MF projections.[10] Abbreviations: HI, two mouse lines selectively bred for high activity in the open field, both of them pigmented; LO, two mouse lines selectively bred for low activity in the open field, both of them albinotic; CTL, two randomly bred mouse lines, except that one was selected for pigmented coat color, the other for albinism.

In the Morris water maze, phenotypic correlations were found, too.[11,12] However, the variation of the IIP-MF was correlated not with the learning of the task, but with the ability of relearning a new platform location, which was superior in mice with extended IIP-MF projections (Figure 26.6). Curiously, the correlation appeared to be better in the left hippocampus. Subsequent studies showed that the asymmetry of the IIP-MF projection was contributing to this kind of performance as well; mice with larger IIP-MF projections at left crossed over the former platform position more during the first day of reversal, but reoriented faster toward the new platform location the following day. The worst relearners were animals with small symmetrical IIP-MF projections.[11,12]

These studies indicated that variations of the IIP-MF projections (or of an intrahippocampally linked structural covariate) were probably associated with a physiological process inside the hippocampus causing behavioral differences in spatial learning and exploration. On the other hand, the results did not blend well into unitary concepts of hippocampal function, such as spatial reference[13] vs. working memory.[14] Depending on the task, they appeared to correlate once with working memory, once with spatial reference memory, and seemed to be also involved in behavioral flexibility and perhaps hemispheric lateralization.

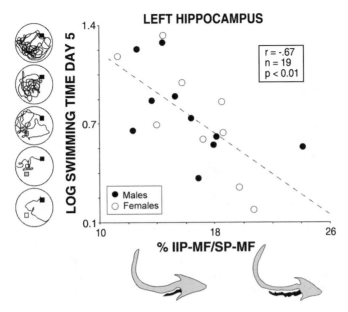

FIGURE 26.6 Correlation between IIP-MF and escape performance from the Morris water maze in the left hippocampus of 19 mice from a randomly bred stock generated by means of a 4-way diallel cross. Day 5 is the second day after platform reversal. Note superior performance of mice with enlarged IIP-MF projections. Along the ordinate typical swim paths (modified after Bernasconi-Guastalla et al. 1994).[12]

26.7 SEARCHING FOR "NON HIPPOCAMPAL" BEHAVIORAL CORRELATES

Initially, the term hippocampus-dependent behavior denoted a behavior sensitive to or abolished by lesions of the hippocampus but not other brain structures. This distinction has never been validated thoroughly because in most lesion studies, the number of control structures (mostly neocortex) was limited. Nonetheless, it was adopted by many studies using targeted disruption of genes and shortened to the term "hippocampal" implying a category of behavior exclusively mediated by the hippocampus. The most frequent example is swimming-navigation learning, which is impaired or abolished after hippocampal lesions, and with it the ability to remember the location of or to find a hidden target platform. Since hippocampal lesions usually spare the capacity of finding a cued (visually marked) platform, it is assumed that this form of escape learning is mediated by nonhippocampal structures and is thus frequently labeled as "nonhippocampal." This distinction is not logical, since there is no evidence that the hippocampus is not participating in comediating "nonhippocampal behavior," such as visually cued water escape learning. The only safe conclusion is that an intact hippocampus is more important for complex learning than for a simpler form. In addition, there is enough evidence that other brain regions can impair swimming navigation, particularly its procedural components. Nevertheless, it has become current practice to contrast behavioral effects of targeted gene deletions as "hippocampal," usually with the connotation "cognitive," vs. "nonhippocampal" and "noncognitive."

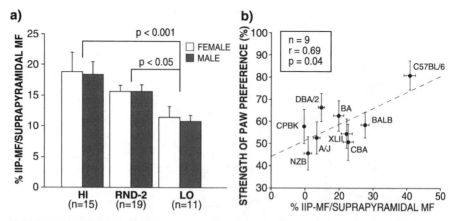

FIGURE 26.7 IIP-MF variations and strength of paw preference. (a) Mice with strong paw preference (HI) show the largest IIP-MF projections, the smallest are found in mice with weak paw preference. Random-bred controls (RND-2) lie in-between; (b) weak genetic (strain mean) correlation between IIP-MF and strength of paw preference in inbred strains of mice (modified after Lipp et al., 1996).[15]

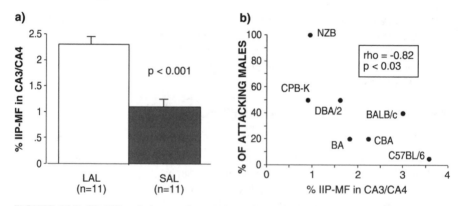

FIGURE 26.8 IIP-MF variations and attack latencies toward an intruder into the home cage. (a) Mice selectively bred for long attack latencies show larger IIP-MF projections;[16] (b) percentage of attacking males is higher in mouse strains with scanty IIP-MF projections (modified after Guillot et al., 1994).[17]

With the appearance of IIP-MF variations as good markers for variations of intrahippocampal physiological processing, it became possible to test whether the hippocampus was also comediating some of the so-called "nonhippocampal" behaviors, which are thought to be nonlearned and nonspatial. Lipp et al.[15] were testing whether the IIP-MF distribution would correlate with paw preference of mice. They found that selective breeding for strong vs. weak paw preference was associated with larger IIP-MF in the mice with consistent paw preference (Figure 26.7) and that asymmetries of the IIP-MF distribution were weakly but significantly correlated with ipsilateral paw preference. Sluyter et al.[16] found that selective breeding for short vs. long attack latencies in the male intruder paradigm was associated with smaller IIP-MF projections and found a similar genetic correlation in inbred strains[17] (Figure 26.8).

26.8 CONFLICTING DATA

This short and incomplete review has focused mainly on our own studies and on some data from the groups of Crusio and Schwegler. Other laboratories have reported partially different results whose detailed evaluation, however, would go beyond the scope of this chapter. There appears to be agreement that variations of the IIP-MF correlate positively but sometimes weakly with exploratory behavior in mice[18,19] Prior et al.[20] found, by comparing two lines, that mice bred for high male aggression levels had larger IIP-MF projections yet were more fearful and less explorative. On the other hand, in the more exploratory line with smaller IIP-MF projections, the size of the IIP-MF projection correlated positively with exploratory-like behavior while no such within-line correlation was found in the aggressive animals. Such strain dependencies of correlations of the IIP-MF projections were also observed in other studies.[21,22]

On the other hand, there were studies reporting missing genetic correlations between IIP-MF and radial maze learning abilities,[23] and mouse strains with differential avoidance learning abilities did not show differences of the IIP-MF distribution.[24] Occasionally, size variations of the other MF fields also were associated or correlated with behavioral scores,[12,18,25] but these relations cannot be discussed here.

The discrepancies found for radial maze learning may reflect procedural variations. Correlations as reported by Crusio and others were observed in a comparatively small radial maze at ground level, while the negative reports from the laboratory of Roullet and Lassalle[23] were based on the use of a much larger elevated radial maze. Possibly, such differences could entail a differential use of cues and favor differential strategies, thus masking or unmasking participation of factors related to the IIP-MF projection. Also, it should be kept in mind that IIP-MF variations are only one cerebral trait among many others that might influence behavior.

In the past decade, there have been several reports of altered MF distributions in genetically engineered mice. In many cases, these alterations appear to be primarily markers of anomalous brain development, since the final adult distribution of hippocampal MFs is the result of proper timing in generation, migration and axonal growth of pyramidal cells, and, separately, of granule cells that are generated in an ectopic proliferation zone in the hilus of the dentate gyrus. Obviously, behavioral phenotypes of such mutant mouse lines may have multiple causations and it is difficult to disentangle the possible contributions from the targeted systems and altered MF projections. For example, mice lacking the mineral corticoid receptor show a strongly extended IIP-MF projection and many behavioral symptoms of hippocampal lesions, but it is difficult to judge whether their rather moderate impairment in swimming-navigation learning reflects partial compensation by the IIP-MF projection, or a minor role of the hormone receptors in this task. Apart from developmental changes being reflected in relative permanent changes of the MF distribution (as observed in normal strain differences), mutant mice may show adult MF sprouting due to epilepsy, apoptosis of target neurons and/or changes in adult neurogenesis of the granule cells themselves that can produce both altered patterns of mossy fiber distribution and behavioral changes. Finally, interpretations should be made carefully when the changes in MF distribution in mutant mice are subtle and have been assessed only in a limited portion of the hippocampus. The extent of the IIP-MF decreases gradually from rostral (septal) to ventral (temporal) levels of the hippocampus, while recurrent MF collaterals in the dentate granule cell layer are abundant in ventral parts and almost missing in the rostral portion of the hippocampus. Thus, subtle mutation-induced changes in hippocampal size and form can fake a difference in MF distribution when compared at one hippocampal plane only.

26.9 EMERGING HYPOTHESES

The findings of correlations between IIP-MF and "noncognitive" behaviors did and do not fit well into prevailing theories about hippocampal functions, which require considerable bending to explain hippocampal contributions to paw preference and attack behavior. A less theoretical explanation is simply to accept multifunctionality of the hippocampus as it is reflected by the diversity of inputs and outputs at the cortical and subcortical levels. Thus, one would assume that the behavioral function of the dorsal hippocampus with its connections to the prefrontal cortex is related to movement planning and spatial processing, while the ventral-most parts that project to the ventromedial nucleus of the hypothalamus and the amygdala might be more relevant to emotional/aggressive behavior. Such parallel processing of multiple inputs and outputs requires temporal stabilization of activity as well as some form of shielding against interference from neuronal activity in neighbored hippocampal zones involved in different activities. Neuroanatomically, the simplest explanation would be that extended IIP-MF projections provide better protection against interference arising from neighboring regions or other afferent subcortical and cortical systems, simply because they occupy a prominent position in controlling the firing behavior of the CA3 pyramidal cells, diminishing the impact of entorhinal and intrahippocampal connections. Thus, activity in a given channel or segment along the septo–temporal axis remains confined according to the amount of MF terminals on both apical and basal dendrites, IIP-MF reflecting the genetic and environmentally more variable part of the CA3-projection. Thus, small mossy fiber projections with entirely missing IIP-MF would characterize a hippocampus in which neuronal activity from any input segment could spread quickly through the entire hippocampal formation, while massive suprapyramidal and IIP-MF projections would prevent such spread.

At the behavioral level, individuals with small IIP-MF projections would be expected to be hyper-reactive, quickly aggressing or fleeing, good two-way avoiders, but inferior in complex tasks and memory because they are easily distracted. On the other hand, individuals with large MF input to CA3 would be characterized by more predictable (longer ongoing) behavior, regardless of what they are doing. For example, they might be expected to use a preferred paw more insistently, showing prolonged attack latency (as mice usually show an ethological "foreplay" before launching an attack). They would also be expected to appear more explorative and less anxious (because of better buffering or even ignoring moderate external or motivational stimuli), and to perform better in complex learning and memory tasks requiring a certain stress tolerance and concentration. On the other hand, they might also show prolonged inappropriate responses such as in two-way avoidance learning (Figure 26.3), or prolonged searching over old target locations in the water maze. Taken together, genetic variations of the MF projection to CA3 would be a mechanism for tuning a basic hippocampal function, namely, the control of parallel processing in various parts of the forebrain that are all interconnected with the hippocampal formation. Lesions of the entire hippocampus must then appear as a combination of both extremes, namely hyper-reactivity and hyperemotionality combined with stereotyped and rigid behavior. This in fact is the case, at least in rodents.

26.10 SOME MORE PROBLEMS

This explanation has an intuitive appeal but poses another methodological problem: how can this be tested behaviorally? Multifunctionality of the hippocampus also implies that any change in this structure probably will be manifested in behavior, but the outcome may depend on the behavioral context, genetic background, and other factors.

Another problem is to explain why genetic variation of the IIP-MF in mice occurs at all. Having scarce IIP-MF projections equivalent to a mild hippocampal lesion syndrome would seem an undesirable property for mice, and one is left wondering why natural selection has not eliminated such an unhealthy feature, even if it were only of partial importance for animal cognitive behavior. Nonetheless, it occurs in a couple of mouse strains, and investigations of wild house mice have consistently revealed individuals with fairly scanty IIP-MF projections (unpublished observations by H.-P. Lipp and H. Schwegler).

A disturbing possibility is that MF variations somehow influence behavior, but the behavioral tests applied to mice fail to measure biologically meaningful abilities of the animals—an argument often proposed by ethologists. Likewise, it can be argued that laboratory mice are unrealistic models because of their long domestication and behavioral degeneration and one might dismiss the observed correlations as laboratory artifacts. One approach to solve this problem would be a broad screening for genes underlying MF variations, and trying to elucidate the path from gene to behavior. But if the former arguments were true, such (costly) approaches would be doomed for failure.

26.11 THE REAL WORLD TEST FOR GENETIC BRAIN–BEHAVIOR RELATIONS: NATURAL SELECTION AND EVOLUTION

The arguments above can only be answered by studying correlations between MF variations and behavior in an ecological context. If MF variations are of any relevance for natural behavior, they must respond to natural selection. Likewise, species with different lifestyles, but comparable brains must show differences in their IIP-MF projections. To study this issue, a field station was established in Russia in collaboration with behavioral geneticists.

In order to observe natural selection within a species, laboratory mice with genetically different IIP-MF projections (C57BL/6, DBA/2, C3H and NZB) were crossed in order to obtain a sample of hybrid mice characterized by large genetic variability of this trait. These were released to two large outdoor pens and left there for 2 years by now. Samples of mice were taken in yearly intervals and compared with control mice from the same crosses that were kept under standard laboratory conditions and random mating. An analysis after 2 years revealed that the hippocampi of the feralized mice were morphologically adapting. In both pens, there was a significant increase of the suprapyramidal MF projection, which was correlated with a general increase in brain weight of about 5%. Independently of this increase, there was a concomitant reduction of the IIP-MF projection (Figure 26.9).

Further follow-up over 4 years showed the divergence of the IIP-MF to be consistent.[26] Moreover, samples of mice caught at the end of the second year outdoors were transferred to the laboratories (Zürich and Moscow) and bred there, the reduction in the IIP-MF projections in Zürich being largely conserved in the seventh generation, even after embryo transfer. Behavioral studies in Zürich showed no parallel selection effects in the water maze and open field,[27] but naturally selected lines showed neophobia toward a novel object in an open field.[28] Studies in Moscow with the naturally selected lines showed increased defecation in the open field, reduced activity in exploratory arenas after repeated exposures, and, interestingly, an improved ability of extrapolating the movements of a reward toward a hidden consumption site.[26] Taken together, this study confirmed the prediction of an association between reduced IIP-MF and behavioral reactivity, and showed, most importantly, an extremely rapid natural selection effect on both hippocampal circuitry and behavior. Obviously, the association between IIP-MF and behavior is of

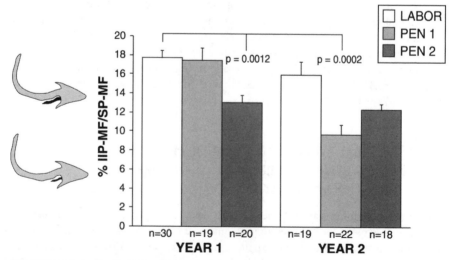

FIGURE 26.9 Reduction of the IIP-MF projection in two demes of feralized mice after 2 years in Russian outdoor pens as compared with control mice kept indoors. The time span includes 7 to 8 generations of mice. Note that the reduction in pen 2 was observed after 1 year, while it took 2 years in the other pen.

ecological relevance. The mice had to live in the shelters throughout the year. Intentional exploration of the pen territory (as expected for carriers with expanded IIP-MF projections) bears a high risk of predation by owls, while the social structure in the shelter (a few highly aggressive dominant males and many quickly escaping subordinates) favors again high behavioral reactivity (either for flight or fight), at least in males. As the lifestyle in these types of pens is not demanding in terms of spatial abilities and complex learning, the lacking selection effects on performance in classic cognitive tests such as the water maze is not unsurprising. Most likely, other genetically dependent brain traits supporting the observed phenotype were subject to selection as well. Nonetheless, the speed of genetic adaptation of the IIP-MF projection implies that selection acted upon a few major loci only.

These results received support from studying the hippocampi of small wild mammals in Russia and elsewhere. From the results of experimental natural selection, it could be predicted that small rodent species living in ecological niches requiring high behavioral reactivity or readiness for flight would show small IIP-MF projections. An example of such adaptation is given in Figure 26.10, which shows large IIP-MF projections in the bank vole (*Clethrionomys glareolus*), a species adapted to a wide range of different habitats, and extremely reduced IIP-MF projections in the root vole (*Microtus oeconomus*), living in homogeneous grassland habitats with small home ranges. Interestingly, the two species showed about equal performance in water maze learning but used very different strategies, the bank voles showing controlled spatial searching behavior, the root voles searching inefficiently but swimming at high speed, reacting to any change in the setup, such as platform reversal, with frantic but ill-directed swimming.[29] Other studies showed that wood mice (*Apodemus* ssp.), known for roaming large territories, have large IIP-MF projections as well, which were also found in patrolling insectivores (*Sorex* ssp.). On the other hand, species living chiefly underground, such as the vole *Microtus subterraneus* and the mole (*Talpa europea*), have shown much reduced or missing IIP-MF projections.

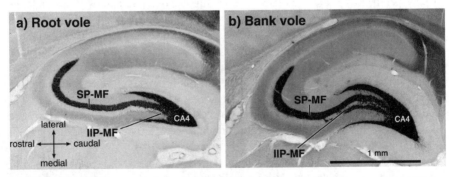

FIGURE 26.10 IIP-MF projections in wild voles. (a) Root vole (*Microtus oeconomus*), living in small monotonic habitats and showing rapid yet chaotic swim strategies in the water maze; (b) Bank vole (*Clethrionomys glareolus*), a species capable of adapting to many different habitats and showing predictable and efficient learning in the water maze (modified after Pleskacheva et al. 2000).[29]

While these findings confirmed the prediction of MF variations underlying behavioral reactivity, we also noted that rodents with relatively large IIP-MF projections had larger home ranges. Thus, a link of the size of the IIP-MF projection with spatial abilities remained an alternative hypothesis. In order to test this question, we were looking for a species with excellent spatial abilities and found it represented by nectar-eating bats that must patrol large territories with constantly changing locations of flowers delivering nectar. If large IIP-MF projections would be a prerequisite for mastery of spatial short- and long-term memory, these bats should show them. If, on the other hand, behavioral reactivity would be the main factor associated with MF variations, freely and rapidly flying species should be characterized by scanty MF projections enabling them to react instantaneously to stimuli encountered during flight. Thus far, three South American species have shown the smallest MF projections among all species observed thus far.[30] Pending further investigations, this would imply that the behavioral dimension associated with the size of the MF projections is indeed the degree of behavioral reactivity, that is, sensitivity to distraction. If so, one would also expect that the species with the highest degree of parallel processing and multitasking should show the most extreme adaptation of the IIP-MF projection. Thus far, this seems to be the case in the human hippocampus, where every pyramidal cell in CA3 receives both suprapyramidal and infrapyramidal MF afferents (unpublished data). Further predictive comparative studies will be necessary to confirm this conclusion. In any case, it is now almost certain that natural selection is operating massively on MF traits, within and between species. Thus, variations of the IIP-MF are not simply epiphenomena irrelevant for behavior but may reflect gene actions important not only for the regulation of individual behavior, but also for the evolutionary selection of traits relevant at the population level.

26.12 CONCLUSIONS AND OUTLOOK

This short review has shown that an analysis of natural genetic variation can lead to new perspectives and hypotheses about hippocampal functions that would be difficult to obtain by any invasive technique. Its power is based on the fact that a correlative approach using available genetically defined mouse and rat populations can identify in the brain natural regulation sites for setting permanent differences in behavioral responsiveness and particular talents. Once discovered, such natural regulation sites can be used to study the function of a brain structure beyond the gross deficits that follow its destruction or inactivation. There is probably no other technique by which brain structures can be varied so elegantly, and it is difficult to see how an interfering technique, tissue destruction or gene targeting, could alter behavior, memory and learning in such a natural way.

On the other hand, the goal of neurogenetics is to understand the pathway from gene to brain and to behavior. The example here is a typical phenotype-to-genotype approach. It models the end-product of the genotype-to-phenotype pathway quite nicely and offers a relevant target to be investigated further. However, it lends itself not easily to the discovery of candidate genes, perhaps best by using quantitative trait locus (QTL) mapping techniques (see Chapter 5). Even then, it will be difficult to identify by molecular techniques those major genes that determine the genetic variation of the IIP-MF projection because these regulatory genes might act locally and their products occur in low concentrations during a limited developmental period. But it is not impossible.

Another point is that such a correlation analysis is time consuming, requiring many control studies. This is because working with brain traits means working with relatively minor behavioral differences as compared with the effects of structural destruction. Minor differences are more sensitive to environmental effects and genetic interactions. One must notice, however, that the natural genetic differences in behavior as found between mouse strains are often much larger than those observed after targeted disruption of genes.[31] Thus, the most straightforward genotype-to-behavioral phenotype approach faces similar problems but tends to overlook them.

Correlative approaches are often dismissed because of the difficulties in disentangling pleiotropic effects of unknown genes causing the traits in brain and behavior. These difficulties should not be denied. Anyone familiar with the effects of gene targeting on memory, learning, and cognition is aware of the fact that there is no way to avoid pleiotropy of the targeted mutation. Even worse, there is a common belief that elimination of a particular locus is revealing the function of that gene. What is really observed, however, is the function of the entire genome without the targeted locus. Again, the basic problems remain always the same—they are just perceived differently.

The unique strength of this approach is that it starts with individual, within-species variations of brain traits and behavior that are sensitive to artificial selection and manipulation by classical breeding and eventually helps predict species-specific adaptations. This is a fundamental difference to the traditional comparative approach, which tries to match a special talent of a species with the particular development of a brain structure. While fruitful in linking sensory abilities to corresponding development of brain parts, this strategy faces problems in matching cognitive abilities with altered patterns of neuronal circuitry. It would seem that the example described in this chapter is, thus far, the only rational approach in finding natural regulatory sites for complex behavior that is not depending on luck and speculation.

Last, one would expect that such a simple and relatively cost-effective approach might win more followers. Unfortunately, the popularity of the mouse as a main species for genetic engineering has created a ballooning veterinarian industrial and administrative complex imposing excessive (and often unnecessary) costs of mouse care that threaten to suffocate traditional behavioral genetic methods in this species. With steadily rising costs, the insight may grow that neither classical Mendelian crosses nor artificial selection for behavioral traits needs pathogen-free conditions and veterinarian management and supervision.

ACKNOWLEDGMENTS

This article was supported by Swiss National Science Foundation and the NCCR "Neural Plasticity and Repair."

REFERENCES

1. Lipp, H.-P., Genetic variability, individuality and the evolution of the mammalian-brain, *Behavioural Processes* 35, 19–33, 1995.
2. Barber, R.P., Vaughn, J.E., Wimer, R.E., and Wimer, C.C., Genetically-associated variations in the distribution of dentate granule cell synapses upon the pyramidal cell dendrites in mouse hippocampus, *Journal of Comparative Neurology* 156, 417–434, 1974.
3. Lipp, H.-P., and Wolfer, D.P. New paths towards old dreams: microphrenology or the study of intact brains in intact worlds. In *Behavioural Brain Research in Naturalistic and Semi-Naturalistic Settings: Possibilities and Perspectives,* Alleva, E., Fasolo, A., Lipp, H.-P., Nadel, L. and Ricceri, L., eds. (Kluwer, Dordrecht, the Netherlands), pp. 1–39, 1995
4. Lipp, H.-P., Schwegler, H., Crusio, W.E., Wolfer, D.P., Heimrich, B., Driscoll, P., and Leisinger-Trigona, M.-C., Using genetically-defined rodent strains for the identification of hippocampal traits relevant for two-way avoidance learning: a non-invasive approach, *Experientia* 45, 845–859, 1989.
5. Gerlai, R., Gene-targeting studies of mammalian behavior: Is it the mutation or the background genotype? *Trends in Neurosciences* 19, 177–181, 1996.
6. Wolfer, D.P.P., Müller, U., Stagliar-Bozizevic, M., and Lipp, H.-P., Assessing the effects of the 129/Sv genetic background on swimming navigation learning in transgenic mutants: a study using mice with a modified beta-amyloid precursor protein gene, *Brain Research* 771, 1–13, 1997.
7. Jamot, L., Bertholet, J.Y., and Crusio, W.E., Neuroanatomical divergence between two substrains of C57BL/6J inbred mice entails differential radial-maze learning, *Brain Research* 644, 352–356, 1994.
8. Crusio, W.E., Schwegler, H., and Brust, I., Covariations between hippocampal mossy fibres and working and reference memory in spatial and non-spatial radial maze tasks in mice, *European Journal of Neuroscience* 5, 1413–1420, 1993.
9. Prior, H., Schwegler, H., and Dücker, G., Dissociation of spatial reference memory, spatial working memory, and hippocampal mossy fiber distribution in two rat strains differing in emotionality, *Behavioural Brain Research* 87, 183–194, 1997.
10. Hausheer-Zarmakupi, Z., Wolfer, D.P., Leisinger-Trigona, M.-C., and Lipp, H.-P., Selective breeding for extremes in open-field activity of mice entails a differentiation of hippocampal mossy fibers, *Behavior Genetics* 26, 167–176, 1996.
11. Schöpke, R., Wolfer, D.P., Lipp, H.-P., and Leisinger-Trigona, M.-C., Swimming navigation and structural variations of the infrapyramidal mossy fibers in the hippocampus of the mouse, *Hippocampus* 1, 315–328, 1991.
12. Bernasconi-Guastalla, S., Wolfer, D.P., and Lipp, H.-P., Hippocampal mossy fibers and swimming navigation in mice: correlations with size and left–right asymmetries, *Hippocampus* 4, 53–64, 1994.
13. O'Keefe, J., and Nadel, L. *The Hippocampus as a Cognitive Map,* Clarendon Press, Oxford, 1978.
14. Olton, D.S., and Feustle, W.A., Hippocampal function required for nonspatial working memory, *Experimental Brain Research* 41, 380–389, 1981.
15. Lipp, H.-P., Collins, R.L., Hausheer-Zarmakupi, Z., Leisinger-Trigona, M.-C., Crusio, W.E., Nosten-Bertrand, M., Signore, P., Schwegler, H., and Wolfer, D.P., Paw lateralization and intra/infrapyramidal mossy fibers in the hippocampus of the mouse, *Behavior Genetics* 26, 167–176, 1996.

16. Sluyter, F., Jamot, L., Van Oortmerssen, G.A., and Crusio, W.E., Hippocampal mossy fiber distributions in mice selected for aggression, *Brain Research* 646, 145–148, 1994.

17. Guillot, P.-V., Roubertoux, P.L., and Crusio, W.E., Hippocampal mossy fiber distributions and intermale aggression in seven inbred mouse strains, *Brain Research* 660, 167–169, 1994.

18. Belzung, C., Hippocampal mossy fibres: implication in novelty reactions or in anxiety behaviours? *Behavioural Brain Research* 51, 149–155, 1992.

19. Roullet, P., and Lassalle, J.-M., Genetic variation, hippocampal mossy fibres distribution, novelty reactions and spatial representation in mice, *Behavioural Brain Research* 41, 61–70, 1990.

20. Prior, H., Schwegler, H., Marashi, V., and Sachser, N., Exploration, emotionality, and hippocampal mossy fibers in nonaggressive AB/Gat and congenic highly aggressive mice, *Hippocampus* 14, 135–140, 2004.

21. Dimitrieva, N.I., Gozzo, S., Dimitriev, Y., and Ammassari-Teule, M., Mossy fiber distribution in four lines of rats: a correlative study with avoidance abilities and excitability thresholds, *Physiological Psychology* 12, 30–34, 1984.

22. Lipp, H.-P., Schwegler, H., Heimrich, B., Cerbone, A., and Sadile, A.G., Strain-specific correlations between hippocampal structural traits and habituation in a spatial novelty situation, *Behavioural Brain Research* 24, 111–123, 1987.

23. Roullet, P., and Lassalle, J.M., Behavioural strategies, sensorial processes and hippocampal mossy fibre distribution in radial maze performance in mice, *Behavioural Brain Research* 48, 77–85, 1992.

24. Hoffmann, H.J., Wenkel, R., Schicknick, H., Bernstein, H.-G., and Schwegler, H., Hippocampal mossy fiber distribution does not correlate with two-way active avoidance performance in backcross lines derived from inbred mouse strains DBA/2 and C3H, *Brain Research* 589, 171–174, 1992.

25. Isgor, C., Slomianka, L., and Watson, S.J., Hippocampal mossy fibre terminal field size is differentially affected in a rat model of risk-taking behaviour, *Behavioural Brain Research* 153, 7–14, 2004.

26. Poletaeva, I.I., Pleskacheva, M.G., Markina, N.W., Perepiolkina, O.W., Scheffrahn, H., Wolfer, D.P.P., and Lipp, H.-P., [Environmental habitat-related pressure: behavioral alterations and morphological changes in the brain of the house mouse] Russian, *Ecologia* 3, 231–236, 2001.

27. Rissi, S. Moosfasersystem und Verhalten bei der Hausmaus nach Embryonentransfer. M.D. thesis, University of Zürich, Zürich (2002).

28. Ceschi, A. Effects of experimental natural selection on exploratory activity and anxiety in mice: assessment by means of three behavioral paradigms. M.D. thesis, University of Zürich, Zürich (2004).

29. Pleskacheva, M.G., Wolfer, D.P., Kupriyanova, I.F., Nikolenko, D.L., Scheffrahn, H., Dell'Omo, G., and Lipp, H.-P., Hippocampal mossy fibers and swimming navigation learning in two vole species occupying different habitats, *Hippocampus* 10, 17–30, 2000.

30. Amrein, I. Functional and neuroanatomical correlates of adult neurogenesis in the dentate gyrus of domesticated and wild rodents. Ph.D. thesis, University of Zürich, Zürich (2004).

31. Lipp, H.-P., and Wolfer, D.P., Genetic background problems in the analysis of cognitive and neuronal changes in genetically modified mice, *Clinical Neuroscience Research* 3, 221–232, 2003.

LIST OF ABBREVIATIONS IN TEXT

CA1, CA3	Subregions of hippocampus, CA stands for cornu ammonis (Ammon's horn)
CA4	Hilus of the dentate gyrus
IIP-MF	Intra/infrapyramidal mossy fibers
MF	Mossy fibers
QTL	Quantitative trait locus
RHA	Roman high avoidance
RLA	Roman low avoidance
SP-MF	Suprapyramidal mossy fibers

27 Expression and Brain Structure: Black Boxes between Genes and Behaviors

Jeremy L. Peirce and Robert W. Williams

CONTENTS

27.1 INTRODUCTION

The operation of the adult mammalian brain involves a large fraction of the genome.[1,2] Upward of 50% of the gene complement is expressed in even single discrete brain regions such as the cerebellum, striatum, and hippocampus. Transcriptional complexity in the brain is unusually high compared with other organs, and roughly 40% of all genes are transcribed in two or more alternative forms.[3] Development adds another major axis of complexity. In small mammals such as mice, assembling a brain is a beautifully choreographed dance that involves the proliferation, migration, differentiation, and interconnection of 75 million neurons, 25 million glial cells, and a web of 10 million or more blood vessel cells over a 30-day period. In humans, this process involves three orders of magnitude more neurons and their supporting cast and takes a decade or two.[4] It seems conservative to estimate that 75% or more of the mammalian genome is engaged at some stage of making or maintaining the brain.

The major topic of this chapter is the study the genetic influences on brain organization at various levels. Variations in genes affect either transcription or function, and both of these can influence brain structure. Brain structure, in turn, affects behavior both in variations in gross organization and variation in the ways circuits are formed and utilized. Studying each level of organization is valuable and gives us the opportunity to treat transcription and morphology in the brain as intermediate phenotypes and not simply a black box between brain and behavior.

27.2 GENES AND DEVELOPMENT

Most of what we have to say is based on data collected exclusively from sexually mature adult mice. We recognize that genes act as part of transitory developmental processes. Much of what we see in adults is therefore the downstream consequence of embryonic expression. This is an important caveat to keep in mind. However, it is equally important to keep in mind that variation in adult brain structure is the final and most important product of development. The variation among brains of adults can be exploited using genetic methods to lead us straight to the causative genes, even when those genes have been in the full-off position for months or years. In fact, the genetic analysis of adult brain variation is now becoming a reasonably effective and objective way to define key developmental genes responsible for variation in brain development, structure, and even function.

There are several important complementary large-scale research efforts, notably GENSAT, that have examined prenatal and early postnatal gene expression patterns in the brain. These resources can give us insight into these transitory developmental expression patterns. In addition, many genes are likely to have selective and/or partially redundant functions in different brain regions.[5,6] At every level, of course, these genetic effects will exist alongside and interact with complex environmental effects.

27.3 BRAINS AND BEHAVIOR

Differences in brain structure generate differences in behavior. Even a crude measurement such as brain weight is associated with behavioral variation, most prominently when the comparisons are made among species.[7–11] Of course, the more specific the region, cell population, or synaptic subcircuit being compared, the more specific the related behavior differences are likely to be. In songbirds, for example, variation in the size of neuron populations in the system is correlated with various features of song learning and production.[12,13] A more specific relationship is the variation in size of the infrapyramidal mossy fibers in the hippocampus, which correlates with differences in open-field exploration and novelty-induced fear,[14] as well as with avoidance learning.[15,16] Other aspects of hippocampal morphology also have behavioral correlates,[17–20] a subset of which is likely to be causal. We expect that numerous other structural variants will be linked to behavioral differences.

27.4 DEALING WITH COMPLEXITY

The most common approaches to studying the remarkable complexities of brain development and cellular architecture are reductionist. Approaches are focused on isolated molecules, single circuits, or nuclei in simple model systems. In the case of mice, we often study effects of single-gene modifications on a single inbred genetic background—often that of the inbred mouse strain C57BL/6J. This approach has led to many impressive stories and has provided numerous insights into key mechanisms. However, reductionist approaches often provide only a partial picture and will not necessarily provide results that are readily applicable to other systems. This is one of the complaints leveled against mouse models—that we can cure a single particular mouse of a disease, but that these results do not mean that we will make equivalent progress in humans. The reason for the failure has nothing to do with the particular mouse, but with the assumption that results from any single, defined system will generalize well. Biological processes and the generation of behavior are somewhat messy and contingent affairs, and behavior in particular is often the result of complex nonlinear interactions that are hard to model or predict. The phrase *behavioral mechanism* is in many cases an oxymoron. Reductionist methods may provide unequivocal answers to specific questions in a specific context. We may be able to understand the role of perturbations of *Camk2a* expression level at one particular age, in one particular sex, and in one particular strain or species. But how successful will this data make us in prediction of structure or function at another stage, in another sex, and on a different genetic background? The unfortunate answer is that our predictive powers are woefully limited. We certainly cannot predict effects of gene knockouts with any assurance before we do the experiment, and once we have results from one strain, we still usually cannot predict the effects of the same allele in another strain. Results of this type may lead to high-impact papers, but may not lead to real understanding of a biological system.

27.5 COMPLEX TRAIT ANALYSIS AND QUANTITATIVE
TRAIT LOCI (QTL) MAPPING

Complex trait analysis is an effective approach for studying sets and systems of central nervous system (CNS) phenotypes. This approach has been commonly used to study behavioral and pharmacological traits of relevance to drug and alcohol use and abuse.[21-27] Complex trait analysis has also been applied to a wide variety of CNS divisions and cell populations: total brain weight,[28,29] neuron and glial cell number,[30] the size and structure of the hippocampus[31] and dentate gyrus,[32] olfactory bulb,[33] striatum,[34] retina,[35] neocortex,[36] cerebellum,[37] as well as cytoarchitectonic fields including the barrel cortex.[36,38]

These and other quantitative trait locus (QTL) mapping studies are intermediaries between purely reductionist tools such as single-background gene knockouts, and large, human-genetic studies, which may have much lower power (certainly per participant!) but whose results, hopefully, are more general. A typical mouse QTL study falls somewhere between these extremes. It addresses the complexity of genetic variation but limits its scope to improve the power to detect differences. When applied to a genetic reference population and to molecular phenotypes, however, QTL mapping provides an invaluable tool for building observations that can bridge different levels of complexity.

Instead of studying a small number of isogenic animals or using a single animal model, we usually study a segregating population that consists of 20 to 2,000 genetically unique individuals or strains. Studies of the genetic basis of common diseases in human populations often exploit this approach. For example, much of the population of Iceland has been harnessed to find common gene variants that contribute to pervasive human diseases. Rodent geneticists have their own, much simpler, version of a population-based approach that usually involves breeding two inbred strains known to differ significantly in a trait of biological interest. These parental strains are used to generate heterozygous F1 offspring. The F1s are used to make an experimental cross—either an intercross (F2) or a backcross to one of the parental strains (usually abbreviated N2). Both F2 and N2 populations are genetically complex—no two animals are alike. We therefore refer to these populations as a segregating cross because gene variants segregate following Mendel's laws. For example, there may be a twofold difference between the parental strains in the number of dopaminergic neurons in the substantia nigra, horizontal cells in the retina, or the volume of the hippocampal dentate gyrus.[32] The question we would like to answer with the F2 and N2 progeny is what set of gene variants contribute to the variation among members of the cross. Can we track down the QTLs that are jointly responsible for the genetic fraction of the variation?

A quantitative trait can generally be measured continuously, often on a ratio scale. While an essentially dichotomous Mendelian trait potentially can be converted into a quantitative measurement, most quantitative traits vary more continuously. Generally, this is because multiple genetic and environmental factors affect the phenotype. QTL mapping is a process in which we tease apart the individual and joint contributions. We test whether there is a statistically significant association between differences in genotype at a particular gene or marker and differences in the phenotype. The question is, "Is there an association between a trait, such as the volume of the dentate gyrus, and any gene (a gene variant can act as a marker) or genetic marker in our experimental cross?" In the case of a Mendelian trait such as albinism in mice or Huntington's disease in humans, we will find an almost perfect linkage between a gene variant (tyrosinase and huntingtin, respectively) and variation in the phenotype.

In the case of QTL mapping when a marker is very near a gene that modulates the trait, there will also be a comparatively high level of association between the phenotype and the genotype, though far less than for a Mendelian trait. When a trait is genuinely polygenic, however, a single QTL may account for only 5 to 10% of the variance. In this case, no single marker will be highly predictive of phenotype.

A QTL map is a simple graph of the degree of association between chromosomal intervals (usually on the x-axis) and trait values. These plots are usually presented with degree of association (often given as likelihood ratio score (LRS), log of odds (LOD), or −log P) plotted on the y-axis as a function of chromosomal location. A chromosomal region with a high degree of phenotypic association marks each QTL. While we call this a genetic linkage or association, a significant linkage is actually a directional claim of causality (though not necessarily of mechanistic interaction): a QTL map is a search for the set of genuine genetic causes of variation in phenotype, and in that respect is no different from a study of Mendelian mutations, although loci with small effects on the phenotype are, of course, harder to detect.

Another way to think of a QTL map is as an association heat map by position, generated when we fit a model that allows for one marker at a time to have an influence on the phenotype. Considered this way, it is easy to imagine extending the model to allow two genes and their interaction to have an effect on the phenotype. This two-gene model, or pair scan, is valuable for finding pairs of QTLs with significant interactions, with or without individual effects. Higher-order models are easy to imagine, but they rapidly become prohibitive due to the large numbers of cases needed to rigorously test more complex models.

Most of the common tools for QTL mapping, including pair scans, are integrated into GeneNetwork (GN, www.genenetwork.org), a Web-accessible resource for systems genetics. Many of the examples in the rest of this chapter are drawn from this resource and can be easily replicated and extended from any browser. GN includes tools for QTL mapping and analysis of large-scale phenotypes and array data, a variety of analyses utilizing genetic correlation, clustering, network building and other exploratory tools, and extensive connections to external systems genetics related tools. The remainder of this chapter will focus on the capabilities of this tool set (and associated external data sets built by other investigators) and approaches to analyzing morphological phenotypes.

27.6 THE VALUE OF GENETIC REFERENCE POPULATIONS

For a highly heritable phenotype with little environmental noise, (noise in this sense includes any nongenetic influence such as maternal environment and technical error of the measurement) a standard F2 intercross (F2; stage marked on Figure 27.1) offers an extremely efficient means of mapping genetic variation. The environmental noise for a behavioral or physiological QTL can often be quite high in a single measurement, however, and unfortunately, the uniqueness of each F2 animal means

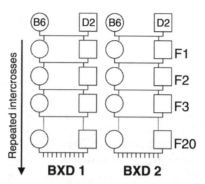

FIGURE 27.1 Recombinant inbred (RI) breeding scheme. Breeding (RI) lines simply requires sequential intercrosses of offspring, usually starting with inbred parental strains, until the resulting offsprings are fully inbred (approximately F20).

that observations on a given F2 are limited. This includes genotype observations, so each F2 population must be separately genotyped for analysis to be possible.

These issues are strong arguments for using recombinant inbred strains (RIs) to map QTLs. RI strains (Figure 27.1) are generated by repeatedly intercrossing progeny from an F2 intercross until the offspring are themselves inbred, with genomes made up of "patches" of each parental genome. Since these strains breed true, observations on different animals within the same strain share a common genome (common genetic influences) and have independent environmental components. Averaging phenotypes among several animals thus serves to reduce environmental noise. This is especially important with relatively noisy measurements such as array-based mRNA phenotypes, because it allows us to pool samples for each array and pool arrays for each strain, all of which dramatically reduce the noise levels of our phenotypes.

In addition to pooling the same observations, using an RI strain set allows us to compare different observations. If we measure two phenotypes using different animals from the same strain set, the common factor between these measurements is their genetic context. Therefore if the two phenotypes are correlated, we know that the correlation must represent common genetic underpinnings of the two phenotypes. (This is known as a genetic correlation. GN provides several tools for exploring this sort of correlation, which will be explored later.) The RI strains serve as an effectively immortal genetic reference population (GRP) about which phenotypic data can be accumulated.

27.7 MICROARRAY DATA: THE CHEAPEST, HIGHEST THROUGHPUT PHENOTYPE

The advent of microarrays gives us a unique opportunity to look at the abundance of the entire steady-state mRNA populations in a tissue *at once*. Essentially this means that we can perform from 10,000 to 1,000,000 measurements of mRNA level in parallel, depending on the precise platform used. If we estimate that each array costs $600 and we pool 4 arrays per strain (2 per sex), our cost is still only 6 cents per phenotype, assuming use of the Affymetrix platform, which measures about

40,000 messages. Array costs continue to decline rapidly even as array quality improves, making mRNA abundance the cheapest phenotype to acquire.[39]

27.8 MICROARRAY DATA: CAVEATS FROM COMPLEX TISSUE

The variation in which we are most interested is that of expression levels of genes and proteins. However, we cannot estimate *per cell* expression level. Instead we are able to estimate the relative expression levels of many transcripts among genetically different samples and the covariance of many transcripts with each other and with other phenotypes among this same set of samples. Variation in cell composition will be one cause of differences between samples. For example, ratios of GABA-ergic and glutamatergic transcripts (*Gad*, *Psd95*, etc) will be under genetic control and will vary either as a result of genetic cell autonomous expression differences, or as a result of genetic differences in ratios of excitatory and inhibitory neurons. The same will be true of neuron:glial ratios.[40] The range of variation in expression levels generally exceeds variation in cell composition, however, so this complication does not typically invalidate simple observations on gross tissue. Because most of our microarray observations are performed using genetic reference population (GRPs), we can always relate QTLs and other relationships back to differences in cell populations. The Mouse Brain Library, described later in this chapter, is an excellent tool with which to characterize cell-type composition differences by strain. (This is not simply a matter of eliminating a nuisance variable. Differences in cell population ratios are an interesting phenotype with interesting potential correlates; for example, variation in ratios of old and new neurons in the hippocampus would be of great interest to researchers studying learning and memory.)

27.9 MRNA ABUNDANCE QTLs

When taken using a GRP or other mapping population, measurements of mRNA abundance can be used as phenotypes for QTL analysis. This crucial insight into the value of molecular phenotypes was made for proteins by Damerval and colleagues[41] seven years before Jansen and Nap resuggested it as an approach for analyzing mRNA abundance using microarrays.[42] In the five intervening years, mRNA abundance has been measured in yeast,[43] mice,[44-46] rats,[47] humans,[48-50] and even eucalyptus trees.[51]

While calculating significance thresholds for a single mRNA trait is equivalent to calculating them for any other complex trait, assigning significance to mRNA abundance QTLs from an array experiment with 39,000 or more mRNA abundance traits is not yet well worked out. A False Discovery Rate[52] (FDR) approach may be helpful but may also be too conservative given our assumption that at some level most transcripts will be genetically modulated by other factors in the genome. In addition, it may be appropriate to apply different significance criteria to *cis*-acting QTLs (those that modulate the transcript maps to the transcript's location on the genome) and *trans*-acting QTLs, since the prior expectation that a *cis*-acting QTL is real (given that it has mapped to the location of the transcript itself and that such occurrences are at a rate much higher than chance in the population of transcripts) is much higher than would be the case for a *trans*-acting QTL.

27.10 *CIS*-ACTING QTLs AND *TRANS*-ACTING QTLs

A useful way to categorize mRNA abundance QTLs is by whether they modulate mRNA abundance in a *cis*-acting or *trans*-acting manner. To put it another way, is a QTL that modulates a particular gene map near the gene itself? If so, it is likely that a variation in the gene's regulatory region is responsible for the variation in gene expression and the presence of the QTL.

Two alternative interpretations prevent us from immediately declaring that we have identified a gene responsible for a QTL when we find apparently *cis*-acting QTLs. The first alternate interpretation is that a closely linked gene is exerting a modulatory influence. The second alternate interpretation is Affymetrix specific. The probes used in the Affymetrix mouse arrays were designed against public mouse sequence, meaning their probes were designed to complement C57BL/6J mRNAs. Single nucleotide polymorphisms (SNPs) and other sequence variants that happen to occur within the probe sequence reduce the probe binding and may be detected as *cis*-acting QTLs. Doss and colleagues[53] applied a *cis*-trans test to apparent *cis*-acting QTLs, reasoning that if the QTL were *cis*-acting the ratio of transcripts in the F1 would differ from 1:1. Using this test they were able to confirm the mode of action for 64% of significant, apparently *cis*-acting QTLs.

There are, however, many apparent *cis*-acting QTLs. Plotting transcript and QTL positions against each other yields a strong diagonal of *cis*-acting QTLs and a number of off-diagonal, potentially *trans*-acting QTLs (Figure 27.2). In addition to

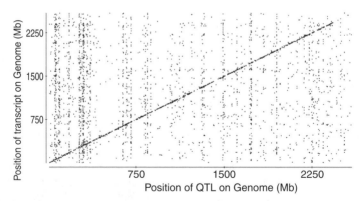

FIGURE 27.2 *cis*- and *trans*-acting QTLs in the BXD brain. This figure plots QTL position (x-axis) vs. transcript genomic position (y-axis) using QTLs from the INIA M430 Brain data set (PDNN transform, Jan. 06 version). The diagonal line indicates *cis*-acting QTLs (the QTL maps to the same position as the transcript) while dots off of the vertical line indicate *trans*-acting QTLs. The precise cause of the light, vertical bands of *trans*-acting QTLs (often called *trans*-bands) is unknown, but caution should be taken interpreting them since they do not replicate well between data sets.

the diagonal line of *cis*-acting QTLs, there are several vertical "lines" where there appear to be a high concentration of *trans*-acting QTLs. Exactly what causes these "*trans*-bands" is unclear, but they may relate to the action of one or more transcription factors with many targets. It is interesting to note, however, that while many *cis*-acting QTLs replicate between our RI array data and F2 array data in the same tissue (whole brain, cerebellum), fewer *trans*-acting QTLs and only one transband replicates.[54]

27.11 MRNA IS NOT PROTEIN

Generally speaking, our interest in mRNA abundance is not fascination with mRNA *per se* but a desire to learn something about the status of proteins in the cell, with the mRNA abundance standing in for protein abundance. Of course, it is important to keep in mind that mRNA is not a protein and that while mRNA and protein levels will generally co-vary together, the relationship between the two varies from transcript to transcript and is often distinctly nonlinear.

27.12 FROM QTL TO GENE

Over the past decade, defining a QTL interval given a reasonably heritable phenotype has become largely routine. The WebQTL module of GeneNetwork has even reduced the barrier to entry for routine QTL mapping to the effort required to input the phenotype, at least for data gathered in RI strains. It is the next stage, determining which gene or genes in the QTL interval are responsible for the influence of the QTL that is not yet a routine process.

One approach to proceeding from QTL to gene is fine mapping, possibly followed by positional cloning of the polymorphism. There are a number of approaches to fine mapping, including generation of large intercross populations (F2s) and congenic animals. This approach could indeed be made routine but involves a great deal in the way of effort and resources.

Researchers may also aggregate phenotypic information in an attempt to intelligently choose a candidate gene from among the tens to hundreds of possible genes in the QTL interval. Techniques to further examine the candidate include analysis of the phenotype of a knockout or transgenic pharmacological interventions to produce phenocopy, evaluating predictions in another appropriate population, and other modes of biological confirmation. Generally these methods are expensive in terms of time and material, so choosing a good candidate is very important. It is also, unfortunately, difficult.

GN is a valuable tool in the process of taking a phenotype from QTL mapping to candidate analysis, and is particularly well suited to the pursuit of genes affecting CNS morphology.

27.13 DATA SETS IN GENENETWORK

There are a wide variety of data sets currently in GenNetwork (GN), and the list is constantly expanding. Because GN used the BXD RI strains as a GRP from early on, the deepest data sets, especially where CNS morphology is concerned, are for these strains. While the crown jewels of the GN data sets are obviously the extensive microarray data, a valuable tool GN provides is easy access to published phenotype data in a useful analysis environment for many common GRPs. Again, the BXDs are the most thoroughly characterized with 830 published phenotypes in the database at time of writing.

For the BXD strain set there are many CNS-related array data sets (all generated using Affymetrix arrays, though GN easily supports data from other array platforms) including data for the forebrain, hippocampus, striatum, cerebellum, and eye. (There is also a data set for liver mRNA abundance generated using the Agilent array and an Affymetrix array-based study of hematopoietic stem cells.) Each of these data sets was generated using at least one array per sex per strain and from 2 to 4 age- and sex-matched littermates pooled per array. We are currently in the midst of expanding on this range of CNS-related tissues. Since the BXD strain set has recently been expanded[55] we can also considerably improve the power of our inquiries by simply expanding our phenotyping effort to the new strains.

For the reader following along on the GN site, these data sets are named according to six characteristics: the institution or group that funded them, tissue the array was applied to, array type, array version, date this version of the data was released, and normalization method. We generally provide pre-normalized and quality controlled array data using at least three normalization procedures: RMA, PDNN, and MAS5. Choosing a good normalization to address with your research question is well beyond the scope of this chapter, but normalization can substantially shift the results of an array analysis, so care is necessary.

A primary advantage of GRPs is never having to genotype again. Over 10,000 SNP genotypes were assayed in many of our mouse GRPs using the Illumina SNP genotyping platform, resulting in very high-density genotypes generated from those SNPs that differed between strains. In the BXD strain set, for instance, there are now 3,795 combined SNP and traditional simple sequence length polymorphism (SSLP) markers that differ between the B6 and D2 strains!

While GN is already set up to analyze data for the AKXD, BXH, CXB, AXB/BXA, and LXS RI strain sets, currently there are only two array data sets available for non-BXD mouse RI strain sets. The first, of considerably more interest for brain structure research, is an array data set for the CXB hippocampus. The second is for mammary tumors in the AKXD strain set. The others are currently limited to genotypes and published phenotypes. In addition, GN also has genotypes and published phenotype data for the HXB/BXH rat strain set.

GN is an easily extensible framework, and adding the most simple data sets to the system is simply a matter of importing phenotypes and genotypes. There are currently several F2 crosses resident on the site, for instance. While they are not an effectively immortal inbred population, like the RIs, integrating them into the GN analytical framework affords us the opportunity to compare and confirm results between crosses and to apply the GN tool set to their analysis. As we continue to expand the variety of data sets available in GN, we look forward to the integration of new tissues, crosses, and treatment conditions into the site's repertoire.

27.14 GENETIC CORRELATION

The use and interpretation of correlations is often regarded with suspicion by biologists with a strong mechanistic or molecular bent. But correlation is one of the conditions for establishing causality—causality without a directional arrow or guarantee of no intermediate steps, as it were. Biologists need to exploit and understand sources of correlation—not shun them. The correlations between traits studied using genetic reference populations are of a special type called *genetic correlations*. The addition of the adjective "genetic" implies that the correlation has an underlying, heritable basis. This is the optimistic interpretation of correlations computed in a GRP for two traits for which we have a reasonably accurate strain, meaning that the correlation is due to neither poor study design nor noise. Consider as an example, the high negative correlation between prepulse inhibition (PPI), a phenotype thought to measure some aspects of schizophrenia in humans,[56] and the expression of the norepinephrine transporter (NET or Slc6a2) in the striatum (Figure 27.3). This correlation is remarkably tight ($r = 0.8$), but does it mean anything?

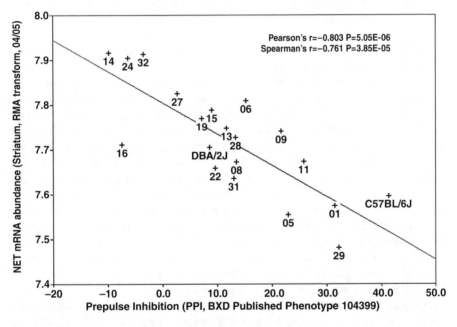

FIGURE 27.3 Prepulse inhibition and NET mRNA abundance. This scatterplot shows the genetic correlation between prepulse inhibition (percent inhibition, BXD published phenotype 10399; x-axis) and expression of the norepinephrine transporter mRNA (NET, Slc6a2; RMA transform 4/05, probe set 1447311; y-axis) in the striatum. Each point in this scatterplot is the average value for an entire strain (BXD24 is labeled 24, etc.). This covariance is quite significant. ($p = 5 \times 10^{-6}$) Relations like these can lead to novel hypotheses or be used to test existing ones.

Let us examine some possibilities. The correlation between PPI and NET may be caused by a true molecular and mechanistic interaction. If the transporter gene has a strong allelic variant (e.g., a premature stop codon in one of the parents of the BXDs), and if variation in PPI just happens to map near the NET locus on Chr 8 at 91.6 Mb, then we would have a good causal hypothesis—that variation in NET protein causes correlation between these two traits (there are at least 73 B vs. D SNPS in NET). In this case, however, there is absolutely no evidence that PPI is modulated by a QTL on Chr 8.

An alternative explanation is that a cascade of protein intermediaries, some of which happen to overlap, generates the correlation between these two traits. While PPI and NET may overlap in this molecular network space, they may not be directly associated. The correlation may be generated by common gene variants that are far upstream of both PPI and NET. If this is the case, then NET and PPI are still in some sense mechanistically coupled, but they do not have a direct causal relationship. If we understood the molecular cascades and networks controlling both traits, we could provide a compelling reason for the correlation, but it might not be possible to induce a change in the PPI simply by changing NET expression as implied by the graph. Over-expression of NET would not improve PPI scores or cure schizophrenia.

A final alternative is that PPI and NET co-vary tightly but that this covariation is entirely due to linkage disequilibrium. This source of correlation is a bit harder to grasp. Linkage disequilibrium is a biological phenomenon, but it usually does not interest nongeneticists. To understand this source of correlation, consider a pair of very different QTLs that both just happen to be located on Chr 11 at 98 Mb. One of these QTLs is actually *NeuroD2*, a gene apparently essential for the normal development of the basolateral amygdala.[57] Mutations in *NeuroD2* can produce fearless, mean mice. Right next to *NeuroD2* is another key developmental gene, namely growth hormone (GH). Imagine that strain X just happened to have alleles at *NeuroD2* and GH that lead to the production of small and unusually aggressive mice and that strain Y just happened to have alleles that lead to gentle and large mice. Imagine also that *NeuroD2* and GH had many other independent downstream molecular targets (which they do). Because these two genes are physically tied together on Chr 11, these two mechanistically separate downstream networks will also be tied together in any experimental cross between strains X and Y. Linkage produces secondary effects that ripple through data sets. Disequilibrium between modifiers that control separate biochemical networks will cause these networks to appear to co-vary, leading to a mixture of mechanistic covariance and linkage covariance.

Fortunately, at least for individual hypotheses where upstream modulators are known, it is easy to control for linkage disequilibrium effects by stratifying the correlation analysis by allele or genotype at the upstream modulator. In the example above, we would stratify our analysis by allele at a marker near *NeuroD2* and GH to disambiguate correlation due to linkage disequilibrium from correlation due to variation in individual networks.

27.15 CLUSTERING AND NETWORKS

GN includes several tools for examining the clustering and network properties of groups of transcripts. Since the pathways involved in higher-order phenotypes are arguably of even greater interest (and applicability to the human condition) than the particular genes varying between parental strains, understanding the organization of transcripts is valuable for understanding larger-scale genome organization.

27.16 CLUSTER TREE

The Cluster Tree tool takes selected traits and builds a hierarchical cluster tree diagram from the distances, measured using 1-r where r is the Pearson product–moment correlation, between pairs of traits, then by successively linking groups of traits to create the hierarchy. The tool then computes the QTL map, using permuted, genome-wide, adjusted p-values for each transcript and plots it as a heat map for each transcript, allowing the user to easily compare QTL maps for related traits.

27.17 NETWORK GRAPH

The Network Graph tool takes selected traits and displays them as nodes, with correlations displayed as edges, given user-specified cutoffs. The user can choose from Pearson or Spearman correlation coefficients as well as literature correlations. Literature correlations are computed using the Semantic Gene Organizer (SGO) software by Homayouni and colleagues.[58] SGO uses latent semantic indexing (LSI) to extract gene–gene relationships from MEDLINE titles and abstracts.

LSI represents the semantic structure of each MEDLINE entry as a vector in a constructed word space and measures the degree of similarity between documents (where genes are mentioned) as the angle between the vectors. Because a direct link between genes in a particular document is not required to calculate the necessary vectors, LSI is a useful tool for proposing novel gene–gene interactions.

27.18 MORE TO DO WITH A GENE LIST: WEBGESTALT AND RELATED TOOLS

Not every analysis starts with a QTL. For physiological and molecular phenotypes, we are frequently as interested in the genes that correlate with the phenotype as in the regions of the genome containing genes that modulate the phenotype. A first step in analyzing these lists of related genes is to order them thematically. GN allows the user to select a list of genes from his or her "shopping cart" and export the list automatically to a variety of tools including WebGestalt, a "WEB-based GEne SeT AnaLysis Toolkit." WebGestalt provides tools to organize a human or mouse gene set by function, tissue expression, chromosome location, and publication co-occurrence. All of these tools are intended to layer themes on gene lists, which is a crucial organizational task for microarray results. For most GN-derived gene lists, however, the primary interest is functional relatedness.

WebGestalt provides four similar tools for gene set analysis by function. The most commonly used is the gene ontology tree tool (GO Tree). GO Tree uses gene ontology categories and compares the observed and expected fraction of genes represented in the gene set across category, returning the gene set in a tree form with over-represented areas marked. It is worth exploring broad categories that are not over-represented because narrower subcategories may still have over-representation, especially in a large gene list. This can be easily done via the Enriched GO DAG tool, which visualizes the GO Tree results as a directed acyclic graph. For convenience this option is directly linked from GN but can also be accessed from WebGestalt.

WebGestalt also provides three other tools for gene set analysis by function. Users can fit gene sets to and perform over-representation analyses on KEGG (Kyoto Encyclopedia of Genes and Genomes; www.genome.jp/kegg) or BioCarta pathways (www.biocarta.com). In these cases over-representation means more genes (in the gene set) that participate in a given pathway than would be expected by chance. These options are particularly good for correlation analyses because genes in the same pathway would be expected to have correlated expression levels. In addition, users can look for over-representation of PFAM (www.sanger.ac.uk/Software/Pfam) protein domains in their gene list.

27.19 BACK TO THE BRAINS—MOUSE BRAIN LIBRARY (MBL)

Behind most morphological measurements of the CNS are collections of high-quality brain sections. The MBL[29] is our way of making such sections accessible to a much broader audience. MBL is a Web-based tool that contains a variety of mouse atlases as well as high-resolution images of more than 800 brains from more than 180 genetically characterized strains of mice. These include several RI strain sets: 28 AXB/BXA, 34 BXD, 11 BXH, 13 CXB, and 57 LXS, and a wide variety of other inbred strains.

GN already contains phenotypes for a variety of CNS morphological traits, especially in the BXD strain set. These can be found in the BXD *Phenotypes* database mentioned earlier, and come from a variety of sources. Well over 100 neuroanatomical phenotypes taken from the MBL have been entered into GeneNetwork: hippocampal volume, dentate gyrus neuron number, neuron:glial ratio in the dorsal lateral geniculate nucleus, neuron number in the basolateral amygdala, and volume of the IGL of the cerebellum. These morphometric and stereological data sets can now be compared and correlated with large behavioral and neuropharmacological data sets. Does the cell population or volume of basolateral amygdala co-vary with important behavioral traits or with the expression of *NeuroD2* or GH? These types of complex multiscale questions can now be addressed very rapidly because the relevant data have been assembled along with appropriate analysis tools in GN.

If you are interested in collecting your own volumetric data using the MBL data set, the MBL includes metadata on the age, sex, body, and brain weight for each animal—in addition to strain identification—and each of these terms can be searched to select animals according to the user's requirements. Images are available at a variety of resolutions, depending on the slide. All slides are available at the base resolution of 24.5 ± 0.5 μm per pixel in the XY plane. Along the z-axis, sections were taken at a 150 μm interval (300 μm between sections on each slide, 2 slides per case). Significantly higher resolution images of single brain sections have been acquired at 4.5 μm per pixel for more than 100 cases (look for the blue, "hi-res" button), and 1 μm/pixel images for the neocortex, hippocampus, and dorsal lateral geniculate nucleus are underway.

Images can be used to calculate volumes using several free programs: NIH Image, Scion Image, or Image/J. (Adobe Photoshop can also be used.) Details on processing, imaging, and calibration using MBL images can be found on procedures, pages, or in papers written based on this data, such as our analysis of hippocampal volume.[31]

27.20 PUTTING THE PICTURES TOGETHER

While array analysis is useful for analyzing mRNA abundance in brain tissues, it does not provide any information on the distribution of expression within that tissue with respect to cell populations. In a homogeneous tissue this is not problematic, but as discussed above, brain and even brain subregions are anything but homogeneous tissues. Fortunately there are two Web-based toolsets that do an excellent job of filling this gap. The Allen Brain Atlas and GENSAT both provide high-resolution images of expression in the brain. The ABA provides these for adult C57BL/6J animals, while GENSAT uses embryonic, neonate, and adult FVB animals; both plan to provide data for nearly all brain-expressed genes.

There are naturally many uses for these data. On the most basic level, presence of expression in the same brain region as indicated in an array result suggests the array is correctly identifying the presence of the gene. While all current array data in GN was gathered using adults, this sort of comparison can eventually be made using animals age-matched to the GENSAT samples. This basic comparison is valuable for analyzing candidate genes—if a gene is not expressed in the target region, it is less likely to be a good candidate.

Putting these expression assays together with MBL images to analyze gene expression in a cell-type specific manner is another exciting use of these databases. Since array expression results are based on the ratio of one mRNA species to the total mRNA population, it is an important caveat that morphological variations include variation in fractions of a given cell type within a structure. In a GRP we can untangle this relationship by directly measuring morphological variations in cell type fraction in the MBL images. Then, if we assume that the cell population type in which genes are expressed is constant, we can normalize the expression ratios from the array by the cell type ratios from the morphological data. This will allow us to make useful statements about cell type populations within morphological regions without the need to purify single cell types for array analysis—an expensive and difficult proposition.

27.21 ALLEN BRAIN ATLAS (ABA)

The ABA (www.brainatlas.org) is a Web-based application that allows researchers to visualize the *in situ* hybridization patterns for genes expressed in the brain. The ABA is the first project of the Allen Institute for Brain Science (AIBS) and will soon include each of the 24,000 genes that AIBS researchers believe is expressed in the brain. The *in situ* data can be searched by expression in a number of brain compartments as well as by intensity and coverage of that expression. Keep in mind, however, that all of these gene expression patterns are specific to the C57BL/6J mice and do not address variation along the genetic axis.

The ABA is a fantastic tool for determining whether QTL candidate genes are indeed expressed in the brain region for which they are expected to modulate expression. While it is possible that the gene responsible for a QTL may exert its influence from some other brain region, it is likely that genes expressed in a brain region are better candidates for QTLs that are related to a particular compartment.

The ABA is also valuable as an internal control on quality-of-brain-region preparations. If you are generating array data, it may be helpful to examine signature expression profiles for your chosen region. In other words, if a gene is unexpressed in your tissue of interest but highly expressed in a neighboring tissue, assaying for the gene expression is a useful way of quality-controlling your dissection protocol.

27.22 GENSAT

One other mouse imaging tool worth mentioning is the GENSAT database[59] (www.gensat.org), which contains a mouse gene expression database for embryos (E15.5), neonates (P7), and adult animals. GENSAT's expression model is based on bacterial artificial chromosomes (BACs) in which coding sequences were replaced by the enhanced green fluorescent protein (EGFP). Gene expression here is a relative and somewhat more indirect measure, since the EGFP mRNA and protein may have different stability than the gene it replaces. GENSAT uses FVB mice to generate its transgenic embryos, so, like ABA, it is not designed to investigate a genetic axis.

The ability to introduce multiple copies of the BAC enhances the ability to detect genes at low levels, however, and may be more sensitive than *in situ* methods. When adult expression is of primary interest, GENSAT is a useful resource to check, either to confirm expression observed in the ABA or for when the ABA shows no expression and there is reason to suspect that the actual expression level may simply be low. Of course, when embryonic or neonatal expression is crucial, as will often be the case with morphological characteristics, GENSAT is an extremely important resource.

27.22.1 EXAMPLE: MAPPING DENTATE GYRUS VOLUME

A sample QTL map from the WebQTL module of GeneNetwork (RecordID/10460) for dentate gyrus volume in the BXD RI strain set is shown in Figure 27.4. The trace of association (given as LOD) between genotype and phenotype is given. The $p = 0.05$ genome-wide adjusted significance threshold (adjusted for the many nonindependent tests for association between marker and phenotype) is shown as a line near the top of the graph. The significance threshold is derived from 1,000 permutations by the method of Churchill and Doerge.[60] For each permutation the relationship between animal ID (strain for RI populations) and phenotype is randomized, the permuted QTL map is generated, and the highest LOD score from each map is recorded. The ordered list of permuted LOD scores serves as the reference for genome-wide adjusted p-value thresholds. Note that the trace of marker-phenotype association only exceeds the genome-wide $p = 0.05$ threshold once, on chromosome (Chr) 13. This locus was named *DGV13a* when it was discovered[32] to indicate its effect on dentate gyrus volume, its location on Chr 13, and that it was the first such QTL to be identified on that chromosome. Having identified a QTL affecting a morphological trait, the next logical question is how to define the interval of the QTL—the range of positions within which we expect to find the gene responsible for the QTL's presence. We know the QTL is on Chr 13, and we can use the GN Map Viewer to generate a view of this chromosome only (Figure 27.5). This expanded view highlights several useful functions of GN. The bars indicate the frequency that, using a bootstrapped sample, the maximum position of the QTL was in the area indicated. This is a method for defining the QTL interval. A more widely used method is to follow the QTL association trace from its highest point to the nearest point 1 LOD down to the left and right. Dupuis and Siegmond[61] discovered, however, that a 1.5 LOD drop is closer to a 95% confidence interval. This method estimates that the gene causing the effect of *DGV13a* is likely to lie between 46 to 64 Mb on Chr 13.

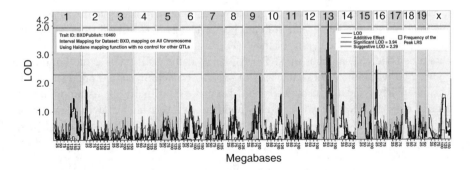

FIGURE 27.4 Whole genome QTL map of dentate gyrus volume. This figure shows a typical QTL map for a morphological phenotype. Dentate gyrus volume has one significant QTL on Chr 13. The line near the top indicates genome-wide adjusted $p = 0.05$ threshold.

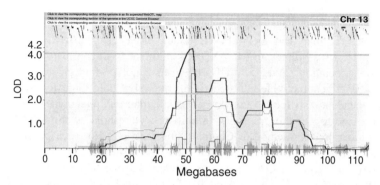

FIGURE 27.5 Chr 13 QTL map of dentate gyrus volume QTL *DGV13a*. This is a closeup of *DGV13a*. Features are discussed further in the text and include the SNP density map at the bottom, bars indicating the frequency with which the best QTL position was in that position in a bootstrapped population, and the light trace indicating the additive effect of the marker on the phenotype. A B6 allele increases dentate gyrus volume.

27.22.2 EXAMPLE: *ROR2* AND *MSX2* AS CANDIDATE GENES FOR *DGV13A*

There are many genes within the 18 Mb *DGV13a* interval, and searching for the gene or genes responsible for the QTL's effect is more art than science. There are two major complementary approaches: narrowing the interval and building evidence for a particular gene. Narrowing the QTL interval reduces the number of potential candidates and generally involves more genetic approaches, such as building con-genics or phenotyping additional animals that are recombinant across the QTL interval. Building evidence for a particular candidate is trickier given the formidable number of possibilities and the capacity of the human mind for constructing plausible stories.

That said, our favorite candidates for *DGV13a* are *Ror2* and *Msx2*. *Ror2* is a receptor tyrosine kinase expressed in the developing nervous system[62] as well as the branchial arches, heart, and limb/tail bud.[63] The gene is located from 51.673 to 51.845 Mb, right in the center of the *DGV13a* interval, and, crucially, in a region with a considerable number of SNPs. A glance at the SNP track at the bottom of Figure 27.5 will indicate how unevenly the SNPs between B6 and D2 are distributed. This is a fairly typical distribution, and if we make the simplifying assumption that a gene with no SNPs is not a likely candidate, we can immediately eliminate a substantial fraction of the genes in the interval. In this case, *Ror2* has 289 SNPs, though none of these are non-synonymous (the apparent mis-sense mutation, mCV23266532, is actually in intron 1).

Ror2 regulates transcription of MSH-like homeo box 2 (Ms × 2) by sequestering Maged1 (Dlxin-1) from the membrane.[64] Since Ms × 2 is located very near *Ror2*, we might expect this to result in what would appear to be a *cis*-acting QTL. While Ror2 seems to be expressed in adult tissue according to the ABA, it has extremely low expression, which is hard to distinguish from noise, in our hippocampus data. This suggests that the effect of the gene may occur during development. This

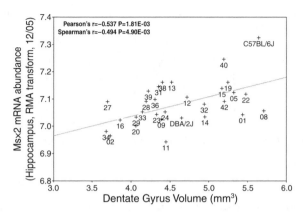

FIGURE 27.6 Correlation of Ms × 2 mRNA abundance and dentate gyrus weight. This scatterplot shows the genetic correlation between Ms × 2 (probe set 1449559, BXD hippocampus data) mRNA abundance and dentate gyrus weight. (BXD published phenotype 10460).

suggestion is bolstered by the expression of *Ror2* in the early nervous system.[62] The peri-natal lethality of a *Ror2* knockout[63] and the association of *Ror2* with dwarfism, cyanosis, and short limbs and tails also underscore its importance in development.

Ms × 2 interacts with *Ror2*, and its expression is strongly correlated with dentate gyrus volume (Figure 27.6; Pearson's $r = 0.53$; $p = 0.0018$), which make it a compelling candidate in the same pathway. Ms × 2 is located from 52.026 to 52.031 Mb on Chr 13 and has an even higher SNP density per Kb than *Ror2*. Maged1, which mediates the interaction between Ms × 2 and *Ror2*, is located on Chr X and has 0 SNPs, so while it interacts with our favorite candidates it is not itself a candidate.

27.23 A SIDE NOTE ON THE PROBLEMS OF SPECIFICITY AND POWER

We discussed above a significant QTL called DGV13a, so named because we found a significant QTL for dentate gyrus volume, but not a significant QTL near that position for volume of the pyramidal cell layer or other aspects of hippocampal morphology. However, lack of detection in a QTL analysis does not imply any definite proof of effect absence.

In addition, it is entirely possible that a QTL may have a large effect on the dentate gyrus and a small effect in nearby areas, so that in a larger population its effect size would be sufficient to generate a QTL independently. In this case the QTL has a real effect on multiple areas yet is likely to be named as affecting only one. When the areas being analyzed are overlapping rather than neighboring any QTL affecting a subregion, the situation will seem similar but will have a different underlying cause. Unless balanced by a corresponding, opposite effect, a QTL affecting a smaller structure should be picked up by a sufficiently sensitive experiment analyzing a larger structure that overlaps the smaller structure.

Also, at least with the smaller BXD RI strain set, (this is to some extent ameliorated by the larger, extended BXD strain set) GN experiments are underpowered for detection of QTLs of small to medium effect. This is not to say that one should not identify a QTL by the designation of the least restrictive population it effects or by the region in which it has the largest effect size if that is where it demonstrated significance. This still seems appropriate in the absence of biological understanding of the QTL's effect. It does, however, suggest liberal interpretation of candidate-gene evidence regarding the gene's effect in larger areas—and cautious interpretation of the importance of having selectively identified a subregion QTL.

27.24 CONCLUSION

The major purpose of understanding biological processes is to be able to make reliable predictions and to modulate outcome in a logical and controlled way. To achieve this level of understanding, we must understand complex systems and networks of phenotypes. We need to understand how combinations of gene variants, networks of molecules, and neuronal circuits work together in different individuals and in many different environments to achieve functional equilibrium. We need to adopt holistic and integrative approaches to experimental design that would have been completely impractical a decade ago. The advent of sophisticated statistical and computational methods and the advent of high-throughput methods to acquire hundreds of thousands of genotypes and phenotypes are rewriting the rules of design. The scientific and statistical chaos that followed the introduction of microarrays is just the technical spearhead of this new experimental order. We should not be surprised if there has been a backlash on the part of those scientists more accustomed to juggling one ball at a time. We are now living through an awkward transition that is moving us in the direction of systems biology for good reason: only this approach will enable us to make reliable predictions of complex outcomes. Animals and humans are not machines. We are stochastic, fluid, willful, and richly redundant cybernetic systems.

REFERENCES

1. Adams MD, Soares MB, Kerlavage AR, Fields C, Venter JC: Rapid cDNA sequencing (expressed sequence tags) from a directionally cloned human infant brain cDNA library. *Nat Genet* 1993, 4(4):373–380.
2. Sutcliffe JG: mRNA in the mammalian central nervous system. *Annu Rev Neurosci* 1988, 11:157–198.
3. Yeo G, Holste D, Kreiman G, Burge CB: Variation in alternative splicing across human tissues. *Genome Biol* 2004, 5(10):R74.
4. Williams RW, Herrup K: The control of neuron number. *Annu Rev Neurosci* 1988, 11:423–453.
5. Gautvik KM, de Lecea L, Gautvik VT, Danielson PE, Tranque P, Dopazo A, Bloom FE, Sutcliffe JG: Overview of the most prevalent hypothalamus-specific mRNAs, as identified by directional tag PCR subtraction. *Proc Natl Acad Sci USA* 1996, 93(16):8733–8738.

6. Usui H, Falk JD, Dopazo A, de Lecea L, Erlander MG, Sutcliffe JG: Isolation of clones of rat striatum-specific mRNAs by directional tag PCR subtraction. *J Neurosci* 1994, 14(8):4915–4926.
7. Lashley K: Persistent problems in the evolution of mind. *Quart Rev Biol* 1949, 24:28–42.
8. Wimer C, Prater L: Some behavioral differences in mice genetically selected for high and low brain weight. *Psychol Rep* 1966, 19(3):675–681.
9. Keverne EB, Martel FL, Nevison CM: Primate brain evolution: genetic and functional considerations. *Proc Biol Sci* 1996, 263(1371):689–696.
10. Aboitiz F: Does bigger mean better? Evolutionary determinants of brain size and structure. *Brain Behav Evol* 1996, 47(5):225–245.
11. Rensch B: Increase of learning capability with increase of brain size. *Amer Natur* 1956, 90(81–95).
12. Ward BC, Nordeen EJ, Nordeen KW: Individual variation in neuron number predicts differences in the propensity for avian vocal imitation. *Proc Natl Acad Sci U S A* 1998, 95(3):1277–1282.
13. Devoogd TJ, Krebs JR, Healy SD, Purvis A: Relations between song repertoire size and the volume of brain nuclei related to song: comparative evolutionary analyses amongst oscine birds. *Proc Biol Sci* 1993, 254(1340):75–82.
14. Crusio WE, Schwegler H, van Abeelen JH: Behavioral responses to novelty and structural variation of the hippocampus in mice. II. Multivariate genetic analysis. *Behav Brain Res* 1989, 32(1):81–88.
15. Schwegler H, Lipp HP: Hereditary covariations of neuronal circuitry and behavior: correlations between the proportions of hippocampal synaptic fields in the regio inferior and two-way avoidance in mice and rats. *Behav Brain Res* 1983, 7(1):1–38.
16. Lipp HP, Schwegler H, Heimrich B, Driscoll P: Infrapyramidal mossy fibers and two-way avoidance learning: developmental modification of hippocampal circuitry and adult behavior of rats and mice. *J Neurosci* 1988, 8(6):1905–1921.
17. Lipp HP, Collins RL, Hausheer-Zarmakupi Z, Leisinger-Trigona MC, Crusio WE, Nosten-Bertrand M, Signore P, Schwegler H, Wolfer DP: Paw preference and intra-/infrapyramidal mossy fibers in the hippocampus of the mouse. *Behav Genet* 1996, 26(4):379–390.
18. Lipp HP, Schwegler H, Crusio WE, Wolfer DP, Leisinger-Trigona MC, Heimrich B, Driscoll P: Using genetically-defined rodent strains for the identification of hippocampal traits relevant for two-way avoidance behavior: a non-invasive approach. *Experientia* 1989, 45(9):845–859.
19. Schwegler H, Crusio WE, Lipp HP, Heimrich B: Water-maze learning in the mouse correlates with variation in hippocampal morphology. *Behav Genet* 1988, 18(2):153–165.
20. Lipp HP, Schwegler H, Heimrich B, Cerbone A, Sadile AG: Strain-specific correlations between hippocampal structural traits and habituation in a spatial novelty situation. *Behav Brain Res* 1987, 24(2):111–123.
21. Patel NV, Hitzemann RJ: Detection and mapping of quantitative trait loci for halo-peridol-induced catalepsy in a C57BL/6J x DBA/2J F2 intercross. *Behav Genet* 1999, 29(5):303–310.
22. Kanes S, Dains K, Cipp L, Gatley J, Hitzemann B, Rasmussen E, Sanderson S, Silverman M, Hitzemann R: Mapping the genes for haloperidol-induced catalepsy. *J Pharmacol Exp Ther* 1996, 277(2):1016–1025.
23. Bucan M, Abel T: The mouse: genetics meets behaviour. *Nat Rev Genet* 2002, 3(2):114–123.

24. Melo JA, Shendure J, Pociask K, Silver LM: Identification of sex-specific quantitative trait loci controlling alcohol preference in C57BL/ 6 mice. *Nat Genet* 1996, 13(2):147–153.

25. Peirce JL, Derr R, Shendure J, Kolata T, Silver LM: A major influence of sex-specific loci on alcohol preference in C57Bl/6 and DBA/2 inbred mice. *Mamm Genome* 1998, 9(12):942–948.

26. Takahashi JS, Pinto LH, Vitaterna MH: Forward and reverse genetic approaches to behavior in the mouse. *Science* 1994, 264(5166):1724–1733.

27. Plomin R, McClearn GE, Gora-Maslak G, Neiderhiser JM: Use of recombinant inbred strains to detect quantitative trait loci associated with behavior. *Behav Genet* 1991, 21(2):99–116.

28. Belknap JK, Phillips TJ, O'Toole LA: Quantitative trait loci associated with brain weight in the BXD/Ty recombinant inbred mouse strains. *Brain Res Bull* 1992, 29(3–4):337–344.

29. Williams RW: Mapping genes that modulate mouse brain development: a quantitative genetic approach. In: *Mouse Brain Development*. Edited by Goffinet A, Rakic P. New York: Springer; 2000: 21–49.

30. Williams RW, Strom RC, Goldowitz D: Natural variation in neuron number in mice is linked to a major quantitative trait locus on Chr 11. *J Neurosci* 1998, 18(1):138–146.

31. Lu L, Airey DC, Williams RW: Complex trait analysis of the hippocampus: mapping and biometric analysis of two novel gene loci with specific effects on hippocampal structure in mice. *J Neurosci* 2001, 21(10):3503–3514.

32. Peirce JL, Chesler EJ, Williams RW, Lu L: Genetic architecture of the mouse hippocampus: identification of gene loci with selective regional effects. *Genes Brain Behav* 2003, 2(4):238–252.

33. Williams RW, Airey DC, Kulkarni A, Zhou G, Lu L: Genetic dissection of the olfactory bulbs of mice: QTLs on four chromosomes modulate bulb size. *Behav Genet* 2001, 31(1):61–77.

34. Rosen GD, Williams RW: Complex trait analysis of the mouse striatum: independent QTLs modulate volume and neuron number. *BMC Neurosci* 2001, 2:5.

35. Zhou G, Williams RW: Eye1 and Eye2: gene loci that modulate eye size, lens weight, and retinal area in the mouse. *Invest Ophthalmol Vis Sci* 1999, 40(5):817–825.

36. Airey DC, Robbins AI, Enzinger KM, Wu F, Collins CE: Variation in the cortical area map of C57BL/6J and DBA/2J inbred mice predicts strain identity. *BMC Neurosci* 2005, 6(1):18.

37. Airey DC, Lu L, Williams RW: Genetic control of the mouse cerebellum: identification of quantitative trait loci modulating size and architecture. *J Neurosci* 2001, 21(14):5099–5109.

38. Li CX, Wei X, Lu L, Peirce JL, Williams RW, Waters RS: Genetic analysis of barrel field size in the first somatosensory area (SI) in inbred and recombinant inbred strains of mice. *Somatosens Mot Res* 2005, 22(3):141–150.

39. Williams RW: Expression Genetics and the Phenotype Revolution. *Mamm Genome* 2006, in press.

40. Seecharan DJ, Kulkarni AL, Lu L, Rosen GD, Williams RW: Genetic control of interconnected neuronal populations in the mouse primary visual system. *J Neurosci* 2003, 23(35):11178–11188.

41. Damerval C, Maurice A, Josse JM, de Vienne D: Quantitative trait loci underlying gene product variation: a novel perspective for analyzing regulation of genome expression. *Genetics* 1994, 137(1):289–301.

42. Jansen RC, Nap JP: Genetical genomics: the added value from segregation. *Trends Genet* 2001, 17(7):388–391.

43. Brem RB, Yvert G, Clinton R, Kruglyak L: Genetic dissection of transcriptional regulation in budding yeast. *Science* 2002, 296(5568):752–755.

44. Chesler EJ, Lu L, Shou S, Qu Y, Gu J, Wang J, Hsu HC, Mountz JD, Baldwin NE, Langston MA et al: Complex trait analysis of gene expression uncovers polygenic and pleiotropic networks that modulate nervous system function. *Nat Genet* 2005, 37(3):233–242.

45. Bystrykh L, Weersing E, Dontje B, Sutton S, Pletcher MT, Wiltshire T, Su AI, Vellenga E, Wang J, Manly KF et al: Uncovering regulatory pathways that affect hematopoietic stem cell function using 'genetical genomics.' *Nat Genet* 2005, 37(3):225–232.

46. Schadt EE, Monks SA, Drake TA, Lusis AJ, Che N, Colinayo V, Ruff TG, Milligan SB, Lamb JR, Cavet G et al: Genetics of gene expression surveyed in maize, mouse and man. *Nature* 2003, 422(6929):297–302.

47. Hubner N, Wallace CA, Zimdahl H, Petretto E, Schulz H, Maciver F, Mueller M, Hummel O, Monti J, Zidek V et al: Integrated transcriptional profiling and linkage analysis for identification of genes underlying disease. *Nat Genet* 2005, 37(3):243–253.

48. Stranger BE, Forrest MS, Clark AG, Minichiello MJ, Deutsch S, Lyle R, Hunt S, Kahl B, Antonarakis SE, Tavare S et al: Genome-Wide Associations of Gene Expression Variation in Humans. *PLoS Genet* 2005, 1(6):e78.

49. Cheung VG, Spielman RS, Ewens KG, Weber TM, Morley M, Burdick JT: Mapping determinants of human gene expression by regional and genome-wide association. *Nature* 2005, 437(7063):1365–1369.

50. Morley M, Molony CM, Weber TM, Devlin JL, Ewens KG, Spielman RS, Cheung VG: Genetic analysis of genome-wide variation in human gene expression. *Nature* 2004, 430(7001):743–747.

51. Kirst M, Basten CJ, Myburg AA, Zeng ZB, Sederoff RR: Genetic architecture of transcript-level variation in differentiating xylem of a eucalyptus hybrid. *Genetics* 2005, 169(4):2295–2303.

52. Benjamini Y, Hochberg Y: Controlling the False Discovery Rate: a practical and powerful approach to multiple testing. *J Royal Statistical Society B* 1995, 57:289–300.

53. Doss S, Schadt EE, Drake TA, Lusis AJ: *Cis*-acting expression quantitative trait loci in mice. *Genome Res* 2005, 15(5):681-691.

54. Peirce J, Lu L, Hongqiang L, Wang J, Manly KF, Hitzemann B, Belknap JK, Rosen GD, Williams RW: How replicable are mRNA expression QTLs. *Mamm Genome* 2006, in press.

55. Peirce JL, Lu L, Gu J, Silver LM, Williams RW: A new set of BXD recombinant inbred lines from advanced intercross populations in mice. *BMC Genet* 2004, 5(1):7.

56. McCaughran J, Jr., Bell J, Hitzemann R: On the relationships of high-frequency hearing loss and cochlear pathology to the acoustic startle response (ASR) and prepulse inhibition of the ASR in the BXD recombinant inbred series. *Behav Genet* 1999, 29(1):21–30.

57. Lin CH, Hansen S, Wang Z, Storm DR, Tapscott SJ, Olson JM: The dosage of the neuroD2 transcription factor regulates amygdala development and emotional learning. *Proc Natl Acad Sci U S A* 2005, 102(41):14877–14882.

58. Homayouni R, Heinrich K, Wei L, Berry MW: Gene clustering by latent semantic indexing of MEDLINE abstracts. *Bioinformatics* 2005, 21(1):104–115.

59. Heintz N, Hatten M: The Gene Expressio Nervous System Atlas (GENSAT) Project, NINDS Contract #N01NS02331 to The Rockefeller University, New York. February 1, 2006. www.gensat.org.

60. Churchill GA, Doerge RW: Empirical threshold values for quantitative trait mapping. *Genetics* 1994, 138(3):963–971.

61. Dupuis J, Siegmund D: Statistical methods for mapping quantitative trait loci from a dense set of markers. *Genetics* 1999, 151(1):373–386.

62. Oishi I, Takeuchi S, Hashimoto R, Nagabukuro A, Ueda T, Liu ZJ, Hatta T, Akira S, Matsuda Y, Yamamura H et al: Spatio-temporally regulated expression of receptor tyrosine kinases, mRor1, mRor2, during mouse development: implications in development and function of the nervous system. *Genes Cells* 1999, 4(1):41–56.

63. Takeuchi S, Takeda K, Oishi I, Nomi M, Ikeya M, Itoh K, Tamura S, Ueda T, Hatta T, Otani H et al: Mouse Ror2 receptor tyrosine kinase is required for the heart development and limb formation. *Genes Cells* 2000, 5(1):71–78.

64. Matsuda T, Suzuki H, Oishi I, Kani S, Kuroda Y, Komori T, Sasaki A, Watanabe K, Minami Y: The receptor tyrosine kinase Ror2 associates with the melanoma-associated antigen (MAGE) family protein Dlxin-1 and regulates its intracellular distribution. *J Biol Chem* 2003, 278(31):29057–29064.

28 Synaptic Mechanisms Involved in Cognitive Function: Cues from Mental Retardation Genes

Guntram Borck, Florence Molinari, Birgit Dreier, Peter Sonderegger, and Laurence Colleaux

CONTENTS

Mental retardation (MR) is likely the result of alterations in molecular pathways underlying neuronal processes involved in cognitive functions; however, the exact nature of these pathways remains unknown. Remarkable progress in understanding the molecular and cellular basis of mental retardation has occurred in recent years. Genes associated with nonsyndromic mental retardation have been evaluated with regard to a role in synaptic transmission. Some of them have been linked to presynaptic vesicle release, whereas others are involved in the regulation of actin cytoskeleton dynamics via Rho GTPases. Detailed investigations of their molecular and cellular roles at the synapse may provide cues for synaptic mechanisms that are essential for cognitive functions.

28.1 INTRODUCTION

Mental retardation (MR) affects about 2% of the general population and is the most frequent handicap in children and young adults. It is characterized by a broad range of deficits in higher brain functions that result in significant limitations in adaptive and cognitive capacities required for competence in daily living, communication, social interaction and integration, self-direction, and work (DSM-IV). An onset of the symptoms in the developmental period is an essential diagnostic trait. The severity of mental retardation is commonly classified on the basis of the intelligence quotient (IQ) although other criteria have also been used. Based on a population mean of 100 and a standard deviation of 15, MR is usually classified as "mild" when the IQ ranges between 50 to 70 and as "severe" when the IQ value is below 50.[1,2]

The causes of MR are diverse and include environmental factors, teratogens, numerical or structural anomalies of chromosomes, gene defects, and metabolic diseases. An estimated fraction of 25 to 40% of the severe MRs and most mild MRs remain unexplained.[3–5] They are currently thought to be due to monogenic or multigenic defects or to a combination of genetic and environmental factors. Only 25 to 50% of the severe forms of MR are estimated to be genetically determined. They include metabolic diseases impairing neuronal function in a nonspecific manner, conditions which alter the normal patterning of the brain, neuromuscular disorders, as well as MRs that do not show any clinical features besides cognitive deficits. The last category is termed nonsyndromic MR (NSMR).

Although NSMR conditions are much difficult for geneticist studies, they are best suited for identifying the molecular basis of cognitive functions, and analysis of these disorders will very likely give new insight into the neurobiology of human cognitive processing.

28.2 AT THE BEGINNING WAS THE X

A linkage of a relevant proportion of nonsyndromic MRs to the X chromosome was suggested already in the 1930s based on the observed predominance of MR in men. The large number of families in which MR is inherited as an X-linked trait provided further, more direct evidence for sex linkage of mental retardation.[6] Taking advantage of the ease of gene mapping on the X chromosome and of the large number of X-linked mental retardation (XLMR) families collected by international consortia,

NS-XLMRs were first investigated. Since 1996, positional and functional candidate gene approaches led to the identification of 11 X-linked nonsyndromic MR genes.[7,8] Despite the extreme variety of processes in which genes responsible for these disorders are involved, several common cellular processes have emerged.

28.2.1 NS-MR Genes Involved in the Regulation of the Actin Cytoskeleton: Possible Roles for the Adaptive Response of Postsynaptic Spines

Three X-linked nonsyndromic MR genes encode components that are directly involved in signal transduction pathways of the Rho family GTPases (Figure 28.1). Small GTPases of the Rho family act as transducers of extracellular signals to the

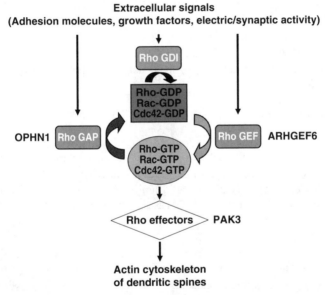

FIGURE 28.1 (SEE COLOR INSERT FOLLOWING PAGE 236) One cluster of nonsyndromic MR genes relates to the Rho family of small GTPases. Rho family GTPases transduce extracellular signals into adaptive responses of the actin cytoskeleton. Actin-dependent processes at the synapse include the regulation of the morphology and the dynamics of the dendritic spines. The Rho family comprises three members, termed Rho, Rac, and Cdc42. They shuttle between an inactive (red) and an active (green) state under the control of three types of regulatory proteins (blue). GEFs (guanine nucleotide exchange factors) promote the release of GDP and its replacement by GTP and, thereby, mediate the transition of Rho, Rac, and Cdc42 from the inactive into the active state. GAPs (GTPase-activating proteins) activate the endogenous GTPase function and, thus, the self-inactivation of Rho, Rac, and Cdc42. GDIs (GDP dissociation inhibitors) bind to the GDP form of the GTPases and prevent their premature activation as long as the GTPase is not in the correct place and situation for a new round of activation. Further downstream activators (yellow) eventually act directly or indirectly on regulatory components of the actin cytoskeleton. The genes affected in MRs are printed red.

cytoskeleton and regulate gene expression. They are involved in neurite outgrowth, axon guidance, dendrite maturation, synapse formation, and the morphogenesis and dynamics of dendritic spines.[9–12]

The family comprises three members, termed Rho, Rac, and Cdc42. They exert their diverse cellular functions via distinct effects on the actin cytoskeleton that are mediated by PAK and ROCK kinases as immediate downstream signaling molecules. Both PAK and ROCK kinases in turn activate LIM kinase, which acts upon and regulates the dynamics of the actin cytoskeleton. Three major classes of upstream regulators control the signaling activity of the Rho family GTPases. First, the guanine nucleotide exchange factors (RhoGEFs) mediate the release of GDP and its replacement by GTP and, thus, drive the conversion from the inactive into the active form. Second, the GTPase activating proteins (RhoGAPs) enhance the conversion of bound GTP into GDP and, thus, drive the conversion from the active into the inactive state. And third, the GDP dissociation inhibitors (RhoGDIs) bind to the GDP-bound form of the GTPase and prevent their activation as long as other prerequisites, such as the correct location for reinitiation of the activation cycle, are not fulfilled.

Over the past years, two regulators and one downstream effector of Rho family GTPases have been found among the genes causing X-linked nonsyndromic MR, described as follows.

28.2.1.1 *Oligophrenin-1*

The *OPHN1* gene, encoding *oligophrenin-1*, maps to Xq12 and is expressed in fetal brain at high levels. It is disrupted in a mildly mentally retarded female carrier of a balanced X;12 translocation. Furthermore, it has been shown to be mutated in affected males of a large X-linked MR family by Billuart et al.[13] Oligophrenin contains a domain typical for Rho-GTPase-activating proteins (RhoGAP). *In vitro* assays demonstrated that it can regulate the activity of the Rho GTPases RhoA, Rac1 and Cdc42Hs.

Recent studies show that *oligophrenin-1* is present in neuronal and astroglial cells and that it colocalizes with actin at the tip of growing neurites.[14] In addition, using siRNAs in organotypic hippocampal slices, Govek and co-workers[15] demonstrated that knocking down oligophrenin-1 significantly decreased dendritic spine length in CA1 pyramidal neurons. Spine morphological changes of the same magnitude have been reported for a mouse model of fragile X,[16] indicating that such changes can compromise synaptic plasticity and potentially lead to learning and memory deficits.

28.2.1.2 *PAK3* (p21 Activated Kinase 3)

Interestingly, soon after the characterization of *OPHN1*, a second NS-XLMR gene implicated in the Rho GTPase pathway was described. Using a candidate gene approach, Allen et al.[17] showed that a mutation in the *PAK3* gene was associated with MR. *PAK3* encodes a serine-threonine kinase involved in the Rac/Cdc42-dependent regulation of actin cytoskeleton dynamics. In neuronal cells, *PAK3* acts downstream of the GTPases in a signaling pathway that drives polarized growth of the actin cytoskeleton in developing neurites. In agreement with such a function,

PAK3 is highly expressed in fetal human brain. In the newborn mouse brain it localizes mainly to cortical and hippocampal dendrites and axons. LIM kinase 1, a downstream effector kinase regulated by PAK kinase, was recently reported to regulate the morphology of dendritic spines of hippocampal neurons.[18]

28.2.1.3 *αPIX* or *Cool-2*

Finally, taking advantage of a balanced X;21 translocation with a breakpoint in Xq26, Kutsche et al.[19] showed that the gene *ARHGEF6*, encoding a Rac1/Cdc42-specific guanine nucleotide exchange factor called a *PIX*, was truncated in a female patient with severe MR and that a null mutation of *ARHGEF6* cosegregated with the MR phenotype in a X-linked MR family. Due to this regulation of Rac and Cdc42 *ARHGEF6* has been implicated in signal transduction pathways involved in cell migration and axonal outgrowth. Although little is known about the biological function of *ARHGEF6* in neuronal processes, its interaction with the focal adhesion molecule beta-Parvin (PARVB) suggests that *ARHGEF6* is involved in integrin-mediated signaling leading to the activation of Rac1 and/or Cdc42 GTPases.[20] A related Rho GEF, termed kalirin-7, has recently been reported to be involved in dendritic spine morphogenesis via activation of Rac1 and its effector PAK.[21]

Although *OPHN1*, *PAK3* and *ARHGEF6* genes affect the signaling of the Rho family members Rho, Rac, and Cdc42 differently, they all affect actin cytoskeleton dynamics, as this is the hallmark of Rho family GTPase function. Dendritic plasticity depends on morphological changes that follow rearrangements of the cytoskeleton in response to neural activity at the synaptic site.[22–24] The observation that mutations in these genes are responsible for NSMR sheds light on the connection between Rho GTPases and cognition and makes a clear case for the central role played by these effectors in neuronal signaling. The importance of the regulation of actin dynamics is further demonstrated by the involvement of downstream effectors of the Rho GTPase signaling pathway. Not only *PAK3*, a direct effector kinase of Rac and Cdc42, but also LIM kinase-1, a kinase downstream of *PAK3*, have been linked to MR. If *PAK3* mutations are responsible for NS-XLMR, a mutation in LIM kinase-1 was found in patients with Williams syndrome, a syndromic form of MR. Finally, the hypothesis that a deficiency in the cytoskeletal dynamics at the synapse may be responsible for cognitive deficit is also supported by the observation of abnormal shapes and numbers of dendritic spines.[25]

28.2.2 NONSYNDROMIC MR GENES INVOLVED IN THE REGULATION OF PRESYNAPTIC NEUROTRANSMITTER RELEASE

Vesicle release from the presynaptic nerve terminal is an essential process for synaptic transmission. For a long time it has been known that synaptic vesicle release is subject to modulations that depend on previous activity patterns of a given synapse. It is known that this modulatory mechanism, termed synaptic facilitation, depends on intracellular calcium, but its exact molecular basis is still under investigation. Neurotransmitter release is essential for basic synaptic function. Its activity-dependent regulation is of critical importance for the coordinated and dynamic

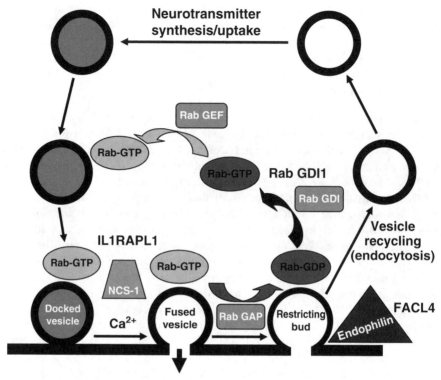

FIGURE 28.2 (SEE COLOR INSERT) A second cluster of nonsyndromic MR genes relates to the release of neurotransmitters from the presynaptic nerve terminal and the regeneration of new releasable vesicles by the budding and endocytosis of vesicles from the presynaptic membrane. The release of neurotransmitters from storage vesicles occurs in several consecutive steps. First, the vesicles are docked to the release site, the presynaptic active zone, a process that involves the interaction of several proteins of the vesicle and the active zone. Docked vesicles fuse with the presynaptic membrane in a process regulated by transiently increased intracellular calcium, and release their content into the synaptic cleft. In order to maintain a constant supply of releasable neurotransmitters, vesicles recycle by budding off the presynaptic membrane and endocytosis. The budding occurs in a clathrin-dependent manner. The restriction of the bud neck, a process that prepares the bud for scission, requires endophilin. Endophilin is a lysophosphatidic acid acyltransferase, which introduces arachidonic acid into the cytoplasmic face of the bud neck. It, thereby, facilitates the formation of a strong negative curvature of the membrane required for the restriction of the bud neck in order to allow the scission process to begin. It is conceivable that the supply of arachidonoyl-CoA requires FACL4. The genes affected in MR are printed in red.

function of synaptic circuits. Three nonsyndromic MR genes play essential roles in the presynaptic vesicle release and recycling machinery, *RabGDI1*, *IL1RAPL1*, and *FACL4* (Figure 28.2). Particularly intriguing in the context of MR is the finding that their function is indispensable for the fine-tuning of the release machinery during activity-dependent adaptive processes. The deficiency of these genes is still compatible with basic synaptic function, but adaptive responses of the secretory machinery, such as those thought to be necessary for learning and memory, may be deficient.

28.2.2.1 *RabGDI1* (Rab GDP-Dissociation Inhibitor 1)

The members of the GDI family (Rab GDP-dissociation inhibitors) play an essential role in the recycling of Rab GTPases required for vesicular transport. Using a candidate gene approach, D'Adamo et al.[26] identified *RabGDI1* mutations in two X-linked MR families linked to the Xq28 region. *RabGDI1* binds Rab3a, a small GTP-binding protein highly enriched in the synapse and involved in neurotransmitter release. First evidence for a role of *RabGDI1* protein in neurite outgrowth and synaptic function came from *in vitro* studies showing that *RabGDI1*-antisense oligonucleotides impaired neurite extension in cultured rat hippocampal neurons. Further support for this role came from histological, behavioral and electrophysiological studies of *RabGDI1*-deficient mice. These mice, which do not have any gross morphological or neuropathological anomalies, display an altered plasticity of hippocampal neurotransmission.[27] In addition, behavioral studies have revealed defects in short-term memory, a reduced aggression level, and altered social behavior.[28] A recent ultrastructural analysis revealed pronounced presynaptic changes in several brain regions. Most strikingly, a marked reduction of the number of synaptic vesicles and a clustering of the residual vesicles were observed. These ultrastructural alterations are in perfect accordance with a role of *RabGDI1* in synaptic vesicle recycling.

28.2.2.2 *IL1RAPL1* (Interleukin-1 Receptor Accessory Protein-Like1)

The *IL1RAPL1* gene has been identified in the analysis of non-overlapping Xp22 deletions and point mutations in X-linked MR families by Carrie et al.[29] The *IL1RAPL1* protein is a member of the IL-1/Toll receptor family. It shows homologies with the IL-1 receptor accessory proteins (IL1RACPs), and localizes to the plasma membrane. Recent functional studies showed that *IL1RAPL1* is not a receptor for IL-1 but a specific interaction partner of neuronal calcium sensor 1.[30] The NCS-1 protein has been shown to be involved in synaptic facilitation at excitatory hippocampal synapses.[31] Synaptic facilitation is a mechanism of short-term plasticity that enhances transmitter release from the presynaptic terminal and increases postsynaptic activation as a consequence of recurrent stimulation. Because synaptic facilitation has long been known to depend on calcium and because NCS-1 was demonstrated to be a sensor of presynaptic calcium, NCS-1 is put into a crucial position for the activity-dependent adaptation of presynaptic transmitter release. By interacting with NCS-1 in the inhibition of exocytosis, *IL1RAPL-1* might exert a regulatory role on the activity-dependent dynamics of presynaptic transmitter release.

28.2.2.3 *FACL4* (Fatty Acid-CoA Ligase 4)

By deletion mapping, *FACL4*, which encodes fatty acid-CoA ligase 4, was identified as an X-linked nonsyndromic MR gene.[32] Acyl-CoA ligases (or synthases) form a family of enzymes that catalyze the formation of acyl-CoA esters from fatty acids, ATP and coenzyme A. Both MR-associated mutations lead to a drastic decrease of enzymatic activity. That fatty acid-CoA ligase 4 has a strong preference for arachidonic acid as a substrate raises interesting speculations about its role in the recycling of synaptic vesicles. Recent reports suggest an essential role of arachidonic acid in the endocytosis of synaptic vesicles from the presynaptic membrane, a process required for the regeneration of releasable neurotransmitter vesicles. A crucial step in this process is the formation of a bud which then constricts and separates from the presynaptic plasma membrane to form a new vesicle. This process requires the remodeling of the lipids on the cytosolic side of the bud neck by endophilin A1.[33] Endophilin A1 acts as a lysophosphatidic acid acyltransferase, i.e., it binds lysophosphatidic acid and fatty acyl-coenzyme A and condenses them to phosphatidic acid. The formation of phosphatidic acid promotes the negative curvature required at the bud neck for bud restriction. Unsaturated fatty acyl-CoAs, such as arachidonoyl-CoA, are most effective in promoting negative curvature when inserted into the cytosolic leaflet. Thus, a deficiency in unsaturated fatty acyl-CoAs, such as arachidonoyl-CoA, that may result from deficient *FACL4* is likely to reduce the efficiency of bud restriction and thus results in reduced synaptic vesicle recycling.

28.3 LESSONS FROM THE X

Some common features emerge from the studies on NS-XLMR. There are now several examples of genes initially described as syndromic MR genes that account for rare cases of NS-XLMR. This is the case for the *RSK2*, *MECP2*, and *FGD1* genes leading to Coffin-Lowry, Rett, and Aarskog-Scott syndromes, respectively, and mutated in rare families with NSMR.[34–36] In addition, the overlap between syndromic XLMR and NS-XLMR exists in both directions: not only can syndromic XLMR genes cause NS-XLMR but recently *OPHN1* mutations were found to cause MR and congenital cerebellar hypoplasia.[37,38] Another common point is that all NS-XLMR genes that have been tested are highly expressed in the fetal brain (from as early as embryonic day E8 on in the mouse) with a high expression in the cortex and the hippocampus, i.e., structures important for higher cognitive functions such as memory and learning. The fact that some XLMR genes are part of common physiological pathways has already been mentioned above. Finally, genetic heterogeneity of this pathology is still growing. None of the NS-XLMR genes accounted for all families linked to a specific region, implying that the number of genes, previously thought to lie between 8 and 12 based on non-overlapping linkage intervals, is much higher.[39]

28.4 NEUROTRYPSIN, A NONSYNDROMIC MR GENE WITH SYNAPTIC LOCALIZATION BUT UNKNOWN FUNCTION

While an autosomal recessive mode of inheritance may account for nearly one fourth of individuals with NS-MR, only X-linked genes have been identified until recently. In fact, the broad genetic heterogeneity of MR and the scarcity of large pedigrees suitable for linkage analyses have hitherto hampered identification of the genes responsible for these diseases. The hope that a careful study of individuals with chromosomal anomalies would lead to the rapid identification of autosomal MR genes has been largely disappointed. However, linkage studies and chromosomal breakpoint analyses recently led to the identification of the first autosomal NS-MR genes.

28.4.1 NEUROTRYPSIN MUTATION IN AUTOSOMAL RECESSIVE MR

The *PRSS-12* gene encoding the synaptic serine protease neurotrypsin was recently identified as the first autosomal-recessive gene involved in nonsyndromic MR. Using homozygosity mapping in a large consanguineous Algerian family, Molinari et al. mapped an autosomal recessive nonsyndromic MR gene on chromosome 4q24-q25.[40] The sibship comprises four mentally retarded and four healthy children born to first-cousin Algerian parents. All affected children (3 girls and 1 boy) exhibited a severe impairment of cognitive functions with an IQ below 50. Analysis of candidate genes mapping to this part of the genome led to the identification of a homozygous 4-bp deletion in the neurotrypsin gene, resulting most likely in a null allele due to the formation of a shortened protein lacking the catalytic domain.[40] Neurotrypsin is predominantly expressed in neurons of the cerebral cortex, the hippocampus and the amygdala.[41] By immuno-electronmicroscopy, neurotrypsin was localized in the pre-synaptic membrane and the presynaptic active zone of both asymmetrical (excitatory) and symmetrical (inhibitory) synapses. *In vitro* studies have demonstrated that it is a secreted protein which remains associated with the presynaptic membrane after its secretion. In search for additional mutations in this gene, Molinari et al. screened 18 inbred families and 30 nonconsanguineous families with individuals affected with nonsyndromic MR.[40] They found the same 4-bp deletion in a child born to first-cousin Algerian parents. The two families appear unrelated, but originate from the same area of eastern Algeria. In both families, the mutation was carried on the background of the same haplotype across the neurotrypsin locus, suggesting a founder effect in the Algerian population.

The pathophysiological phenotype and the age of disease onset in the affected individuals are consistent with the idea that neurotrypsin might regulate adaptive synaptic functions, such as synapse reorganization, during later stages of neurode-velopment and postnatal synaptic plasticity. In all affected children the course of the disease was similar. They reached the milestones of normal psychomotor devel-opment in the first 18 months. Signs of MR were first observed by their parents when they were around 2 years of age. This suggests that neurotrypsin is not involved critically in the formation of synapses, but rather plays a crucial role for adaptive synaptic functions, such as those subserving higher cognitive functions.

Altogether, these results provide the first evidence for an association between cognitive impairment and defects in extracellular proteolytic activity at the synapse, opening a novel field in the pathophysiology of MR. The generation of animal models such as mice deficient in the catalytic domain of neurotrypsin in CNS neurons or mice overexpressing neurotrypsin will provide further insight into the function of this protein.

28.5 WHAT DO NONSYNDROMIC MR GENES TELL US ABOUT THE CELLULAR MECHANISMS THAT ARE IMPORTANT FOR COGNITIVE FUNCTIONS?

The currently available data about monogenic defects resulting in nonsyndromic MR indeed reveal a relatively high proportion of genes that encode synaptic proteins. The best-studied among them contribute to presynaptic release of neurotransmitter and the recycling of transmitter vesicles (*IL1RAPL1, RabGDI1, FACL4*) or the regulation of the cytoskeletal dynamics of postsynaptic spines (*OPHN1, ARHGEF6, PAK3*). Both the presynaptic release of neurotransmitters and the morphology of the postsynaptic spines are elements of the adaptive response of synapses that is summarized under the term synaptic plasticity. The first gene involved in autosomal recessive NS-MR, *PRSS-12*, encodes the synaptic serine protease neurotrypsin, whose role in synaptic structure, function, and plasticity has not yet been elucidated. However, its almost exclusive synaptic localization strongly suggests a regulatory role of synaptic function and thus makes it an interesting candidate for a gene involved in cognitive processes.

A common feature that makes nonsyndromic MR genes particularly interesting is the observation that their deficiency is compatible with life, but results in severe deficits in cognitive functions while synapse formation and basic synaptic transmission is not affected. Thus, one can conclude that cognitive functions need more than basic synaptic activity. They depend on the coordination of extended synaptic circuits. This in turn depends on synaptic plasticity, the capacity of individual synapses to adapt their transmission on demand to fit the preconditions of the entire circuit. Intact mechanisms of synaptic plasticity are thought to be an important prerequisite for the development of these complex and dynamic circuits in the brain that underlie cognitive functions. Although at present, none of these genes has been unequivocally assigned to a specific synaptic function that when compromised would result in a selective loss of a cognitive function, e.g., memory and learning, their identification provides the basis for a detailed characterization of cognitive functions at the molecular level.

Many more NS-MR genes remain to be characterized because, for most of the genes currently known, mutations turn out to be very rare with only few mutations identified in affected families. The observation of common or closely related pathways dysfunction underlying NS-XLMR led to some excitement, but it should not result in a dangerous oversimplification. Many other molecular pathways are likely to be involved in NS-MR as well, and in most cases the relationship between these proteins and cognitive functions remains elusive.[42–47]

This observation is obviously bad news with respect to diagnosis and counseling, but opens up the possibility that there are still many other genes that play a major role in NS-MR. A better understanding of the function of MR genes will ultimately yield not only deeper insight into brain function, but may also provide targets for future therapies.

REFERENCES

1. Penrose, L.S., A clinical and genetic study of 1280 cases of mental defects (The Colchester Survey). Special report series, Medical Research Council No. 229, Her Majesty's Stationary Office, London, 1938.
2. Roeleveld, N., Zielhuis, G.A., and Gabreels, F. The prevalence of mental retardation: a critical review of recent literature, *Dev Med Child Neurol*, 39, 125, 1997.
3. Turner, G. An etiological study of 1,000 patients with an I.Q. assessment below 51. *Med J Aust*, 2, 927, 1975.
4. Lamont, M.A. and Dennis, N.R., Etiology of mild mental retardation. *Arch Dis Child*, 63, 1032, 1988.
5. Kahler, S.G. and Fahey, M.C., Metabolic disorders and mental retardation. *Am J Med Genet, Part C (Sem Med Genet)*, 117C, 31, 2003.
6. Stevenson, R.E., Schwartz, C.E., and Schroer, R.J., Emergence of the concept of X-linked mental retardation, In *X-Linked Mental Retardation*, Stevenson, R.E., Schwartz, C.E., and Schroer, R.J.. New York: Oxford University Press, 2000:23–67.
7. Chelly, J. and Mandel, JL. Monogenic causes of X-linked mental retardation. *Nature Rev Genet*, 2, 669, 2001.
8. Ramakers, G.J.A. Rho proteins, mental retardation and the cellular basis of cognition. *Trends in Neurosci*, 25, 191, 2002.
9. Mackay, D.J., Nobes, C.D., and Hall A. The Rho's progress: A potential role during neuritogenesis for the Rho family of GTPases. *Trends Neurosci*, 18, 496, 1995.
10. Luo, L. Rho GTPases in neuronal morphogenesis. *Nat Rev Neurosci*, 1, 173, 2000.
11. Ishizaki, H. et al. Role of rab GDP dissociation inhibitor alpha in regulating plasticity of hippocampal neurotransmission. *Proc Natl Acad Sci USA*, 97, 11587, 2000.
12. Negishi, M. and Katoh, H. Rho family GTPases as key regulators for neuronal network formation. *J Biochem.* (Tokyo) 132, 157, 2002.
13. Billuart, P. et al. Oligophrenin-1 encodes a rhoGAP protein involved in X-linked mental retardation. *Nature*, 392, 923, 1998.
14. Fauchereau, F. et al. The RhoGAP activity of OPHN1, a new F-actin-binding protein is negatively controlled by its amino-terminal domain. *Mol. Cell. Neurosci.*, 23, 574, 2003.
15. Govek, E.E. et al. The X-linked mental retardation protein oligophrenin-1 is required for dendritic spine morphogenesis. *Nat Neurosci.*, 7, 364, 2004.
16. Comery, T.A. et al. Abnormal dendritic spines in fragile X knockout mice: Maturation and pruning deficits. *Proc. Natl. Acad. Sci. USA*, 94, 5401, 1997.
17. Allen, K.M. et al. PAK3 mutation in nonsyndromic X-linked mental retardation. *Nat Genet.*, 20: 25–30, 1998
18. Meng, Y. et al. Abnormal spine morphology and enhaced LTP in LIMK-2 knockout mice. *Neuron* 35: 121–133, 2002.
19. Kutsche, K. et al. Mutations in ARHGEF6, encoding a guanine nucleotide exchange factor for Rho GTPases, in patients with X-linked mental retardation. *Nat Genet.* 26, 247–50, 2000.
20. Rosenberger, G. et al. Interaction of alphaPIX (ARHGEF6) with beta-parvin (PARVB) suggests an involvement of alphaPIX in integrin-mediated signaling. *Hum Mol Genet* 12: 155–167, 2003.
21. Penzes, P. et al. Rapid induction of dendritic spine morphogenesis by trans-synaptic ephrinB-EphB receptor activation of the Rho-GEF kalirin. *Neuron* 37: 263–274, 2003.

22. Matus, A., Brinkhaus, H., and Wagner, U. Actin dynamics in dendritic spines: a form of regulated plasticity at excitatory synapses. *Hippocampus*, 10, 555, 2000.

23. Matus, A. Postsynaptic actin and neuronal plasticity. *Curr Opin Neurobiol*, 9, 561, 1999.

24. Yuste, R. and Bonhoeffer, T. Morphological changes in dendritic spines associated with long-term synaptic plasticity. *Annu Rev Neurosci*, 24, 1071, 2001.

25. Purpura, D.P. Dendritic spine 'dysgenesis' and mental retardation. *Science*, 186, 1126, 1974.

26. D'Adamo, P. et al. Mutations in GDI1 are responsible for X-linked non-specific mental retardation. *Nat Genet*, 19,134, 1998.

27. Ishizaki, H. et al. Role of rab GDP dissociation inhibitor alpha in regulating plasticity of hippocampal neurotransmission. *Proc Natl Acad Sci USA*, 97, 11587, 2000.

28. D'Adamo, P. et al. Deletion of the mental retardation gene Gdi1 impairs associative memory and alters social behavior in mice. *Hum Mol Genet*, 11, 2567, 2002.

29. Carrie, A. et al. A new member of the IL-1 receptor family highly expressed in hippocampus and involved in X-linked mental retardation. *Nat Genet* 23 , 25, 1999.

30. Bahi, N. et al. IL1 receptor accessory protein like, a protein involved in X-linked mental retardation, interacts with Neuronal Calcium Sensor-1 and regulates exocytosis. *Hum Molec Genet*, 12, 1425, 2003.

31. Sippy, T. et al. Acute changes in short-term plasticity at synapses with elevated levels of neuronal calcium sensor-1. *Nature Neurosci*, 6, 1031, 2003.

32. Meloni, I. et al. FACL4, encoding fatty acid-CoA ligase 4, is mutated in nonspecific X-linked mental retardation. *Nat Genet*, 30, 436, 2002.

33. Huttner, W.B. and Schmidt, A. Lipids, lipid modification and lipid-protein interaction in membrane budding and fission — insights from the roles of endophilin A1 and synaptophysin in synaptic vesicle endocytosis. *Curr Opin Neurobiol*, 10, 543, 2000.

34. Merienne, K. et al. A missense mutation in RPS6KA3 (RSK2) responsible for non-specific mental retardation. *Nat Genet*, 22, 13, 1999.

35. Couvert, P. et al. MECP2 is highly mutated in X-linked mental retardation. *Hum Mol Genet*, 10, 941, 2001.

36. Lebel, R.R. et al. Non-syndromic X-linked mental retardation associated with a missense mutation (P312L) in the FGD1 gene. *Clin Genet.*, 61, 139, 2002.

37. Philip, N. et al. Mutations in the oligophrenin-1 gene (OPHN1) cause X linked congenital cerebellar hypoplasia. *J Med Genet.*, 40, 441, 2003.

38. Bergmann, C. et al. Oligophrenin 1 (OPHN1) gene mutation causes syndromic X-linked mental retardation with epilepsy, rostral ventricular enlargement and cerebellar hypoplasia. *Brain*, 126, 1537, 2003.

39. Ropers, HH. et al. Nonsyndromic X-linked mental retardation: where are the missing mutations? *Trends Genet.*, 19, 316, 2003.

40. Molinari, F. et al. Truncating neurotrypsin mutation in autosomal recessive nonsyndromic mental retardation. *Science*, 298, 1779, 2002.

41. Gschwend, T.P. et al. Neurotrypsin, a novel multidomain serine protease expressed in the nervous system. *Mol Cell Neurosci*, 9, 207, 1997.

42. Fukami, M. et al. A member of a gene family on Xp22.3, VCX-A, is deleted in patients with X-linked nonspecific mental retardation. *Am J Hum Genet*, 67, 563, 2000.

43. Zemni, R. et al. A new gene involved in X-linked mental retardation identified by analysis of an X;2 balanced translocation. *Nat Genet*, 24, 167, 2000.

44. Freude, K. et al. Mutations in the FTSJ1 Gene Coding for a Novel S-Adenosylme-thionine-Binding Protein Cause Nonsyndromic X-Linked Mental Retardation. *Am J Hum Genet.* 2004.
45. Kleefstra, T. et al. Zinc finger 81 (ZNF81) mutations associated with X-linked mental retardation. *J Med Genet.*, 41, 394, 2004.
46. Shoichet, S.A. et al. Mutations in the ZNF41 gene are associated with cognitive deficits: identification of a new candidate for X-linked mental retardation. *Am J Hum Genet.*, 73, 1341, 2003.
47. Gecz, J. The FMR2 gene, FRAXE and non-specific X-linked mental retardation: clinical and molecular aspects. *Ann Hum Genet.*, 64, 95–106, 2000.

29 Pharmacogenetics

Byron C. Jones

CONTENTS

29.1 INTRODUCTION

Pharmacogenetics is the study of individual differences in drug response. The field has been recognized since the early 1950s, but the biomedical research devoted to it has been relatively slow in development. Nevertheless, pharmacogenetics has manifold, important implications for health. For some individuals, for example, what would otherwise be termed a drug's "side effect" may in fact be the main drug effect. Alternatively, an individual may be predisposed to insensitivity to a particular drug, with implications for dose or to make administration of the drug futile. In this chapter we will explore the genetic bases for individual differences in drug actions (although there are environmental effects as well).

29.2 THE TWO TRADITIONS IN PHARMACOGENETICS

There have been two, somewhat distinct paths that the study of pharmacogenetics has taken over the past half century. The first, historically is the pharmacokinetic tradition and the second, the pharmacodynamic tradition. In pharmacology, these distinctions are somewhat arbitrary but focus the attention to different aspects of drug action.

29.2.1 THE PHARMACOKINETIC TRADITION

Pharmacokinetics deals with drug administration, absorption, distribution, biotransformation and elimination. The early pharmacokinetic studies focused on the last two of these, biotransformation and elimination.

29.2.1.1 Inborn Errors of Metabolism

The pharmacokinetic tradition is related to inborn errors of metabolism and is based largely on work in humans. In 1902, Archibald Garrod[1] published an article, "The Incidence of Alkaptonuria: A Study in Chemical Individuality," in *The Lancet*. Alkaptonuria is a condition diagnosed when an individual's urine, exposed to air, turns black. The cause is a problem with the catabolism of tyrosine to produce homogentisic acid in urine, which is responsible for the color change. Garrod described the pattern of inheritance as being consistent with an autosomal recessive allele. In a later work, Fölling[2] reported on phenylketonuria in a family, a work that led to phenylketonuria being described as following the same pattern of inheritance as alkaptonuria. With these pioneering studies, the groundwork was laid to show that many enzymes that are responsible for biotransformation of autacoidal substances come in alternate forms (isoenzymes) produced by genes that vary in their DNA (alleles) and hence in their amino acid sequence.

29.2.1.2 Pharmacogenetics of Drug Metabolism

Not surprisingly, the isoenzymes that catalyze reactions of endogenous hormones, neurotransmitters, etc., are also involved in the biotransformation of drugs. The usual consequence is in individual differences in reaction rates based on abundance, activity or differing affinities among the enzymes for the drug. In the 1950s, Kalow[3] described in a family study, individual differences in pseudocholinesterase. This enzyme is responsible for biotransformation of local anesthetics, such as procaine and biotransformation of succinylcholine, a muscle paralyzer used as adjunct to general surgery. Individuals who were slow metabolizers of succinylcholine had greater difficulty in recovering from its effects on skeletal muscles following general surgery.

A particularly compelling example of the importance of genetic differences in pharmacokinetics is in alcohol metabolism. The major metabolic pathway for alcohol biotransformation is through alcohol dehydrogenase which converts ethyl alcohol to acetaldehyde (and methyl alcohol to formaldehyde) followed by the action of acetaldehyde dehydrogenase to produce acetate which goes into intermediate metabolism. There are two prominent alleles of acetaldehyde dehydrogenase, ALDH2*1 and ALDH2*2. Individuals who are homozygous for the ALDH2*1 allele are able to oxidize acetaldehyde much faster than those homozygous for the ALDH2*2 allele or who are heterozygotes.[4] The consequences of consuming alcohol for those carrying the ALDH2*2 allele include rapid intoxication, facial flushing, headache and nausea—all as a result of the toxic effects of accumulated acetaldehyde. About 50% of East Asians carry the ALDH2*2 allele and flush after consuming alcohol. The ALDH2*2 allele is rare among Caucasians, Africans, and South Asians. Carrying the ALDH2*2 allele is thought to protect against alcoholism; however, there are some carrying the allele and who consume large amounts of alcohol. These individuals incur a many-fold increased risk for esophageal cancer and liver disease compared to those who can metabolize acetaldehyde rapidly.[5]

29.2.1.3 The Special Case of the Cytochrome P 450 Superfamily of Phase I Drug Metabolizing Enzymes

Drug metabolizing enzymes are classified into Phase I and Phase II enzymes. Phase I enzymes by oxidation, reduction or hydrolysis increase the polarity of a drug and in some cases make the drug more active for example, changing codeine to morphine. Phase II enzymes act by conjugation, adding for example acetyl group, to further increase polarity and thus enable elimination by the kidneys. Genetic polymorphisms may occur in either type, however, some of the most important occur in the Phase I, especially in the major group of oxidizing enzymes, the Cytochrome P 450 enzymes (CYP). The CYP enzymes consist of more than 150 classified into subfamilies. The CYP2 family is particularly interesting for psychopharmacology, because many of the drugs used to treat neuropsychiatric illnesses are metabolized by CYP2D6. Drugs so affected include tricyclic and specific serotonin uptake inhibitor antidepressants (fluoxetine, paroxetine) and haloperidol, thoridizine, perphenazine and similar antipsychotics. It is further estimated that in all therapeutic categories, CYP2D6 is involved in the metabolism of more than 20% of prescribed drugs.[6] More than 50 allelic variants of this enzyme are known and are used to categorize individuals into four classes of metabolizers: poor, intermediate, efficient, and ultrarapid. Of particular interest is that the development of Parkinsonian symptoms in individuals taking antipsychotic medication is associated with the poor metabolizing phenotype. It is obvious, therefore, that drug dosing recommendations should be made based on the allelic configuration of the individual and one researcher believes that the differential distribution of the polymorphisms would make genotyping worthwhile prior to prescription of medications metabolized by CYP2D6.[6]

29.2.2 THE PHARMACODYNAMIC TRADITION

The study of individual differences in drug action at target tissues began in the 1950s. Initially, the almost sole focus of this line of research was on individual differences in alcohol response and sensitivity and in rats and mice predominantly. More recent work has focused on other drugs, albeit for the most part on drugs of human misuse. In 1951, Mardones[7] reported research on successful selective breeding for rats that preferred and for rats that did not prefer alcohol (vs. water). This work showed that at least some aspects of alcohol consumption were under genetic influence. One of the most salient demonstrations of genetic influence on drug actions is the work of G.E. McClearn and colleagues.[8] In the 1960s this team screened a large number of genetically heterogeneous mice for hypnotic sensitivity to ethanol. The animals were injected intraperitoneally with 3.3 g/kg ethanol in saline and then observed for loss of righting response by placing the animals in a supine position. The time from loss of righting response to time of regain was recorded as the dependent variable. Large individual differences were observed among these animals and mass selection over several generations produced relatively rapid differentiation between the two lines. The more sensitive line, as determined by sleep time was named long sleep (LS) and the less sensitive line was named short sleep (SS). The pharmacogenetics of alcohol is more completely covered in the next chapter by J.C. Crabbe. The example of the LS and SS mice is used here to illustrate two important principles in pharmacogenetics. First, what is the nature of the phenotype? In the example of the LS and SS mice, does sleep time reflect a pharmacodynamic effect (i.e., target tissue sensitivity) or is the effect a result of different rates of alcohol disappearance between the two lines, i.e., do SS mice eliminate alcohol more rapidly than LS? Tabakoff and Ritzmann addressed this question directly by measuring blood alcohol concentrations (BEC) at loss and regain of the righting response[9] and they reported that the BEC for the SS was nearly twice that of the LS both at loss and at regain of righting. Thus, this gives us good evidence that pharmacodynamic processes account for most of the differential effect of ethanol in these two lines of mice. The second principle illustrated by the LS-SS example is that differential response progressed at nearly steady rate across generations, rather than within one or two generations of selection. This gives good evidence for hypnotic sensitivity to ethanol as being influenced by multiple genes (alleles) with algebraic additive effects (some increase effect, some decrease) and that across generations, each line accumulates more of the alleles that push the phenotype in the direction of selection. In the pharmacokinetic approach, most of the attention is focused on single genes and their alleles that alter the function of metabolizing enzymes. In the pharmacodynamic approach, most of the phenotypes measured show continuous variation, respond to selection very much like the LS-SS, thus indicating polygenic influence.

29.2.3 LET US NOT FORGET THE INFLUENCE OF THE ENVIRONMENT

Individual differences in drug response are not entirely caused by differences in genetic makeup. Environment is broadly defined and can include the immediate external or interior milieu, rearing conditions during infancy, nutrition and even prior exposure to drugs. The environment may also include hormonal or other humoral status. Early handling, i.e., removal of pups from the nest for brief periods, can alter preference for alcohol in some inbred mouse strains but not others,[10] individual vs. group housing can alter ethanol sensitivity[11] and iron deficiency early in life can alter sensitivity to and self administration of cocaine.[12,13] It behooves the researcher, therefore, to know how environmental conditions can affect the phenotype and how any environmental perturbation may interact with the genetic makeup of the subject.

29.3 SINGLE GENES AND PHARMACOGENETICS

In Chapter 13, the relationship among the short allele of the 5HT transporter promoter, stressful life events and depression was discussed. A recent study has reported that individuals who carry at least one long allele of 5-HTTLPR fare better in long-term antidepressant treatment than those who are homozygous for the short allele.[14]

In animals, approaches to single-gene effects on drug sensitivity include induced mutations by irradiation, chemical mutagenesis, and creation of targeted null mutants. Genes may also be amplified, i.e., by insertion of multiple copies of desired alleles into the genome. These models can be useful for elucidating the role of specific proteins, receptors, etc. in drug action. For example, in the serotonin 1b receptor null mutant mouse, treatment with paroxetine, a selective serotonin reuptake inhibitor used to treat depression causes a twofold increase in extracellular serotonin compared with treatment with the same drug in wild-type mice. In humans, allelic variants of the $5HT_{1B}$ receptor gene promoter region have been described.[15] The allelic variants are associated with transcriptional activity of the receptor gene, showing a more than twofold difference between the highest efficiency and lowest efficiency alleles. Although the allelic configuration may not affect the action of specific serotonin reuptake inhibitors, the fact that they do exist, in light of the work in the null mutant mice should give cause for further investigation into possible ramifications for possible individual differences in drug treatment for depression.

29.4 PHARMACOGENETICS AND COMPLEX TRAITS ANALYSIS

A rapidly developing approach in pharmacogenetics is complex traits analysis. In this approach, we recognize the multiple gene influence on most of our phenotypes of interest and how these genes interact with each other and with the environment. The major features of complex traits analysis include quantitative trait analysis, gene network analysis, and gene expression. The object of complex traits analysis is to identify genes that influence target tissue sensitivity to drugs and to identify related gene networks. The usual place to start is to specify precisely the phenotype of interest and perhaps related phenotypes. Quantitative trait analysis is then performed on data collected from a genetic reference population. Quantitative trait locus (QTL) analysis gives general locations on chromosomes containing phenotype-related genes. The next task is to narrow the chromosomal areas, nominate candidate genes and then verify whether the candidate gene is truly the gene of interest. As an example of preliminary QTL analysis, Jones et al.[16] conducted a QTL analysis of cocaine-related behaviors and neurochemistry in the BXD/Ty recombinant inbred mouse panel. The behavioral phenotypes measured were spontaneous activity (loco-motion, exploration, stereotyped movements and thigmotaxis) under 4 doses of cocaine, 5, 15, 30 and 45 mg/kg (vs. saline) administered intraperitoneally. In a separate set of animals, density of dopamine receptors (D_1 and D_2) and the dopamine transporter were measured by ligand binding in homogenates derived from the prefrontal cortex, caudate-putamen, nucleus accumbens, and ventral midbrain. The purpose of measuring drug response and then measuring putatively related neuro-biological parameters was to find QTL for the behavioral measures, QTL for the neurobiological measures and QTL common to both. The last of these would provide more robust evidence for common genetic mechanisms than would the individual

QTL. Of course the related behavioral–neurochemical parameters would have to "make sense" in terms of what we know about the neurobiology of the drug action. Numerous QTL were identified for the behavioral effects of cocaine and neurochemical indices and at least four of these QTLs were common between cocaine-related behavior and dopamine neurobiology. Of particular interest were markers on chromosome 15 that correlated with both dopamine receptor types in the caudate-putamen and with locomotion and stereotyped movements. This "makes sense" in that the caudate-putamen is involved in movement and is sensitive to psychostimulants and that dopamine actions in this part of the striatum affect movement. Thus, we have identified QTL that indicate common genetic mechanisms for the expression of dopamine receptors and for the behavioral actions of cocaine.

29.5 QTL TO QTGENE

QTL analysis is useful for pointing to chromosomal locations containing genes that modulate drug actions and may also indicate how many genes are involved. The next step is to move from QTL to actually identifying the genes in question. This can be a lengthy process, beginning with narrowing the region of the chromosome to an area that contains dozens rather than hundreds to thousands of genes. QTL analysis of neurobehavioral traits has a short history of a little more than a dozen years. In the initial stages, very few genes were identified; however, as analytical techniques, including high throughput sequencing, have evolved, the pace of QTGene identification has picked up dramatically. At this writing, more than 30 genes have been identified following QTL mapping and the number is growing as more refined techniques are brought into play.[17] Indeed such an approach has been undertaken in the laboratory of Buck who identified *Mpdz* on mouse chromosome 4 as one QTGene underlying susceptibility to alcohol and barbiturate withdrawal-induced seizures.[18]

29.6 PHARMACOGENOMICS—THE NEXT STEP

For the most part, pharmacogenetics has been concerned with identifying individual genes involved with drug safety and sensitivity. Pharmacogenomics casts a wider net by probing the entire genome for genes and gene products that influence safety, sensitivity and efficacy of drugs.[19] Indeed, QTL analysis has provided one approach toward developing the methods in multiple-gene identification. Evolving bioinformatics techniques will help us to identify pleiotropic effects of genes, gene–gene interactions and gene networks that affect drug actions. The interested reader is directed to http://www.genenetwork.org for further information. This is a user-friendly Web site that contains phenotypic data of many kinds, including published pharmacological data from genetic reference populations. An effective tutorial guides the user through QTL analysis, genetic correlations, gene networks, epistasis and more. The site is continually updated and refined.

29.7 SUMMARY

Pharmacogenetics is a small, but important component of the study of individual differences in response to environmental challenges to homeostatic biological systems. It is not different in principle from inborn errors of metabolism, dietary problems such as lactose intolerance, disease resistance, or individual differences in sensitivity to toxins. What makes pharmacogenetics worthy of study is the great importance of its impact on human health and the eventual development of drug treatment strategies based on an individual's genetic constitution.

REFERENCES

1. Garrod, A.E. The incidence of alkaptonuria: a study in chemical individuality. *The Lancet* ii:1616–1620, 1902.
2. Fölling, A. Excretion of phenylpyruvic acid in urine as a metabolic. anomaly in connection with imbecility. *Nord Med Tidskr.* 8: 1054–1059, 1934
3. Kalow, W. Familial incidence of low pseudocholinesterase level. *The Lancet* 2:576–577, 1956
4. Crabb, D.W., Edenberg, H. J., Bosron, W. F., Li, T.-K. Genotypes for aldehyde dehydrogenase deficiency and alcohol sensitivity: the inactive ALDH2*2 allele is dominant. *J. Clin. Invest.* 83: 314–316, 1989.
5. Yokoyama, A., Kato, H., Yokoyama, T., Tsujinaka, T., Muto, M., Omori, T., Haneda, T., Kumagai, Y., Igaki, H., Yokoyama, M., Watanabe, H., Fukuda, H., Yoshimizu, H. and Ingelman-Sundberg, M. Genetic polymorphisms of cytochrome P4502D6 (CYP2D6): clinical consequences, evolutionary aspects and functional diversity. *The Pharmacogenetics Journal* 5:6–13, 2005.
6. Mardones, J. On the relationship between deficiency of B vitamins and alcohol intake in rats. *Q J Stud Alcohol* 12:563–575, 1951.
7. McClearn, G.E. and Kakihana, R. Selective breeding for ethanol sensitivity: short-sleep and long-sleep mice. In *Development of Animal Models as Pharmacogenetic Tools* G.E. McClearn, R. A. Deitrich and V.G. Erwin, eds., USDHHS-NIAAA Research Monographs No 6, Washington, pp. 147–159, 1981.
8. Tabakoff, B. and Ritzmann, R.F. Acute tolerance in inbred and selected lines of mice *Drug Alcohol Depend* 4:87–90, 1979.
9. Jones, B., Goldstine, R., Kegel, M., Gurley, M., and Reyes, E. The influence of infantile handling, age and strain on alcohol selection in mice. *Alcohol*, 2:327–331, 1985.
10. Jones, B. C., Connell, J. M. and Erwin, V. G. Isolate housing affects ethanol sensitivity in long-sleep and short-sleep mice. *Pharmacol Biochem Behav*, 35:469–472, 1990.
11. Erikson, K.M., Jones, B.C., and Beard, J.L. Iron deficiency alters dopamine transporter functioning in rat striatum. *Journal of Nutrition*, 130:2831–2837, 2000.
12. Jones, B.C., Wheeler, D.S., Beard, J.L. and Grigson, P.S. Iron deficiency decreases acquisition of and suppresses responding for cocaine. *Pharmacol Biochem Behav*, 73:813–819, 2002.
13. De Groote, L., Olivier, B. and Westenberg, G.M. The effects of selective serotonin uptake inhibitors on extracellular 5-HT levels in the hippocampus of 5-HT$_{1B}$ receptor knockout mice. *Eur J Pharm* 439:93–100, 2002.

14. Duan, J., Vander Molen, J.E., Martinolich, L., Mowry, B.J., Levinson, D.F., Crowe, R.R., Silverman, J.M. and Gejman, P.V. Polymorphisms in the 5'-untranslated region of the human serotonin receptor 1B (HTR1B) gene affect gene expression. *Mol Psychiat* 8:901–910, 2003.
15. Jones, B.C., Tarantino, L.M., Rodriguez, L.A., Reed, C.L., McClearn, G.E., Plomin, R. and Erwin, V.G. Quantitative trait loci analysis of cocaine-related behaviours and neurochemistry. *Pharmacogenetics* 9:607–617, 1999.
16. DiPetrillo, K., Wang, X., Stylianou, I.M. and Paigen, B. Bioinformatics toolbox for narrowing rodent quantitative trait loci. *TRENDS in Genetics*, 21:683–92, 2005.
17. Shirley, R.L., Walter, N.A., Reilly, M.T., Fehr, C. and Buck, K.J. Mpdz is a quantitative trait gene for drug withdrawal seizures. *Nat Neurosci* 7:699–700, 2004.
18. Kalow, W., Meyer, U.A. and Tyndale, R.F., eds. *Pharmacogenomics*, Marcel Decker, New York, 1991.

30 Alcohol Psychopharmacogenetics

John C. Crabbe

CONTENTS

30.1 INTRODUCTION

Of all drugs of abuse studied to determine genetic contributions to susceptibility to their effects, alcohol has been by far the most frequent focus. Both animal model and human genetic studies will be mentioned here. A historically rich literature with genetic animal models has explored the general contribution of genetics to alcohol responsiveness, and has been helpful in elucidating the drug's mechanism of action on the nervous system. Some studies using genetic animal models have attempted to identify specific genes that increase or decrease responsiveness for a number of alcohol's effects. Most human studies have been of alcoholics and their relatives, and have compared relative risks for alcohol dependence disorders in twins, adoptees, or other relatives. More recently, human studies have also addressed the goal of identifying individual genes that might contribute to alcoholism risk, or to individual differences in endophenotypes, which also are associated with alcoholism risk.

A comprehensive review of these studies is beyond the scope of this chapter, but a few examples will be discussed below, and readers will be referred to relevant reviews.

30.2 ANIMAL PHARMACOGENETIC STUDIES

Two standard pharmacogenetic methods have been employed to investigate the genetic contribution to alcohol sensitivity in rodents. Some studies have been conducted with standard inbred strains, and others have used lines selectively bred for enhanced or diminished alcohol response.

30.2.1 STUDIES WITH INBRED STRAINS

Inbred strains have been created through generations of brother–sister matings, and each strain gradually has lost its genetic diversity such that all same-sex members of a fully inbred strain share two copes of the same allele for each gene. The genotype is faithfully recreated each generation, with the result that data collected in an inbred strain many years (and generations) ago can be compared with data collected now. Since 1959, we have known that inbred mouse strains were characterized by large and reproducible differences in their drinking preference of weak alcohol solutions.[43] C57BL/6 and related strains have a high preference for alcohol, DBA/2 and related strains are extreme avoiders, and other strains tend to fall somewhere between. This literature has been systematically reviewed.[14,16,18,40,51]

Ignoring for the moment the more specialized recombinant inbred (RI) mouse strains, which will be discussed in a later section, systematic studies with multiple inbred mouse strains have now been performed for many responses to alcohol. These include responses representing low-dose stimulant effects,[50] moderate-dose effects such as ataxia and hypothermia, and high-dose effects such as hypnosis and acute and chronic withdrawal severity.[13]

One of the features of studies with inbred strains is that pleiotropic gene influences can be detected. That is, if a gene or genes act to increase sensitivity to one effect of ethanol in certain strains, and if those strains are also found to be more sensitive to another effect of ethanol, this provides evidence for the influence of the same genes on the second response as well.[17] In general, the genetic influences on responses to ethanol appear to be discrete, and there are few striking genetic correlations among different behavioral response domains across inbred strains.[13]

It is also interesting to note that there are striking strain differences in ethanol metabolism.[19,35,58] However, studies with inbred mice have failed to detect common genetic influences across strains between ethanol metabolism and behavioral sensitivity.[13] This suggests that the strain differences in behavioral sensitivity are mediated by physiological differences in brain circuits responsive to the alcohol rather than in the availability of alcohol in the brain.

The stability and availability of inbred mouse strains has led to the recent development of a systematic database comprising physiological, neurobiological and behavioral phenotypes of many sorts, including drug and alcohol responses, for

40 common inbred strains. This effort, the Mouse Phenome Project, can be assessed to search for evidence of pleiotropic gene effects.[48]

30.2.2 STUDIES WITH SELECTIVELY BRED LINES

There are more than a dozen sets of mouse and rat lines bidirectionally selected for their enhanced or diminished response to alcohol.[1,7,19,40,51] Selective breeding for increased intensity of a trait enriches the selected line for alleles that increase the trait, and when such selected lines are compared, pleiotropic effects of those alleles can be identified when they produce concomitant increases in a "correlated response to selection."[17] The most frequently selected trait is high- or low-alcohol preference drinking when animals are offered a choice between alcohol and water: at least six sets of selectively bred rat lines and one set of mouse lines have been selected for variants of this trait.[7,40] The preponderance of studies have been conducted with preferring (P) and nonpreferring (NP) rat lines, and information characterizing these lines is extensive.

The similarity of the several selection studies offers a unique opportunity to identify traits that are reliably differentiated in drinking and nondrinking lines. Many of the older selection studies were not replicated, which increases the risk that differences between the selected lines are due to chance fixation of genes unrelated to ethanol preference.[17] However, a few behavioral and neurochemical traits have been systematically sampled in multiple preference-selected lines, and some apparent commonalities seem likely to exist. P rats show more alcohol tolerance development, and it lasts longer, than NP rats. This difference is shared by both pairs of HAD vs. LAD, and by AA vs. ANA rats as well.[40] This could contribute to the relatively higher drinking of the P, HAD, and AA lines.

The neurotransmitter serotonin (5-HT) and its metabolite 5-HIAA is relatively depleted in several brain areas in P rats vs. NP rats. This difference is also seen in HAD vs. LAD rats, but has not been found to characterize the AA vs. ANA rat lines.[40] A possible reason for the discrepant findings with the AA/ANA rats is that, in contrast to the NP and LAD rats, ANA rats accumulate more acetaldehyde than AA rats after administration of alcohol, so the toxicity of acetaldehyde may serve to limit alcohol consumption in these selected lines. Another potentially important difference between genetically high- and low-drinking lines is in the mesolimbic dopamine system. P rats will self-administer ethanol directly into the ventral tegmental and current studies are exploring the pharmacological control of the circuits regulating this response.[56,57]

Several of the preference selected lines were compared in a set of interesting operant studies that characterized their behavioral patterns of drinking with the goal of understanding how alcohol's rewarding effects were experienced by these rats.[29,59] The genotypes selected for high preference tended to self-administer more ethanol under most conditions studied, suggesting that some genes affect both operant self-administration and two-bottle preference drinking. However, the high-preference lines did not always behave similarly, and the overall picture was that operant and two-bottle preference methods for studying self-administration were influenced mostly by rather different genes.

Studies with lines selected for low-dose sensitivity to ethanol, measured by increases in locomotor activity, have been reviewed elsewhere,[50] as have those for alcohol's sedative effects, measured as a decrease in core body temperature after injection.[51] A long-term project has generated lines of mice (long- and short-sleep, or LS and SS) and more recently rats (high and low alcohol sensitive, or HAS and LAS) for sensitivity to the high-dose effect of ethanol to cause loss of righting reflex. Studies with these animals number in the hundreds, and have been periodically reviewed.[1,19,51] These and other genetic studies implicate the neurotransmitter GABA as an important determinant of the genetic differences in sensitivity to high-dose ethanol.[8]

Finally, another long-term project has created mice that are withdrawal seizure-prone or -resistant (WSP and WSR) by exposing them to ethanol vapor for 72 hrs and measuring handling-induced convulsion severity during withdrawal. Studies with these animals have also been discussed in detail elsewhere, and have revealed many interesting features about chronic neuroadaptation to alcohol.[7,11,46] Perhaps most interesting is the finding that a genetic susceptibility to ethanol withdrawal severity is shared with susceptibility to withdrawal from other depressant drugs such as barbiturates and benzodiazepines,[12] a finding also substantiated in inbred strains.[47]

30.2.3 GENE IDENTIFICATION AND TARGETING STUDIES

Three general approaches have been taken to explore the role of specific genes in modulating alcohol sensitivity. The first approach has been to identify novel genes by using quantitative trait locus (QTL) gene mapping to discover the presence of chromosomal markers near genes affecting alcohol responses. QTLs have been reported for ethanol drinking, sensitivity to various acute effects of alcohol, tolerance, and dependence liability, which is inferred from studies of withdrawal when alcohol is discontinued. The second strategy is employed where specific candidate genes have been identified as potentially important mediators of susceptibility to alcohol. Classically, such studies employ Northern or Western analyses to study mRNA or protein levels, respectively, for a gene of interest. The use of antisense oligodeoxynucleotide treatments to reduce gene translation, or, more recently, RNA inhibitors to modulate transcription, has been reported in only a few studies. Many studies have now used gene targeting technologies to create overexpression or null-mutant (knockout) lines of mice by targeting specific genes of interest. The success of many recent candidate gene studies with animals has been noted.[62]

The third, and most recent, method is a hybrid of the first two. Analogously to QTL mapping, microarray chips can simultaneously study many thousands of genes. However, where QTL mapping seeks to associate degree of alcohol response with a specific variation in gene sequence mapped to a specific chromosomal location, expression microarrays focus on differences in expression in any gene probed.

30.2.3.1 QTL Mapping Studies

Most alcohol QTL studies have been carried out using the BXD RI strains of mice as a starting point for rough QTL identification, followed by verification tests in other populations, such as the F2 cross of C57BL/6J and DBA/2J, lines selectively bred from the F2 cross, or other selected populations. The experiments that have progressed the furthest have shown that three loci for acute alcohol withdrawal severity have been definitively mapped with very high LOD scores to regions of mouse chromosomes 1, 4, and 11.[10] Genes encoding the α_1, α_6, and $\gamma 2$ subunits of the neuroinhibitory $GABA_A$ receptor map near the chromosome 11 QTL. Studies using inbred strains, lines selectively bred from the B6D2F2 population, and a number of so-called congenic strains were then used to reduce the area of genome surrounding the chromosome 4 QTL until it contained only a handful of candidate genes.[28] The remaining candidate genes were gradually eliminated because they did not differ in sequence, expression, or function in ways that systematically co-varied with alcohol withdrawal severity. In the end, a single gene remained, *Mpdz*, which must be the QT Gene for this QTL. MPDZ is a multiple PDZ domain protein that affects the coupling of neurotransmitters with their receptors, including the binding of serotonin with the 5-HT2 receptor. This gene also affects the severity of withdrawal from pentobarbital, another nervous system depressant that interacts with GABA receptors. This provides another example of pleiotropic gene effects, and *Mpdz* may indeed also play a role in susceptibility to seizures more generally, in addition to its specific role in alcohol and pentobarbital withdrawal.[28,63]

For sensitivity to high-dose ethanol, five significant QTLs with high LOD scores have been identified on chromosomes 1, 2, 8, 11, and 15.[41] Starting with the LS and SS selected lines and using RI strains and crosses derived from them, the QTLs were nominated and subsequently verified in a multistage procedure much like that used for acute alcohol withdrawal.[24] Some interesting potential candidate genes near the QTL regions include *Ntsr* on chromosome 2, which codes for the high-affinity neurotensin receptor.

Several different research groups have reported mapping QTLs for ethanol preference drinking and related traits.[3,44,45,52,65] The significant QTLs are found on chromosomes 2, 9, 11 and 15, and several other QTLs have been suggested on chromosomes 1, 3, 4, 5, 7, and 13. The chromosome 9 QTL is near genes encoding both the dopamine D2 and serotonin 5-HT_{1B} receptor, *Drd2*, and *Htr1b*, respectively. The results of these studies are not currently in complete agreement, probably because different crosses, sexes, strains, and mapping strategies were employed, as well as different measures of preference-related drinking. However, a meta-analysis revealed a remarkable degree of convergence of findings for several QTLs.[6]

A number of other alcohol-related traits are also being mapped, although the projects are in earlier stages. One set of studies mapping the effect of ethanol to stimulate locomotor activity in mice has used a variety of mapping methods.[20,34] Other traits being mapped include: ethanol-conditioned place preference and taste aversion; hypothermia; traits related to ataxia; tolerance to some of these effects; and ethanol's anxiolytic effects. Progress in ethanol QTL mapping has been reviewed recently.[9,15,30,49]

30.2.3.2 Transgenic, Null Mutant, Antisense, and Other Gene-Targeted Expression Studies

$GABA_A$ receptors are known to be important modulators of many behavioral effects of ethanol. Ethanol enhances GABA stimulated function *in vitro*, and when antisense oligodeoxynucleotides to several GABA$_A$ receptor subunits were tested in an early study, ethanol enhancement was prevented only by an antisense probe to receptors containing the γ_{2L} subunit. This subunit differs from γ_{2S} only by bearing an additional 8 amino acids which contain a consensus site for phosphorylation by protein kinase C. Site-directed mutagenesis within this phosphorylation site blocked ethanol's effects.[66] As technology developed, subsequent studies showed that null mutant mice that lack the α_6 subunit did not differ from wild types in sensitivity to loss of righting reflex after ethanol.[36] As more subunit null mutants have been studied and more specific genetic tools have become available, it has become possible to see how alcohol's effects can be understood by comparison with other anesthetic agents with strong effects on GABA systems.[37]

Gene targeting strategies have also been employed to study the role of transforming growth factor α, insulin-like growth factor 1, and many other candidate genes in alcohol responses. Gene targeting studies must be interpreted cautiously to avoid attributing the influence of "background" genes to the specific gene that has been transgenically altered.[31] For example, the effect of a gene deletion may only be detectable on certain genetic backgrounds.[53] The strengths and weaknesses of the various gene targeting approaches applied to alcohol research have been discussed.[14,68] Some studies also have explored the induction of the immediate early gene, c-fos, as a marker for intracellular changes following ethanol administration.[9,69] Initial identification of brain regions activated by ethanol can then be followed by studies identifying responsive circuits.[4]

30.3 HUMAN STUDIES

30.3.1 TWIN AND ADOPTION STUDIES

Genetic influence on developing alcoholism has been supported by more than 20 years of controlled studies using twins and adoptees.[23,25,32] Early studies used samples from Scandinavia because the national record keeping systems made the process of ascertaining subjects and finding their relatives fairly straightforward. Although these experiments reported a substantial genetic component to risk, the definitions of alcoholism and alcohol-related problems may have been somewhat idiosyncratic to Scandinavian populations, possibly because of the strong cultural determinants of drinking patterns that differ among countries.

Recent studies with populations in the United States have used more modern diagnostic criteria, and have largely yielded similar estimates of the importance of inheritance. Such studies have always faced the difficulty of defining the phenotype. The standard diagnoses of alcoholism are generally based on possession of a number of symptoms from a list of possible criteria, which range from medical complications (e.g., compromised liver function) to social (e.g., loss of significant relationship attributable to drinking) or functional problems (e.g., loss of job). To give an example, different diagnoses are available for alcohol abuse vs. alcoholism. Are these related disorders on a continuum, or are they distinct (the latter would imply that separate genetic influences should be discernible)? A related complexity is that there is extensive overlap, or comorbidity, among alcoholism and other individual traits, such as: other psychiatric disorders (e.g., antisocial personality, depression, attention deficit hyperactivity disorder); personality characteristics (e.g., impulsivity, aggressiveness); and substance abuse.[54] Disentangling the chains of cause and effect among these different diagnostic aspects makes this a very difficult area of research to interpret. Finally, each typology for conceptualizing alcohol dependence disorders leads to emphasis on different interactions between heritable risk factors and the modulating effects of environmnental factors that are also important for turning risk into diagnosis.[33,38]

Recently, much attention has turned to analyzing identifiable aspects of alcoholism and alcohol responses for their separable genetic determinants. By using other heritable traits as potential indicators, it is hoped that the genetic basis for such traits will make it easier to determine gene–behavior relationships than is trying to study the diagnoses directly. Such studies are meeting with some success.[21,26]

30.3.2 CANDIDATE GENE STUDIES

The human alcoholism literature offers a clear example of the influence of a specific gene on alcoholism susceptibility. Human populations possess several variant isoforms of the metabolic enzymes, alcohol dehydrogenase (ADH) and aldehyde dehydrogenase (ALDH). Asian populations have a high frequency of specific ALDH2 variants that are associated with slow acetaldehyde metabolism, and consequently, facial flushing, nausea, and anhedonia after alcohol ingestion. Possession of these ALDH2 alleles has been shown to reduce risk for alcoholism.[67] Other polymorphisms in ADH genes have a similar but smaller protective effect.[21]

30.3.3 IDENTIFICATION OF RISK MARKERS AND QTLS

Association and linkage studies can only genotype individuals for a limited subset of their total DNA sequence, and therefore must compromise between two choices— genome-wide screens using markers spread widely (which may therefore detect most signals but provide little information about exactly where the responsible gene is) or focused screens using densely spaced markers, which can localize the signal better but may miss signals that are not very near the targeted regions. Many biomarkers of genetic risk have been suggested in studies with human populations, such as monoamine oxidase and adenylyl cyclase. Generally, these studies remain controversial.[2] As studies with genetic animal models increasingly suggest specific candidate genes, it was to be expected that studies with human populations would increasingly examine allelic variation in detail for areas surrounding those genes. Some such studies are now providing evidence for influential genomic regions.[21]

An interesting "risk marker" for alcoholism has been revealed by studies of individuals who have are family history positive (FHP) vs. negative (FHN) for alcoholism. FHP individuals when tested as young men showed lower subjective response to alcohol than FHN controls, and a follow-up study showed that low initial responders (whether of FHP or FHN background) went on to develop alcoholism at higher rates than high responders.[60,61] Another such risk marker that is not identified with a particular gene is the lower P300 event-related potential that characterize FHP subjects.[5]

The National Institute on Alcohol Abuse and Alcoholism is currently sponsoring a large multisite study called the Collaborative Study on the Genetics of Alcoholism (COGA). The goal of this project is to perform a large, population-based association study looking for particular genetic polymorphisms that indicate QTLs associated with alcoholism diagnosis, P300 event-related potentials, and other traits potentially associated with alcoholism and alcohol dependence. This study is described elsewhere.[55] Some genomic regions have appeared from these analyses as having potential linkages, including those containing GABA receptor subunit genes.[22,64]

30.4 GENE EXPRESSION ANALYSES

A few studies have now applied microarray technology to explore tissue from human alcoholics for potential gene expression differences vs. controls. These studies are quite preliminary, but as the technology matures, have the potential to find important genes in addition to those that differ by virtue of their sequence polymorphisms.[27,39,42]

30.5 CONCLUSIONS

The field of alcohol pharmacogenetics offers an interesting viewpoint on the explosive impact modern molecular biological methods have had on genetic analyses. The rich history of standard genetic animal model research into alcohol's effects has provided many cases where clear genetic influence was known to exist. Thus, the move to identification of specific candidate genes and markers occurred quite rapidly after the advent of dense genetic maps and the ability to target specific genes for deletion and overexpression. Because there are readily available populations of human alcoholics with well-defined pedigrees, it will be possible for the animal model and human genetic findings to be related rather directly. This offers the hope that in addition to improving our understanding of how alcohol works, genetic studies will be able to provide novel methods for identifying at-risk individuals and for developing new pharmacotherapies to attack this major disease.

ACKNOWLEDGMENTS

Preparation of this chapter was supported by a grant from the Department of Veterans Affairs and by NIH Grants from NIAAA and NIDA.

REFERENCES

1. Allan AM, Harris RA. Neurochemical studies of genetic differences in alcohol action, In: Crabbe JC, Harris RA, editors. *The Genetic Basis for Alcohol and Drug Actions.* New York: Plenum; 1991. pp. 105–52.
2. Anthenelli RM, Tabakoff B. The search for biological markers. *Alcohol Health and Research World* 1996;19:176–81.
3. Bachmanov AA, Reed DR, Li X, Li S, Beauchamp GK, Tordoff MG. Voluntary ethanol consumption by mice: genome-wide analysis of quantitative trait loci and

their interactions in a C57BL/6ByJ x 129P3/J F2 intercross, *Genome Res.* 2002;12:1257–68.

4. Bachtell RK, Tsivkovskaia NO, Ryabinin AE. Strain differences in urocortin expression in the Edinger-Westphal nucleus and its relation to alcohol-induced hypothermia. *Neuroscience* 2002;113:421–34.

5. Begleiter H, Porjesz B. Neurophysiological phenotypic factors in the development of alcoholism, In: Begleiter H, Kissin B, editors. *Genetic Factors and Alcoholism*. New York: Oxford University Press; 1996. pp. 269–93.

6. Belknap JK, Atkins AL. The replicability of QTLs for murine alcohol preference drinking behavior across eight independent studies. *Mamm Genome* 2001;12:893–99.

7. Browman, K. E., Crabbe, J. C., and Li, T.-K. Genetic strategies in preclinical substance abuse research. Bloom, F. E. and Kupfer, D. J. *Psychopharmacology: A Fourth Generation of Progress* (CD-ROM version 3). 2000. Lippincott, Williams & Wilkins. Ref Type: Electronic Citation (url = http://www.acnp/org/citations/GN400000077).

8. Buck KJ. Molecular genetic analysis of the role of GABAergic systems in the behavioral and cellular actions of alcohol. *Behav. Genet.* 1996;26:313–23.

9. Buck KJ, Crabbe JC, Belknap JK. Alcohol and other abused drugs, In: Pfaff DW, Berrettini WH, Joh TH, Maxson SC, editors. *Genetic Influences on Neural and Behavioral Functions*. Boca Raton FL: CRC Press; 2000. pp. 159–83.

10. Buck KJ, Metten P, Belknap JK, Crabbe JC. Quantitative trait loci involved in genetic predisposition to acute alcohol withdrawal in mice. *J Neurosci* 1997;17:3946–55.

11. Crabbe JC. A genetic animal model of alcohol withdrawal. *Alcohol Clin. Exp. Res.* 1996;20(8):96A–100A.

12. Crabbe JC, Merrill CD, Belknap JK. Acute dependence on depressant drugs is determined by common genes in mice. *Journal of Pharmacology and Experimental Therapeutics* 1991;257:2:663–67.

13. Crabbe, J. C., Metten, P., Cameron, A. J., and Wahlsten, D. An analysis of the genetics of alcohol intoxication in inbred mice. *Neuroscience and Biobehavioral Reviews,* 2005;28:785–802.

14. Crabbe JC, Phillips TJ. Pharmacogenetic studies of alcohol self-administration and withdrawal. *Psychopharmacology* 2004;174:539–60.

15. Crabbe JC, Phillips TJ, Buck KJ, Cunningham CL, Belknap JK. Identifying genes for alcohol and drug sensitivity: recent progress and future directions. *Trends Neurosci* 1999;22:173–79.

16. Crabbe JC, Phillips TJ, Cunningham CL, Belknap JK. Genetic determinants of ethanol reinforcement. *Annals NYAS* 1992;654:302–10.

17. Crabbe JC, Phillips TJ, Kosobud A, Belknap JK. Estimation of genetic correlation: interpretation of experiments using selectively bred and inbred animals. *Alcohol Clin. Exp. Res.* 1990;14:2:141–51.

18. Crawley JN, Belknap JK, Collins A, Crabbe JC, Frankel W, Henderson N et al. Behavioral phenotypes of inbred mouse strains: implications and recommendations for molecular studies. *Psychopharmacology* 1997;132:107–24.

19. Deitrich RA, Baker RC. Genetic influences on alcohol metabolism and sensitivity to alcohol in animals, In: Begleiter H, Kissin B, eds. *The Genetics of Alcoholism*. New York: Oxford University Press; 1995. p. 139–64.

20. Demarest K, Koyner J, McCaughran JJ, Cipp L, Hitzemann R. Further characterization and high-resolution mapping of quantitative trait loci for ethanol-induced locomotor activity. *Behav. Genet.* 2001;31:79–91.

21. Dick DM, Foroud T. Candidate genes for alcohol dependence: a review of genetic evidence from human studies. *Alcohol Clin Exp Res* 2003;27:868–79.

22. Dick DM, Nurnberger J, Jr., Edenberg HJ, Goate A, Crowe R, Rice J et al. Suggestive linkage on chromosome 1 for a quantitative alcohol-related phenotype. *Alcohol Clin. Exp. Res.* 2002;26:1453–60.

23. Ehringer MA, Thompson J, Conroy O, Goldman D, Smith TL, Schuckit MA et al. Human alcoholism studies of genes identified through mouse quantitative trait locus analysis. *Addiction Biology* 2002;7:365–71.

24. Ehringer MA, Thompson J, Conroy O, Yang F, Hink R, Bennett B et al. Fine mapping of polymorphic alcohol-related quantitative trait loci candidate genes using interval-specific congenic recombinant mice. *Alcohol Clin. Exp. Res.* 2002;26:1603–08.

25. Enoch MA, Goldman D. The genetics of alcoholism and alcohol abuse. *Curr Psychiatry Rep* 2001;3:144–51.

26. Enoch MA, Schuckit MA, Johnson And BA, Goldman D. Genetics of alcoholism using intermediate phenotypes. *Alcohol Clin. Exp. Res.* 2003;27:169–76.

27. Fan L, Bellinger F, Ge YL, Wilce P. Genetic study of alcoholism and novel gene expression in the alcoholic brain. *Addict. Biol.* 2004;9:11–18.

28. Fehr C, Shirley RL, Belknap JK, Crabbe JC, Buck KJ. Congenic mapping of alcohol and pentobarbital withdrawal liability loci to a <1 centimorgan interval of murine chromosome 4: Identification of *Mpdz* as a candidate gene. *J Neurosci* 2002;22: 3730–38.

29. Files FJ, Samson HH, Denning CE, Marvin S. Comparison of alcohol-preferring and nonpreferring selectively bred rat lines. II. Operant self-administration in a continuous-access situation. *Alcohol Clin. Exp. Res.* 1998;22:2147–58.

30. Flint J. Analysis of quantitative trait loci that influence animal behavior. *J Neurobiol.* 2003;54:46–77.

31. Gerlai R. Molecular genetic analysis of mammalian behavior and brain processes: caveats and perspectives. *Seminars in the Neurosciences* 1996;8:153–61.

32. Heath AC. Genetic influences on alcoholism risk: a review of adoption and twin studies. *Alcohol Health and Research World* 1995;19:166–71.

33. Heath AC, Todorov AA, Nelson EC, Madden PAF, Bucholz KK, Martin NG. Gene-environment interaction effects on behavioral variation and risk of complex disorders: The example of alcoholism and other psychiatric disorders. *Twin Research* 2004;5:30–37.

34. Hitzemann R, Demarest K, Koyner J, Cipp L, Patel N, Rasmussen E et al. Effect of genetic cross on the detection of quantitative trait loci and a novel approach to mapping QTLs. *Pharmacol. Biochem. Behav.* 2000;67:767–72.

35. Holmes RS. Genetic variants of enzymes of alcohol and aldehyde metabolism. *Alcohol Clin. Exp. Res.* 1985;9:535–38.

36. Homanics GE, Ferguson C, Quinlan JJ, Daggett J, Snyder K, Lagenaur C et al. Gene knockout of the alpha6 subunit of the gamma-aminobutyric acid type A receptor: lack of effect on responses to ethanol, pentobarbital, and general anesthetics. *Mol. Pharmacol.* 1997;51:588–96.

37. Homanics GE, Xu Y, Tang P. Integrated approaches to the action of general anesthetics and alcohol. *Physiol Behav.* 2002;77:495–99.

38. Johnson EO, van den Bree M, Gupman AE, Pickens RW. Extension of a typology of alcohol dependence based on relative genetic and environmental loading. *Alcohol Clin. Exp. Res.* 1998;22:1421–29.

39. Lewohl JM, Wang L, Miles MF, Zhang L, Dodd PR, Harris RA. Gene expression in human alcoholism: microarray analysis of frontal cortex. *Alcohol Clin. Exp. Res.* 2000;24:1873–82.

40. Lumeng L, Murphy JM, McBride WJ, Li T-K. Genetic influences on alcohol preference in animals, In: Begleiter H, Kissin B, eds., *The Genetics of Alcoholism.* New York: Oxford University Press; 1995. p. 165–201.
41. Markel PD, Bennett B, Beeson M, Gordon L, Johnson TE. Confirmation of quantitative trait loci for ethanol sensitivity in long-sleep and short-sleep mice. *Genome Res.* 1997;7:92–99.
42. Mayfield RD, Lewohl JM, Dodd PR, Herlihy A, Liu J, Harris RA. Patterns of gene expression are altered in the frontal and motor cortices of human alcoholics. *J. Neurochem.* 2002;81:802–13.
43. McClearn GE, Rodgers DA. Differences in alcohol preference among inbred strains of mice. *Quarterly Journal of Studies on Alcohol* 1959;20:691–95.
44. McClearn GE, Tarantino LM, Rodriguez LA, Jones BC, Blizard DA, Plomin R. Genotypic selection provides experimental confirmation for an alcohol consumption quantitative trait locus in mouse. *Mol. Psychiat* 1997;2:486–89.
45. Melo JA, Shendure J, Pociask K, Silver LM. Identification of sex-specific quantitative trait loci controlling alcohol preference in C57BL/6 mice. *Nat. Genet.* 1996;13:147–53.
46. Metten P, Crabbe JC. Dependence and withdrawal, In: Deitrich RA, Erwin VG, eds. *Pharmacological Effects of Ethanol on the Nervous System.* Boca Raton FL: CRC Press; 1996. p. 269–90.
47. Metten P, Crabbe JC. Genetic determinants of severity of acute withdrawal from diazepam in mice: commonality with ethanol and pentobarbital. *Pharmacol Biochem Behav* 1999;63:473–79.
48. Grubb SC, Churchill GA, Boque MA. A collaborative database of infred mouse strain characteristics. *Bioinformatics* 2004; 20: 2857–2859.
49. Palmer AA, Phillips TJ. Quantitative trait locus (QTL) mapping in mice, In: Liu Y, Lovinger DM, eds. *Methods in Alcohol-Related Neuroscience Research.* Boca Raton FL: CRC Press; 2002. p. 1–30.
50. Phillips TJ. Behavior genetics of drug sensitization. *Crit Rev. Neurobiol.* 1997;11:21–33.
51. Phillips TJ, Crabbe JC. Behavioral studies of genetic differences in alcohol action, In: Crabbe JC, Harris RA, editors. *The Genetic Basis of Alcohol and Drug Actions.* New York: Plenum Press; 1991. p. 25–104.
52. Phillips TJ, Crabbe JC, Metten P, Belknap JK. Localization of genes affecting alcohol drinking in mice. *Alcohol Clin. Exp. Res.* 1994;18:931–41.
53. Phillips TJ, Hen R, Crabbe JC. Complications associated with genetic background effects in research using knockout mice. *Psychopharmacology* 1999;147:5–7.
54. Pickens RW, Svikis DS, McGue M, LaBuda MC. Common genetic mechanisms in alcohol, drug, and mental disorder comorbidity. *Drug and Alcohol Dependence* 1995;39:129–38.
55. Reich T. A genomic survey of alcohol dependence and related phenotypes: results from the Collaborative Study on the Genetics of Alcoholism (COGA). *Alcohol Clin. Exp. Res.* 1996;20:133A–7A.
56. Rodd ZA, Bell RL, Melendez RI, Kuc KA, Lumeng L, Li TK et al. Comparison of intracranial self-administration of ethanol within the posterior ventral tegmental area between alcohol-preferring and wistar rats. *Alcohol Clin. Exp. Res.* 2004;28:1212–19.
57. Rodd-Henricks ZA, McKinzie DL, Melendez RI, Berry N, Murphy JM, McBride WJ. Effects of serotonin-3 receptor antagonists on the intracranial self-administration of ethanol within the ventral tegmental area of Wistar rats. *Psychopharmacology* 2003;165:252–59.

58. Rout UK, Holmes RS. Alcohol dehydrogenases and aldehyde dehydrogenases among inbred strains of mice: multiplicity, development, genetic studies and metabolic roles. *Addiction Biology* 1996;1:349–62.
59. Samson HH, Files FJ, Denning C, Marvin S. Comparison of alcohol-preferring and nonpreferring selectively bred rat lines. I. Ethanol initiation and limited access operant self-administration. *Alcohol Clin. Exp. Res.* 1998;22:2133–46.
60. Schuckit MA. Self-rating of alcohol intoxication by young men with and without family histories of alcoholism. *J. Stud. Alcohol* 1980;41:242–49.
61. Schuckit MA, Smith TL. An 8-year follow-up of 450 sons of alcoholic and control subjects. *Archives of General Psychiatry* 1996;53:202–10.
62. Schumann G, Spanagel R, Mann K. Candidate genes for alcohol dependence: animal studies. *Alcohol Clin Exp Res* 2003;27:880–88.
63. Shirley RL, Walter NA, Reilly MT, Fehr C, Buck KJ. Mpdz is a quantitative trait gene for drug withdrawal seizures. *Nat. Neurosci.* 2004;7:699–700.
64. Song J, Koller DL, Foroud T, Carr K, Zhao J, Rice J et al. Association of GABAA receptors and alcohol dependence and the effects of genetic imprinting. *Am. J. Med. Genet.* 2003;117B:39–45.
65. Vadasz C, Saito M, Balla A, Kiraly I, Gyetvai B, Mikics E et al. Mapping of quantitative trait loci for ethanol preference in quasi-congenic strains. *Alcohol* 2000;20:161–71.
66. Wafford KA, Burnett D, Harris RA, Whiting PJ. GABAA receptor subunit expression and sensitivity to ethanol. *Alcohol Suppl* 1993;2:327–30.
67. Wall TL, Ehlers CL. Genetic influences affecting alcohol use among Asians. *Alcohol Health and Research World* 1995;19:184–89.
68. Wehner JM, Bowers BJ. Use of transgenics, null mutants, and antisense approaches to study ethanol's actions. *Alcohol Clin. Exp. Res.* 1995;19:811–20.
69. Weitemier AZ, Woerner A, Backstrom P, Hyytia AP, Ryabinin AE. Expression of c-Fos in Alko Alcohol rats responding for ethanol in an operant paradigm. *Alcohol Clin. Exp. Res.* 2001;25:704–10.

Index

A

Aarskog-Scott syndrome, 442
ABA (Allen Brain Atlas), 426
Abiola studies, 116
Abkevich studies, 237
Absence, genetic activity, 23
Acceptance of association, 175
Accessory olfactory bulb (AOB), 22–23
Acetaldehyde toxicity, 451, 459
Acoustic startle reflex, 295
Acoustic stimuli, 320–321
Actin cytoskeleton, *437*, 437–439
Activators, regulation of transcription, 81
Active avoidance paradigms, 295, 298
Active genotype-environment correlation, 190
Addictions, 12, 87, *see also specific substance*
Additive genetic effects and variations, 33–34, *34*,
    40, 184
Additive model, gene-environment interactions,
    179
ADHD (attention-deficit hyperactivity disorder),
    269
Adobe Photoshop, 425
Adoption studies, *see also* Family and twin
    studies; Human studies; Twins studies
    alcohol psychopharmacogenetics, 462–463
    bipolar disorders, 229
    historical research developments, 4, 12
    schizophrenia, 214–216
Advanced intercross lines (AILS), 60, 133
Affected cohorts requirement, 174
Affective disorders, major
    bipolar disorders, 228–236
    family studies, 228, 236
    fundamentals, 227, 237–238
    linkage disequilibrium studies, 231–236,
        233
    molecular linkage studies, 230–231,
        232–233, 237
    recurrent unipolar disorders, 236–237
    twins studies, 228–229, *229*, 237
Affymetrix system, 90–91, 381, 416, 418, 420
Aggression
    fundamentals, 281–282
    gene-environment interactions, 195
    genetic vulnerability, 178

historical research developments, 11
humans, 286–287
interim solution, 25
mice, 282–285, *283, 285–286,* 287
phenotypes, 30–31
rodents, 9
types, 282
voles, 12
AILS (advanced intercross lines), 60, 133
Akaike's information criterion (AIC), 256
AKR/J strains, 134
Albino strains and albinism, 138, 414
Alcohol, alcohol-related disorders, and abuse,
    see also Ethanol and ethanol resistance
    bipolar disorders, 234
    brain dopamine systems, 371–372
    Caenorhabditis elegans, 10
    expression variation sources, 87
    false positives, 175
    gene-environment interactions, 35, 191
    historical research, 7
    inbreeding, 136
    metabolism, 450
    multi-gene disorder, 30
    pharmacodynamics, 451–452
    QTLs, 301
    recurrent unipolar disorders, 236
    response and sensitivity, 451–452
    rodents, 9, 11
    serotonin, 24–25
Alcohol psychopharmacogenetics
    adoption studies, 462–463
    animal studies, 458–462
    candidate gene studies, 463
    fundamentals, 457–458, 464
    gene expression analysis, 464
    gene identification, 460–462
    human studies, 462–464
    inbred strains studies, 458–459
    QTLs identification, 463–464
    risk marker identification, 463–464
    selectively bred lines studies, 459–460
    targeting studies, 460–462
    twins studies, 462–463
ALDH2 variants, 450–451, 463
Algerian family study, 443
Alkaptonuria, 4, 450